FUNDAMENTALS OF MICROELECTRONICS PROCESSING

McGraw-Hill Chemical Engineering Series

BUILDING THE LITERATURE OF A PROFESSION

Fifteen prominent chemical engineers first met in New York more than 60 years ago to plan a continuing literature for their rapidly growing profession. From Industry came such pioneer practitioners as Leo H. Baekeland, Arthur D. Little, Charles L. Reese, John V. N. Dorr, M. C. Whitaker, and R. S. McBride. From the universities came such eminent educators as William H. Walker, Alfred H. White, D. D. Jackson, J. H. James, Warren K. Lewis, and Harry A. Curtis. H. C. Parmelee, then editor of *Chemical and Metallurgical Engineering*, served as chairman and was joined subsequently by S. D. Kirkpatrick as consulting editor.

After several meetings, this committee submitted its report to the McGraw-Hill Book Company in September 1925. In the report were detailed specifications for a correlated series of more than a dozen texts and reference books which have since become the McGraw-Hill Series in Chemical Engineering and which became the cornerstone of the chemical engineering curriculum.

From this beginning there has evolved a series of texts surpassing by far the scope and longevity envisioned by the founding Editorial Board. The McGraw-Hill Series in Chemical Engineering stands as a unique historical record of the development of chemical engineering education and practice. In the series one finds the milestones of the subject's evolution: industrial chemistry, stoichiometry, unit operations and processes, thermodynamics, kinetics, and transfer operations.

Chemical engineering is a dynamic profession, and its literature continues to evolve. McGraw-Hill and its consulting editors remain committed to a publishing policy that will serve, and indeed lead, the needs of the chemical engineering profession during the years to come.

The Series

Bailey and Ollis: *Biochemical Engineering Fundamentals*
Bennett and Myers: *Momentum, Heat, and Mass Transfer*
Beveridge and Schechter: *Optimization: Theory and Practice*
Brodkey and Hershey: *Transport Phenomena: A Unified Approach*
Carberry: *Chemical and Catalytic Reaction Engineering*
Constantinides: *Applied Numerical Methods with Personal Computers*
Coughanowr and Koppel: *Process Systems Analysis and Control*
Douglas: *Conceptual Design of Chemical Processes*
Edgar and Himmelblau: *Optimization of Chemical Processes*
Fahien: *Fundamentals of Transport Phenomena*
Finlayson: *Nonlinear Analysis in Chemical Engineering*
Gates, Katzer, and Schuit: *Chemistry of Catalytic Processes*
Holland: *Fundamentals of Multicomponent Distillation*
Holland and Liapis: *Computer Methods for Solving Dynamic Separation Problems*
Katz, Cornell, Kobayashi, Poettmann, Vary, Elenbaas, and Weinaug: *Handbook of Natural Gas Engineering*
Katz and Lee: *Natural Gas Engineering: Production and Storage*
King: *Separation Processes*
Lee: *Fundamentals of Microelectronics Processing*
Luyben: *Process Modeling, Simulation, and Control for Chemical Engineers*
McCabe, Smith, J. C., and Harriott: *Unit Operations of Chemical Engineering*
Mickley, Sherwood, and Reed: *Applied Mathematics in Chemical Engineering*
Nelson: *Petroleum Refinery Engineering*
Perry and Chilton (Editors): *Chemical Engineers' Handbook*
Peters: *Elementary Chemical Engineering*
Peters and Timmerhaus: *Plant Design and Economics for Chemical Engineers*
Reid, Prausnitz, and Rolling: *The Properties of Gases and Liquies*
Resnick: *Process Analysis and Design for Chemical Engineers*
Satterfield: *Heterogeneous Catalysis in Practice*
Sherwood, Pigford, and Wilke: *Mass Transfer*
Smith, B. D.: *Design of Equilibrium Stage Processes*
Smith, J. M.: *Chemical Engineering Kinetics*
Smith, J. M., and Van Ness: *Introduction to Chemical Engineering Thermodynamics*
Treybal: *Mass Transfer Operations*
Valle-Riestra: *Project Evolution in the Chemical Process Industries*
Van Ness and Abbott: *Classical Thermodynamics of Nonelectrolyte Solutions: With Applications to Phase Equilibria*
Wei, Russell, and Swartzlander: *The Structure of the Chemical Processing Industries*
Wentz: *Hazardous Waste Management*

FUNDAMENTALS OF MICROELECTRONICS PROCESSING

Hong H. Lee

Professor of Chemical Engineering
University of Florida, Gainesville

McGraw-Hill Publishing Company

New York St. Louis San Francisco Auckland Bogotá Caracas
Hamburg Lisbon London Madrid Mexico Milan Montreal New Delhi
Oklahoma City Paris San Juan São Paulo Singapore Sydney Tokyo Toronto

This book was set in Times Roman.
The editors were B. J. Clark and John M. Morriss;
the production supervisor was Denise L. Puryear.
The cover was designed by Marie Christine Lawrence.
Project supervision was done by Santype International Ltd.
R. R. Donnelley & Sons Company was printer and binder.

FUNDAMENTALS OF MICROELECTRONICS PROCESSING

1234567890 DOC DOC 89432109

ISBN 0-07-037056-7

Acknowledgments appear on pages 507–510;
Copyrights included on this page by reference.

Library of Congress Cataloging-in-Publication Data

Lee, Hong H.
Fundamentals of microelectronics processing/Hong H. Lee.
p. cm.—(McGraw-Hill chemical engineering series)
Includes bibliographies and index.
ISBN 0-07-037056-7
1. Integrated circuits—Design and construction. 2. Integrated circuits—Materials. I. Title. II. Series.
TK7874.L34 1990
621.381′5—dc20 89-12114

ABOUT THE AUTHOR

HONG H. LEE is currently Professor of Chemical Engineering at the University of Florida. He received his B.S. degree from Seoul National University and his Ph.D. from Purdue University. After working in industry for six years, he joined the Department in 1977. He authored a textbook on catalytic reactions and reactors and holds six patents. He is also an author or coauthor of more than sixty technical papers, about one-third of which are in the area of microelectronics processing. His current research interests in the area are GaAs surface passivation, heteroepitaxy, and development of chemical vapor deposition reactors. He is a consultant to chemical and semiconductor industries.

To Sukie, Caroline, Iris, and Audrey
and the family

CONTENTS

PREFACE

This book is on the processing involved in the fabrication of microelectronics, from raw materials preparation to integrated circuit fabrication and to packaging of the integrated circuits. The text is intended primarily for a senior-level course in microelectronics or semiconductor processing but also for an introductory course at the first-year graduate level. The book has been written for chemical engineers. However, those in materials, chemistry, and electrical engineering may find it useful as a textbook when the course emphasis is on the fundamentals of the processing.

Microelectronics processing is perhaps the most interdisciplinary field by its nature. As such, there are certain aspects of the processing that are relatively unique to each discipline. Therefore, the aspects that are unique to chemical engineering have been emphasized. They are also the subjects that flow naturally from the traditional chemical engineering education. Nontraditional subjects that are necessary to interface with other disciplines are included but are given less emphasis. The processing steps in microelectronics are given different names depending on the materials involved. Each of these steps is treated separately in the books currently available, when in fact the same physical phenomena are involved in many cases. Efforts have been made, wherever possible, to treat the subjects according to their physical nature.

Much thought has been given to properly reflecting the progression of chemical engineering education in writing this book. Our education has progressed from mere descriptions of processing to understanding and using basic physical principles governing the processes, and from empiricism to quantitative description based on engineering sciences. Efforts have been made to reflect the progress to the extent possible.

An introduction to and an overall view of microelectronics processing are given in Chapter 1. Sections 1.5 and 1.6 that deal with device physics can be skipped without loss of continuity; however, some instructors may find them essential. Chapter 2 is on raw materials used in microelectronics and Chapter 3 is on the process of making wafers on which devices are fabricated. Fabrication conditions are related to critical aspects of electronic device performance in Chapter 4 to pave the way for the chapters to follow. This chapter, however, may be covered at the end of a course.

Chapters 5 through 10 deal with the processes leading to integrated circuits. The approach taken here is to combine various processing steps into chemical and physical rate processes. The chemical rate processes and the reactors are treated in Chapters 5 and 6. The rationale is that the same physical principles apply whether a metal or a passivating layer is deposited. The fabrication steps unique to microelectronics are doping (Chapter 7), which enables a semiconductor to function as a circuit, and lithography (Chapter 8), which specifies the order and manner in which the fabrication is carried out. The physical and physico-chemical processes are then treated in Chapters 9 and 10. Packaging of fabricated integrated circuits (Chapter 11) involves diverse phenomena. As such, only the major performance factors are touched upon.

The subjects of microelectronics necessarily involve concepts and terminology often unfamiliar to chemical engineers, although many aspects of the processing are essentially of chemical engineering. Thus, numerous examples are provided in the book to help the students to grasp the material presented: some for familiarization and others for further development. There are also ample problems at the end of each chapter. A solutions manual for these problems is available. The problems are structured in such a way that they become more challenging toward the latter part of the problem section. The first part, however, covers all the subjects treated in the text. A notation section is provided at the end of each chapter.

Many in the profession were gracious enough to review one or two chapters or offer comments. Their comments and suggestions are gratefully acknowledged. They are D. L. Flamm, Bell Laboratories; Rudolf G. Frieser, University of North Carolina; Dennis W. Hess, University of California; Y. H. Lee, from academia; J. W. Moody, Monsanto Electronics Company from industry; R. Pollard, from academia; Gregory B. Raupp, Arizona State University; Harvey Stenger, Lehigh University; L. F. Thompson, Bell Laboratories; and Matthew Tirrell, University of Minnesota. Present students of the author contributed directly and indirectly. They are: J. V. Cole, D. C. Koopman, E. C. Stassinos, and J. S. Yoo. Special thanks are due to D. C. Koopman for editing the entire manuscript from the viewpoint of a student whose research is not in the area.

Enthusiastic support from our department has made this endeavor possible. The need for a book of this kind was recognized and shared by our department chairmen and colleagues on our faculty. Initiation of the work was made possible by A. L. Fricke. Enthusiastic support from our current chairman, D. O. Shah, provided the continuity. Advice and encouragement from S. A. Svoronos, whose enthusiasm for teaching is contagious, carried the hard days often encountered in the course of the writing. Dialogues with T. J. Anderson on the microelectronics area in general have been valuable. Finally, the author is indebted to the class of '85.

My special thanks are due to Caroline and Iris Lee for typing the bulk of the manuscript, some parts of it several times. The author is indebted to Debbie Hitts for taking care of numerous correspondences.

Hong H. Lee

FUNDAMENTALS OF MICROELECTRONICS PROCESSING

CHAPTER 1

INTEGRATED CIRCUITS AND FABRICATION

1.1 INTEGRATED CIRCUITS

An integrated circuit (IC) is a collection of electronic circuits made by simultaneously forming individual transistors, diodes, and resistors on a small chip of semiconductor material, typically silicon, that are intraconnected to one another with a metal, such as aluminum, deposited on the chip surface. The metal serves as "wireless wires" for the intraconnections. The concept of the monolithic (meaning literally "one stone") integrated circuit (all components fabricated within a solid block of semiconductor material) was first suggested by Dummer in 1952 who stated: "It seems now possible to envisage electronic equipment in a solid block with no connecting wires." The integrated circuit was invented six years later by Kilby (1976). Integrated circuits are grouped into two major categories: analog (or linear) and digital (or logic). Analog integrated circuits amplify or respond to variable voltages. Typical examples are amplifiers, timers, oscillators, and voltage regulators. Digital integrated circuits produce or respond to signals having only two voltage levels. Typical examples are microprocessors, memories, and microcomputers. Although integrated circuits can also be grouped into monolithic and hybrid circuits according to fabrication, only the monolithic circuits will be considered here, as discussed in the Preface.

The scale of integration for silicon integrated circuits has increased geometrically since the first commercial integrated circuits, introduced by Texas

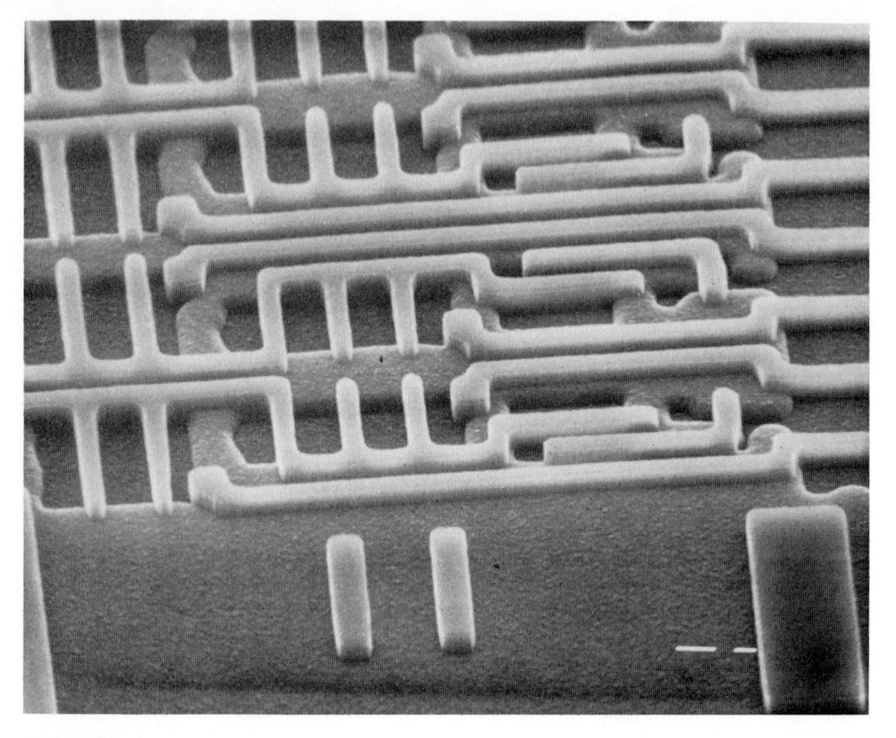

FIGURE 1-1
A silicon wafer containing 89 chips and 7 test structures (Sze, 1983). (Courtesy A. Kornblit and T. Giniecki, AT&T Bell Laboratories)

Instruments in 1960, from small scale integration (SSI) to present-day very large scale integration (VLSI). Although the scale of integration is usually restricted to digital circuits with somewhat nebulous dividing lines associated with the number of logic gates per circuit, one can only marvel at the processing technologies that have made it possible to squeeze 10^5 or more equivalent gates into a chip with an area much less than 1 cm^2 for the VLSI. Shown in Fig. 1-1 is a silicon wafer containing eighty-nine 256-kilobyte dynamic random access memory chips (256 K DRAM) and seven test structures. Each memory chip has more than 600,000 components (gates). A typical component fabricated by metal-oxide semiconductor (MOS) technology is shown in Fig. 1-2 for both planar and sectional views. The minimum device dimension, in this case the gate-to-gate distance, is 2 μm.

The dominant trend for device miniaturization in integrated circuit (IC) technology owes its *impetus* to lower costs per function, lower power consumption, and higher device speed. Another major trend is the use of continually larger wafers, which also has led to lower costs per chip. To prevent a decrease in yield every time a larger substrate size was introduced, processing technology had to undergo improvements to maintain the same uniformity across these larger wafers. The economics of scale for mass production processing applies to chips as

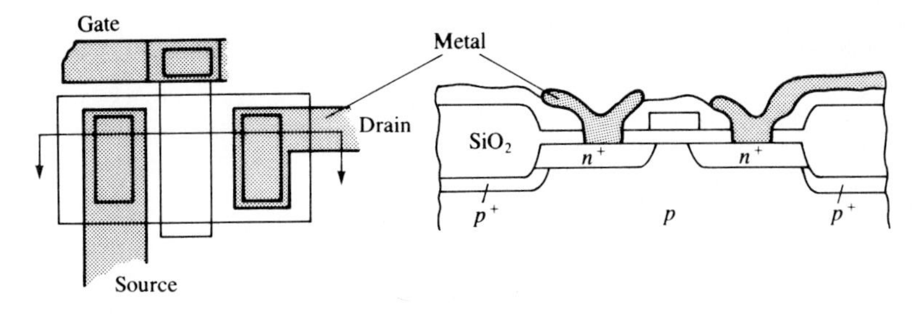

FIGURE 1-2
Planar and sectional view of a MOS structure (Hodges and Jackson, 1983).

well. Therefore processing technology has had to keep up with ever-decreasing device sizes and increasing wafer sizes.

The major current IC technologies in terms of device structure are MOS and bipolar junction transistor (BJT) technologies for silicon and metal semiconductor (MES) technology for gallium arsenide. The MOS and MES transistors are usually in the form of field-effect transistors, and are called MOSFET and MESFET, respectively. One needs to have an understanding of semiconductors and charge carriers before studying the basics (fundamentals) and operating characteristics of transistors.

1.2 SEMICONDUCTORS AND CHARGE CARRIERS

The best-known semiconductor is undoubtedly silicon. Single-crystal silicon is the base material of silicon ICs. A silicon atom has only four electrons in its outermost shell, but the outermost shell can accommodate eight electrons. Therefore, a silicon atom links up with four of its neighbors to share electrons, leading to covalent bonds. A cluster of silicon atoms sharing the outermost electrons forms a regular solid arrangement called a *crystal.* A planar view of the silicon crystal with only the outermost electrons of each atom is shown in Fig. 1-3*a.* A semiconductor can be defined as a material with controllable conductivities, intermediate between insulators and conductors. A greater freedom for controllability of the conductivity over orders of magnitude can be gained by introducing either atoms of three outermost electrons such as boron or atoms of five outermost electrons such as phosphorus into the silicon structure. Incorporating these electron-deficient (relative to the host semiconductors) atoms or excess-electron atoms, which are called *dopants* or *intentional impurities*, into the pure semiconductor crystal structure is called *doping.* Pure semiconductors are *intrinsic* semiconductors; doped semiconductors are *extrinsic* semiconductors. As shown in Fig. 1-3*c*, an aluminum atom that has only three electrons in a cluster of silicon atoms leads to the outermost shell occupied by only seven electrons, leaving a vacant electron opening called a *hole.* An electron from a nearby atom can "fall" into the hole. Therefore, the hole has moved to a new location. This type of

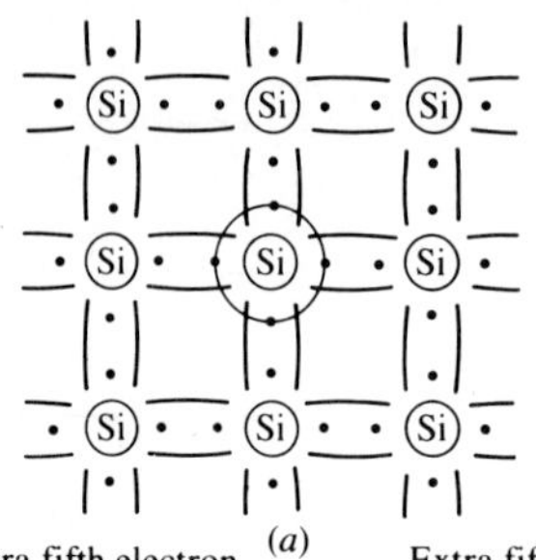

Extra fifth electron of phosphorus atom

Extra fifth electron is removed from phosphorus atom

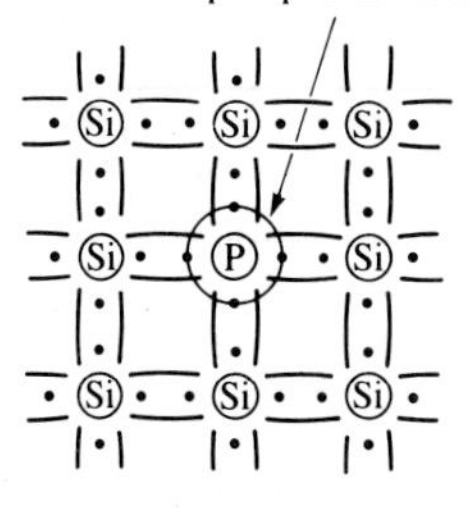

(b)

Fourth bonding electron of aluminum atom is missing and creates a hole

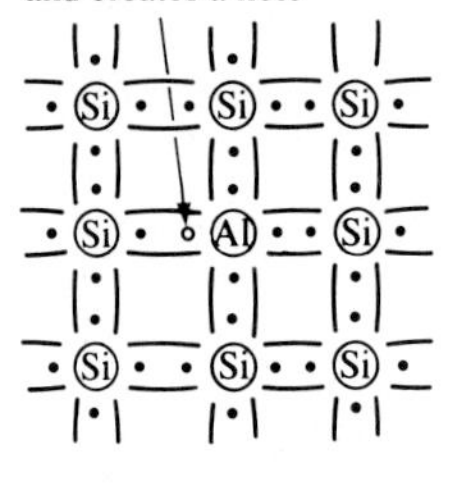

Electron is attracted from a silicon atom to fill hole in aluminum-silicon bond

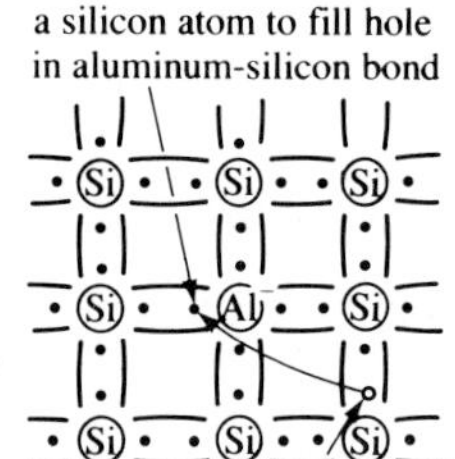

New hole is created here

(c)

FIGURE 1-3
Intrinsic and extrinsic (*n*- and *p*-type) silicon (Smith, 1986).

semiconductor is called a *p-type* semiconductor. A phosphorus atom that has five outermost electrons in a cluster of silicon atoms, as shown in Fig. 1-3*b*, has one excess electron after sharing its electrons with the surrounding silicon atoms, donating the extra "free" electron to its surrounding. Such an extra electron can move through the crystal with comparative ease. This type of semiconductor is called an *n-type* semiconductor. All substrates used in IC fabrication are either *p*- or *n*-type semiconductors. Electrons and holes carry with them negative and positive charges and thus are called *charge carriers;* that is, electrons and holes that are "free" to move about are the charge carriers. The movement of charge carriers leads to the flow of (electrical) current.

A semiconductor can be either an element or a compound. A compound semiconductor has two or more elements. The average number of shared electrons per atom for two-element compound semiconductors is four with the exception of V-VI compound semiconductors. Given in Table 1.1 are the portion of the periodic table from which elemental and compound semiconductors are

TABLE 1.1
Periodic table and semiconductors

	Group IIB	Group IIIA	Group IVA	Group VA	Group VIA
1 1*s*					
2 2*s*2*p*		5 B	6 C		
3 3*s*3*p*		13 Al	14 Si	15 P	16 S
4 4*s*3*d*4*p*	30 Zn	31 Ga	32 Ge	33 As	34 Se
5 5*s*4*d*5*p*	48 Cd	49 In	50 Sn	51 Sb	52 Te

Element	IV compound	III-V compound	II-VI compound	V-VI compound
Si(1.1†)	SiC (2.86 for α structure)	AlP (2.85), GaP (2.26), InP (1.28)	ZnS (3.6), CdS (2.42)	PbS (0.37), PbSe (0.27), PbTe (0.29)
Ge (0.67)		AlAs (2.16), GaAs (1.43), InAs (0.36), AlSb (1.6), GaSb (0.7), InSb (0.18)	ZnSe (2.7), CdSe (1.73), ZnTe (2.25), CdTe (1.58)	

† Energy gap in electronvolts (eV).

formed and a list of elemental and two-element compound semiconductors. Group III and group V elements in the table have three and five outermost electrons, respectively. For silicon, therefore, it follows that a *p*-type silicon results when the group III elements (B, Al, Ga, In) in the table are used as dopants; an *n*-type silicon results when the group V elements (P, As, Sb) are used as dopants. The former are often referred to as *acceptor* atoms for silicon since they accept electrons; the latter as *donor* atoms since they donate electrons.

An intrinsic semiconductor is an insulator unless it is excited thermally or optically. If the excitation is high enough, a semiconductor becomes a conductor. Since the electron energy levels are discrete, any excitation will bring the energy levels of electrons to higher discrete levels. Since a semiconductor can be an insulator or a conductor depending on the magnitude of the excitation, it can behave as a conductor only when a certain energy barrier is overcome. This energy barrier E_g is called an *energy gap*. The values of the energy gap are also given in Table 1.1 for the semiconductors listed there. It is seen that the energy gap ranges from 0.18 eV for InSb to 3.6 eV for ZnS. Conductors such as metals do not have any energy gap and therefore they conduct with or without excitation; insulators do not conduct even when excited, which means that the energy

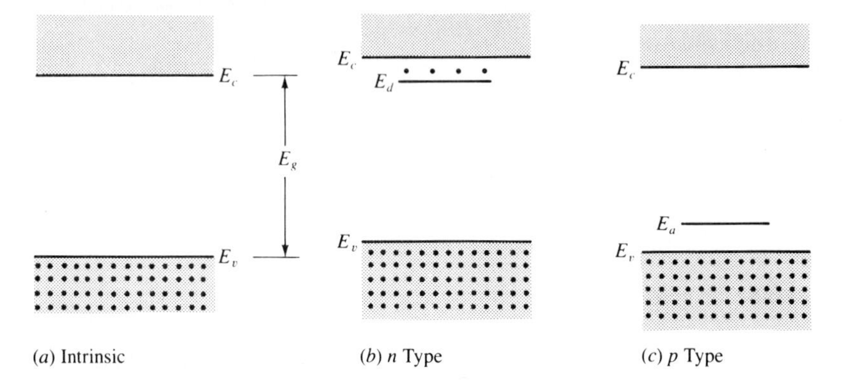

FIGURE 1-4
Energy states at 0 K for intrinsic, *n*-type and *p*-type semiconductors for the same semiconductor (Streetman, 1980).

gap is prohibitively high. With no excitation, all electrons of a semiconductor occupy the low energy levels in the valence states. Although energy levels are discrete, they would look like a band since there are so many energy levels. This band is called a *valence band.* The highest energy level of the valence band is usually denoted by E_v. Across the energy gap, there is another band of energy levels for conduction, called the *conduction band.* The lowest energy level of the conduction band is usually denoted by E_c. This energy level configuration at 0 K is shown in Fig. 1-4*a* for an intrinsic semiconductor. When the intrinsic semiconductor is doped with donor atoms, the extra donor electrons of the atoms, which cannot occupy the valence band that is already full, occupy the energy levels somewhere in between E_v and E_c. The lowest level is called the *donor level* and is denoted by E_d, as shown in Fig. 1-4*b*. The donor atoms are usually chosen to have the donor level closer to E_c, as will be apparent shortly. When the electron-deficient acceptor atoms are introduced to the intrinsic semiconductor, holes are created that can accommodate electrons of energy levels slightly higher than E_v, as shown in Fig. 1-4*c*. The highest of these energy levels is called the *acceptor level,* denoted by E_a.

The intrinsic semiconductor in Fig. 1-4*a* would give the energy band diagram of Fig. 1-5*a* when it is thermally excited. As shown, some of the electrons in the valence band are excited across the band gap to the conduction band, creating holes in the valence band and an equal number of electrons in the conduction band, which leads to electron-hole pairs (EHP). Since the donor level for the *n*-type semiconductor is quite close to E_c, slight excitation is sufficient to cause all donor electrons to jump into the conduction band, as shown in Fig. 1-5*b*. Thus, an *n*-type semiconductor can have a considerable number (concentration) of electrons in the conduction band, even when the temperature is too low for the intrinsic EHP concentration to be appreciable. For *p*-type semiconductors, the acceptor energy level is close to the valence band and slight excitation can cause some of the electrons in the valence band to jump and

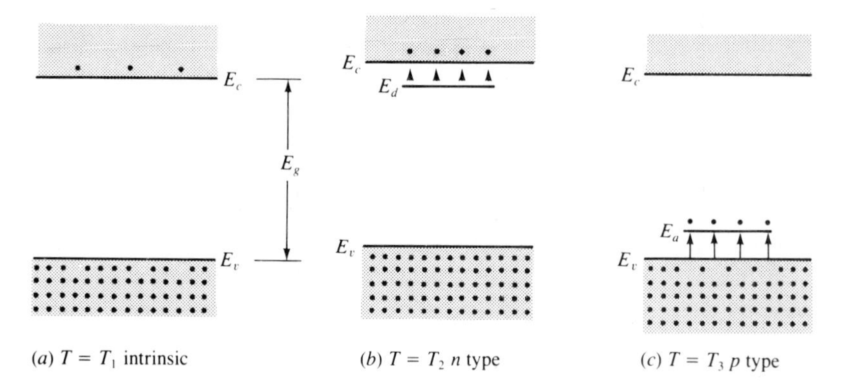

FIGURE 1-5
Energy states at high temperatures for intrinsic, *n*-type and *p*-type semiconductors. Temperature T_1 is higher than T_2 and T_3.

occupy the acceptor energy levels, as shown in Fig. 1-5*c*. This leads to creation of holes in the valence band. Therefore, a *p*-type semiconductor can have a considerable concentration of holes even when the temperature is too low for the intrinsic EHP concentration to be appreciable.

When a semiconductor is doped *n* or *p* type, one type of charge carrier dominates. The dominant carrier in terms of its concentration is referred to as the *majority carrier*; the other carrier is called the *minority carrier*. It is clear from the above discussion that the dominant charge carriers for *n*-type semiconductors are electrons. Therefore, the majority carriers are electrons and the minority carriers are holes. For *p*-type semiconductors, the majority carriers are holes and the minority carriers are electrons.

1.3 BASIC RELATIONSHIPS AND CONDUCTIVITY

The motion of charge carriers is electrical conductance. Therefore, the charge carrier concentration is a quantity of prime interest in any IC device applications. For intrinsic semiconductors, electrons and holes are created in pairs. Therefore, the electron concentration (in the conduction band) n (electrons per cubic centimeter) is equal to the hole concentration (in the valence band) p (holes per cubic centimeter):

$$n = p = n_i \tag{1.1}$$

where each of the intrinsic carrier concentrations has been denoted by n_i in accordance with the usual notation practice. A few basic relationships are useful in considering the carrier concentrations for an extrinsic semiconductor under equilibrium conditions, i.e., at steady state.

Suppose that the dopants (impurities) are distributed uniformly. In order to satisfy the charge neutrality (electrostatic neutrality) in an extrinsic semiconductor, the positive charge must be equal to the negative charge. For silicon, the dopants are either one electron deficient or in excess relative to the silicon. Therefore, one has

$$p + N_D = n + N_A \tag{1.2}$$

where N_D is the concentration of the donor atoms and N_A is the concentration of the acceptor atoms. Equation (1.2) is one form of the basic relationship known as *space-charge neutrality*. One assumption made here is that all donor electrons and acceptor holes are fully activated such that the donor and acceptor levels are fully occupied by electrons. At room temperature, this assumption is in general correct for silicon unless the dopant atoms are present in relatively high concentrations ($>10^{18}$ cm^{-3}). Otherwise N_D should be replaced by the ionized concentration N_D^+ and N_A by N_A^-. As will be shown shortly, the product of electron and hole concentrations in thermal equilibrium is equal to the square of the intrinsic carrier concentration n_i:

$$pn = n_i^2 \tag{1.3}$$

Note that this relation holds for any type of semiconductor under equilibrium.

In the light of the above two basic relationships, consider the electron and hole concentrations in an extrinsic semiconductor. For an *n*-type semiconductor, the electron concentration n_n can be obtained by substituting Eq. (1.3) into Eq. (1.2) and solving for n (n_n):

$$n_n = \tfrac{1}{2}\{N_D - N_A + [(N_D - N_A)^2 + 4n_i^2]^{1/2}\} \tag{1.4}$$

Likewise, the hole concentration in a *p*-type semiconductor, p_p, can be derived to give

$$p_p = \tfrac{1}{2}\{N_A - N_D + [(N_A - N_D)^2 + 4n_i^2]^{1/2}\} \tag{1.5}$$

The intrinsic carrier concentration n_i for silicon is 1.45×10^{10} cm^{-3} at 27 °C; it is approximately 9×10^6 for GaAs. The magnitude of the net impurity (dopant) concentration $|N_D - N_A|$ is in general much larger than n_i. Therefore, the majority carrier concentrations can be approximated by

$$n_n \approx N_D - N_A \tag{1.6}$$

$$p_p \approx N_A - N_D \tag{1.7}$$

which follow from Eqs. (1.4) and (1.5). The approximate concentrations of the minority carriers also follow from Eqs. (1.4) and (1.5) and the *pn* product relationship [Eq. (1.3)]:

$$p_n \approx \frac{n_i^2}{N_D - N_A} \tag{1.8}$$

$$n_p \approx \frac{n_i^2}{N_A - N_D} \tag{1.9}$$

The subscript refers to the type of semiconductor such that p_n is the hole concentration in an n-type semiconductor, which is the minority carrier concentration in this case.

Example 1.1. In general, both acceptor and donor impurities may be present. For this example, however, consider a silicon wafer containing 10^{16} cm^{-3} boron atoms and no donor impurity. What are the majority and minority carrier concentrations at 27 °C?

Solution. Only the acceptor impurity is present. Therefore, $N_A = 10^{16}$, $N_D = 0$; holes are the majority carrier. Since $n_i = 1.45 \times 10^{10}$, Eq. (1.5) yields

$$p_p = \tfrac{1}{2}\{10^{16} + [(10^{16})^2 + 4(1.45 \times 10^{10})^2]^{1/2}\}$$
$$= 10^{16} \text{ cm}^{-3}$$

It can be seen that the value of p_p calculated from Eq. (1.7) is essentially the same as N_A. For the minority carrier (electron) concentration n_p, Eq. (1.9) yields

$$n_p = \frac{(1.45 \times 10^{10})^2}{10^{16}} = 2.1 \times 10^4 \text{ cm}^{-3}$$

It is seen that $p_p \gg n_p$ but that $p_p n_p = n_i^2$.

Another basic relationship central to semiconductor behavior is a probability function that electrons in solids obey, known as the *Fermi-Dirac* probability function. The probability $f(E)$ that an electronic state with energy E is occupied by an electron is given by

$$f(E) = \frac{1}{1 + e^{(E - E_F)/kT}} \tag{1.10}$$

where T is the temperature, k is the Boltzmann constant (8.62×10^{-5} eV/K = 1.38×10^{-23} J/K) and E_F is called the *Fermi level.* The Fermi level is essentially the chemical potential of electrons in a solid but may be considered as the energy at which the probability of occupation of an energy state by an electron is exactly one half, which can be verified by setting E equal to E_F in Eq. (1.10). An examination of Eq. (1.10) should reveal that at 0 K the distribution takes the simple rectangular form shown in Fig. 1-6. At temperatures higher than 0 K, some probability exists for states above the Fermi level to be filled. For example, at $T = T_1$ in Fig. 1-6 there is some probability $f(E)$ that states above E_F are filled, and there is a corresponding probability $[1 - f(E)]$ that states below E_F are empty. The Fermi function is symmetrical about E_F for all temperatures. Thus, if the number of energy states in the conduction and valence bands is the same, and if the number of electrons in the conduction band and that of holes in the valence band is also the same, the Fermi level must be located in the middle of the energy gap. This is approximately the case in an intrinsic semiconductor. The Fermi level in an intrinsic semiconductor is often referred to as the *intrinsic Fermi level* and is denoted by E_i.

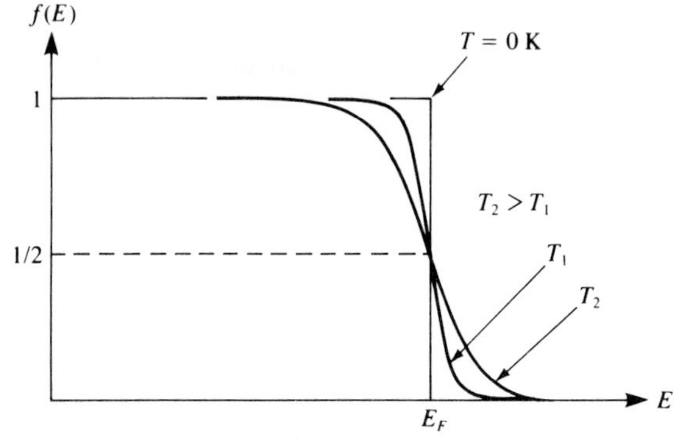

FIGURE 1-6
Fermi-Dirac distribution function.

Since the Fermi distribution function represents the probability of the occupancy of an energy state by an electron, the (probable) number of electrons in a semiconductor with a certain energy level can be calculated from the probability density function $N(E)$. If the number of energy states per unit volume (or the density of states) in the energy range dE is $N(E)\,dE$, the number of electrons per unit volume or the electron density in the conduction band, n, is given by

$$n = \int_{E_c}^{\infty} f(E)N(E)\,dE \tag{1.11}$$

since those with energy larger than E_c are in the conduction band. In principle, $N(E)$ can be calculated from quantum mechanics and the Pauli exclusion principle (e.g., Streetman, 1980). However, it is more convenient to represent all of the distributed electron states in the conduction band by an *effective density of states* N_c located at the conduction band edge E_c. Then the conduction band electron concentration is simply the effective density of states at E_c times the probability of occupancy at E_c:

$$n = N_c f(E_c) \tag{1.12}$$

where N_c is given by

$$N_c = 2\,\frac{2\pi m_n^* kT^{3/2}}{h^2}$$

$$= \begin{cases} 2.8 \times 10^{19}(T/300)^{3/2}\ \text{cm}^{-3} \text{ for Si} \\ 4.7 \times 10^{17}(T/300)^{3/2}\ \text{cm}^{-3} \text{ for GaAs} \end{cases} \tag{1.13}$$

Here, m_n^* is the effective electron mass and h is the Planck constant. The effective mass is a quantity that takes the place of the mass of a free electron m_n in various calculations. It essentially corrects for the effect of the presence of the semiconductor crystal lattice on the behavior of the electron. Since $f(E_c)$ can be approximated by

$$f(E_c) = \frac{1}{1 + e^{(E_c - E_F)/kT}} \approx e^{-(E_c - E_F)/kT} \tag{1.14}$$

when $(E_c - E_F)$ is at least several times kT, which is often the case (note that at room temperature $kT = 0.026$ eV), Eq. (1.12) can be rewritten as

$$n = N_c e^{-(E_c - E_F)/kT} \tag{1.15}$$

Similarly, the valence band hole concentration can be derived to yield

$$p = N_v e^{-(E_F - E_v)/kT} \tag{1.16}$$

where

$$N_v = 2\,\frac{2\pi m_p^* kT^{3/2}}{h^2}$$

$$= \begin{cases} 1.04 \times 10^{19}(T/300)^{3/2}\ \text{cm}^{-3} \text{ for Si} \\ 7.0 \times 10^{18}(T/300)^{3/2}\ \text{cm}^{-3} \text{ for GaAs} \end{cases} \tag{1.17}$$

Here m_p^* is the effective mass of hole. It is noted that Eqs. (1.15) and (1.16) are valid for both intrinsic and extrinsic material provided E_F is replaced with E_i for the intrinsic material.

Example 1.2. Derive Eq. (1.3) and show the pn product to be equal to $N_c N_v e^{-E_g/kT}$.

Solution. From Eqs. (1.15) and (1.16) written for the intrinsic semiconductor,

$$n_i = N_c e^{-(E_c - E_i)/kT}$$

and

$$p_i = N_v e^{-(E_i - E_v)/kT}$$

Thus,

$$n_i p_i = N_c e^{-(E_c - E_i)/kT} N_v e^{-(E_i - E_v)/kT} = N_c N_v e^{-E_g/kT}$$

From Eqs. (1.15) and (1.16) written for the extrinsic semiconductor,

$$np = N_c N_v e^{-(E_c - E_v)/kT} = N_c N_v e^{-E_g/kT}$$

Since N_c and N_v are the same for both intrinsic and extrinsic semiconductors one has

$$pn = p_i n_i = n_i^2 = N_c N_v e^{-E_g/kT}$$

Consider the flow of current now that the charge carrier concentrations can be calculated from the relationships developed so far. The current flows in the direction that the charge carriers move. More precisely, the current flows in the opposite direction to the electron flow and in the same direction as the holes. When charge carriers are uniformly distributed, the carriers will move only when an external force is applied. In view of the fact that the average velocity attained by electrons is different from that of holes for the same external force, consider the current flux J (usually referred to as current density, in units of amperes per square centimeter) due solely to electrons in one dimension (direction). If the average velocity is $\bar{v}_n$, then the flux is simply the velocity times the charge concentration (charge times the carrier concentration, qn):

$$J_n = -qn\bar{v}_n \tag{1.18}$$

where the subscript n is for electron. The current flux is equivalent to the convective mass flux except that the charge on the electron $(-q)$ is multiplied. If, in addition, the carrier concentration is not uniform, the carrier diffuses due to the concentration gradient. Therefore, the current flux for electrons, in general, can be written as

$$J_n = -qn\bar{v}_n - D_n(-q)\frac{dn}{dx} \tag{1.19}$$

Note that $(-q)n$ can be considered as the (electric) concentration for electrons. The first term is usually referred to as the *drift* term. It is often more convenient to express the drift term in terms of the electric field ε (in units of volts per centimeter), which is the electric force when multiplied by q. An important quantity called *mobility* μ results when the electric field is related to the average velocity:

$$\bar{v} = \mu\varepsilon; \qquad \mu[\text{cm}^2/(\text{V}\cdot\text{s})] \tag{1.20}$$

Note that electrons drift opposite to the field so that, for electrons, $\bar{v}_n = -\mu_n\varepsilon$. The electron mobility μ_n is different from hole mobility μ_p: in general, the electron mobility is higher than the hole mobility. They are dependent on the impurity concentration, as shown in Fig. 1-7, and also on temperature. The linear relationship of Eq. (1.20) holds for relatively small values of ε, say less than 0.2 V/cm. At larger values, the velocity increases slower with ε, eventually reaching a

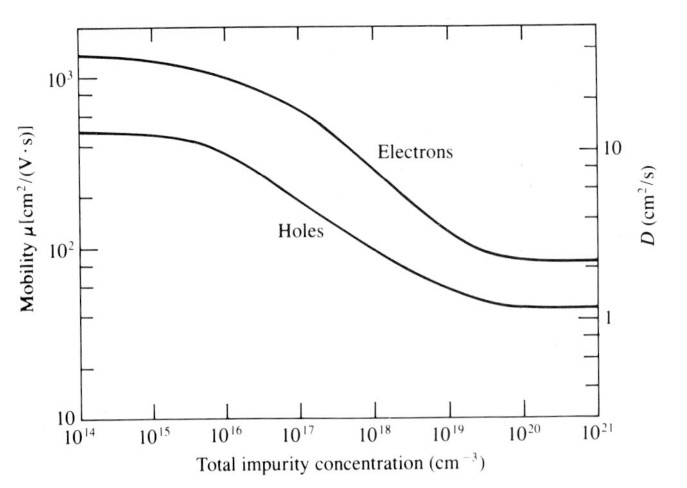

FIGURE 1-7
Mobility as a function of total dopant concentration (Grove, 1967).

plateau. The total current is the sum of the current due to electrons and that due to holes:

$$J = J_n + J_p \tag{1.21}$$

with

$$\begin{aligned} J_n &= q\mu_n n\varepsilon + qD_n \frac{dn}{dx} \\ J_p &= q\mu_p p\varepsilon - qD_p \frac{dp}{dx} \end{aligned} \tag{1.22}$$

where Eq. (1.20) and an equation similar to Eq. (1.19) for holes have been used.

The conductivity $\sigma(\Omega\cdot\text{cm})^{-1}$ is the (electric) concentration times the mobility:

$$\sigma = q(n\mu_n + p\mu_p) \tag{1.23}$$

The resistivity, which is the inverse of the conductivity and is denoted by ρ, is a quantity often determined to characterize a homogeneous semiconductor sample. If the semiconductor sample has a length L (from contact to contact) and a cross-sectional area A (width times height) with voltage V applied across the sample, the resistivity is simply

$$\rho = \frac{RA}{L} \tag{1.24}$$

where R is the resistance. The resistivity of both p- and n-type silicon at room temperature is shown in Fig. 1-8 as a function of dopant concentration.

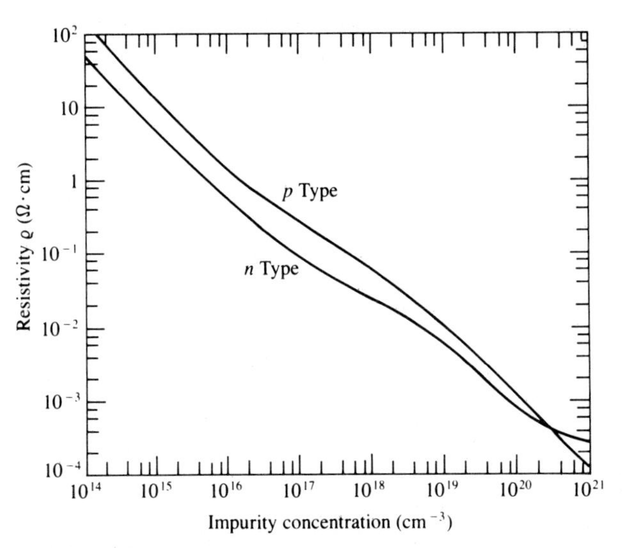

FIGURE 1-8
Resistivity of silicon at room temperature (Grove, 1967).

The diffusivity in the current flux relationships is related to the mobility by Einstein's relationship:

$$D = \frac{kT}{q}\mu \tag{1.25}$$

which is applicable to both electrons and holes.

Example 1.3. What fraction of the conduction in intrinsic silicon is due to electrons? For pure silicon, the conductivity is $5 \times 10^{-6} (\Omega\cdot\text{cm})^{-1}$. What concentration of boron atoms is required to increase the conductivity to $1 (\Omega\cdot\text{cm})^{-1}$? What fraction of the conduction in this extrinsic silicon is due to holes, electrons? Use the following information:

$$\mu_n = 1900\ \text{cm}^2/(\text{V}\cdot\text{s}) \qquad \mu_p = 425\ \text{cm}^2/(\text{V}\cdot\text{s}) \qquad \text{for Si at } 20\,^\circ\text{C}$$

Solution. The total conductivity in the intrinsic silicon is given by

$$\sigma = qn_i(\mu_n + \mu_p)$$

which follows from Eq. (1.23) since $n = p = n_i$. Therefore, the fraction of the conductance due to electrons is

$$\frac{\sigma_n}{\sigma} = \frac{\mu_n}{\mu_n + \mu_p} = \frac{1900}{1900 + 425} = 0.82$$

With regard to the boron concentration required to increase the conductivity, one may consider only the majority carrier in this case (holes) since the increase is from 5×10^{-6} to $1 (\Omega\cdot\text{cm})^{-1}$. From Eq. (1.23),

$$\sigma = qp\mu_p$$

or

$$p = \frac{\sigma}{q\mu_p} = \frac{1}{(0.16 \times 10^{-18}\ \text{A}\cdot\text{s})[425\ \text{cm}^2/(\text{V}\cdot\text{s})]}$$

$$= 1.5 \times 10^{16}\ \text{cm}^{-3}$$

Each boron atom contributes one acceptor site and hence one hole. Therefore 1.5×10^{16} boron atoms are required per cubic centimeter. This is, of course, a large number. However, it is still small (0.3 ppm) when compared with the number of silicon atoms per cubic centimeter, which is $5 \times 10^{22}\ \text{cm}^{-3}$. Needless to say, the conduction is almost entirely by holes since the electron concentration, which is n_i in this case, is $1.45 \times 10^{10}\ \text{cm}^{-3}$.

Example 1.4. Calculate the electron and hole concentrations, the resistivity, and the position of the Fermi level of a silicon wafer containing 10^{16} boron atoms/cm^3 and 0.8×10^{16} phosphorus atoms/cm^3 at 27 °C.

Solution. According to the problem, $N_A = 10^{16}$ and $N_D = 0.8 \times 10^{16}$ cm^{-3}. Also, $n_i = 1.45 \times 10^{10}$ cm^{-3}. The majority carrier here is holes since $N_A > N_D$. From Eq. (1.5),

$$p = \tfrac{1}{2}\{10^{16} - 1.8 \times 10^{16} + [(10^{16} - 0.8 \times 10^{16})^2 + (1.45 \times 10^{10})^2]^{1/2}\}$$

$$= 2 \times 10^{15} \text{ cm}^{-3}$$

From Eq. (1.9),

$$n = \frac{n_i^2}{N_A - N_D} = 1.05 \times 10^5 \text{ cm}^{-3}$$

The conductivity is given by Eq. (1.23):

$$\sigma = q(n\mu_n + p\mu_p)$$

$$= (1.6 \times 10^{-19} \text{ A·s})\{(1.05 \times 10^5)(1900) + (2 \times 10^{15})[425 \text{ cm}^2/(\text{V·s})]\}$$

$$= 0.136\ (\Omega\text{·cm})^{-1}$$

Thus, the resistivity is

$$\rho = \frac{1}{\sigma} = 7.35\ \Omega\text{·cm}$$

For the position of the Fermi level, one can divide Eq. (1.16) by the equation written for p_i to get

$$\frac{p}{p_i} = \frac{N_v e^{-(E_F - E_v)/kT}}{N_v e^{-(E_i - E_v)/kT}} = e^{(E_F - E_i)/kT}$$

where the subscript i is for the intrinsic semiconductor. The equation can be rewritten as

$$E_F = E_i - kT \ln \frac{p}{p_i}$$

$$= E_i - (8.62 \times 10^{-5} \text{ eV/K})(300 \text{ K}) \ln \frac{2 \times 10^{15}}{1.45 \times 10^{10}}$$

$$= E_i - 0.3 \text{ eV}$$

It is seen for this p-type semiconductor that the Fermi level is shifted down (closer to E_v or the valence band edge) by 0.3 eV compared to the intrinsic Fermi level E_i. For n-type semiconductors, the Fermi level is shifted upward closer to E_c than the intrinsic Fermi level.

Example 1.5. (*a*) Show that the minimum conductivity of doped silicon is smaller than the conductivity of the intrinsic silicon at 300 K. (*b*) Write a relationship for the calculation of the dopant concentration required for the extrinsic silicon to have a higher conductivity than the intrinsic silicon. (*c*) Comment on temperature effects. (*d*) Would a silicon doped with 10^{15} atoms/cm^3 of phosphorus be useful as an n-type semiconductor at 300 K?

Solution

(*a*) The conductivity of the intrinsic silicon is

$$\sigma_i = qn_i(\mu_n + \mu_p) \tag{A}$$

which follows from Eq. (1.23). For the minimum conductivity of an extrinsic silicon, Eq. (1.23) is rewritten as follows with the aid of Eq. (1.3):

$$\sigma = q\left(n\mu_n + \frac{n_i^2}{n}\mu_p\right) \tag{B}$$

Differentiating with respect to n and setting the derivative equal to zero yields

$$n_{\min} = n_i\left(\frac{\mu_p}{\mu_n}\right)^{1/2}; \qquad n_{\min}\text{: } n \text{ at which } \sigma \text{ is the minimum}$$

Using this in Eq. (B) yields, after some rearrangement, the following minimum σ, $\sigma_{\min}$:

$$\sigma_{\min} = 2qn_i(\mu_p \mu_n)^{1/2} \tag{C}$$

Taking the ratio of $\sigma_{\min}$ to σ_i yields

$$\frac{\sigma_{\min}}{\sigma_i} = \frac{2(\mu_p \mu_n)^{1/2}}{\mu_n + \mu_p} = \frac{2(1900 \times 425)^{1/2}}{1900 + 425}$$

$$= 0.77 < 1$$

(*b*) The relationship can be written as $\sigma/\sigma_i > 1$. With the aid of Eqs. (A) and (B), the relationship can be written as

$$\frac{\sigma}{\sigma_i} = \frac{(n/n_i)\mu_n + (n_i/n)\mu_p}{\mu_n + \mu_p} > 1$$

or

$$\frac{n}{n_i}\mu_n + \frac{n_i}{n}\mu_p > \mu_n + \mu_p \tag{D}$$

(*c*) Following the same procedures as in Example 1.4 for p/p_i, one can show that

$$\frac{n}{n_i} = e^{(E_F - E_i)/kT}$$

The mobilities, μ_n and μ_p, are also a function of temperature although it is of the order of $T^{\pm 1.5}$. Thus, it is dominated by the exponential temperature dependence of the concentrations.

(*d*) In order for an extrinsic semiconductor to be useful, Eq. (D) has at least to be satisfied since otherwise no advantage in conductance can be gained. Now $n = 10^{15}$ and $n_i = 1.45 \times 10^{10}$. Using these and the mobility values at 300 K, Eq. (D) becomes

$$\frac{10^{15}}{1.45 \times 10^{10}}(1900) + \frac{1.45 \times 10^{10}}{10^{15}}(425) = 1.31 \times 10^{8} \text{ cm}^2/(\text{V·s})$$

This value is much larger than the left-hand side of Eq. (D), which is 2325 $\text{cm}^2/(\text{V·s})$. It is seen that a five orders of magnitude increase in the conductivity has been gained.

1.4 BASIC UNITS OF INTEGRATED CIRCUITS

The building blocks for the current integrated circuits are metal-oxide semiconductor field-effect transistor (MOSFET) and bipolar junction transistor (BJT) for silicon-based ICs and metal semiconductor field-effect transistor (MESFET) for gallium arsenide–based ICs. In order to examine these basic units, let us consider a diode first.

A diode is similar to a one-way valve for fluid flow in that it conducts in one direction (forward) while blocking the flow of current in the opposite direction (reverse). The diode can be fabricated, for instance, by forming a *p*-type region in an *n*-type semiconductor. The interface between the *p* and *n* regions is called the *pn* junction. Since an *n*-type semiconductor contains a large concentration of electrons and a *p*-type semiconductor a large concentration of holes, one might think that electrons and holes would diffuse to even out the large concentration gradients existing between the regions across the metallurgical (*pn*) junction. Diffusion does take place. However, as holes leaving the *p* region leave behind uncompensated acceptor ions (N_A^-) and electrons diffusing from the *n* to the *p* region leave behind uncompensated donor ions (N_D^+) in the *n* material, there develops a region of negative space charge near the *p* side of the junction and a region of positive space charge near the *n* side, as shown in Fig. 1-9*a*. In this *space-charge region* (or sometimes called the *transition region*), therefore, an electric field develops in the direction opposite to that of diffusion current for each type of carrier. The field creates the drift component of the current opposing and exactly canceling the diffusion current at equilibrium ($V = 0$). Therefore, the net current is zero when no external field is applied. When a forward bias of magnitude V_f is applied to the diode by placing a battery, for instance, with the positive side of the battery connected to the *p* side (Fig. 1-9*b*), the charge from the battery repels holes and attracts electrons toward the *n* side of the junction. This results in a smaller length of the space-charge region and a lowering of the electrostatic potential (barrier) at the junction by qV_f. As the potential is lowered, more diffusion current develops and the current flows. When the diode is reverse biased, as shown in Fig. 1-9*c*, the charge from the battery attracts holes and electrons away from the junction. This results in a widening of the space-charge

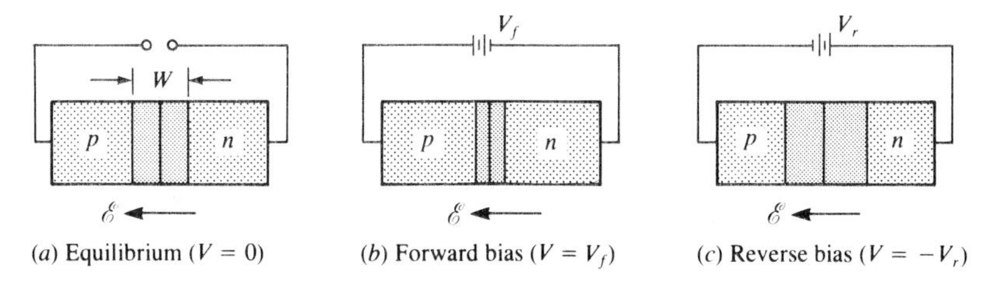

(*a*) Equilibrium ($V = 0$) (*b*) Forward bias ($V = V_f$) (*c*) Reverse bias ($V = -V_r$)

FIGURE 1-9
Effects of bias at a *pn* junction: transition region width and electric field (Streetman, 1980).

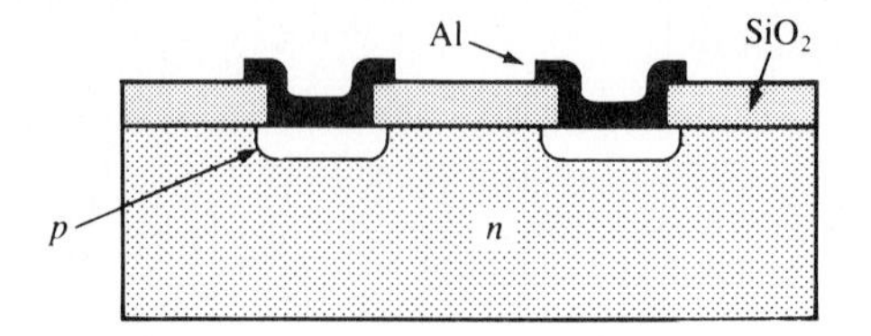

FIGURE 1-10
A typical side view of diodes. Two diodes are shown (Streetman, 1980).

region and an increase of the potential barrier by qV_r. Because of the large potential barrier, the diffusion current becomes negligible and almost no current flows when the diode is reverse biased. The bias condition corresponding to easy conduction is called forward bias; the condition corresponding to almost no conduction is called reverse bias. A typical side view of fabricated semiconductor diodes is shown in Fig. 1-10. The insulator SiO_2 isolates one diode from another and the metal Al contact is used for electrical intraconnection for the "wireless" wiring.

An important application of the *pn* junction in IC fabrication is its use for electrical isolation of many different types of active components. It should be clear from the discussion of the potential energy barrier at the junction that the junctions should be either reverse or zero biased with respect to the substrate if the junction is to be used as an isolation junction. The potential energy barrier at zero bias increases with increasing dopant concentrations.

Transistors are semiconductor devices with three leads. A very small current or voltage at one lead can control a much larger current flowing through the other two leads. This means transistors can be used as amplifiers and switches. There are two main families of transistors: bipolar and field effect.

A bipolar transistor results when a second junction is added to a *pn* junction diode. As shown in Fig. 1-11, the bipolar transistor can be *npn* or *pnp*. For both, the middle layer acts as a faucet or gate that controls the current moving through the three layers. The three layers of a bipolar transistor are the emitter,

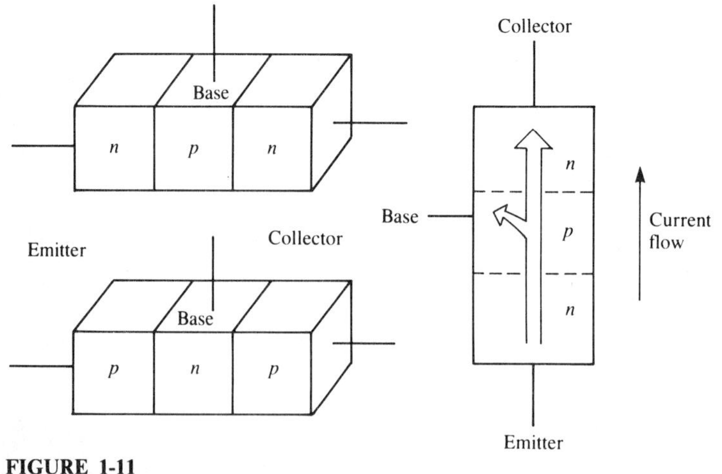

FIGURE 1-11
The *npn* and *pnp* bipolar transistors.

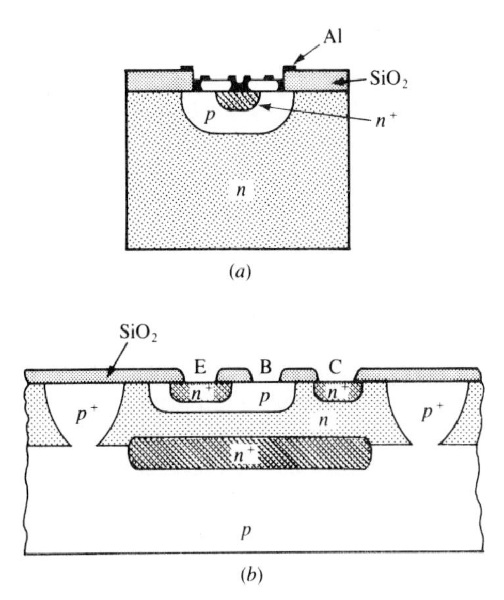

FIGURE 1-12
An *npn* transistor arrangement and a typical basic unit of *npn* bipolar junction IC transistors (Streetman, 1980; Hodges and Jackson, 1983).

base, and collector. The base is very thin and has fewer dopant atoms than the emitter and collector. Therefore, a very small emitter-base current will cause a much larger emitter-collector current to flow.

An *npn* transistor can be fabricated as shown in Fig. 1-12*a*. A typical *npn* junction isolated bipolar transistor for integrated circuits is also shown in Fig. 1-12*b*. Since all three contacts must be on the top surface of the chip, the collector current must pass along a high-resistance path in the lightly doped *n* material (Fig. 1-12*b*) while flowing from the active part of the collector to the contact. One common method of decreasing the collector resistance is to include a heavily doped *n* layer (n^+) just below the collector. This n^+ layer is referred to as the buried layer. Since a *pn* junction can be used to isolate one unit of transistor from another because of the electrostatic potential barrier at the interface, the isolation between units is done by forming *p* layers as shown in Fig. 1-12*b*. Bipolar junction transistors are usually *npn* type rather than *pnp* type because of the simple fabrication involved.

Field-effect transistors (FETs) have become more important than bipolar transistors. In fact, FETs based on metal-oxide semiconductor technology has become the dominant IC technology, particularly for logic ICs. Most microcomputer and memory ICs are arrays of thousands of MOSFETs on a small sliver of silicon. In FETs, an output current is controlled by a small input voltage and practically no input current.

The MOSFETs can be either *n* (*n* channel) or *p* (*p* channel) type depending on whether electrons or holes are responsible for the conductance. Because of higher electron mobility, *n* type or NMOS is preferred over *p* type or PMOS. A combination of NMOS and PMOS is called complimentary MOS or CMOS. A simple side view of an *n*-channel MOSFET is shown in Fig. 1-13. If no voltage is applied to the gate, no current flows from the drain to the source since the drain-

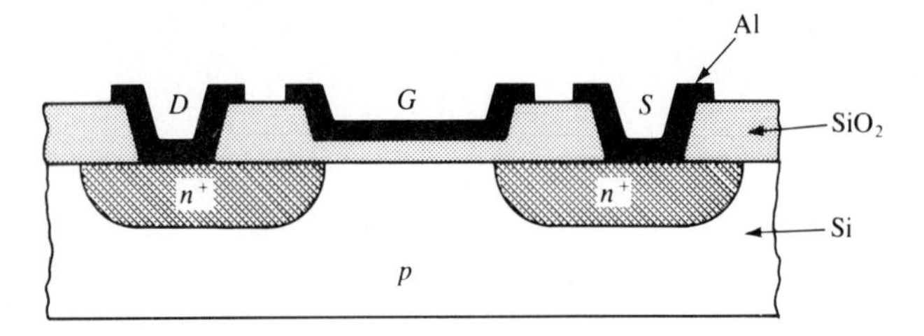

FIGURE 1-13
An n-channel MOSFET basic unit of integrated circuits (Streetman, 1980).

substrate-source combination includes oppositely directed *pn* junctions (isolation) in series. In order for the current to flow, there should be a conducting n channel between the source and the drain (both n^+). When a positive gate voltage is applied to the gate, positive charges are in effect deposited on the gate metal. In response, negative charges are induced in the underlying silicon. Thus, electrons are attracted to the region below the gate, creating a thin n-type channel in the p-type silicon between the source and the drain. This field effect allows the current to flow from the drain to source. To create the channel that connects the source to the drain, the applied voltage has to be above a certain threshold, called threshold voltage. For logic (digital) circuits, this means that the MOSFET can be operated at two states of "on" or "off" depending on the voltage applied.

Example 1.6. Complimentary MOS is a combination of NMOS and PMOS. Draw a simplified side view of a CMOS for an n-type substrate.

Solution. The side view can be drawn from the NMOS in Fig. 1-13. If an n-type substrate is used, there should be a p region or p-type tub in the substrate in the n-channel part:

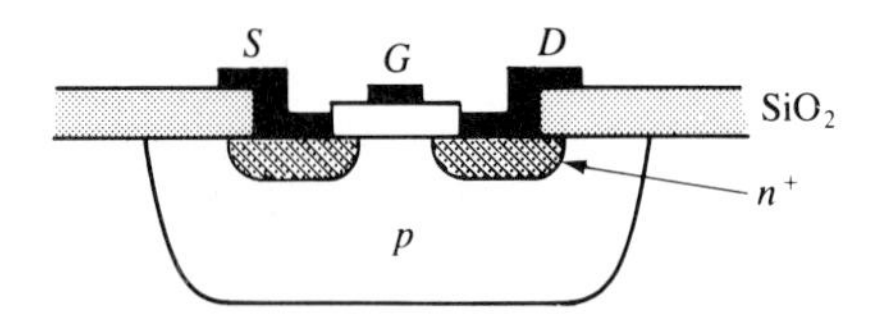

One can then add on the p channel alongside the n channel:

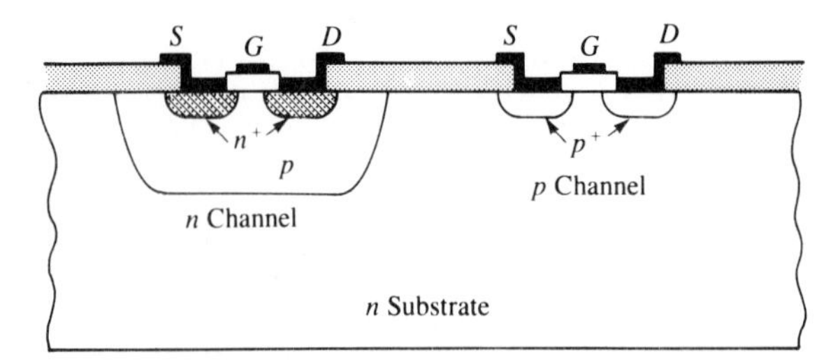

This should be sufficient for the example, although additional isolation may be added by introducing p and n^+ regions between the channels.

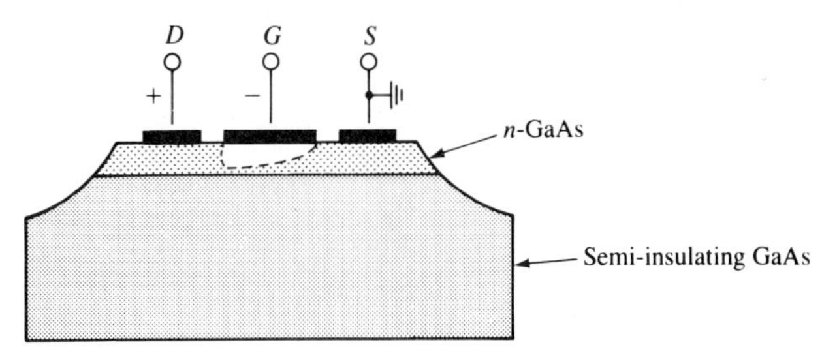

FIGURE 1-14
A simple MESFET structure (Streetman, 1980).

The metal semiconductor field-effect transistor (MESFET) is the dominant building block of gallium arsenide–based integrated circuits. There are a number of difficulties in the processing of gallium arsenide for MOS structure with respect to diffusion, formation of insulator, and passivation (for protection from the environment). On the other hand, gallium arsenide offers several advantages in the MESFET structure such as higher speed and packing density, and wider bandgap (E_g). These are the reasons for the MESFET structure for gallium arsenide.

The MESFET works in much the same way as the junction field-effect transistor (JFET). A simple MESFET in GaAs is shown in Fig. 1-14. The MESFET is operated with reverse-biased metal Schottky gate and ohmic contacts for the drain and source. The substrate is a semi-insulating gallium arsenide made by introducing a suitable impurity such as chromium so that the Fermi level is pinned near the center of the gap, resulting in a very high resistivity material. This is called semi-insulating gallium arsenide. When no voltage is applied, current flows through the *n* channel of the *n*-GaAs shown in Fig. 1-14. When the gate is reverse biased, a depletion region (holes attracted to the region under the gate) develops and an induced *pn* junction forms when the depletion extends to the semi-insulating GaAs. Therefore, no current will flow. The MESFET is usually *n* type because of higher electron mobility. As shown in Fig. 1-14, isolation between device elements is done by etching through the *n* region to the semi-insulating substrate, which is known as mesa isolation.

1.5 ELEMENTARY DEVICE PHYSICS

In addition to the current-density equations given by Eqs. (1.21) through (1.23), there are two other basic sets of equations for semiconductor device operation. Among the set for Maxwell equations, the most important is the Poisson equation:

$$-\frac{\partial^2 V}{\partial x^2} = \frac{\partial \varepsilon}{\partial x} = \frac{\rho}{\varepsilon_s} \tag{1.26}$$

where V is the voltage, ε_s is the permittivity, ρ is the total electric charge density, and ε is the electric field. Continuity equations constitute the other set:

$$\frac{\partial J_n}{\partial x} = q(R_n - G_n) \tag{1.27}$$

$$\frac{\partial J_p}{\partial x} = q(R_p - G_p) \tag{1.28}$$

where R is the recombination rate, G is the generation rate, and the subscripts n and p, respectively, are for electrons and holes. The basic equations have been written in one dimension in the direction of current flow at steady state. All device descriptions follow from the three basic sets of equations, although they are reduced to specific forms depending on the type of devices. A constitutive equation is the Fermi-Dirac probability given by Eq. (1.10). This can often be approximated by the Boltzmann probability:

$$f(E) = \exp\left(\frac{E_F - E}{kT}\right) \tag{1.29}$$

In this section, only the bare minimum relationships will be considered for semiconductor device operation. These include electrical characteristics of a *pn* junction diode, bipolar junction transistor, and MOSFET (MIS) transistor. Details on these devices can be found elsewhere (e.g., Streetman, 1980; Sze, 1981). A *pn* junction diode is considered first.

Suppose an *n*-type material and a *p*-type material are joined together forming a junction, as shown in Fig. 1-15. Before they are joined, the *n* material has a large concentration of electrons and few holes whereas the converse is true for the *p* material. Upon joining, however, diffusion of carriers takes place because of the large carrier concentration gradients at the junction. Thus holes diffuse from the *p* side to the *n* side and electrons diffuse from the *n* side to the *p* side. As the holes and electrons diffuse, they leave behind uncompensated ions near the junction: uncompensated acceptors (N_A^-: negative acceptor ions) in the *p* region and uncompensated donors (N_D^+: positive donor ions) in the *n* region. Because of the development of this space charge and the corresponding electric field, the diffusion current (the current due to diffusion) cannot build up indefinitely. The electric field opposing the diffusion determines the behavior of the space-charged region, also known as the transition or depletion region, along with the impurity concentrations N_A and N_D.

The minority carrier concentration on each side of a *pn* junction varies with the applied bias because of variations in the diffusion of carriers across the junction. It can be shown (see Prob. 1.8) that the ratio of the minority carrier hole concentration at the edge of the transition region on the *n* side, $p(x_{n0})$, to that at equilibrium (no bias), p_{ne}, is given by

$$\frac{p(x_{n0})}{p_{ne}} = \exp\left(\frac{qV}{kT}\right) \tag{1.30}$$

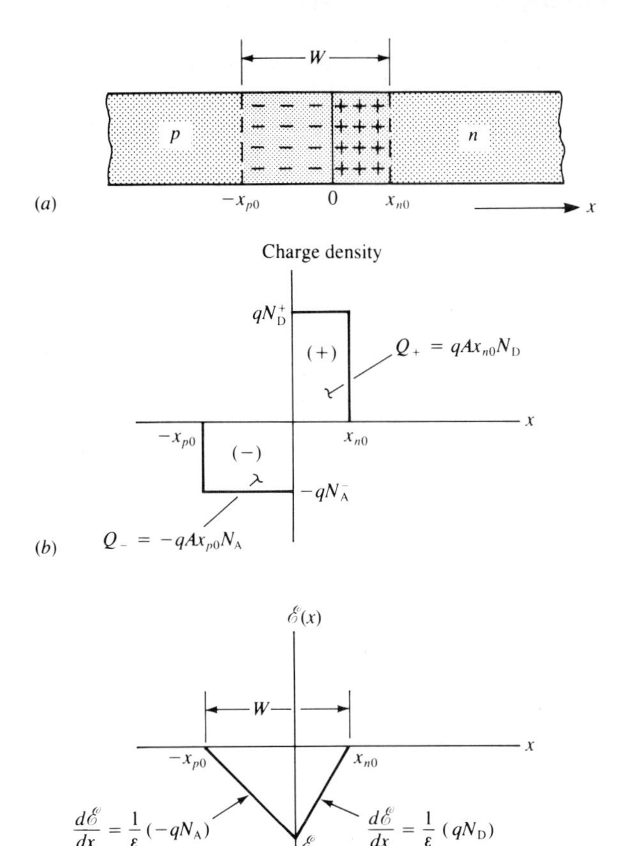

FIGURE 1-15
Space-charge and electric field distribution within the transition region of a *pn* junction with $N_D > N_A$: (*a*) transition region with $x = 0$ at metallurgical junction, (*b*) charge density within the transition region, and (*c*) electric field distribution (Streetman, 1980).

where V is V_f for forward bias and $-V_r$ for reverse bias. It can be seen that the minority carrier concentration at the edge of the transition region increases exponentially with forward bias. This increase in the carrier concentration due to the applied bias is termed (minority) carrier injection. The excess carrier concentration of holes Δp_n on the n side follows directly from Eq. (1.30):

$$\Delta p_n = p(x_{n0}) - p_{ne} = p_{ne}(e^{qV/kT} - 1) \tag{1.31}$$

Similarly, for excess electrons on the p side, one has

$$\Delta n_p = n(-x_{p0}) - n_{pe} = n_{pe}(e^{qV/kT} - 1) \tag{1.32}$$

The excess holes in the n side as given by Eq. (1.31) with forward bias represent the amount of injection of holes from the p side to the n side. Similarly, the excess electrons in the p side represent the amount of electron injection from the n side to the p side. Under reverse bias, Δp_n approaches $-p_{ne}$ and Δn_p approaches $-n_{pe}$ for $V_r \gg kT/q$, which means that the charge carriers are depleted from the transition region. This reverse-bias depletion of minority carriers can be thought of as minority carrier extraction.

Outside the transition region or in the neutral region, there is no electric field and no regeneration taking place. Thus, the steady-state continuity equation [Eq. (1.27)] for holes reduces to

$$D_p \frac{\partial^2 p_n}{\partial x^2} = -qR_p = -\frac{q[p(x_{n0}) - p_n]}{\tau_p} \tag{1.33}$$

where τ_p is the hole lifetime and the recombination rate for low injection levels is given by $[p(x_{n0}) - p_n]/\tau_p$. The solution of Eq. (1.3) with the boundary conditions of Eq. (1.31) and $p_n(x_n = \infty) = p_{ne}$ is

$$\begin{aligned} p_n - p_{ne} &= p_{ne}(e^{qV/kT} - 1)e^{-(x_n - x_{n0})/L_p} \\ L_p &= (D_p \tau_p)^{1/2} \end{aligned} \tag{1.34}$$

The flux at $n_n = x_{n0}$ from Eq. (1.34) is

$$J_p = -qD_p \frac{\partial p_n}{\partial x}\bigg|_{x_{n0}} = \frac{qD_p p_{ne}}{L_p}(e^{qV/kT} - 1) \tag{1.35}$$

Similarly, one has for the p side:

$$J_n = qD_n \frac{\partial n_p}{\partial x}\bigg|_{-x_{p0}} = \frac{qD_n n_{pe}}{L_n}(e^{qV/kT} - 1) \tag{1.36}$$

The total current is the sum of J_p and J_n multiplied by the cross-sectional area A perpendicular to the current flow:

$$I = A(J_p + J_n) = I_0(e^{qV/kT} - 1) \tag{1.37}$$

$$I_0 \equiv \left(\frac{qD_p p_{ne}}{L_p} + \frac{qD_n n_{pe}}{L_n}\right)A \tag{1.38}$$

Equation (1.37) is the Shockley equation for ideal diodes, also known as the diode equation. This I-V characteristic of an ideal *pn* junction is shown in Fig. 1-16. When V is positive (forward biased), the exponential term is much greater than unity ($kT/q = 0.0259$ V at room temperature) and thus the current increases exponentially with forward bias. When V is negative (reverse biased), the exponential term approaches zero and the current is $-I_0$, which is in the n to p (negative) direction. This is called the reverse saturation current. This reverse saturation current depends on the rate at which holes arrive at x_{n0} (and electrons at x_{p0}) by diffusion from the neutral region, which can be seen from Eqs. (1.35) and (1.37) for $V_r \gg kT/q$ or Eq. (1.38). As seen in the figure, current flows relatively freely in the forward direction of the diode, but almost no current flows in the reverse direction.

The function of the transistor as a signal amplifier or for switching can be achieved by a *pnp* or *npn* junction, better known as a bipolar junction transistor

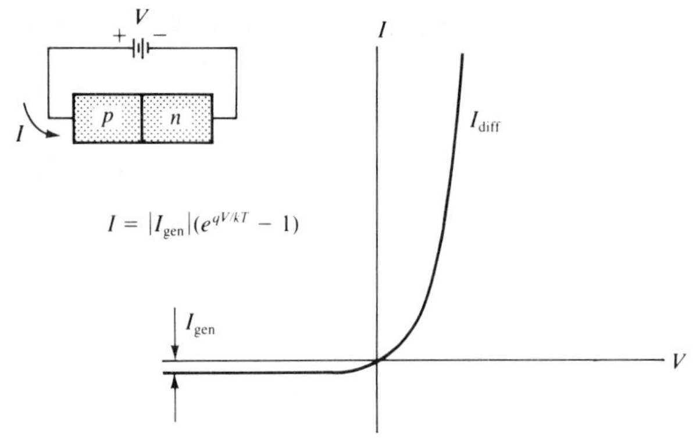

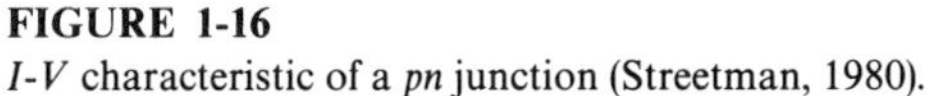

FIGURE 1-16
I-V characteristic of a *pn* junction (Streetman, 1980).

(BJT). Consider a *pnp* transistor connected in the common-base configuration, shown in Fig. 1-17, for an amplifier application. The *pn* junction on the emitter side or the emitter junction is forward biased whereas the *np* junction on the collector side or the collector junction is reverse biased. The object of a BJT is to produce a controllable constant output (the collector current I_C in Fig. 1-17) for a given input of the emitter current I_E. In a *pn* junction under reverse bias, the constant reverse saturation current from *n* to *p* depends on the hole injection rate and is essentially independent of the bias voltage, as Eq. (1.38) shows. The rate of hole injection, on the other hand, can be controlled by a *pn* junction under forward bias. Thus, in the arrangement of the *pnp* junctions shown in Fig. 1-17, the emitter side *pn* junction, which is forward biased, injects holes to the base. In the collector side *pn* junction (*np*), which is reverse biased, the holes injected into

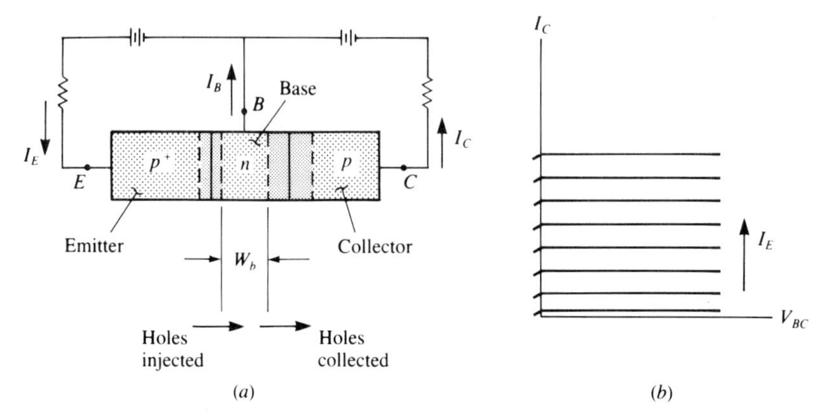

FIGURE 1-17
A *pnp* transistor: (*a*) schematic representation with forward-biased emitter and reverse-biased collector, and (*b*) *I-V* characteristic of the reverse-biased *np* junction as a function of emitter current (Streetman, 1980).

the base are then collected to yield the collector current. This structure of a *pnp* BJT gives the output (I_C) as shown in Fig. 1-17 for the base-collector bias V_{BC}.

For a good *pnp* transistor, almost all the holes injected by the emitter into the base should be collected. Thus the *n*-type base region should be narrow such that the neutral length of the base W_b is much less than the hole diffusion length. Only then would an average hole injected at the emitter junction diffuse to the transition region of the collector junction without recombining in the base. Further, the current crossing the emitter junction should be composed almost entirely of holes injected into the base rather than electrons crossing from base to emitter. This can be satisfied by doping the base region lightly compared with the emitter. Thus the usual arrangement is to dope the emitter heavily so that the p^+n emitter junction results.

For the performance of a BJT, several quantities can be defined. These are the emitter injection efficiency γ, current transfer ratio α, and base-collector current amplification factor β:

$$\gamma \equiv \frac{I_{Ep}}{I_{En} + I_{Ep}} \tag{1.39}$$

$$\alpha \equiv \frac{I_C}{I_E} = \frac{I_{Ep}}{I_E}\frac{I_C}{I_{Ep}} = \gamma \frac{I_C}{I_{Ep}} \tag{1.40}$$

$$\beta \equiv \frac{I_C}{I_B} = \frac{\alpha}{1-\alpha} \tag{1.41}$$

where I_{Ep} is the emitter current due to holes and I_{En} is that due to electrons. The last part of Eq. (1.41) follows from Eq. (1.40) and $I_E = I_B + I_C$. For an efficient BJT, α should be close to unity, which in turn means that the amplification factor β should be very high. Also, the emitter efficiency should be close to unity.

Continuity equations can be solved as in *pn* junctions for ideal *I-V* characteristics. Although there are two junctions in a BJT, it is sufficient to consider only the continuity equations for the neutral region in the base, since the currents are determined by the charge carrier behavior in the two transition regions about the base, which serve as the boundary conditions for the continuity equations. When both forward and reverse biases are large and the emitter is heavily doped, the solutions can be reduced to the following approximate results:

$$I_B = a_1 \tanh\left(\frac{W_b}{2L_p}\right) \tag{1.42}$$

$$I_C = a_1 \operatorname{csch}\left(\frac{W_b}{L_p}\right) \tag{1.43}$$

$$I_E = a_1 \coth\left(\frac{W_b}{L_p}\right) \tag{1.44}$$

$$a_1 = \frac{qAD_p\Delta p_E}{L_p} \qquad \Delta p_E = p_{Be}(e^{qV_{EB}/kT} - 1) \tag{1.45}$$

where L_p is the diffusion length of holes in the base region and p_{Be} here is the equilibrium hole concentration in the base.

Example 1.7. The continuity equation of holes in the neutral region is the same as Eq. (1.33). The boundary conditions are

$$p = \begin{cases} p_{Be}[\exp(qV_{EB}/kT) - 1] = \Delta p_E & \text{at } x = 0 \\ p_{Be}[\exp(qV_{CB}/kT) - 1] = \Delta p_C & \text{at } x = W_b \end{cases}$$

Similarly for electrons, one has

$$D_n \frac{d^2 n}{dx^2} = \frac{n - n_{pe}}{\tau_n}$$

with

$$n = \begin{cases} n_{Ee}[\exp(qV_{EB}/kT) - 1] = \Delta n_E & \text{at } x = -x_E \\ n_{Ce}[\exp(qV_{CE}/kT) - 1] = \Delta n_C & \text{at } x = x_C \end{cases}$$

The electron continuity equation is applicable in the neutral regions of the emitter and collector. The solutions are

$$p(x) = p_{Be} + \left[\frac{\Delta p_C - \Delta p_E e^{-W_b/L_p}}{2\sinh(W_b/L_p)}\right] e^{x/L_p} - \left[\frac{\Delta p_C - \Delta p_E e^{W_b/L_p}}{2\sinh(W_b/L_p)}\right] e^{-x/L_p}$$

$$n(x) = \begin{cases} n_{Ee} + \Delta n_E \exp[(x + x_E)/L_E] & x < -x_E \\ n_{Ce} + \Delta n_C \exp[-(x - x_C)/L_C] & x > x_C \end{cases}$$

where L_E and L_C are the diffusion lengths defined by $(D_n \tau_{nE})^{1/2}$ and $(D_n \tau_{nC})^{1/2}$, respectively. τ_{nC} is the electron lifetime in the collector and τ_{nE} is the same in the emitter. Show that Eqs. (1.42) through (1.44) result when $V_{EB} \gg kT/q$, $V_{CB} \ll 0$, and the emitter is heavily doped compared with the base doping level.

Solution. The emitter current is given by

$$I_E = A\left(-qD_p \frac{dp}{dx}\bigg|_{x=0}\right) + A\left(-qD_n \frac{dn}{dx}\bigg|_{x=-x_E}\right)$$

$$= Aq \frac{D_p p_{Be}}{L_p} \coth\left(\frac{W_b}{L_p}\right)\left[(e^{qV_{EB}/kT} - 1) - \frac{1}{\cosh(W_b/L_p)}(e^{qV_{CB}/kT} - 1)\right]$$

$$+ Aq \frac{D_n n_{Ee}}{L_E}(e^{qV_{EB}/kT} - 1) \qquad \text{(A)}$$

The collector current is given by

$$I_C = A\left(-qD_p \frac{dp}{dx}\bigg|_{x=W_b}\right) + A\left(-qD_n \frac{dn}{dx}\bigg|_{x=x_C}\right)$$

$$= Aq \frac{D_p p_{Be}}{L_p} \frac{1}{\sinh(W_b/L_p)}\left[(e^{qV_{EB}/kT} - 1) - \coth\left(\frac{W_b}{L_p}\right)(e^{qV_{CB}/kT} - 1)\right]$$

$$- Aq \frac{D_n n_{Ce}}{L_c}(e^{qV_{CB}/kT} - 1) \qquad \text{(B)}$$

The base current is then obtained from

$$I_B = I_E - I_C \tag{C}$$

When $V_{EB} \gg kT/q$, $e^{qV_{EB}/kT} - 1 = e^{qV_{EB}/kT}$. Also, $e^{qV_{CB}/kT} \approx 0$ for $V_{CB} \ll 0$. Further, $p_{Be} \gg n_{Ee}$ for a heavily doped emitter. Using these in Eq. (A) yields

$$I_E \approx Aq \frac{D_p p_{Be}}{L_p} \coth\left(\frac{W_b}{L_p}\right)\left[(e^{qV_{EB}/kT} - 1) + \frac{1}{\cosh (W_b/L_p)}\right]$$

When $W_b/L_p \gg 1$, $\cosh (W_b/L_p) \gg 1$. Thus, the above equation reduces to Eq. (1.44). Using the same conditions in Eq. (B) leads to Eq. (1.43). Neglecting the last term in Eqs. (A) and (B), which are the currents due to electrons ($p_{Be} \gg n_{Ee}, n_{Ce}$), and substituting them into Eq. (C) yields

$$I_B = qA \frac{D_p}{L_p}\left[(\Delta p_E + \Delta p_C)\left(\text{ctnh} \frac{W_b}{L_p} - \text{csch} \frac{W_b}{L_p}\right)\right]$$

$$= qA \frac{D_p}{L_p}\left[(\Delta p_E + \Delta p_C) \tanh \frac{W_b}{2L_p}\right]$$

This reduces to Eq. (1.42) since $\Delta p_E \gg \Delta p_C$.

The emitter efficiency γ and the current transfer ratio α follow from the currents given by Eqs. (A) and (B) in Example 1.7. From Eq. (1.39), one has

$$\gamma = \frac{1}{1 + I_{En}/I_{Ep}}$$

$$= \frac{1}{1 + (D_n n_{Ee} L_p/D_p p_{Be} L_E) \tanh (W_b/L_p)} \tag{1.46}$$

where the second term within the bracket has been neglected in comparison with the first in Eq. (A). Since I_E is almost entirely due to I_{Ep}, Eqs. (1.43) and (1.44) can be used to obtain I_C/I_{Ep}, which is known as the base transport factor for use in Eq. (1.40). It follows then from Eqs. (1.43) through (1.45) that

$$\alpha = B\gamma = \left(\cosh \frac{W_b}{L_p} + \frac{L_p D_n n_{Ee}}{L_E D_p p_{Be}} \sinh \frac{W_b}{L_p}\right)^{-1} \tag{1.47}$$

where B is the base transport factor.

The fundamental relationships for a BJT are essentially determined from three considerations. First, the applied voltages control the boundary carrier densities through the term exp (qV/kT). Second, the emitter and collector currents are given by the minority density gradients at the junction boundaries. Third, the base current is the difference between the emitter and collector currents. As opposed to the *pnp* transistor just considered the three current directions are reversed in an *npn* transistor since electrons flow from emitter to collector and holes must be supplied to the base. However, the operation can be understood simply by reversing the roles of electrons and holes in the *pnp* discussion.

Just as a *pn* junction constitutes a basis for many electronic devices, so is a basic structure known as a metal-insulator semiconductor (MIS), particularly for

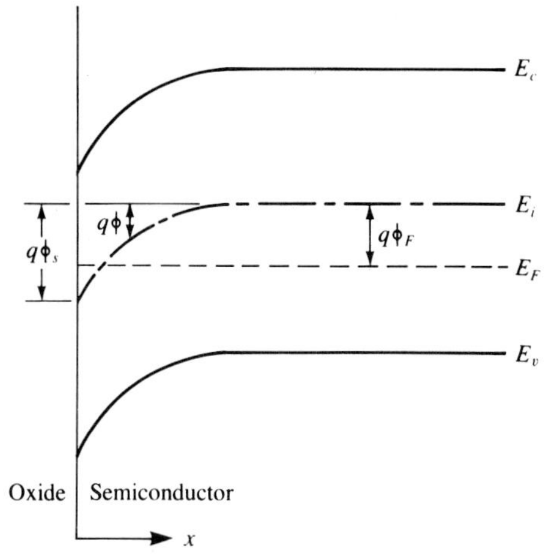

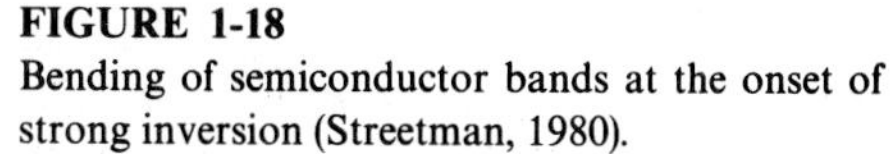

FIGURE 1-18
Bending of semiconductor bands at the onset of strong inversion (Streetman, 1980).

digital ICs. When the insulator is an oxide, it is known as a MOS (metal-oxide semiconductor). The MIS structure is a capacitor arrangement. An *n*-channel MOSFET is shown in Fig. 1-13. When a positive voltage is applied to the gate (*G* in Fig. 1-13), it deposits positive charge on the metal, which in turn induces a corresponding net negative charge at the surface of the semiconductor. This inverted charge region in the *p* semiconductor under the insulator becomes a conducting channel for electrons and allows the current to flow from the drain to the source. The characteristics of a MOSFET, therefore, are largely determined by the behavior near the insulator-semiconductor interface.

Consider the energy bands near the insulator (oxide)-semiconductor interface shown in Fig. 1-18. When there is no voltage applied to the gate, the energy bands in the semiconductor are flat. When the voltage is applied, however, the semiconductor energy bands bend near the interface to accommodate the accumulation of electrons. If one defines a potential ϕ at any point x, relative to the equilibrium position of the intrinsic Fermi level E_i, the energy $q\phi$ represents the extent of band bending at x, and $q\phi_s$ is the bending at the surface. For the p semiconductor being considered, the bands bend up at the surface when $\phi_s < 0$ (negative gate voltage) and holes accumulate at the interface. When $\phi_s > 0$ (positive gate voltage) as shown in Fig. 1-18, depletion of holes occurs from the region near the surface, leaving behind uncompensated ionized acceptors, and the bands bend down. When ϕ_s is larger than ϕ_F, the bands at the surface are bent down such that E_i at the surface lies below the semiconductor Fermi level E_F, and inversion is obtained in the sense that an n region is created in the p-type semiconductor. A true n-type conducting channel exists at the surface when the surface is as strongly n type as the substrate is p type. This strong inversion occurs when

$$(\phi_s)_{\text{inv}} = 2\phi_F = 2\frac{kT}{q}\ln\frac{N_A}{n_i} \tag{1.48}$$

An important quantity for MOSFET operation is threshold voltage (or turn-on voltage), which is defined as the voltage required for strong inversion. To obtain an expression for the threshold voltage, three factors have to be considered in addition to the potential given by Eq. (1.48). One of them has to do with the assumption that the semiconductor bands are flat when the applied voltage is zero. This is valid only when the work function of the metal is the same as that of the semiconductor (work function here may be defined as the energy required to move an electron from the Fermi level to outside the material) or when the metal Fermi level is the same as the semiconductor Fermi level. Such is usually not true. First, the semiconductor Fermi level changes with doping. Invariably, therefore, bending of energy bands occurs just as in Fig. 1-18 because of the difference in the Fermi levels and the corresponding electric field. This in turn means that the band bending occurs even when the gate voltage is zero. Another factor has to do with the voltage drop due to the charge at the insulator-semiconductor interface induced by interface states. The voltage required for flat bands or the flat-band voltage V_{FB} is therefore given by

$$V_{FB} = \phi_{ms} - \frac{Q_i}{C_i} \qquad \phi_{ms} \equiv \phi_m - \phi_s \tag{1.49}$$

where ϕ_m and ϕ_s are the work functions of the metal and semiconductor, respectively, Q_i is the charge per unit area at the interface, and C_i is the insulator capacitance per unit area. The last factor has to do with the charge in the depleted region, which is required for strong inversion in addition to the surface potential. Therefore, the threshold voltage V_T is given by

$$V_T = V_{FB} - \frac{Q_d}{C_i} + 2\phi_F \tag{1.50}$$

where Q_d is the charge per unit area in the depletion region. The applied gate voltage consists of the voltage required to achieve the flat band, the induced charge in the semiconductor Q_s, and the surface potential:

$$V_G = V_{FB} - \frac{Q_s}{C_i} + \phi_s \tag{1.51}$$

In a MOSFET, the applied drain voltage V_D has to be taken into consideration for the surface potential $\phi_s(x)$. Thus the surface potential ϕ_s is that required for strong inversion plus voltage $V_x(x)$ due to the applied drain voltage, for which $V_x = V_D$ at $x = 0$ (Fig. 1-19). The induced charge Q_s in the semiconductor consists of the mobile charge Q_n and fixed charge in the depletion region Q_d. Substituting $Q_s = Q_d + Q_n$ and $\phi_s = 2\phi_F + V_x$ into Eq. (1.51) and solving for the mobile charge, one has

$$Q_n = -C_i(V_G - V_T - V_x) \tag{1.52}$$

where Eq. (1.50) has been used and the variation of Q_d with V_x has been neglected for approximate relationships. The conductance of the differential element dx in

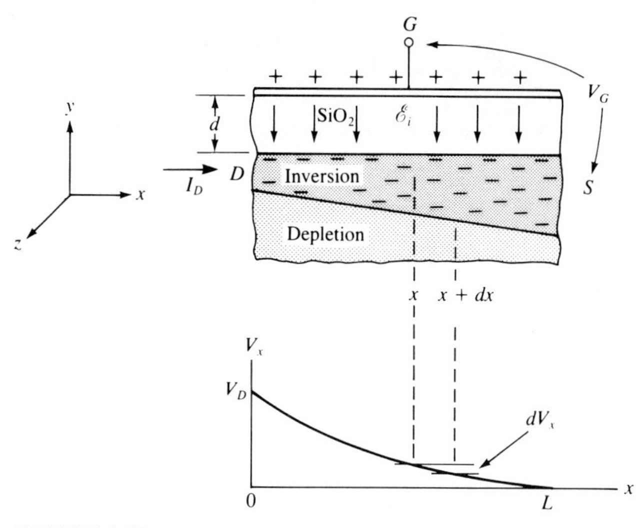

FIGURE 1-19
The *n*-channel region of a MOS transistor under bias below pinch-off, and variation of voltage V_x along the conducting channel (Streetman, 1980).

Fig. 1-19 is $\bar{\mu}_n Q_n(x)Z/dx$, where Z is the depth of the channel and $\bar{\mu}_n$ is the surface electron mobility. Thus, one has at point x,

$$I_D\, dx = \bar{\mu}_n Z Q_n(x)\, dV_x \tag{1.53}$$

Integrating from drain to source, one has

$$\int_D^L I_D\, dx = -\mu_n Z C_i \int_{V_D}^{0} (V_G - V_T - V_x)\, dV_x$$

$$I_D = \frac{\bar{\mu}_n Z C_i}{L} [(V_G - V_T)V_D - \tfrac{1}{2}V_D^2] \tag{1.54}$$

which is an approximate *I-V* characteristic of a MOSFET. Here L is the channel length. More accurate results for the *I-V* characteristics (Streetman, 1980) are shown in Fig. 1-20. An approximate expression for the saturation drain current (the asymptotic current value in Fig. 1-20) can be obtained by setting $\partial I_D/\partial V_D$ equal to zero, which yields an approximate relationship for the saturation drain voltage:

$$(V_D)_{\text{sat}} = V_G - V_T \tag{1.55}$$

Use of this equation in Eq. (1.54) yields

$$(I_D)_{\text{sat}} = \frac{Z}{2L} \bar{\mu}_n C_i (V_D)_{\text{sat}}^2 \tag{1.56}$$

As shown in Fig. 1-20, the voltage across the oxide decreases near the drain and Q_n becomes smaller there as the drain voltage is increased [see Eq. (1.53) with

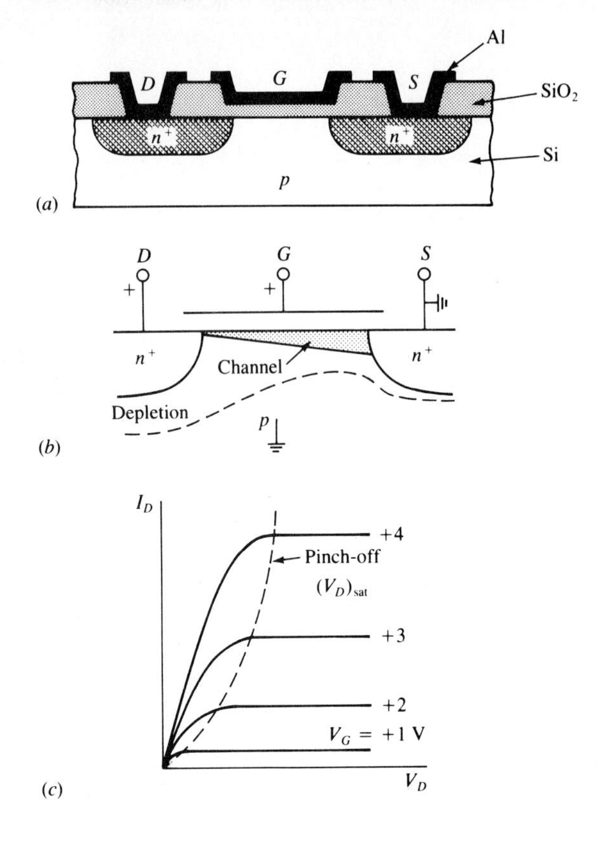

FIGURE 1-20
An enhancement-type n-channel MOS transistor: (*a*) cross section of the device, (*b*) induced n channel and depletion region near pinch-off, and (*c*) drain I-V characteristics as a function of gate voltage (Streetman, 1980).

$V_x = V_D$]. As a result the channel becomes pinched off at the drain end and the current saturates. Equation (1.56) gives an approximate value of this saturation current. The region in which I_D is proportional to V_D is called the linear region as opposed to the saturation region.

1.6 BIPOLAR JUNCTION TRANSISTOR DESIGN: AN EXAMPLE

Design of an IC (integrated circuit) for fabrication of a bipolar junction transistor (BJT) with a buried layer is considered in this section. The design and fabrication sequence does not necessarily represent the standard design but is simply being used as an example to illustrate how fabrication can be specified up to the final step of circuit layout. This example is due to Colclaser (1980), and more details can be found there.

Consider the n^+pn^+ transistor shown schematically in Fig. 1-21. As discussed in the previous section, one major design consideration is the base-collector current amplification factor, or current gain, β. Another major consideration is the cutoff frequency, which can be defined as the frequency at which the ac amplification for the device (base-collector gain that is essentially

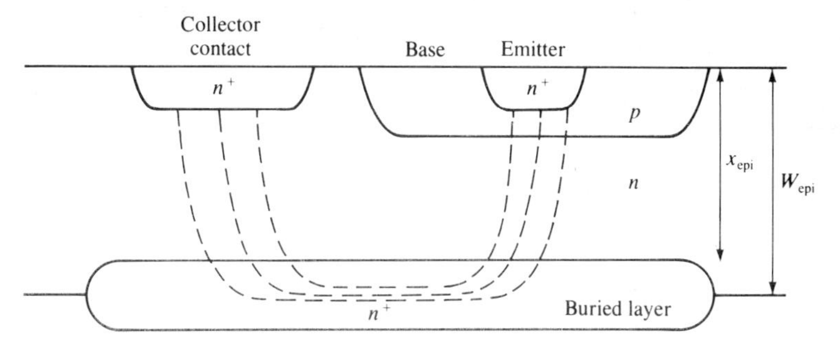

FIGURE 1-21
Schematic of an n^+pn^+ transistor.

the same as the dc gain β) drops to unity. The alpha cutoff frequency is similar to the cutoff frequency but is based on the transit time of the minority carrier through the bases, τ_B. The alpha cutoff frequency, f_α, is defined as the frequency at which the gain has fallen to $1/\sqrt{2}$ of its low-frequency value. The value of f_α is given by

$$f_\alpha = \frac{1}{2\pi\tau_B} \tag{1.57}$$

The time delay τ_B is given by

$$\tau_B = \frac{W_b^2}{\eta D_n} \tag{1.58}$$

where η is dependent on the doping level ($\eta = 2$ for uniformly doped base) and the applied field. In addition, there are two constraints that have to be satisfied for proper operation of the transistor. The first is the constraint on supply voltages that can be applied to the circuit without causing junction breakdown. Under reverse-bias conditions, there are two causes of junction breakdown. The first is that caused by field-induced tunneling, usually between two heavily doped regions. This breakdown is known as the Zener effect. The other breakdown, known as avalanche multiplication, is due to electron-hole pair generation by carriers accelerated by the electric field. The junction breakdown voltage (BV) is often correlated to the collector multiplication factor M as follows:

$$M = \frac{1}{1 - [V_{CB}/(\mathrm{BV})_{CBo}]^n} \tag{1.59}$$

where n is a constant and $(\mathrm{BV})_{CB_o}$ is the breakdown voltage for an open circuit. Although the current gain α was expressed as $B\gamma$, a more rigorous way of writing the gain is as follows:

$$\alpha = \frac{I_C}{I_E} = \frac{I_C}{I_{Cn}}\frac{I_{Cn}}{I_{En}}\frac{I_{En}}{I_E}$$

$$= M\alpha_T\gamma \tag{1.60}$$

where α_T is the base transport factor. Thus, $B = M\alpha_T$. Note that an *npn* rather than a *pnp* is being considered so that $\gamma = I_{En}/I_E$ instead of $\gamma = I_{Ep}/I_E$ since the minority carrier for *npn* transistors is the electron. Following the procedures outlined in Example 1.7, one can arrive at the following results:

$$\alpha_T \approx \operatorname{sech} \frac{W_b}{L_n} \approx \frac{1}{1 + W_b^2/2L_n^2} \tag{1.61}$$

$$\gamma \approx \frac{1}{1 + G_B/G_E} \tag{1.62}$$

where L_n is the electron diffusion length, the Gummel number is defined by

$$G_B = \frac{1}{D_{nB}} \int_{\text{base}} N_B(x)\, dx = \frac{Q_{Bo}/q}{D_{nB}} \tag{1.63}$$

and G_E is defined similarly for the emitter. Q_{Bo} is the built-in charge of the base. The approximate relationships are for $W_b \gg L_n$. In the expression for the emitter efficiency, the depletion approximation for constant charge that was used for the *pnp* transistor [Eq. (1.46)] has been replaced by the Gummel numbers. Using Eqs. (1.59) through (1.62) in Eq. (1.41) and neglecting terms higher than first order leads to the following expression for the gain β:

$$\beta \approx \frac{1}{G_B/G_E + W_b^2/2L_n^2 - [V_{CB}/(\text{BV})_{CBo}]^n} \tag{1.64}$$

If one assumes that a Gaussian dopant profile results when the doping is carried out thermally with a surface concentration of N_{Bo}, one has

$$N_B = N_{Bo} \exp\left(-\frac{x^2}{4Dt}\right) - N_C \tag{1.65}$$

where N_C is the dopant concentration in the collector. Referring to Fig. 1-21, N_C is applicable only on the base side of the collector-base transition region. From the last equality of Eq. (1.63), one obtains after transforming variables

$$\frac{Q_{Bo}}{q} = \left(\frac{Dt}{2}\right)^{1/2} N_{Bo}\left(\int_0^{t_2} e^{-y^2/2}\, dy - \int_0^{t_1} e^{-y^2/2}\, dy\right) - \int_{x_{EB}}^{x_{CB}} N_C\, dx \tag{1.66}$$

where

$$t_1 = \frac{x_{EB}}{(2Dt)^{1/2}} \qquad t_2 = \frac{x_{CB}}{(2Dt)^{1/2}} \tag{1.67}$$

and x_{EB} and x_{CB} are the distances from the surface to the base edges of the emitter-base and collector-base transition regions, respectively (see Fig. 1-22). Finally, the open circuit breakdown voltage for premature punch-through (Chap.

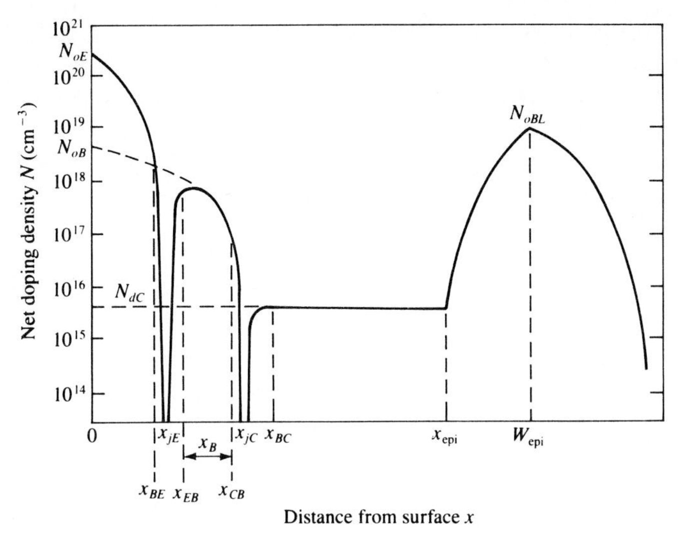

FIGURE 1-22
A transistor doping profile defining the distances used in the example (Colclaser, 1980).

4) can be estimated by

$$(\mathrm{BV})_{CBo} = \frac{q}{\varepsilon_s}\left(\frac{Q_{Bo}x_1}{q} - \frac{N_C x_1^2}{2} + \frac{Q_{Bo}W_b}{2q}\right) \tag{1.68}$$

$$x_1 = x_{\mathrm{epi}} - x_{BC}$$

where x_{epi} and x_{BC} are shown in Fig. 1-21. Equations (1.63) through (1.68) with Eq. (1.57) can now be used as the specifications for the device fabrication. The problem can be stated as follows: given the desired values of $\beta, f_T, (\mathrm{BV})_{CBo}$, and

TABLE 1.2
Design specifications for the example (Colclaser, 1980)

Design values:	$\beta_F = 45$	$\omega_\alpha = 2\pi \times 5 \times 10^9$ rad/s
	$(\mathrm{BV})_{CBo} = 25$ V	$R_{sB} = 200\ \Omega/\square$
Dimensions:	$W_{\mathrm{epi}} = 6.4\ \mu\mathrm{m}$	$x_{\mathrm{epi}} = 3.1\ \mu\mathrm{m}$
	$x_{jC} = 1.45\ \mu\mathrm{m}$	$x_{jE} = 0.67\ \mu\mathrm{m}$
	$x_B = 0.49\ \mu\mathrm{m}$	Buried layer $R_s = 25\ \Omega/\square$

Diffusion processes	***T*, °C**	***t*, min**	**R_s, $\Omega/\square$**	**N_o, cm^{-3}**
Isolation				
Predeposit	950	39		1.55×10^{20}
Drive-in	1200	60	50	5.5×10^{18}
Base				
Predeposit	900	17.4		1.2×10^{20}
Drive-in	1100	23.4	200	6.5×10^{18}
Emitter	950	47.8	~20	8.4×10^{20}

R_{SB} (sheet resistance in the base), specify the dimensions and doping schedules for the fabrication of the BJT in Fig. 1-21. The problem and the solution are summarized in Table 1.2.

The specifications for the problem can follow the steps given in Fig. 1-23 in flowchart form. The step-by-step procedures will be followed in the order indicated on the chart.

1. Pick a typical Gummel number for the emitter G_E of 5×10^{13} cm^{-4}. If it is assumed that β dominates in Eq. (1.64), which can be checked later, Eq. (1.64) can be used to calculate G_B:

$$G_B = \frac{G_E}{\beta} = \frac{5 \times 10^{13}}{45} = 1.11 \times 10^{12} \text{ cm}^{-4}\cdot\text{s}$$

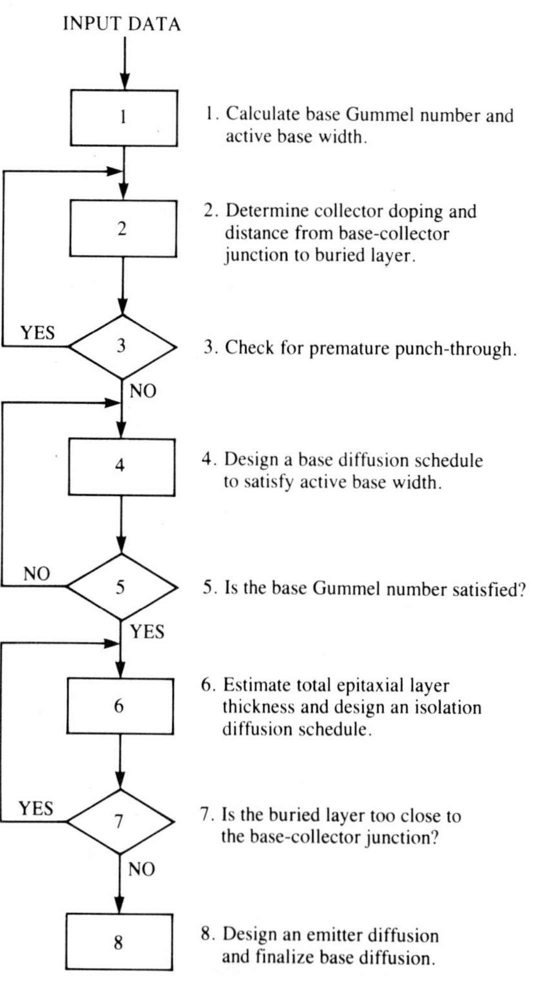

FIGURE 1-23
A flowchart for the design of a processing sequence for the fabrication of a bipolar transistor (Colclaser, 1980).

The average base doping concentration in a transistor is typically 10^{17} cm^{-3}. From Fig. 1-7, the corresponding value of D_{nB} is 15 cm^2/s. It follows from Eq. (1.63) that

$$\frac{Q_{Bo}}{q} = D_{nB} G_B = 1.67 \times 10^{13} \text{ cm}^{-2}$$

Combining Eqs. (1.57) and (1.58) with $\eta = 4$ and solving the result for W_b yields

$$W_b = \left(\frac{\eta D_{nB}}{2\pi f_\alpha}\right)^{1/2} = 0.49 \ \mu\text{m}$$

2. Assuming that the base-collector junction is located approximately 2 μm from the surface (x_{jC} in Fig. 1-24), the breakdown voltage specification [25 V for $(BV)_{CBo}$] and x_{jC} of 2 μm, when used in Fig. 1-24, yield

$$x_{\text{epi}} - x_B = x_1 \approx 1.2 \ \mu\text{m}$$

The maximum allowable collector doping = 8×10^{15} cm^{-3} for N_C.

3. The possible punch-through voltage can now be checked from the values calculated so far using Eq. (1.68):

$$(\text{BV})_{CBo} = \frac{q}{\varepsilon_s}\left(\frac{Q_{Bo} x_1}{q} - \frac{N_C x_1^2}{2} + \frac{Q_{Bo} w_b}{2q}\right) = 372 \text{ V}$$

Since the specification of 25 V is well below 372 V, one can proceed to step 4.

4. For the assumed value of x_{jC} (2 μm), $R_{SB} x_{jC} = 400$ $\Omega\cdot\mu$m. For the thermal diffusion schedule based on a Gaussian profile, Fig. 1-25 can be used to obtain

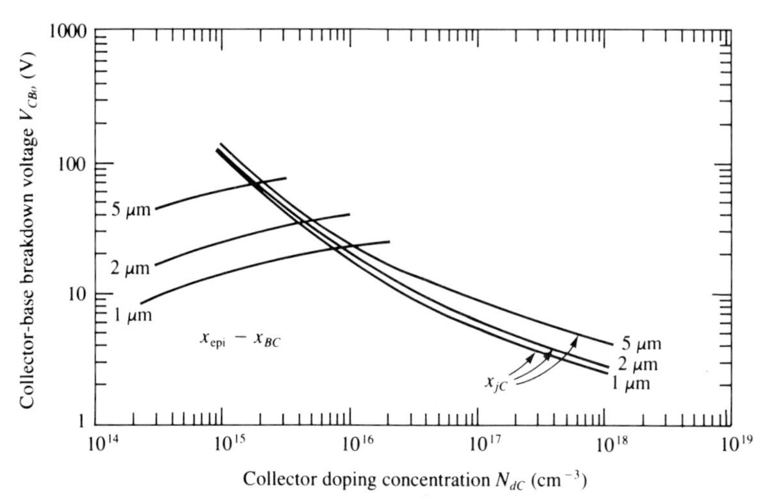

FIGURE 1-24
Collector-base breakdown voltage as a function of collector doping concentration with collector-base junction depth as a parameter (Colclaser, 1980; Wolf, 1969; Allen *et al.*, 1966).

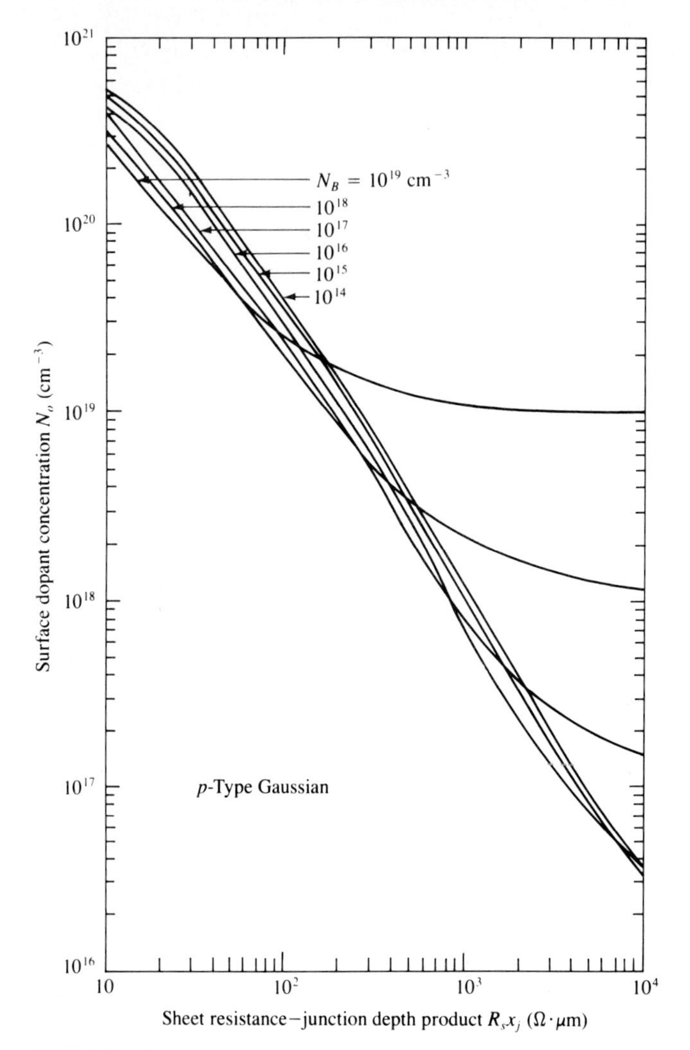

FIGURE 1-25
Surface dopant density of a *p*-type Gaussian diffusion in uniformly doped *n*-type silicon as a function of average resistivity at 300 K (Irvin, 1962).

N_{Bo}. For the curves in the figure, the $N_B = 10^{16}$ cm^{-3} curve is the closest to the value of N_C calculated in step 2. The value of N_{Bo} corresponding to the curve at $R_{SB}x_j = 400$ Ω·μm is 4×10^8 cm^{-3}. Assuming a two-step diffusion, the junction occurs when

$$N(x_j, t) = N_{Bo} \exp\left(\frac{-x_{jC}^2}{4D_{2B}t_{2B}}\right) = N_C$$

or $D_{2B}t_{2B} = 1.6 \times 10^{-9}$ cm^2.

5. Now the value of Q_{Bo}/q previously calculated can be checked with the aid of Fig. 1-26 for the base-collector space-charge region width (x_T) and the respective widths in heavily (x_1) and lightly (x_2) doped regions. If the value calculated with the aid of Fig. 1-26 is larger than Q_{Bo}/q previously calculated, x_{jC} must be increased and step 4 repeated until this constraint is satisfied. For the unbiased junction, the value of V to use in the figure is 0.7 V. For $V = 0.7$ V and $N_B = 10^{16}$, the figure yields the following:

$$\text{Total space-charge region} = 0.5\ \mu\text{m}$$

$$\text{Base side} = 0.19\ \mu\text{m}$$

$$\text{Collector side} = 0.31\ \mu\text{m}$$

The boundaries of active base, x_{CB} and x_{EB} for use in Eq. (1.66) follow from the assumed value of x_{jC} (2 μm):

$$x_{CB} = 2 - 0.19 = 1.81\ \mu\text{m}$$

$$x_{EB} = x_{CB} - W_b = 1.81 - 0.49 = 1.32\ \mu\text{m}$$

Use of these in Eq. (1.66) yields

$$\frac{Q_{Bo}}{q} = 5.3 \times 10^{12}\ \text{cm}^{-2}$$

which is smaller than the value of 1.67×10^{13} previously obtained in step 1. Repeating steps 4 and 5 yields

$$x_{jC} = 1.45\ \mu\text{m}$$

$$D_{2B}t_{2B} = 7.85 \times 10^{-10}\ \text{cm}^2$$

$$N_{Bo} = 6.5 \times 10^{18}\ \text{cm}^{-3}$$

These values give Q_{Bo}/q of 1.66×10^{13} cm^{-2}, which is very close to the value calculated in step 1.

6. The epitaxial layer thicknesses W_{epi} and x_{epi} in Fig. 1-21 can now be calculated. From step 2, one has

$$x_{\text{epi}} - x_{BC} = 1.2\ \mu\text{m}$$

and x_{BC} is

$$x_{BC} = x_{jC} + \text{collector side of the base-collector transition region}$$

$$= 1.45 + 0.31 = 1.76\ \mu\text{m}$$

Therefore, $x_{\text{epi}} = 1.2 + 1.76 = 2.96\ \mu$m. If it is assumed that the predeposit for the buried layer penetrates into the metallurgical epilayer during the epitaxial

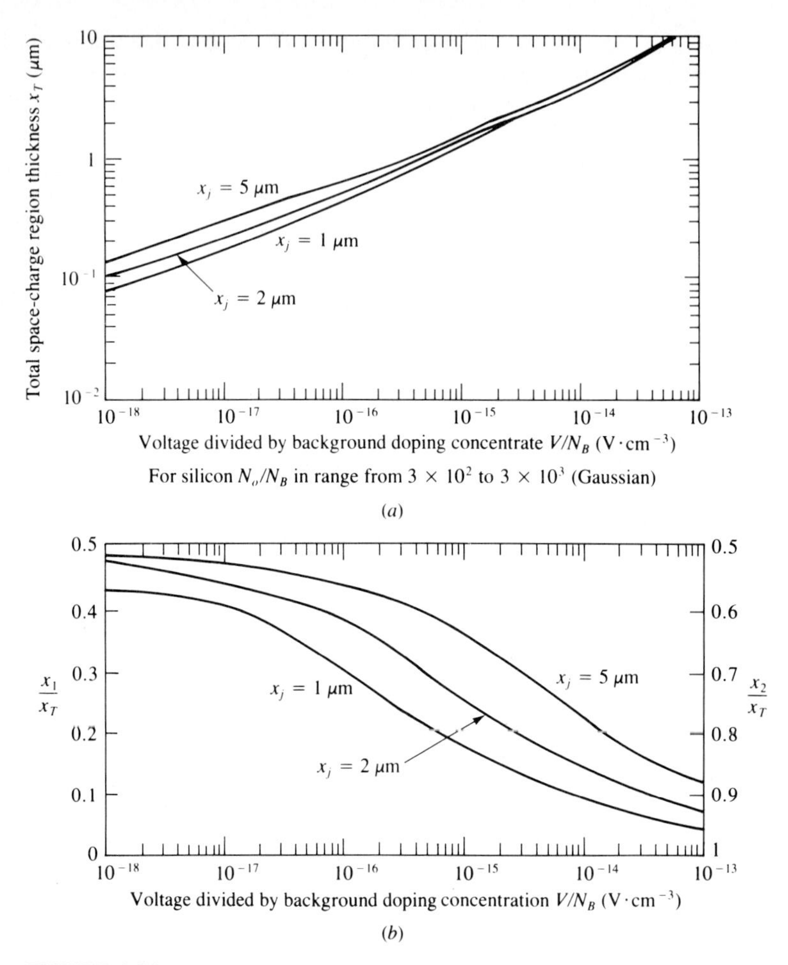

FIGURE 1-26
Space-charge region thickness as a function of voltage for a *pn* junction formed by a Gaussian diffusion into a constant background concentration: (*a*) total width x_T and (*b*) ratios x_1 and x_2 where x_1 is the portion in the heavier doped side and x_2 is that in the lighter doped side (Lawrence and Warner, 1960).

growth, one has

$$0.9W_{\mathrm{epi}} = x_{\mathrm{epi}} + x_2$$

where x_2 is the penetration of the buried layer into the epilayer during the isolation diffusion. For trial and error, set $x_2 = 2$ μm, which yields W_{epi} equal to 5.5 μm. Assuming an isolation diffusion sheet resistance of 50 Ω/□, the final surface concentration, $N_{Io}(N_{Co})$ is found to be 7×10^{18} cm^{-3} from Fig. 1-25. Using this value plus the Gaussian distribution, one can calculate x_2 to be 2.44 μm.

7. The calculated value of 2.44 μm is larger than the trial value of 2 μm. This indicates that the buried layer is too close to the base-collector junction and so step 6 must be repeated. The new results are

$$W_{\text{epi}} = 6.4\ \mu\text{m}$$

$$x_{\text{epi}} = 3.1\ \mu\text{m}$$

$$N_{Io} = 5.5 \times 10^{18}\ \text{cm}^{-3}$$

which in turn yields the doping schedule (refer to Sec. 7.3) as well.

8. The position from the surface of the emitter junction, x_{jE}, has to be specified. The base portion of the base-emitter transition region under forward bias is approximately 0.1 μm. Then,

$$x_{jE} = x_{jC} - x_{CB} - W_b - 0.1$$
$$= 1.45 - 0.19 - 0.49 - 0.1 = 0.67\ \mu\text{m}$$

Finally, the net impurity concentration at x_{jE} must be zero. This in turn implies that

$$N_E = (x_{jE}) = N_B(x_{jE}) - N_C(x_{jE})$$

or $$N_E = N_{Bo} \exp\left(\frac{-x_{jE}^2}{4D_{2B}t_{2B}}\right) - N_C = 1.55 \times 10^{18}\ \text{cm}^{-3}$$

As will be shown in Chap. 7, the equation can be used to specify the doping schedule. The specifications obtained so far are summarized in Table 1.2.

The surface geometry required for composite layout is determined by specifying the surface dimensions of the smallest component (emitter) and then by tolerance for multilevels. For BJT, the emitter contact metal strip of length h is determined by the maximum allowed voltage drop and its width l by a limit on the switching frequency:

$$l(\mu\text{m}) = \frac{2 \times 10^{10}}{f_T(\text{Hz})}$$

$$h(\mu\text{m}) = 4I_E\ (\text{mA})$$

The composite layout resulting from the smallest dimensions and the tolerances, and the corresponding individual masks are given in Fig. 8-2.

1.7 MICROELECTRONICS PROCESSING

In this book, a broad view is taken of semiconductor processing. Therefore, the fabrication of a basic unit such as a MOSFET structure, which is sufficient for consideration of an integrated circuit fabrication, is just a part of microelectronics processing. Crystal growth and preparation of wafers on which the ICs are fabricated is another part of the processing. Integrated circuits come in a certain package. After chips are sorted and tested, the back of a chip (called a die)

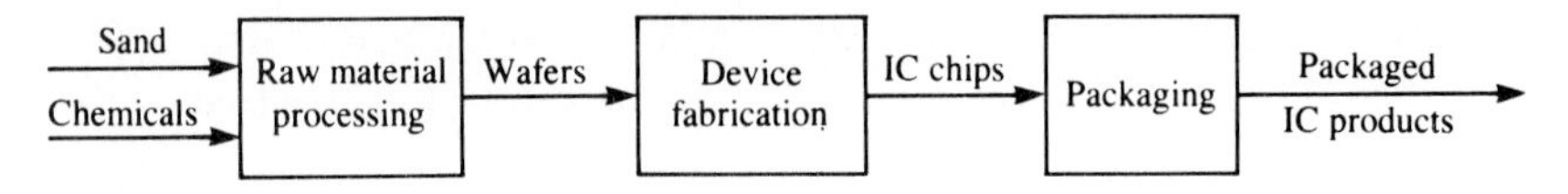

FIGURE 1-27
A broad view of microelectronics processing.

is mechanically attached or bonded to an appropriate mount media, typically plastic or ceramic material. Then the bond pads on the circuit side of the chip are electrically interconnected to the package. The whole package may be sealed and the electrical communication between the chip and outside world is through the interconnected outlets. This packaging is the last part of microelectronics processing. Therefore, there are three major parts: processing leading to wafers, processing for IC fabrication, and package processing or simply packaging. Figure 1-27 shows the three major parts of microelectronics processing.

Wafers are thin plates (0.5 mm, for instance, for silicon) of single-crystal material doped with either donor or acceptor atoms. The need for very high compositional purity and near-perfect crystal structure requires arduous processing steps. Typical processing steps are shown in Fig. 1-28 for silicon.

The starting material can be either sand or metallurgical grade silicon that is available from ore processing, e.g., iron refining. A molten electric arc furnace is used to reduce the sand to metallurgical grade silicon (MGS). The MGS has a silicon purity of about 98 percent. Fine particles of MGS are fed to a fluidized-bed reactor with a carrier gas containing hydrochloric acid to convert the MGS to silicon-containing gases such as silane and chlorosilanes. The gases are then separated and purified in a series of separators such as distillation columns. Two main methods are used to produce electronic grade silicon (EGS). One method involves deposition of silicon from the silicon-containing gas onto a resistance-heated rod of silicon, which provides the surface for nucleation sites for the silicon deposition. The rod grows rapidly in size up to diameters of 20 cm. When

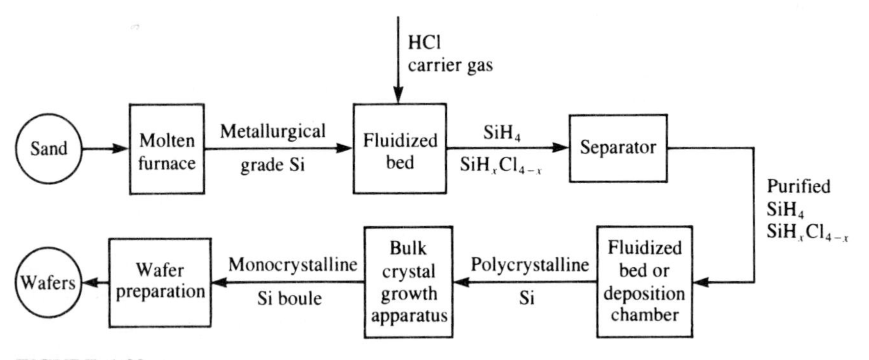

FIGURE 1-28
Processing steps for silicon wafers.

trichlorosilane is used, the overall reaction is

$$SiHCl_3(gas) + H_2(gas) \longrightarrow Si(solid) + 3HCl(gas)$$

The second, more advanced EGS preparation method uses a fluidized-bed reactor to which silicon-containing gas and seed particles of pure silicon are fed. The seed particles grow in size as silicon deposits from the gas phase. The EGS has a polycrystalline atomic structure and contains impurities in the ppm range. The total impurity level at this point is less than 20 ppm (Sze, 1983). Both the rods, as single chucks or crushed into nuggets, and the grown polycrystalline particles from the fluidized bed are then melted to grow monocrystalline silicon boule.

There are three major methods for growing monocrystalline silicon boule from the melted EGS. The most widely used is a growing technique called Czochralski growth. A small seed crystal of monocrystalline structure is dipped into a molten EGS, and the seed crystal is then gradually pulled up so that monocrystalline boule of diameters as large as 15 cm can be formed upon being cooled in the process of pulling. Another method is floating-zone growth. A single vertical polycrystalline rod is locally melted from the bottom up as a local heating element such as an rf (radiofrequency) heater traverses upward. The melted zone is recrystallized into a monocrystalline structure from the bottom using a seed crystal. In the Bridgeman method, used mostly for gallium arsenide, the polycrystalline material is melted in a long closed boat and is cooled from one end, which is attached to a seed crystal. A traversing heater is also used in this method. These growth processes accomplish two objectives: transformation of polycrystalline structure into monocrystalline structure and further removal of the undesired impurities that takes place at the liquid-solid interface. In the crystal growth process, dopants are introduced to make p- or n-type monocrystalline semiconductors. The grown monocrystalline boule is then cut into slices, better known as wafers. The thickness of a wafer is about 0.5 mm. However, only a tiny fraction of that depth (say, of the order of micrometers) is used for actual device fabrication. The need for mechanical stability is the determining factor for the wafer thickness.

Device fabrication for integrated circuits is not as clear-cut as the material processing for wafers. The technology for device fabrication is changing and progressing rapidly and constantly. Some specifics of device fabrication that are current at the time of writing may become obsolete before publication. While the specifics may change, the fundamentals behind the specifics and the basic general goals of the fabrication do not.

Given the circuit requirements and specifications, an IC can be designed in the form of circuit layout, which specifies the width and depth of each element in the basic unit such as those considered in the previous section. For the simple MOSFET in Fig. 1-13, for instance, this involves specification of the dimensions of the n^+ drain and source, silicon dioxide layer, metal contacts, and so on. The composite layout is then converted into a set of oversized drawings with a schematic (for individual layout) for each masking level. The final masks are then

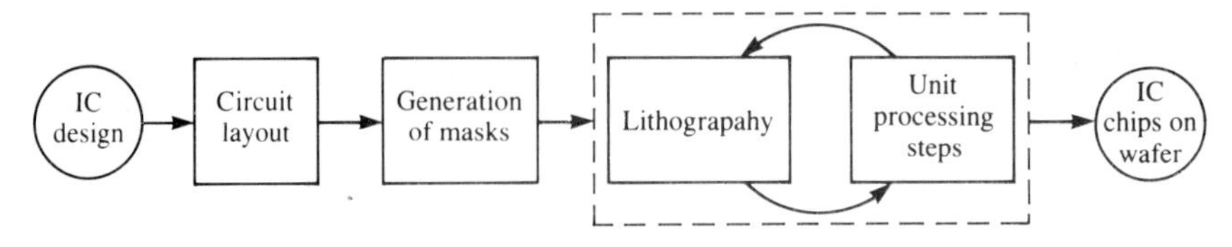

FIGURE 1-29
Device fabrication sequence.

made from the reduced versions of the drawings. For VLSI circuits, however, designers can completely describe the circuit layout electronically in place of the drawings through computer-aided design (CAD) systems. The design is then converted into digital form and stored on magnetic tapes. The whole structure, such as the one in Fig. 1-13, is not fabricated in one step but rather in many steps to separately create the doped regions, to deposit silicon dioxide in a particular pattern, and so on. Different masks are required for different fabrication steps. Once the masks are made for different masking levels, the processing can be carried out in accordance with the pattern specified by the mask for each level, in the order prescribed at the composite layout stage. Masks are made from emulsion-coated glass plates or glass covered with a hard surface material such as chromium, chromium oxide, iron oxide, or silicon. As shown in Fig. 1-29, the next steps in device fabrication are repetitions of lithography and the corresponding processing steps in accordance with the masks specifications until the whole desired structure is fabricated on the wafer in the form of identical IC chips on a wafer. Lithography is the process of transferring geometric shapes on a mask to the surface of a wafer.

Each cycle of lithography-unit processing steps usually involves etching to "open windows" or delineate certain patterns for the next step, which is either deposition of a desired film or placement of a dopant species into the open area through diffusion or by ion implantation. One example of a cycle based on photolithography is shown in Fig. 1-30. The mask is used to open windows through the silicon dioxide so that gaseous dopant species can be diffused into the openings to provide doped regions in an *n*-type silicon substrate. As shown in the figure, a photoresist (PR) of light-sensitive polymeric material is coated onto the silicon dioxide surface. Upon placing the mask and exposing it to an ultraviolet light, the PR is polymerized everywhere except where the pattern (window in this case) is to appear. The unpolymerized areas are then dissolved in a solution, leaving the pattern as shown in Fig. 1-30. The wafer is put into a solution containing buffered hydrofluoric acid (BHF) that etches the oxide but not the PR. The PR is then stripped, yielding the desired windows for the diffusion step. The wafer is then put in a furnace into which dopant species such as B_2H_6 or PH_3 in a carrier gas are fed so that the diffusion of the dopant through the windows can take place. This gives the desired doped regions. Although wet etching, in which wet chemicals are used, is described here, dry etching involving ion bombardment can also be used. Typical unit processing steps include epitaxial (monocrystalline)

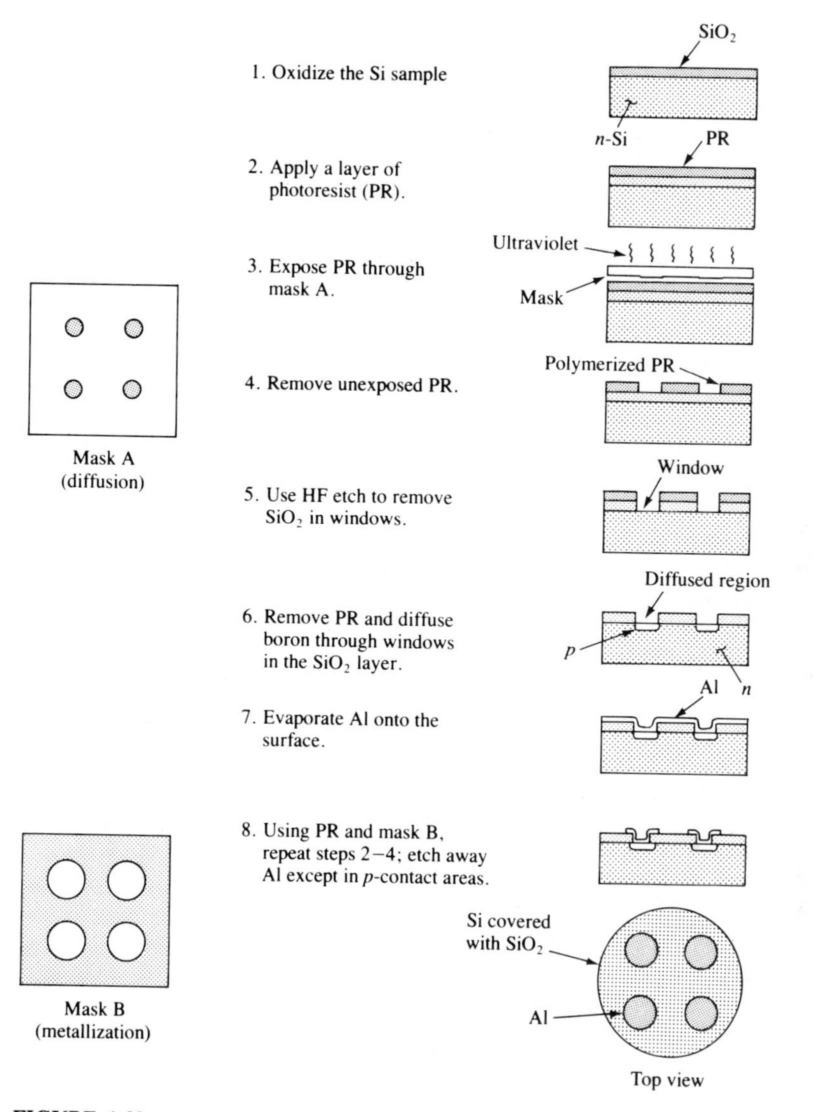

FIGURE 1-30
An example cycle of lithography processing in Fig. 1-29 (Streetman, 1980).

and nonepitaxial film deposition, oxidation, ion implantation that is a method of physically forcing dopant atoms into substrate, and metallization that involves deposition of metals.

Processing for the device fabrication essentially involves creating active (doped) regions and isolating one active region from another. The main factor here is really the distribution of dopants and clear delineation of the distribution. Since the wafer undergoes many processing steps and each step is carried out at high temperature, redistribution of dopants does take place at each step. This redistribution problem is one of the reasons that constant efforts are made to carry out the unit processing at ever lower temperatures.

Integrated circuit packaging involves attaching a die (chip) to a package material, connecting the device externally by attaching wires between the device pad and the package, and fabricating the package that houses the die. The usual package materials are plastics, metals, and ceramics. A detailed manufacturing process for an NMOS silicon-gate IC from wafer to packaged IC, although outdated, is shown in Fig. 1-31 as an illustration.

The rapid increase in the number of devices per chip and the enhanced performance of these devices has led to chips with higher input and output counts and, therefore, more connections to the die and more package pins. With

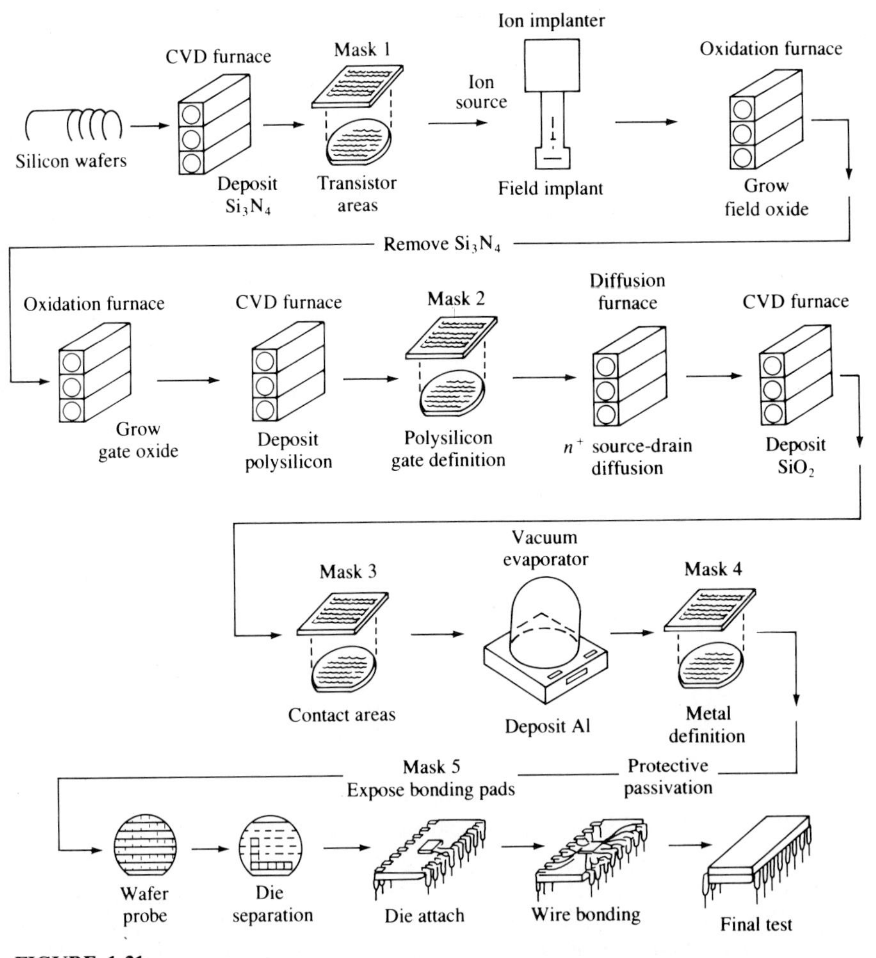

FIGURE 1-31
A manufacturing process for NMOS silicon-gate ICs (Integrated Circuit Engineering Company).

the higher pin count, the packaging has become one limitation in IC miniaturization since higher counts require larger packages and more heat dissipation. Higher input-output counts also causes more signal noise. The lifetime of an IC, as limited by corrosion, has also become an important part of packaging design considerations.

It is instructive at this point to have another look at the three major steps in sequences of microelectronics processing (Fig. 1-27), in terms of interacting effects and integration of the sequences for the final packaged ICs. The raw material processing, in particular the processing for monocrystalline substrate material, has direct effects on device fabrication, since the material properties such as intrinsic mobility and the defects/undesired impurities in the material directly limit the extent to which device fabrication options can be explored. On the other hand, the processability of the substrate material is a limitation that the device fabrication can place on the usefulness of the material in certain device applications. A good example would be the unsuitability of gallium arsenide for MOSFET (MIS) devices, although further research may resolve the processing difficulties. The continuous push for IC miniaturization has to deal with the limitations imposed by packaging in terms of smaller packages, more efficient heat removal, and lower signal noise. Although emphasis in the IC technologies has shifted from material processing to circuit design and then from circuit design to material processing, depending on when the IC design reached maturity for a given material, eventually one has to have an integrated approach to all IC technologies. For microelectronics processing, this means that all three sequences in Fig. 1-27 should be integrated for the advancement of the technologies.

NOTATION

a_1	Constant defined in Eq. (1.45)
A	Cross-sectional area (width times height) (L^2)
B	Base transport factor defined in Eq. (1.47)
BV	Junction breakdown voltage
C_i	Interface capacitance per unit area (C/V)
D	Diffusivity of charge carrier (L^2/t)
D_n	Diffusivity of electron
D_p	Diffusivity of holes
E_A	Acceptor energy level (E)
E_c	Energy level for conduction band edge
E_D	Donor energy level
E_F	Fermi energy level
E_g	Energy gap
E_i	Fermi level for intrinsic material
E_v	Energy level for valence band edge
f_α	Alpha cutoff frequency given by Eq. (1.57) (t^{-1})
f	Fermi-Dirac probability density function (dimensionless)
G	Generation rate; Gummel number defined by Eq. (1.63)

h	Planck's constant = 6.626×10^{-34} J·s
i, I	Current (I)
I_B, I_C, I_D, I_E	Base, collector, drain, emitter current, respectively
I_0	Total current given by Eq. (1.38)
J	Current flux ($IL^{-2}t^{-1}$)
J_n	J due to electrons
J_p	J due to holes
k	Boltzmann constant = 1.38×10^{-23} J/K
L	Distance between contacts; channel length (L)
L_n	Electron diffusion length
L_p	Hole diffusion length defined in Eq. (1.34) (L)
m_e^*	Effective electron mass (M)
m_p^*	Effective hole mass
M	Collector multiplication factor defined in Eq. (1.59)
n	Electron concentration (electrons/L^3)
n_i	Intrinsic value of n
n_n	n in n-type material
n_p	n in p-type material
n_{Ee}	Equilibrium electron concentration in emitter
n_{pe}	n_p at equilibrium
N	Dopant concentration
N_A	Acceptor concentration (atoms/L^3)
N_B	Background or base dopant concentration
N_C	Effective density of states at E_c (states/L^3)
N_D	Donor concentration (atoms/L^3)
N_v	Effective density of states at E_v (states/L^3)
p	Hole concentration (holes/L^3)
p_n	p in n-type material
p_{Be}	Equilibrium hole concentration in base
p_{ne}	p_n at equilibrium
p_p	p in p-type material
q	Elementary charge = 1.60×10^{-19} C
Q_d	Charge per unit area in depletion region (C/L^2)
Q_i	Charge per unit area at interface (C/L^2)
Q_n	Mobile charge (C/L^2)
Q_s	Induced charge (C/L^2)
R	Resistance; recombination rate (R)
R_s	Sheet resistance ($R/\square$)
t_1, t_2	Quantities defined in Eq. (1.67)
T	Temperature
$\bar{v}_n$	Average electron velocity
V	Voltage
V_D	Drain voltage
V_f	Forward-bias magnitude
V_G	Applied gate voltage

V_r	Reverse-bias magnitude
V_T	Threshold voltage [Eq. (1.50)]
V_{CB}	Collector-base voltage
V_{FB}	Flat-band voltage given by Eq. (1.49)
W_b	Base width (L)
x	Axial coordinate (L)
x_j	Junction depth or length
Z	Depth of channel

Greek letters

α	Current transfer ratio defined by Eq. (1.40)
β	Current amplification factor defined by Eq. (1.41)
γ	Injection efficiency defined by Eq. (1.39)
ε	Electric field (V/L)
ε_s	Permittivity (F/L)
μ	Mobility ($L^2V^{-1}t^{-1}$)
μ_n	Electron mobility
$\bar{\mu}_n$	μ_n at surface
μ_p	Hole mobility
ρ	Resistivity; electron charge density (RL); (C/L^3)
σ	Conductivity ($R^{-1}L^{-1}$)
σ_n	σ due to electron
τ_B	Time delay defined by Eq. (1.58) (t)
τ_p	Hole lifetime (t)
ϕ	Work function potential (V)
ϕ_F	$(E_i - E_F)/q$ (V)

Subscripts

B	Base
C	Collector
E	Emitter
n	Electron
m	Metal
p	Hole
s	Semiconductor

Units

C	Charge
E	Energy
F	Farad (C/V)
I	Current
L	Length
M	Mass
R	Resistance

t	Time
T	Temperature
V	Voltage

PROBLEMS

1.1. List acceptor and donor atoms for gallium arsenide based on the periodic table in Table 1.1. Comment on the group IV atoms as dopants.

1.2. What is the number of electrons in the conduction band of intrinsic silicon at 27 °C? For phosphorus in silicon, the energy difference between E_c and E_d is about 0.05 eV. What is the number of electrons in the conduction band in this case at 27 °C if the dopant concentration is 10^{16} cm^{-3}?

1.3. Calculate the approximate maximum dopant concentration that can be incorporated into gallium arsenide. Calculate the dopant concentration in an n-type silicon that will give the same conductivity as the n-type gallium arsenide when it has the maximum dopant concentration. Use the following information:

$$n_i = 9 \times 10^6 \text{ cm}^{-3}, \qquad \mu_n = 8600 \text{ cm}^2/(\text{V·s}) \text{ for GaAs and}$$

$$n_i = 1.45 \times 10^{10} \text{ cm}^{-3}, \qquad \mu_n = 1350 \text{ cm}^2/(\text{V·s}) \text{ for Si}$$

1.4. A silicon wafer is doped with 10^{16} arsenic atoms per cubic centimeter. Calculate the resistivity of the wafer. Would it be sufficient to calculate the resistivity based solely on the majority carrier of electrons? Use the information given in Prob. 1.3. $\mu_p =$ 480 cm^2/(V·s).

1.5. A pn junction is usually made by diffusing donor atoms into a p-type semiconductor or acceptor atoms into an n-type semiconductor. Suppose that boron is diffused into a silicon wafer doped with arsenic at 10^{15} cm^{-3} such that the boron concentration in the p region is 6×10^{15} cm^{-3}. What are the electron and hole concentrations in the p region? What about the hole concentration in the n region?

1.6. Two types of isolation in IC fabrication have been discussed in the text. The first is the pn junction isolation. The second is what is called mesa isolation and is shown in Fig. 1-14. The third method of isolation uses an insulator such as SiO_2. Draw two possibilities of isolating a repeating MOSFET structure in Fig. 1-13 using SiO_2.

1.7. The resistivity of a semiconductor is often determined by a "four-point probe" method shown in Fig. P1-7. The quantity measured is the "sheet resistance," denoted by R_s in ohms per square. The resistance of a resistor made up of n squares laid in a row is nR_s. The formula relating sheet resistance to current and voltage is

$$R_s = 4.53 \frac{V}{I}$$

The relationship is valid when the thickness of the layer being measured is much less than the spacing between the probes and the size of the piece of material is much greater in length and width than the probe spacing. When Eq. (1.24) is used per sheet (or the spacing of S in Fig. P1-7) for R_s, it follows for a thin layer of material that

$$\rho = R_s t; \qquad t\text{: thickness}$$

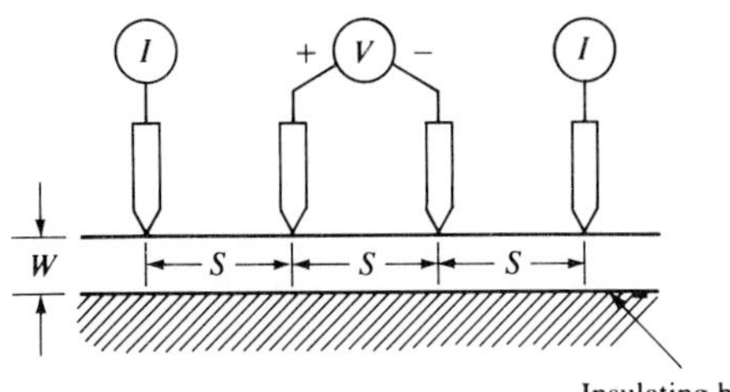

FIGURE P1-7
Four-point probe method.

Suppose that a four-point probe measurement on a uniformly doped silicon wafer (125 mm diameter) of thickness 0.5 mm results in

$$V = 5 \times 10^{-3}\ \text{V} \qquad I = 4.5 \times 10^{-3}\ \text{A}$$

Calculate the sheet resistance and resistivity. Assume that S is 1 cm. When the thickness of the sample is much greater than the probe spacing S, the formula is

$$\rho = \frac{2\pi S V}{I}$$

1.8. At equilibrium, the drift and diffusion components of the hole current (J_p) just cancel:

$$J_p(x) = q\left(\mu_p p\varepsilon - D_p \frac{dp}{dx}\right) = 0$$

Since $\varepsilon = -dV/dx$, one can rewrite the above as

$$-\frac{q}{kT}\frac{dV}{dx} = \frac{1}{p}\frac{dp}{dx} \tag{A}$$

where the Einstein relation for μ_p/D_p has been used. Let the potential on either side of a pn junction be V_p and V_n, and the hole concentration just at the edge of the transition region on either side be p_p and p_n. Integrating Eq. (A) gives

$$-\frac{q}{kT}\int_{V_p}^{V_n} dV = \int_{p_p}^{p_n} \frac{1}{p}\, dp$$

or

$$-\frac{q}{kT}(V_n - V_p) = \ln \frac{p_n}{p_p} \tag{B}$$

For a step junction, it is reasonable to take the electron and hole concentration in the neutral regions outside the transition region as their equilibrium values, p_{pe} and p_{ne}. Thus, one can rewrite Eq. (B) as follows:

$$\frac{p_{pe}}{p_{ne}} = e^{qV_0/kT} \tag{C}$$

since $V_n - V_0$, the electrostatic potential.

Show that Eq. (1.30) results when the junction is biased, in which case the barrier is $V_0 - V$ instead of V_0. Assume that Eq. (A) is still applicable to a junction with external bias V and that under low injection conditions, $p(-x_{p0}) = p_{pe}$.

REFERENCES

Allen, C. C.: *J. Electrochem. Soc.*, vol. 113, p. 5, 1966.

Colclaser, R. A.: *Microelectronics Processing and Device Design*, Wiley, New York, 1980.

Dummer, G. W. A.: Proceedings of the Symposium of the IRE-ALEE-RTMA, Washington, D.C., May 1952, in R. A. Colclaser (ed.), *Microelectronics Processing and Device Design*, Wiley, New York, 1980.

Grove, A. S.: *Physics and Technology of Semiconductor Devices*, Wiley, New York, 1967.

Hodges, D. A., and H. G. Jackson: *Analysis and Design of Digital Integrated Circuits*, McGraw-Hill, New York, 1983.

Irvin, J. C.: *Bell Syst. Tech. J.*, vol. 41, p. 2, 1962.

Kilby, J. S.: *IEEE Trans. Electron Devices*, vol. ED-23, p. 648, 1976.

Lawrence, H., and R. M. Warner Jr.: *Bell Syst. Tech. J.*, vol. 39, p. 3, 1960.

Smith, W. F.: *Principles of Materials Science and Engineering*, McGraw-Hill, New York, 1986.

Streetman, B. G.: *Solid State Electronic Devices*, 2d ed., Prentice-Hall, New York, 1980.

Sze, S. M.: *Physics of Semiconductor Devices*, 2d ed., Wiley, New York, 1981.

——— (ed.): *VLSI Technology*, McGraw-Hill, New York, 1983.

Wolf, H. F.: *Silicon Semiconductor Data*, Pergamon Press, Oxford, 1969.

CHAPTER 2

SILICON REFINING AND OTHER RAW MATERIALS

2.1 INTRODUCTION

There are a number of materials used in the fabrication of integrated circuits. The most important is silicon. The starting material is sand (SiO_2), which is eventually made into very pure silicon wafers after it is refined and shaped through a number of processing steps. Most of the gases used for various types of film growth and doping are also produced in the process of silicon refining. Materials of interest for II-VI and III-V compound semiconductors are metal organics. The metal organics in gaseous or liquid state are used for film growth of the compound semiconductors. Bulk growth of these semiconductors involves similar processing as for silicon refining. However, epitaxial films grown on the bulk crystalline material are mostly used in device fabrication.

Polymers are a major group of materials used in the fabrication of integrated circuits. They are used as a resist material in lithography, which is a pattern transfer specifying the order and manner in which a device is to be fabricated. Polymers are also used as a material for packaging finished integrated circuits. Ceramics are another major group of material of importance. They are used for device packaging, usually for better performance than polymeric material can deliver. These two groups of material are treated in the chapter on lithography (Chap. 8) and on packaging (Chap. 11).

This chapter describes the chemistry and production techniques used to obtain electronic grade silicon, silicon source gases, and metal organics. Preparation of these materials involves reaction engineering and separation problems that are normally encountered in chemical engineering. Most gases and metal organics used in the processing are in general either pyrophoric, toxic, or both. Because of the safety hazards presented by the currently used compounds, synthesis techniques that diminish the safety problems are included and explored.

2.2 METALLURGICAL GRADE SILICON AND SOURCE GASES

Electronic grade silicon is produced from quartzite, SiO_2, in a four-step process (Fig. 2-1). First, the quartzite and sources of carbon such as wood chips, coke, and coal are reduced in a submerged-electrode arc furnace to produce metallurgical grade silicon (MGS). The MGS is then reacted with HCl in a fluidized bed to form silane, SiH_4, and the chlorosilanes, SiH_xCl_{4-x}. These silicon gases are then distilled to reach the high level of purity required for electronic grade silicon (EGS). Finally, polycrystalline EGS is produced by either chemical vapor deposition or decomposition of the chlorosilanes in a fluidized bed.

The primary source of silicon, quartzite, is readily available in quartz rock, quartz sands, and veins of quartz crystals. Silicon is the second most abundant element in the earth's crust and SiO_2 has been concentrated in the growth deposits during nearly all geologic periods. The amount of impurities in the various sources range from <0.03 weight percent in quartz crystals to approximately 1.0 weight percent in quartzite rock (Dietl, 1987). The quartz sand that is commonly used is the least expensive silicon source and has a moderate level of impurities.

The use of submerged-electrode arc furnaces to reduce silicon dioxide with carbon is a large-scale, technologically mature process that has provided MGS for the iron and aluminum industries since the turn of the century (Fig. 2-2). Typical commercial furnaces have electrodes about one meter in diameter and 10

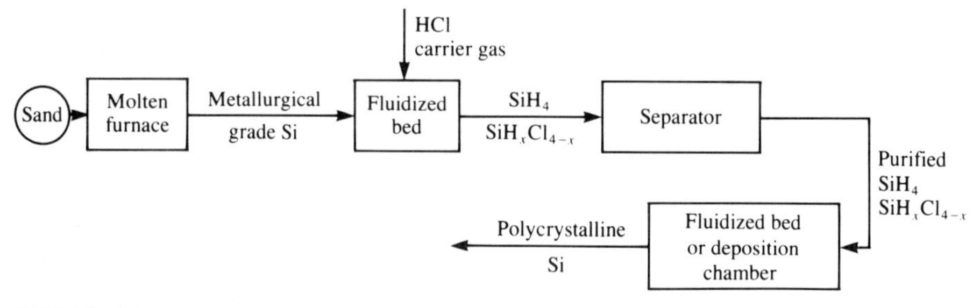

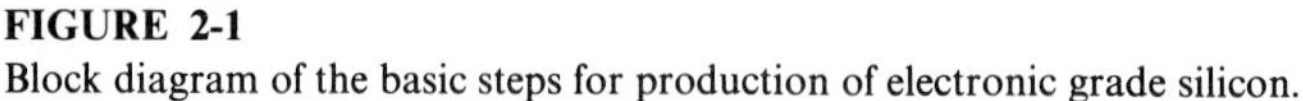
FIGURE 2-1
Block diagram of the basic steps for production of electronic grade silicon.

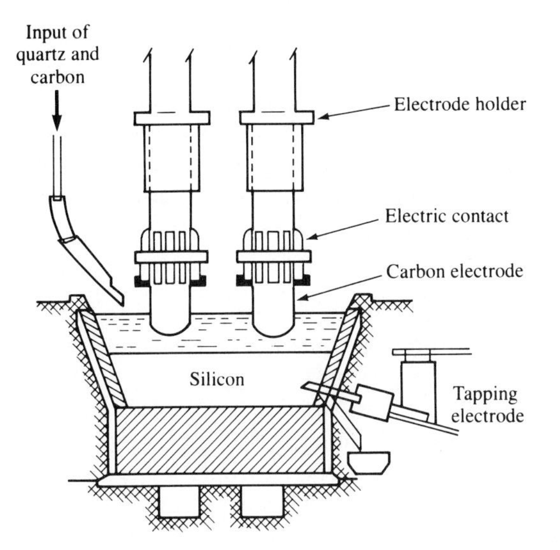

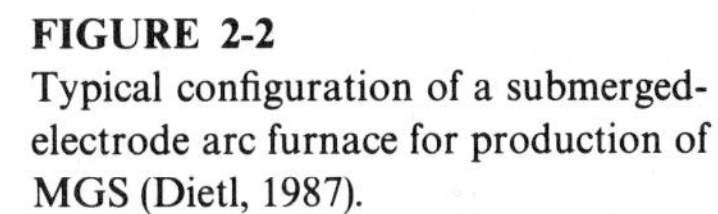
FIGURE 2-2
Typical configuration of a submerged-electrode arc furnace for production of MGS (Dietl, 1987).

to 30 MW power sources. A medium-sized furnace has a production capacity of approximately 20,000 tons of metallurgical grade silicon per year.

The chemical and transport processes occurring in a working furnace are very complex, but for the purposes of this discussion the furnace may be divided into three zones as shown in Fig. 2-3. The reactants are largely converted to silicon carbide by the reaction:

$$SiO_2 + 3C \longrightarrow SiC + 2CO(g) \tag{2.1}$$

when they reach zone II, where the temperature is about 1800 °C. The SiO_2 and SiC continue to descend into the hottest part of the reactor, zone III, where

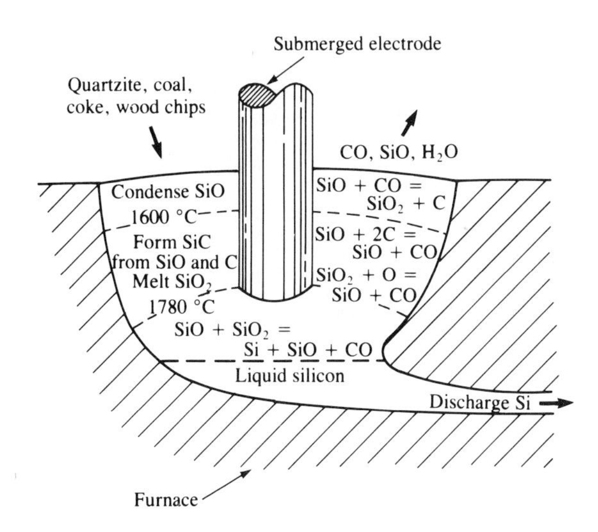

FIGURE 2-3
Schematic of furnace showing the three reaction zones (Crossman and Baker, 1977).

silicon is formed by the reactions:

$$SiO_2 + 2SiC \longrightarrow 3Si + 2CO(g) \tag{2.2}$$

and

$$2SiO_2 + SiC \longrightarrow 3SiO(g) + CO(g) \tag{2.3}$$

The gas produced has a high proportion of SiO, which is cooled by rising through the descending charge of reactants. The majority of SiO is retained in the charge by reactions such as

$$SiO(g) + CO(g) \longrightarrow SiO_2 + C \tag{2.4}$$

The quality, quantity, and structure of the carbon sources used are of the utmost importance. The kinetics of the reduction process is strongly dependent on the form and quality of the reductants. The carbon sources are primarily coal and coke, with wood chips added to increase the porosity and allow the SiO and CO gases to flow through the bed (Crossman and Baker, 1977). Maintaining the balance between carbon and quartzite is very important as well. If the reacting mixture becomes carbon-deficient, the SiO_2 can only be melted. The result is a low-viscosity quartz slag that is impossible to reduce with additional carbon.

The reduction process adds impurities to the silicon, resulting in a contamination level of approximately two percent (see Table 2.1). The impurities already present in the quartzite are reduced and are joined by contaminants from the carbon sources, electrodes, and other materials. The principal contaminants are iron, aluminum, and titanium at approximately 100 ppm, and the doping elements boron and phosphorus below 100 ppm. Another impurity present above its saturation limit is carbon, which is usually at a concentration in the 100 to 1000 ppm range.

The silane and chlorosilanes used as intermediate compounds in silicon purification are formed by the hydrochlorination of metallurgical grade silicon in a fluidized-bed reactor. The operating conditions are set to produce the chlorosil-

TABLE 2.1
Impurity levels in metallurgical grade silicon from several sources (Dietl, 1987)

	Concentration in MGS, ppm wt		
Impurity	**A**	**B**	**C**
Mn	260	500	50
Cr	25	20	50
Cu	25	50	20
Ni	110	30	10
Fe	3800	3500	5000
Al	1600	2400	2500
Ca	2700	2200	500
Mg	60	50	70
Ti	150	250	150
B	10	20	15
P	40	30	20

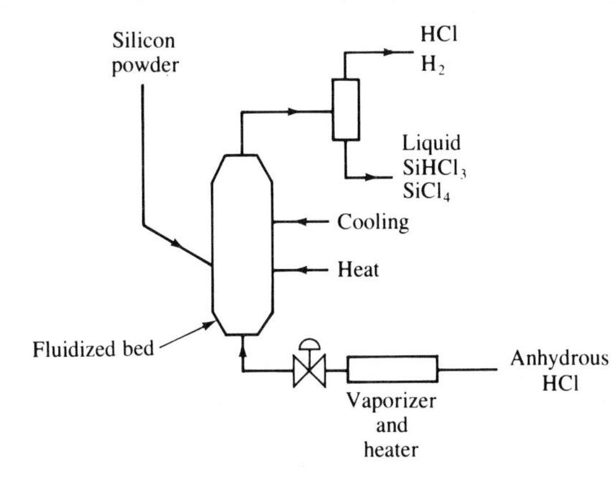

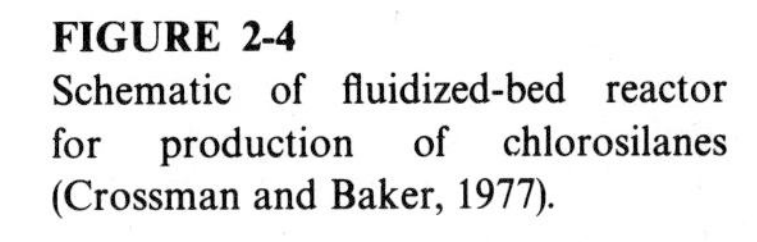

FIGURE 2-4
Schematic of fluidized-bed reactor for production of chlorosilanes (Crossman and Baker, 1977).

anes, $SiCl_4$ and $SiHCl_3$, for several reasons. The chlorosilanes may be produced easily at reasonable temperatures (200 to 400 °C), are liquid at room temperature, may be handled and transported with relative ease, and can be reduced at atmospheric pressure in the presence of hydrogen (Crossman and Baker, 1977).

The production of the chlorosilanes is begun by grinding MGS into a fine powder. The powder is then charged to a fluidized bed where it reacts with anhydrous HCl to produce $SiHCl_3$, $SiCl_4$, and H_2 along with metal chlorides such as $AlCl_3$, BCl_3, and PCl_3 from the impurities in the MGS. The principle reaction is

$$\text{Si(MGS)} + 3\text{HCl} \longrightarrow \text{SiHCl}_3 + \text{H}_2 \tag{2.5}$$

to produce $SiHCl_3$. A schematic of the reactor is shown in Fig. 2-4.

The chlorosilanes are then purified and separated to provide the extremely high purity required. The purification process begins with condensation of the product gas to remove the hydrogen and any light (i.e. BCl_3) or heavy (Al_2Cl_6, PCl_5) impurities. The resulting product stream then enters a series of distillation columns where the remaining impurities are removed and the $SiCl_4$ and $SiHCl_3$ are separated, if necessary (see Table 2.2). A typical procedure is to purify the $SiHCl_3$ by removing essentially all of the less volatile components in a second column, as in Fig. 2-5.

TABLE 2.2
Normal boiling points of silicon compounds and some potential impurities

Compound	Normal boiling point, °C
$SiCl_4$	56.8
$SiHCl_3$	31.8
SiH_3Cl	−30.4
SiH_4	−111.5
BCl_3	12.7
PCl_3	74.2
PCl_5	162.0
Al_2Cl_6	180.2

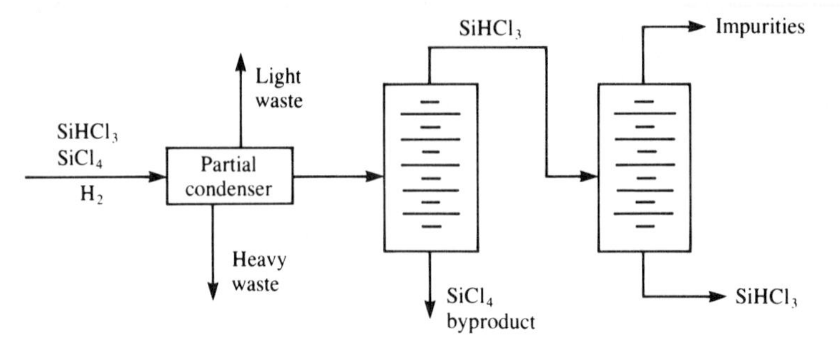

FIGURE 2-5
Schematic of purification process for electronic grade $SiHCl_3$.

The other chlorosilanes used for deposition of silicon are produced by catalytic redistribution of trichlorosilane:

$$2SiHCl_3 \longrightarrow SiH_2Cl_2 + SiCl_4 \tag{2.6}$$

$$3SiH_2Cl_2 \longrightarrow SiH_4 + 2SiHCl_3 \tag{2.7}$$

The purification of trichlorosilane, the redistribution reactions for production of dichloro- and chlorosilanes, and the decomposition reactions to produce silicon provide a variety of chlorosilanes and possibly H_2, HCl, and $SiCl_4$ (McCormick, 1985). The silicon gases are purified and used as precursors for chemical vapor deposition of silicon in device fabrication.

2.3 ELECTRONIC GRADE SILICON

Electronic grade polycrystalline silicon (EGS) is produced by either hydrogen reduction of chlorosilanes or pyrolysis of silane. The reactions for these processes are

$$SiHCl_3 + H_2 \xrightarrow{1000^\circ C} Si + \text{other products} \tag{2.8}$$

$$H_2SiCl_2 + H_2 \xrightarrow{900^\circ C} Si + \text{other products} \tag{2.9}$$

$$SiH_4 \xrightarrow{600^\circ C} Si + 2H_2 \tag{2.10}$$

The decomposition is performed in either a Siemens-type reactor or a fluidized bed.

The Siemens process for reduction of trichlorosilane by hydrogen on heated silicon substrates has dominated the industry for the last two decades. The trichlorosilane, which has impurities from the periodic table groups III and V of less than one part per billion, is vaporized and fed to the reactor in a hydrogen carrier stream. The reactor, shown in Fig. 2-6, has several U-shaped silicon substrates contained in a quartz bell jar. The substrates, which are of purity comparable to the produced silicon, are resistively heated to 1000 to 1100 °C and deposition occurs on the surface of the substrates (McCormick, 1985).

The actual deposition is a much more complicated reaction than that shown by Eq. (2.8). The intermediate $SiCl_2$ forms at the surface of the rods and

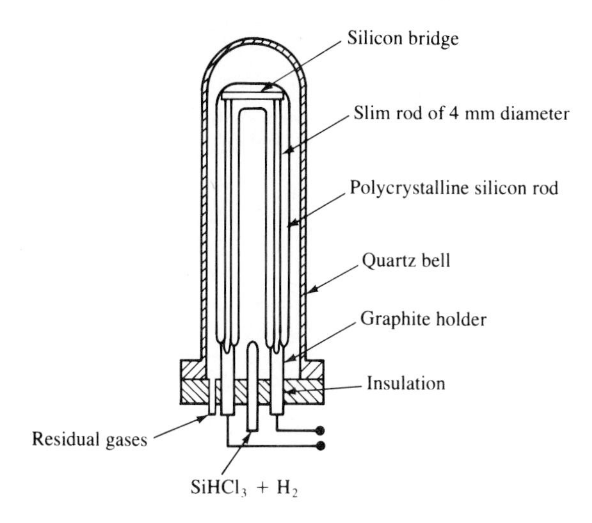

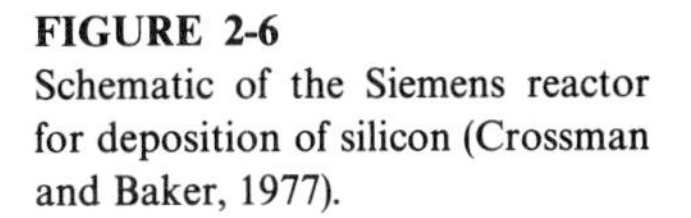
FIGURE 2-6
Schematic of the Siemens reactor for deposition of silicon (Crossman and Baker, 1977).

strongly affects both the decomposition rate of silicon and the production of $SiCl_4$. The byproducts of this process, H_2, $SiCl_4$, and unreacted $SiHCl_3$, are recovered and purified to be either sold or reused (Crossman and Baker, 1977) (see Fig. 2-7).

The Siemens process has several shortcomings, including low silicon and chlorine efficiency because of the large amount of $SiCl_4$ produced, the high labor cost associated with a batch process, and high power consumption and capital cost due to the low deposition rate. Improvements upon the basic processes have

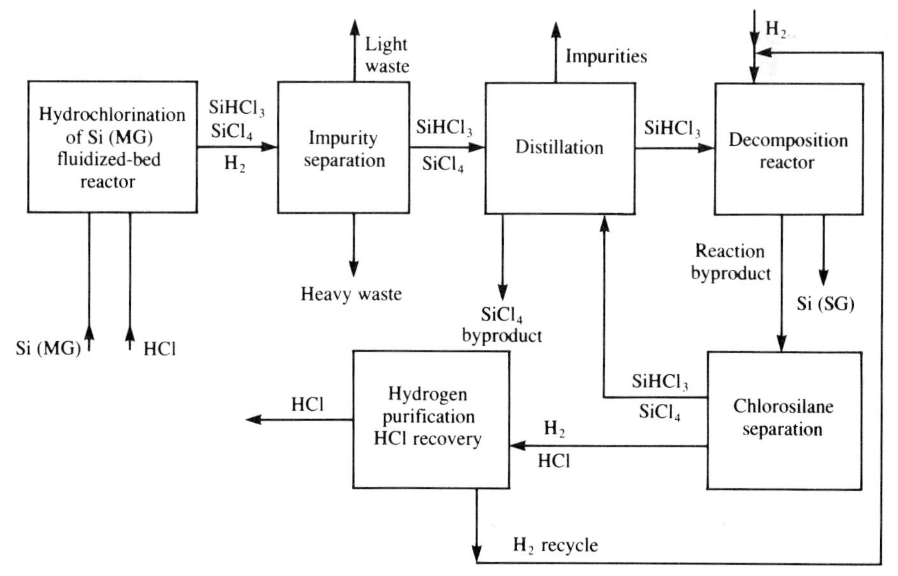

FIGURE 2-7
Block diagram of the Siemens process for production of polycrystalline electronic grade silicon (McCormick, 1985).

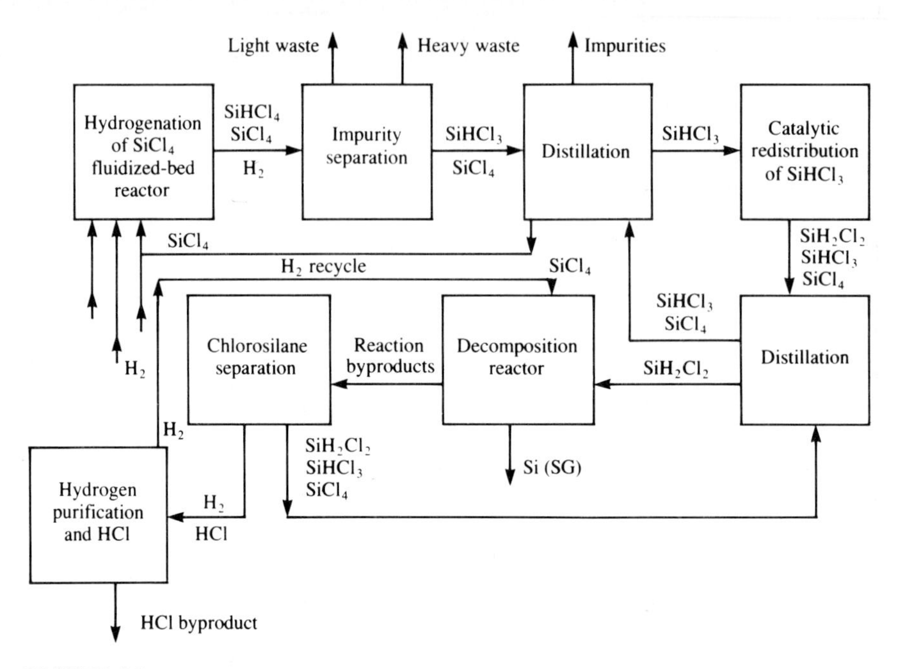

FIGURE 2-8
Block diagram of a process for production of electronic grade silicon by dichlorosilane decomposition in Siemens-type reactors (McCormick, 1985).

included processes for the hydrogenation of $SiCl_4$, eliminating the large byproduct stream, and hydrogenation of SiH_2Cl_2. The dichlorosilane process eliminates the chlorosilane byproduct streams by recycling $SiCl_4$ to the fluidized bed (Fig. 2-8). There, $SiHCl_3$ is produced and recycled to the catalytic redistribution reactor for production of SiH_2Cl_2. The dichlorosilane is better suited for use as a decomposition reactor feed than trichlorosilane because it is less stable, giving an increased reaction rate and reduced power consumption. The disadvantage of this process is the extra reaction and purification steps required for production of dichlorosilane (McCormick, 1985).

The other method by which silicon is produced from the chlorosilane and silane intermediates is decomposition in a fluidized-bed reactor. The reactions are the same as those that occur in the Siemens-type reactors, but the mechanisms are often different since homogeneous nucleation can occur along with heterogeneous deposition. The fluidized beds should eventually replace the Siemens-type reactors for EGS production because of their lower power consumption and continuous operation.

The gases used as silicon sources for fluidized-bed decomposition are $SiCl_4$, $SiHCl_3$, and SiH_4 (Fig. 2-9). The presence of $SiCl_4$ in the feed stream reduces the amount produced in the reactor and it is claimed that at properly selected feed conditions the $SiCl_4$ remains in a closed loop and silicon only leaves the system as EGS (McCormick, 1985).

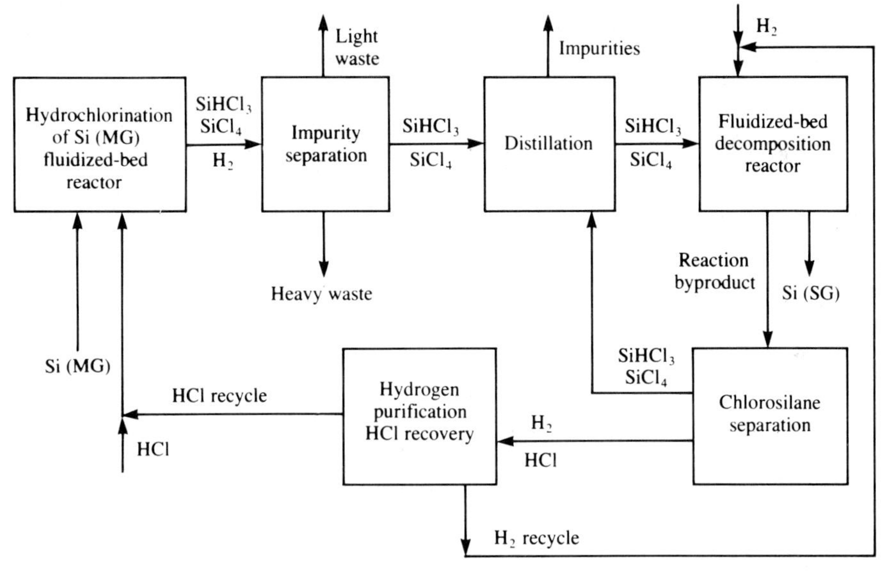

FIGURE 2-9
Block diagram of a process for production of EGS from $SiHCl_3$ and $SiCl_4$ by fluidized-bed decomposition (McCormick, 1985).

A process using silane as the feed gas for pyrolysis in a fluidized bed is being investigated by Union Carbide. The silane is produced by catalytic redistribution of dichlorosilane and then purified by cryogenic distillation, producing a silicon source of purity equivalent to that of $SiHCl_3$. The first approach investigated for the decomposition was a "free-space" reactor which took advantage of the decomposition and homogeneous nucleation of pure silane (Fig. 2-10). Unfortunately, this scheme had to be abandoned because the resulting powder of very small silicon particles was difficult to handle. The introduction of hydrogen into the silane feed and proper selection of operating conditions allows heterogeneous deposition to occur, which increases the resulting silicon particle size and eliminates the handling difficulties of the free-space reactor.

The decomposition of silane in a fluidized bed is a complex process involving heat transfer, homogeneous nucleation, heterogenous deposition, and interaction between particles. The effects which may occur can be examined by considering a simplified example of an isothermal bed (Fitzgerald, 1984).

As the gas enters the bed, it forms a small jet and then disperses through the bed particles and is heated to the bed temperature. To determine how rapidly the gas reaches the bed temperature the heat capacities of the gas and solids should be compared. The heat capacity per unit volume of the solids is on the order of 10,000 times that of the gas. The result of this large difference and the typical measured values of the Nusselt number for particle-to-gas heat transfer (approximately 0.1) is that the gas reaches the particle temperature very quickly. For the small particle size ($\sim 100\ \mu m$) and the low gas flow rate considered here,

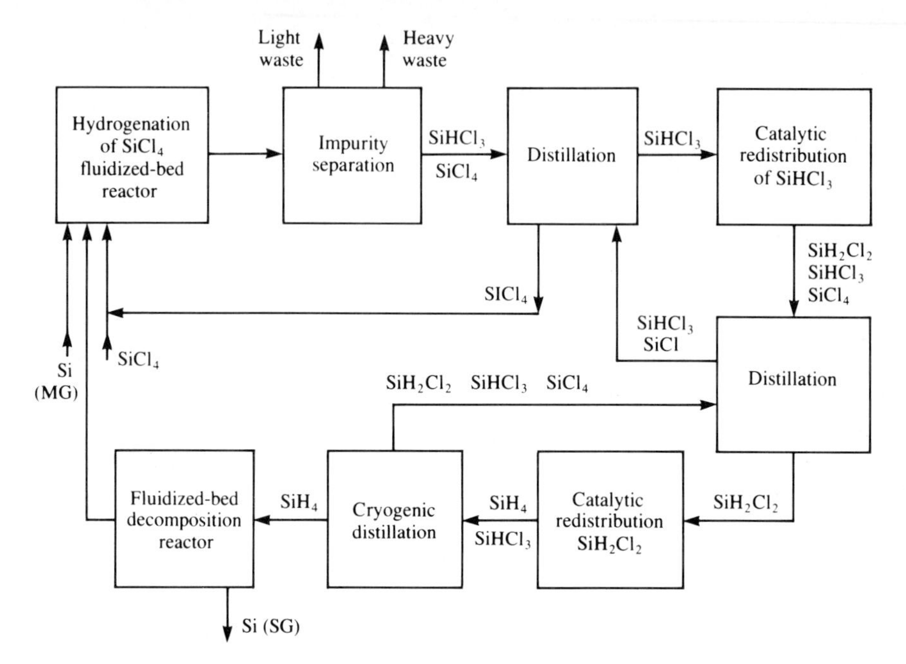

FIGURE 2-10
Block diagram of a process for production of EGS from silane by fluidized-bed decomposition (McCormick, 1985).

the gas reaches the particle temperature in a distance less than the particle diameter with essentially no change in the particle temperature.

As the silane enters the reactor and is heated, homogeneous nucleation begins and a suspension of microscopic particles of silicon is formed in the gas phase. Collisions resulting from Brownian motion will cause these particles to grow to diameters greater than 0.1 μm. At this point, the surface area of the microparticles is nearly equivalent to that of the surrounding bed particles. Since the microparticles are dispersed throughout the gas phase and offer a comparable surface area, any heterogeneous deposition that occurs near the bed entrance should take place on the microparticles.

There are two mechanisms that may account for capture of the microparticles by bed particles. In the entrance region, the deposition of silicon on microparticles causes the temperature of the microparticles to rise. The resulting temperature gradients create a drift velocity toward the cooler bed particles and cause a fraction of the bed particles to be captured. The fraction of microparticles that will be captured because of the thermophoresis, the thermally induced drift velocity, can be approximated by

$$\frac{T_{\text{hot}} - T_{\text{cold}}}{T_{\text{cold}}} = \frac{\Delta T}{T_{\text{bed}}} \tag{2.11}$$

The adiabatic temperature rise which results from the silicon deposition is approximately 350 °C and the bed temperature is approximately 1000 K. Therefore, approximately 35 percent of the microparticles will be captured in the region near the reactor entrance.

The remaining microparticles are captured by impact with the existing bed particles. Before the microparticles travel the length of the bed, they may grow as large as 0.1 μm in diameter and the effect of Brownian motion becomes negligible. The particles follow gas streamlines and are captured by impaction when the streamlines come within one particle radius of a bed particle.

Although the mechanism for particle growth and agglomeration described above is not verified, it does serve to demonstrate the complexity of silane decomposition in a fluidized bed. Both homogeneous nucleation and heterogeneous deposition can occur at typical operating conditions, but the mechanisms are not known well enough to determine which, if either, of the reactions dominate. There are also several other processes that were not included, such as attrition and elutriation, unwanted carryover of fine particles.

In order to control and predict the behavior of a fluidized bed in which particles are growing, the size distribution of the particles should be known. The design equations which will be presented here, or their discrete analogs, provide the basis for such calculations.

Consider a bed such as that shown in Fig. 2-11. The solid feed rate, removal rate, and elutriation rate are labelled F_0, F_1, and F_2, respectively, with units of mass per unit time. The size distribution in each stream and that of the bed is given by the corresponding distribution function $P(R)$ and the mass of the particles in the bed is denoted W. For a bed operating at steady state the design equations may be developed as given in Kunii and Levenspiel (1969).

It is assumed that the solid density, ρ_s, is constant, the bed is sufficiently mixed for $P_B(R) = P_1(R)$, and the rate of particle growth may be described by

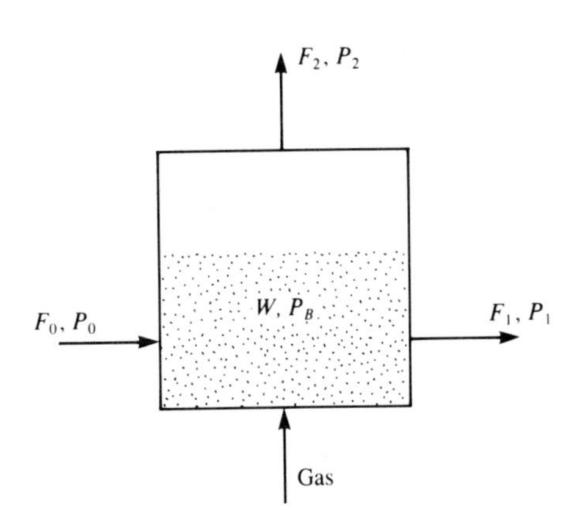

FIGURE 2-11
Schematic diagram of a fluidized bed.

$G(R) = dR/dt$. The number of particles in a size interval R to $R + \Delta R$ is given by

$$N(R) = \frac{WP_B(\bar{R})\Delta R}{\rho_s 4\pi \bar{R}^3/3} \tag{2.12}$$

where $\bar{R}$ is the average size in the interval. Therefore, the rate of solid generation in the interval is

$$\begin{matrix}\text{Solid generation}\\ \text{in the interval}\end{matrix} = \rho_s N(R)\left(\frac{dV}{dt}\Delta t\right) \tag{2.13}$$

in a time interval Δt. By taking the limit as $\Delta R \to dR$ and $\Delta t \to dt$, there results:

$$\begin{matrix}\text{Solid generation}\\ \text{in the interval}\end{matrix} = \frac{3WP_B(R)G(R)\,dR}{R} \tag{2.14}$$

A mass balance for the solids of size between R and $R + \Delta R$ in unit time is expressed as

$$\begin{matrix}\text{Rate at which}\\ \text{solids in } \Delta R\\ \text{enter the feed}\end{matrix} - \begin{matrix}\text{rate at which}\\ \text{solids in } \Delta R\\ \text{leave with the}\\ \text{outlet stream}\end{matrix} - \begin{matrix}\text{rate at which}\\ \text{solids in } \Delta R\\ \text{are lost by}\\ \text{elutriation}\end{matrix}$$

$$+ \begin{matrix}\text{rate at which}\\ \text{solids enter } \Delta R\\ \text{by growth from}\\ \text{a smaller size}\end{matrix} - \begin{matrix}\text{rate at which}\\ \text{solids in } \Delta R\\ \text{leave with the}\\ \text{outlet stream}\end{matrix} + \begin{matrix}\text{rate at which}\\ \text{solids in } \Delta R\\ \text{are lost by}\\ \text{elutriation}\end{matrix} = 0 \tag{2.15}$$

By taking the limit as ΔR goes to zero of the corresponding difference equation, the general expression becomes

$$F_0 P_0(R) - F_1 P_1(R) - WE(R)P_1(R) - W\frac{d}{dR}[G(R)P_1(R)] + 3\frac{W}{R}P_1(R)G(R) = 0 \tag{2.16}$$

where $E(R)$ is the elutriation constant defined by

$$E(R) = \frac{F_2}{W}\frac{P_2(R)}{P_1(R)} \tag{2.17}$$

for a steady-state, well-mixed bed. Integration of Eq. (2.17) over all sizes of particles present yields the overall mass balance:

$$F_1 + F_2 - F_0 = 3W\int_{R_{\min}}^{R_{\max}} \frac{P_1(R)G(R)}{R}\,dR \tag{2.18}$$

The particle size distribution in the bed for a feed of wide size distribution can be computed by an iterative scheme. The equations that must be solved for a bed in which attrition fines are ignored and the particles are growing are

$$P_1(R) = \frac{F_0 R^3 I(R_1 R_{\min})}{W\,|\,G(R)\,|} \int_{R_{\min}}^{R} \frac{P_0(R_i)\,dR_i}{R_i^3 I(R_i, R_{\min})} \tag{2.19}$$

and

$$\frac{W}{F_0} = \int_{R_{\min}}^{R_i \to \infty} \frac{R^3 I(R, R_{\min})}{|\,G(R)\,|} \int_{R_{\min}}^{R} \frac{P_0(R_i)\,dR_i}{R_i^3 I(R_i, R_{\min})}\,dR \tag{2.20}$$

with

$$I(R, R_i) = \exp\left[-\int_{R_i}^{R} \frac{F_1/W + E(R)}{G(R)}\,dR\right] \tag{2.21}$$

These equations are solved for $P_1(R)$ and W by the following scheme:

1. Tabulate values for $I(R, R_{\min})$ at suitable increments of R.
2. Use the tabulated values in Eq. (2.20) to iteratively calculate W.
3. Find $P_1(R)$ at the desired values of R from Eq. (2.19).
4. Determine F_2 from Eq. (2.18).
5. Determine $P_2(R)$ from Eq. (2.17).

These equations may be discretized for calculations in which a finite number of size ranges are used to represent the overall distribution. Unfortunately, significant errors may arise from this approach and it is better to derive the equations using the discrete distribution initially (Kayihan, 1984). For N discrete size ranges $\Delta R_1, \Delta R_2, \ldots, \Delta R_N$, with N corresponding mean sizes $R_1, R_2, \ldots, R_N$, the equations are

$$F_0 P_0(R_i)\Delta R_i - F_1 P_1(R_i)\Delta R_i - WE(R_i)P_1(R_i)\Delta R_i + WG(R_{i-1})P_1(R_{i-1}) - WG(R_i)P_1(R_i) + \frac{3W}{R_i} P_1(R_i)G(R_i)\Delta R_i = 0 \tag{2.22}$$

with $P_1(R_{i-1}) = 0$ for $i = 1$, for growing particles. The constraints are

$$\sum_{i=1}^{N} m_i = \sum_{i=1}^{N} P_1(R_i)\Delta R_i = 1 \tag{2.23}$$

$$F_1 + F_2 - F_0 = 3W \sum_{i=1}^{N} \frac{m_i G(R_i)}{R_i} \tag{2.24}$$

where the m_i are the mass fractions of particles in the size range ΔR_i.

Example 2.1. Particles (single size) of radius $R_i = 1.5 \times 10^{-2}$ cm are fed continuously to a fluidized-bed reactor. The particles grow due to deposition, and there is no elutriation. Determine the feed rate necessary to maintain a bed weight of 20,000 kg and a 1 kg/s product stream.

Data: Particle growth is modeled by

$$G(R) = dR/dt = k = 3.14 \times 10^{-7} \text{ cm/s}$$

Solution. The performance equations for a bed of growing particles and single size feed are

$$P_1(R) = \frac{F_0}{W\,|G(R)|}\frac{R^3}{R_i^3} I(R, R_i)$$

$$\frac{W}{F_0} = \int_{R_i}^{R_{t\to\infty}} \frac{R^3}{G(R)R_i^3} I(R, R_i)\, dR$$

where $I(R, R_i)$ is given by Eq. (2.21):

$$I(R, R_i) = \exp\left[-\int_{R_i}^{R} \frac{F_1/W + E(R)}{G(R)}\, dR\right]$$

Substitution of the known values gives

$$\frac{W}{F_0} = \int_{R_i}^{R_{t\to\infty}} \frac{x^3}{kR_i^3} \exp\left(-\int_{R_i}^{x} \frac{F_1/W}{k}\, dy\right) dx$$

Evaluation of the integral representing $I(R, R_i)$ results in

$$\frac{W}{F_0} = \int_{R_i}^{R_{t\to\infty}} \frac{x^3}{kR_i^3} \exp\left[-Z(x - R_i)\right] dx$$

where $Z = F_1/Wk$. Since $R_{t\to\infty} = \infty$ for the given growth mechanism, integration by parts gives

$$\frac{W}{F_0} = \frac{1}{kR_i^3} \lim_{R\to\infty} \left\{\frac{R^3}{Z \exp[Z(R - R_i)]} + \frac{R_i^3}{Z} - \frac{3R^2}{Z^2 \exp[Z(R - R_i)]} + \frac{3R_i^2}{Z^2} - \frac{6R}{Z^3 \exp[Z(R - R_i)]} + \frac{6R_i}{Z^3} - \frac{6}{Z^4 \exp[Z(R - R_i)]} + \frac{6}{Z^4}\right\}$$

Application of l'Hopital's rule gives the result

$$\frac{W}{F_0} = \frac{1}{kR_i^3}\left(\frac{R_i^3}{Z} + \frac{3R_i^2}{Z^2} + \frac{6R_i}{Z^3} + \frac{6}{Z^4}\right)$$

$$\frac{W}{F_0} = \frac{20{,}000 \text{ kg}}{F_0} = 7.4960 \times 10^4 \text{ s}$$

$$F_0 = 0.2668 \text{ kg/s}$$

2.4 METAL ORGANIC COMPOUNDS

Hydrides and metal organic compounds are used as source gases for film deposition in device fabrication. These compounds are not only used for the growth of compound semiconductor films such as GaAs but also for the doping of silicon films and for other deposition such as metallization (deposition for providing device interconnections). Since high-purity source gases are required to avoid unintential doping of the deposited films, synthesis techniques that diminish the possibility of such contamination are highly desired.

As mentioned in the introduction, most of the metal hydrides and metal organics described in this section are potentially dangerous and must be handled accordingly. Both phosphine and arsine are very toxic gases and their use requires a carefully designed system for delivery to the reactor and extensive scrubbing of the reactor exhaust gases (Pearce, 1983). The metal alkyls used for metal organic chemical vapor deposition (MOCVD) are extremely reactive towards electron donors such as O, S, N, or P because of the electron deficiency of the metal atom (Bradley *et al.*, 1986). The reactions of trimethyl gallium demonstrate the danger this presents, as trimethylgallium ignites with oxygen at $-76\,^{\circ}C$ and explodes if its vapor is mixed with air at room temperature (Sheka *et al.*, 1966). The safety hazards and contamination possibilities caused by the reactivity of metal alkyls have spurred interest in a possible replacement, volatile adducts.

The metal halides are often used as an intermediate between pure or combined metals and the desired metal organic compounds. For this reason, the facility with which metal halides can be produced has a direct bearing on the purity and cost of the metal organic compound. The halides at the group III elements aluminum, gallium, and indium can all be obtained by direct combination of the elements. The procedure for production of gallium trichloride, $GaCl_3$, from molten gallium and chlorine gas provides an example of such a process.

The direct combination of gallium and chlorine is begun by placing the gallium metal in a quartz reactor. The reactor is evacuated and sealed to prevent contamination; then the reaction is begun by bubbling chlorine gas through the molten metal. The dichloride, Ga_2Cl_4, is formed first, with continued reaction causing the trichloride to form as follows:

$$2Ga + 2Cl_2 \longrightarrow Ga_2Cl_4 \tag{2.25}$$

$$Ga_2Cl_4 + Cl_2 \longrightarrow 2GaCl_3 \tag{2.26}$$

The completion of the reaction is marked by the color of the melt changing from white to the yellowish green of chlorine. At this point the flow of chlorine is stopped and an inert gas such as helium is bubbled through the melt to remove any of the dichloride that remains (Goldsmith *et al.*, 1962).

There are several other methods for production of gallium trichloride: the reaction of hydrogen chloride and gallium metal at 200 to $800\,^{\circ}C$ followed by distillation in carbon dioxide; the reaction of carbon tetrachloride and chlorine with gallium oxide with $SOCl_2$ at $200\,^{\circ}C$; and the reaction of gallium metal with silver, lead, copper, or mercury chlorides. The advantage of direct combination of gallium and chlorine over these methods is the high purity of the product. For this reason, very pure gallium trichloride is usually produced by either the sublimation of gallium in a stream of chloride as described above or by zone refining (Sheka *et al.*, 1966).

Arsenic trichloride, $AsCl_3$, may be used as the arsenic source for GaAs deposition. Although $AsCl_3$ may be formed by direct combination of the elements, it is also formed when arsenic (III) oxide is heated with concentrated hydrochloric acid (Sharpe, 1986):

$$As_4O_6 + 6H_2O + 12H^+ + 12Cl^- \longrightarrow 4AsAl_3 + 12H_2O \tag{2.27}$$

Hydrides of main interest are PH_3 and AsH_3. The trihydrides of the group V elements are used as sources for both compound semiconductors containing the elements and for doping of silicon with these elements. Phosphine, arsine, and stibine, SbH_3, are all covalently bonded hydrides which can be made by the hydrolysis of metal salts, i.e.,

$$Ca_3P_2 + 6H_2O \longrightarrow 3Ca(OH)_2 + 2PH_3 \tag{2.28}$$

Arsine and stibine are also prepared by reducing arsenic or antimony compounds with zinc in acid solution as follows:

$$As_4O_6 + 12Zn + 24H^+ \longrightarrow 12Zn^{2+} + 6H_2O + 4AsH_3 \tag{2.29}$$

Metal halides are used as source gases for chemical vapor deposition (CVD) metallization as well as intermediates in the production of metal organics. The compound that has attracted the most interest for this application is tungsten hexafluoride because of the relatively low resistivity of tungsten metal and its refractory nature. Tungsten may be deposited by either pyrolitic or reduction reactions as the following reactions:

$$WF_6 + \text{thermal, plasma, or optical energy} \longrightarrow W + 3F_2 \tag{2.30}$$

$$WF_6 + 3H_2 \longrightarrow W + 6HF \tag{2.31}$$

Other metals are also in use for this application, such as

$$2MoCl_5 + 5H_2 \longrightarrow 2Mo + 10HCl \tag{2.32}$$

$$2TaCl_5 + 5H_2 \longrightarrow 2Ta + 10HCl \tag{2.33}$$

$$2TiCl_5 + 5H_2 \longrightarrow 2Ti + 10HCl \tag{2.34}$$

for the deposition of molybdenum, tantalum, and titanium. Because of the multiple valence states of these transition metals, there are a number of reaction sequences for production of the desired chlorides, including reduction or oxidation of the MX_6 (M for metal and X for halogen atom) and MX_4 compounds.

Metal alkyls have become the main source material for film deposition of III-V and II-VI compound semiconductors. The trimethyl and triethyl metal organic compounds are the usual source gases for MOCVD or the group III metals while hydrides are the usual source gases for group V metals. The reactions used to produce trimethyl gallium (TMG) will be emphasized here since TMG is the most commonly used metal source gas and the reactions undergone by other alkyl groups and trivalent metals are similar. Because of the need for consistant high purity of the metal source for growth of good-quality semiconductor films, several reaction schemes for production of high-purity TMG have been developed.

As mentioned previously, gallium trichloride is often used as an intermediate between gallium metal and trimethyl gallium. Although the production of $GaCl_3$ introduces an extra step into the preparation sequence, the purity of $GaCl_3$ is very good provided the pure gallium and chlorine are used. The standard procedure for preparing main group alkyls of metals with greater electronegativity than magnesium is through the Grignard reagent (Jones *et al.*, 1984):

$$6MeMgI + 2GaCl_3 \xrightarrow{OEt_2} 2GaMe_3 \cdot OEt_2 + 3MgCl_2 \quad (2.35)$$

Unfortunately, the TMG is a strong Lewis acid and forms a stable adduct with the diethylether solvent, OEt_2. Because of the stability of this adduct, it is difficult to produce pure TMG free of a Lewis base by this method unless a more volatile ether, which allows thermal decomposition of the adduct, is used.

Reactions similar to the reaction of gallium trichloride with the Grignard reagent are also possible with alkyl lithium, dialkyl zinc, and trialkyl aluminum as the alkyl source (Jones *et al.*, 1984; Gaines *et al.*, 1974). These reactions are

$$GaCl_3 + 3MeLi \longrightarrow Me_3Ga + 3LiCl \quad (2.36)$$

$$2GaCl_3 + 3Me_2Zn \longrightarrow 2Me_3Ga + 3ZnCl_2 \quad (2.37)$$

$$GaCl_3 + 3Me_3Al \longrightarrow Me_3Ga + 3Me_2AlCl \quad (2.38)$$

Since these reagents are soluble in hydrocarbons, the presence of Lewis bases in the product is no longer a problem. The major disadvantage of these routes is the possibility of unintentional doping, particularly with the dimethyl zinc source. The reaction with trimethyl aluminum is the most commercially viable of the three. The reaction is carried out without a solvent by adding trimethyl aluminum to the $GaCl_3$ dropwise and then distilling the product. To completely remove the trimethyl aluminum from the TMG, the vapor is condensed onto sodium fluoride. An involatile adduct, $Na \cdot Me_3AlF$, is formed from the trimethyl aluminum while the TMG is unreactive and can be distilled off.

There are also several routes to TMG directly from gallium metal. The most straightforward is probably the metal exchange reaction between dimethyl mercury and gallium (Jones *et al.*, 1984):

$$2Ga + 3Me_2Hg \longrightarrow 2GaMe_3 + 3Hg \quad (2.39)$$

Unfortunately, the reaction is slow, taking several days to complete under typical conditions, and the diethyl mercury may introduce contaminants. There is also the possibility of the dimethyl mercury itself contaminating the TMG and thereby introducing a *p*-type dopant.

The electrolysis of Grignard reagents with a sacrificial gallium anode has received some attention because of the high yields and current efficiencies ($\sim$100 percent) that may be obtained (Jones *et al.*, 1984). By adding excess alkyl halide to the mixture, deposition of magnesium at the cathode may be avoided and the Grignard reagent is regenerated. The resulting overall reaction is

$$3MeMgX + 3MeX + 2Ga \xrightarrow{S} 2Me_3Ga \cdot S + 3MgX_2 \qquad (2.40)$$

where X is the iodine or chlorine and S is the solvent. As discussed previously, diethyl ether will produce a stable adduct when used as the solvent. However, the diethyl ether allows higher current densities and correspondingly shorter reaction times than the higher boiling ethers which dissociate more readily. There is evidence that the diethyl ether may be displaced by a higher boiling ether via the equilibrium:

$$Me_3Ga \cdot OEt_2 + R_2O \rightleftharpoons Me_3Ga \cdot OR_2 + OEt_2 \qquad (2.41)$$

The equilibrium normally lies to the left, but the diethyl ether is the most volatile component of the mixture and may be removed to form a solution of $[Me_3Ga \cdot OR_2]$. The TMG can then be released by heating.

Aluminum alkyls may be prepared by synthesis techniques such as those presented for TMG, particularly the mercury alkyl and aluminum reaction and the reaction of Grignard reagents with the chloride. For industrial scale production, a direct synthesis from alkenes, aluminum, and hydrogen is usually employed (Sharpe, 1986):

$$2Al + 3H_2 + 6C_2H_4 \longrightarrow 2Al(C_2H_5)_3 \qquad (2.42)$$

Indium alkyls may be prepared by reaction of the appropriate metal alkyls and indium chloride, as described previously for gallium (Sharpe, 1986). Another synthesis technique of interest is the electrolysis of Grignard reagents with an indium anode in diethyl ether, as outlined for TMG (Jones *et al.*, 1984). The diethyl ether–trimethyl indium adduct that results may be dissociated by dissolution in benzene, removal of the ether, and fractional distillation.

The use of volatile adducts of the metal alkyls as MOCVD source gases was initiated by problems with the deposition of InP from Me_3In and PH_3 (Jones *et al.*, 1984). When the attempts were made to prepare InP by this method an involatile polymer, $[MeInPh]_n$, was formed. Although the growth conditions can be adjusted to prevent polymer formation, some growers found it simpler to use adducts such as $Me_3In \cdot PMe_3$, $Me_3In \cdot PEt_3$, and $Me_3In \cdot NEt_3$ as the indium source instead. The Lewis base blocks formation of the polymer but the adduct compound still allows deposition of InP when it is combined with phosphine.

The electrolysis of Grignard reagents that has been discussed previously is an effective method for synthesis of the volatile adducts. By using tetrahydrofuran (thf) as the solvent, the adduct $Me_3M \cdot thf$ (M = Ga or In) is formed. The thf may be displaced by Lewis bases such as NEt_3, PMe_3, and PEt_3 by the ligand exchange reaction:

$$Me_3M \cdot thf + L \longrightarrow Me_3Ga \cdot L + thf \qquad (2.43)$$

where L represents the Lewis base. The main disadvantage of the volatile adducts as source gases is that the source must be heated, as opposed to the cooling required for metal alkyls. The adducts have several advantages over the metal alkyls, including simpler purification procedures because of the decreased reactivity and easier composition control for lattice-matched ternary or quaternary layers (Bradley *et al.*, 1986).

There are also non-volatile adducts based on the Lewis base "diphos" which dissociate at approximately 100 °C to give very pure metal alkyls:

$$\underset{\displaystyle Me_3In\qquad\quad InMe_3}{(C_6H_5)_2PCH_2CH_2P(C_6H_5)_2} \longrightarrow (C_6H_5)_2PCH_2CH_2P(C_6H_5)_2 + 2Me_3In \quad (2.44)$$

These adducts are prepared, using the ligand exchange reaction, from the metal alkyl diethyl ether adduct and diphos. The nonvolatile adducts can be used as a purification stage for preparation of metal alkyls or as the source of metal alkyl vapor by heating a container filled with the adduct.

NOTATION

E	Elutriation rate (M/t)
F	Mass rate (M/t)
G	Growth rate (L/t)
I	Quantity defined by Eq. (2.21)
M_i	Mass fraction in ith size range
N	Number of particles
P	Size distribution function (number$/L$)
R	Particle radius (L)
$\bar{R}$	Average value of $R(L)$
t	Time (t)
V	Particle volume (L^3)
w	Mass of particles

Greek letter

ρ_s	Solid density (M/L^3)

Subscripts

0	Inlet
1	Outlet
2	Elutriation outlet
B	Bed

Units

L Length
M Mass
t Time

PROBLEMS

2.1. Solids of a wide size distribution are fed continuously to a fluidized-bed reactor where reaction causes them to shrink. There are no fines entrained in the gas outlet stream ($F_2 = 0$), but unreacted solids exit in an overflow stream. For a feed rate of 1.44 kg/s and a bed weight of 10,000 kg, what is the mass flow rate of the overflow stream?

Data:

$$G(R) = dR/dt = -k = -1.24 \times 10^{-6} \text{ cm/s}$$

Size R, cm × 10^2	$P_0(R)$, cm^{-1}
0.5	0
1.0	3.15
2.0	8.40
2.5	10.50
3.0	12.25
3.5	13.65
4.0	14.70
5.0	15.75
6.0	15.40
7.0	13.65
8.0	10.50
8.5	8.40
9.0	5.95
9.5	3.15
9.6	2.55
9.8	1.30
10.0	0

Answer: $F_1 = 0.62$ kg/s

2.2. Determine the weight of the bed for the system in Prob. 2.1 if no solids are allowed to leave the bed.

2.3. Derive the design equations analogous to Eqs. (2.19) and (2.20) for a fluidized bed with a single feed size, R_i, in which (*a*) the particles are growing and (*b*) the particles are shrinking. Describe a procedure for calculation of the weight of particles and outlet particle size distribution.

2.4. Derive Eqs. (2.19) and (2.20) by using the results of Prob. 2.3 and treating the outflow as the sum of outflows from narrow size ranges ΔR.

2.5. Derive the equations analogous to Eqs. (2.19) and (2.20) for a bed in which the particles are shrinking. Describe how these equations may be applied to a fluidized bed in which MGS is decomposed to form $SiHCl_3$.

REFERENCES

Bradley, D. C., M. M. Faktor, M. Scott, and E. A. D. White: *J. Crystal Growth*, vol. 75, p. 101, 1986.

Crossman, L. D., and J. A. Baker: in H. R. Huff and E. Sirtl (eds.), *Semiconductor Silicon 1977*, p. 18, The Electrochemical Society, Inc., Pennington, N.J., 1977.

Dietl, J.: in C. P. Khattak and K. V. Ravi (eds.), *Silicon Processing for Photovoltaics*, vol. II, p. 301, North-Holland, New York, 1987.

Fitzgerald, T. J.: in R. Lutwack and A. Morrison (eds.), *Silicon Material Preparation and Economical Wafering Methods*, p. 129, Noyes Publications, Park Ridge, N.J., 1984.

Gaines, D. F., J. Borlin, and E. P. Fody: *Inorganic Synthesis*, vol. 15, p. 203, 1974.

Goldsmith, N., A. Mayer, and L. Vieland; *J. Less Common Methods*, vol. 4, no. 6, p. 564, 1962.

Jones, A. C., A. K. Holliday, D. J. Cole-Hamilton, M. M. Ahmad, and N. D. Gerrard: *J. Crystal Growth*, vol. 68, p. 1, 1984.

Kayihan, F.: in R. Lutwack and A. Morrison (eds.), *Silicon Material Preparation and Economical Wafering Methods*, p. 146, Noyes Publications, Park Ridge, N.J., 1984.

Kunii, D., and O. Levenspiel: *Fluidization Engineering*, Wiley, New York, 1969.

McCormick, J. R.: in C. P. Khattak and K. V. Ravi (eds.), *Silicon Processing for Photovoltaic Applications*, vol. I, p. 1, North-Holland, New York, 1985.

Pearce, C. W.: in S. N. Sze (ed.), *VLSI Technology*, chap. 2, McGraw-Hill, New York, 1983.

Sharpe, A. G.: *Inorganic Chemistry*, 2d ed., Longman, New York, 1986.

Sheka, I. A., I. S. Chans, and T. T. Mityureva: *The Chemistry of Gallium*, Elsevier, New York, 1966.

CHAPTER 3

BULK CRYSTAL GROWTH

3.1 INTRODUCTION

Bulk single-crystal semiconductors are grown in ingot form from their polycrystalline counterparts. A monocrystalline material, or single crystal, has a uniform crystal structure throughout. A polycrystalline material is a composite of monocrystalline grains. Each grain has the same monocrystalline structure but the grains are randomly joined to one another in the polycrystalline material. The grains' crystal lattices are also randomly oriented from grain to grain.

The ingot or boule is then sectioned into thin slices, which are commonly known as wafers. Single-crystal silicon is made by melting polycrystalline silicon in a furnance and then solidifying it. The silicon ingot is then sectioned into thin wafers. Integrated circuits are fabricated on the wafer surface. The wafer thickness actually used in making the circuit features is only a tiny fraction of the overall thickness.

Although polycrystalline semiconductor materials are very pure in the usual sense, they must be refined further before making single-crystal materials since the acceptable impurity content in the wafer is at or below the ppb level. When solidification takes place, the impurities are usually rejected at the (liquid-solid) interface into the liquid, and a purer solid material forms. To form a single-crystal structure, there has to be a single crystalline material into which atoms can grow in an orderly fashion. This is accomplished by providing a single seed crystal and contacting it with the melt. Various methods are used for the cooling that accompanies solidification.

The typical bulk crystal growth techniques in practice are the Bridgeman method, the Czochralski method, and the floating-zone method. Bulk crystal growth of silicon is carried out mostly by the Czochralski method. The floating-zone method constitutes a small fraction of silicon crystal growth today. The other methods are usually employed for the bulk crystal growth of compound semiconductors such as gallium arsenide. Usually, the other methods are considered only when the Czochralski method cannot be used for one reason or another. The apparatus used for the various methods is shown in Fig. 3-1.

Perhaps the simplest technique is the Bridgeman method (Hannay, 1959; Goodman, 1974). The technique includes melting polycrystalline semiconductor after a tube is evacuated and sealed. As shown in Fig. 3-1, the melt is then cooled from the seed crystal end so that the melt can grow into the desired single crystalline structure. For instance, a specific crystallographic orientation can be established by providing a seed crystal with the desired orientation. As illustrated, gallium arsenide is often grown by this technique, since the excess arsenic pressure required for growth can easily be obtained with the arrangement. Excess arsenic pressure is needed because of the much lower boiling point of arsenic compared with gallium. Without the excess pressure, preferential evaporation of arsenic occurs from the solid and molten gallium arsenide. One major problem with the technique is contamination from the container wall. Ideally, no wetting of the wall should occur. Otherwise, in addition to the contamination problem, the container may crack or crystalline defects can develop because of the expansion of the semiconductor material upon solidification. Almost all semiconductors expand upon freezing, including the major semiconductors of silicon and gallium arsenide. To minimize the wetting, vitreous carbon and carbon-coated quartz boats are used. For gallium arsenide, fused quartz and boron nitride containers are usually used. Wetting of the walls can be reduced by making the surface rough, e.g., by abrasion, which reduces the effective wall contact area due to microscopic hills and valleys.

The Czochralski technique (Teal and Little, 1950) is by far the most preferred growth method for silicon and is often used for gallium arsenide. As shown in Fig. 3-1*b*, the melted polycrystalline semiconductor is contained in a suitable container. Resistive or inductive heating is used for melting. After the desired temperature of the melt is reached, the growth process commences following the dipping of the seed crystal. The seed crystal (about 1 cm in diameter) is then pulled according to the Dash technique (1959) in the case of silicon, to minimize formation of dislocations and/or slips. Initially, the crystal is pulled relatively fast to form a thin neck (about 0.3 cm in diameter). In such a thin neck, the strains caused by cooling are too low to cause slip due to the small radial temperature difference. After establishing a dislocation-free growth for the neck, the crystal diameter is enlarged by slower pulling. Just before the desired diameter is reached, the pull rate is raised again so that after a short transition, the diameter remains constant and the cylindrical crystal grows axially. At the end of the

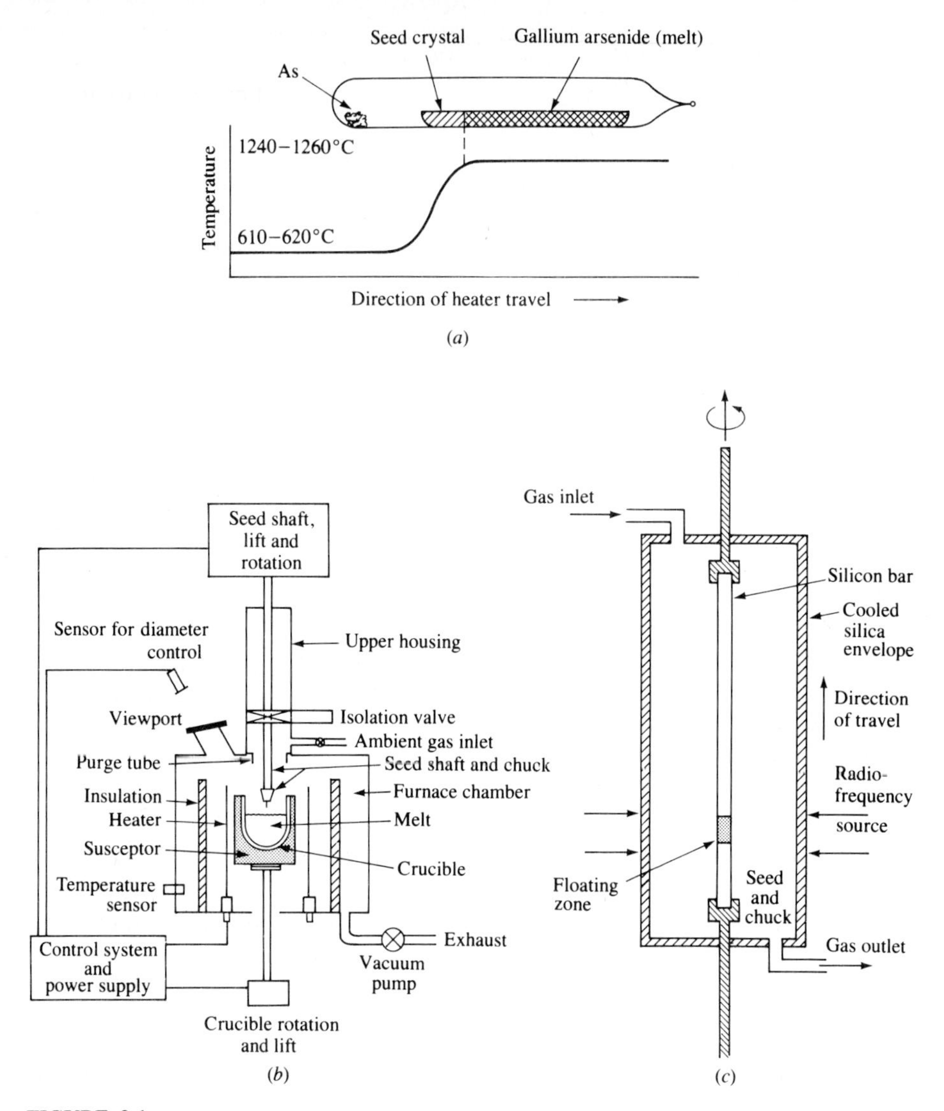

FIGURE 3-1
Apparatus for bulk crystal growth: (*a*) Bridgeman (Ghandhi, 1983), (*b*) Czochralski (Pearce, 1983), and (*c*) floating zone (Ghandhi, 1983).

pulling period, a second cone must be grown at the bottom end of the crystal to complete dislocation-free growth, prior to removing the crystal from the melt. Thus the diameter of the solid-liquid interface is nearly zero at separation. Otherwise, dislocations can be generated and propagated. Cooling is accomplished by radiation primarily and by passing an inert gas through the enclosure that envelopes the whole apparatus. As opposed to the Bridgeman technique, the Czochralski method does not involve any contacting of growing crystal with vessel walls. Although the Czochralski-grown crystals are very low in defects, a large amount of oxygen is incorporated into the melt and consequently into the crystal due to dissolution of the quartz crucible. The oxygen incorporated into the crystal in this way is present primarily in clumps and is electronically inactive. However, the oxygen incorporated into interstitial sites is active. The Czochralski growth of gallium arsenide suffers from the need to maintain arsenic overpressure. A method typically used to overcome the problem is to use a cap layer of inert liquid to cover the melt (Mullin *et al.*, 1965). The cap prevents decomposition of gallium arsenide provided the pressure on its surface is higher than one atmosphere. This can be realized by using an inert gas such as argon. A typical capping material for this liquid encapsulation method is B_2O_3 (about 0.5 cm thick). The boron oxide also remains on the gallium arsenide surface as the crystal is pulled. Since B_2O_3 dissolves SiO_2, the crucible is made of silicon nitride. While the crystal is pulled, the crystal is rotated to have a more uniform dopant distribution in the grown crystal. The crucible is also rotated for the same purpose.

The floating-zone process shown in Fig. 3-1*c* is one form of zone process (Pfann, 1958). A rod of polycrystalline semiconductor is held in a vertical position, and rotated during the growth. A small zone less than a few centimeters long is kept molten by a radio-frequency (rf) heater, which is moved so that this floating zone traverses the length of the rod. A seed crystal is provided at the starting point where the molten zone is initiated. As the zone traverses the rod, single-crystal semiconductor freezes at its retreating end and grows as an extension of the seed crystal. As with the Czochralski process, the rod and coils are enclosed in the cooled silica envelope in which an inert atmosphere is maintained (Ghandi, 1983). Unlike the Czochralski method, there is no contact of semiconductor material with the crucible walls. Therefore, the crystal purity is better and the oxygen content is lower by almost two orders of magnitude. The use of a rod starting material makes the process more costly. Thus, the crystals grown by this method are used for wafers intended for high-performance device applications.

Another form of (bulk) crystal growth is to use a seed crystal in a melt solution of two components and let one of the desired components precipitate onto the seed crystal. This technique, which is often referred to as liquid phase epitaxy, is essentially a liquid phase crystallization. One advantage here is that the temperatures required for the crystallization are much lower than those for single-component melts, due to the lowering of the freezing point by the presence of a solute in the liquid. One disadvantage is that the growth rate is much lower than other bulk crystal growth techniques (Small and Crossley, 1974; Pamplin, 1980). Because of the relatively high melting points of oxide, the technique is suitable for single-crystal growth of oxides.

After the boule (ingot) is made, it is first shaped to the desired form by grinding. The external surface of the acceptable portion of the boule (typically the middle section with the ends cut off) is ground to obtain the desired wafer diameter. The external surface is also ground as shown in Fig. 3-2 to form flats for wafer identification of doping type and crystallographic orientation. The boule is

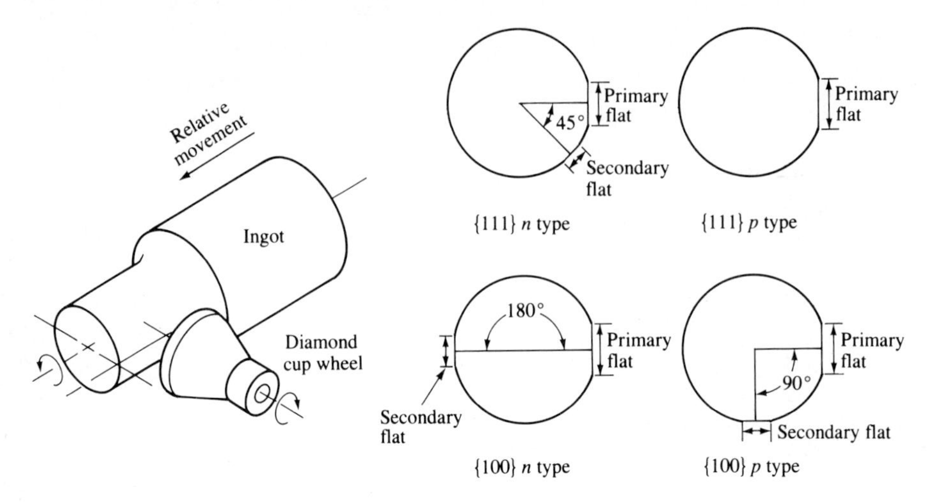

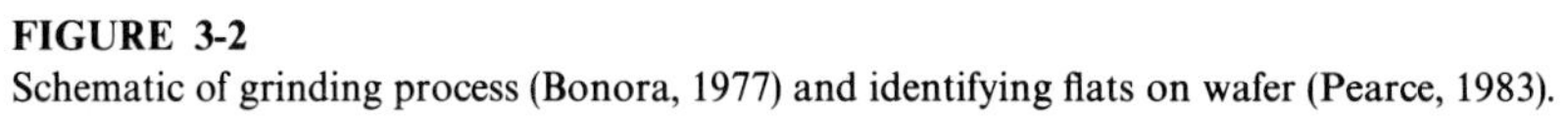

FIGURE 3-2
Schematic of grinding process (Bonora, 1977) and identifying flats on wafer (Pearce, 1983).

then sawed by a stainless steel blade into slices (wafers). A mechanical, two-sided lapping operation is performed to obtain uniformly thick wafers. The surface and edge of the wafers thus prepared are damaged and contaminated. The damaged and contaminated region, which is about 20 μm deep on each side, can be removed by chemical etching (Pearce, 1983). Normally, mixtures of HF, HNO_3 and acetic acid are used as etchants but alkali hydroxides are also used. The overall reaction for the oxidation-reduction taking place in water or acetic acid (Robbins and Schwartz, 1976; Kern, 1978) can be written as follows:

$$3Si + 4HNO_3 + 18HF \longrightarrow 3H_2SiF_6 + 4NO + 8H_2O$$

where the oxidant is HNO_3. The oxidized products are dissolved by HF. Since HF is a very strong solvent, the reaction is limited by the oxidation step when a HF-rich solution is used. The resulting etching is anisotropic, i.e., dependent on crystallographic direction. The use of HNO_3-rich mixtures yields isotropic etching, and the rate-limiting step is the dissolution process. The final step in wafer preparation is polishing of the etched wafer to provide a smooth, specular (reflective) surface. A colloidal suspension of fine SiO_2 particles (10 nm) in aqueous NaOH is used for the polishing step. Frictional heat causes NaOH to oxidize silicon via the OH^- ion. The oxidized silicon is abraded away by the silica particles in the slurry. Only one side of wafer is polished. After polishing, cleaned wafers are then ready for use as the building substrate for integrated circuits.

Although one side of the wafer is polished, the other side is usually damaged intentionally as a means of collecting undesired impurities. This is one form of a gettering treatment. Gettering is a process in which undesired impurities are removed from the device region, which is near the surface of the polished side of the wafer. The intentionally damaged back surface provides a wafer

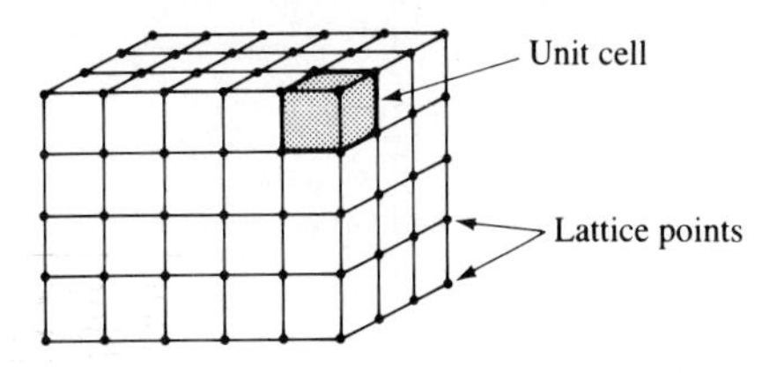

FIGURE 3-3
Lattice and unit cell.

with sinks that can absorb impurities. Impurities diffuse faster along defects and thus the impurities tend to diffuse out to the damaged back side rather than the front side during processing at elevated temperatures. A controllable process for the back side damaging can be accomplished by a focused laser beam (Pearce and Zaleckas, 1979).

The major concern in bulk crystal growth is with crystal defects that can be generated during the growth process. The temperature gradients generated in the cooling process are the major cause of crystal defects. Oxygen precipitation is another source of defects. Crystal structures and defects are treated in the next section before examining growth processes.

3.2 CRYSTAL STRUCTURES AND DEFECTS

Crystal structure refers to the size, shape, and atomic arrangement of the element(s) of the material. The atoms form a regular, repetitive gridlike pattern, or lattice. The unit cell is the smallest subdivision of the lattice that still retains the overall characteristics of the entire lattice. As shown in Fig. 3-3, the entire lattice of the crystal can be constructed by stacking identical unit cells. The surroundings of each lattice point are identical. One or more atoms are associated with each lattice point. Depending on the three angles each lattice point makes with the three axes and the three lengths of the unit cell, there result 14 Bravais lattices (Kittel, 1971). For semiconductors, it is sufficient to consider the simple cubic, face-centered cubic and hexagonal lattices, which are shown in Fig. 3-4 along with their characteristics. For the cubic lattice including the face-centered

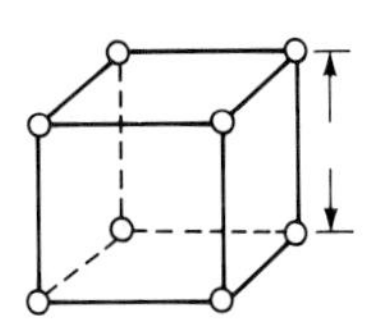

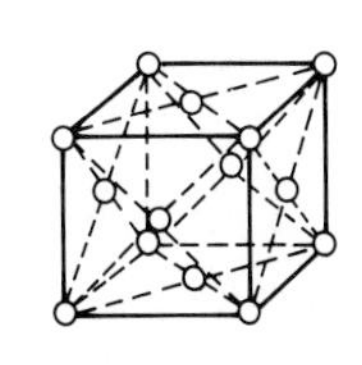

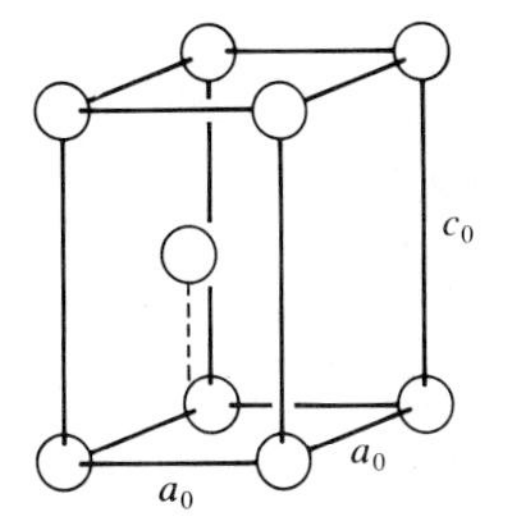

FIGURE 3-4
Three Bravis lattices of interest to semiconductor crystals.

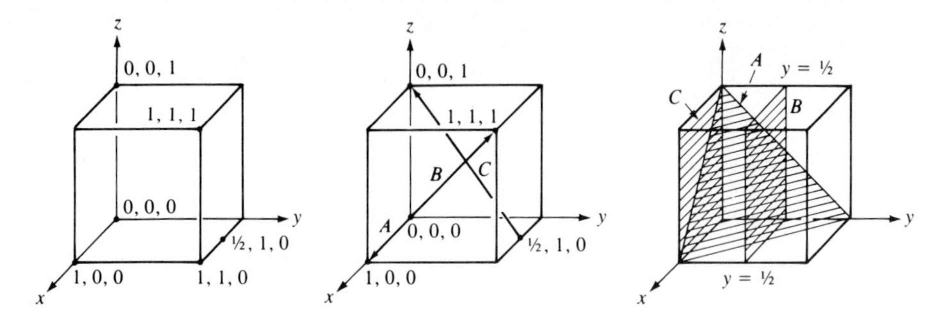

FIGURE 3-5
Crystallographic points, directions, and planes (Askeland, 1985).

cubic lattice, all three lengths are identical to the lattice parameter (lattice constant), a, and all angles are 90°. The hexagonal lattice has its unit cell with one angle of 120° and the other two of 90°, with two axis lengths of a and the third of c, which is different from a (Fig. 3-4).

A right-handed coordinate system is used to locate points, directions, and planes in a unit cell. Distance is measured in terms of the number of lattice parameters (unit cell lengths) one must move in each of the x, y, and z coordinates to go from the origin to the point of interest. The coordinates are written as the three unit cell lengths, with commas separating the numbers, as shown in Fig. 3-5. Directions in the unit cell are important.

Semiconductors dislocate, for example, in certain directions. Miller indices for directions are the shorthand notation used to describe these directions. The Miller indices for directions are determined by first locating and determining the coordinates of two points that lie on the direction. The "tail" point coordinates are then subtracted from those of the "head" point and the resulting numbers are made into lowest integers. The result is then put into a square bracket [ ], with the convention that a negative number is denoted by a bar over the number.

Example 3.1. Determine the Miller indices of directions A and C in Fig. 3-5.

Solution

Direction A. The head point coordinate is 1, 0, 0 and the tail point coordinate is 0, 0, 0. The subtraction yields 1, 0, 0. Thus, the direction is simply [100].

Direction C. The head point coordinate is 0, 0, 1 and the tail point coordinate is $\frac{1}{2}$, 1, 0. The substitution yields $-\frac{1}{2}$, -1, 1. For the lowest integer coordinates, one has -1, -2, 2. Thus the direction is $[\bar{1}\bar{2}2]$.

Certain groups of directions are equivalent. For example, [110] and $[\bar{1}\bar{1}0]$ are equivalent directions. The special bracket $\langle\ \rangle$ is used to indicate a group of equivalent directions. For the example, $\langle 110 \rangle$ represents a group of directions equivalent to [110].

Crystallographic planes are also represented by Miller indices, this time with parenthesis ( ) to denote planes. The Miller indices of planes are determined by first identifying the point at which the plane intercepts the x, y, and z coordinates and then taking the reciprocals of these intercepts. The rest of the procedures are the same as those for the Miller indices for directions.

Example 3.2. Determine the crystallographic planes A, B, and C in Fig. 3-5 (Askeland, 1985).

Solution

Plane A. The plane A intercepts the axes at $x = 1$, $y = 1$, and $z = 1$. The reciprocals are 1, 1, 1. Thus the plane is denoted by (111).

Plane B. The plane B never intercepts the x and z axes. So $x = \infty$, $y = \frac{1}{2}$, and $z = \infty$. The reciprocals are 0, 2, 0. Thus the plane is (020).

Plane C. The plane passes through the origin, or 0, 0, 0. In such cases, the origin of the coordinate system must be moved. One way is to move the origin one lattice parameter in the y direction. Then $x = \infty$, $y = -1$, and $z = \infty$. The reciprocals are 0, -1, 0. Thus the plane is $(0\bar{1}0)$.

It is also noted that there are groups of identical planes. For example, (010) is identical to (010). These equivalent planes are denoted by { }. The planes belonging to {110}, for example, are $(\bar{1}10)$, (110), (101), $(10\bar{1})$, (011), $(01\bar{1})$.

Close-packed directions and planes, where atoms are in continuous contact, are of importance in semiconductors. The stacking sequence of close-packed planes in Fig. 3-6 produces the face-centered cubic (FCC) structure. The close-packed directions are ⟨110⟩ and the planes are {111} for the FCC structure, which is applicable to silicon and gallium arsenide.

Because of the differences in atomic arrangement in the planes and directions within a crystal, the properties depend on direction. A material is anisotropic if its properties depend on the crystallographic direction along which the property is measured. A crystal is isotropic if the properties are the same in all directions. Semiconductors in general are anisotropic. A crystal usually has one

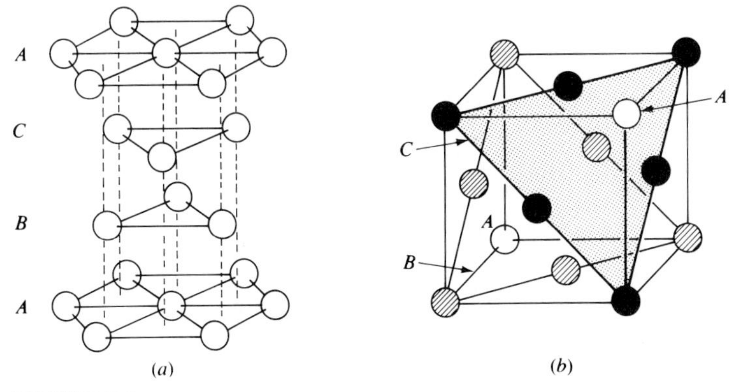

FIGURE 3-6
ACBACBACB stacking sequence of close-packed FCC structure. Note the close-packed planes *A*, *B*, and *C* in the [111] direction in (*b*).

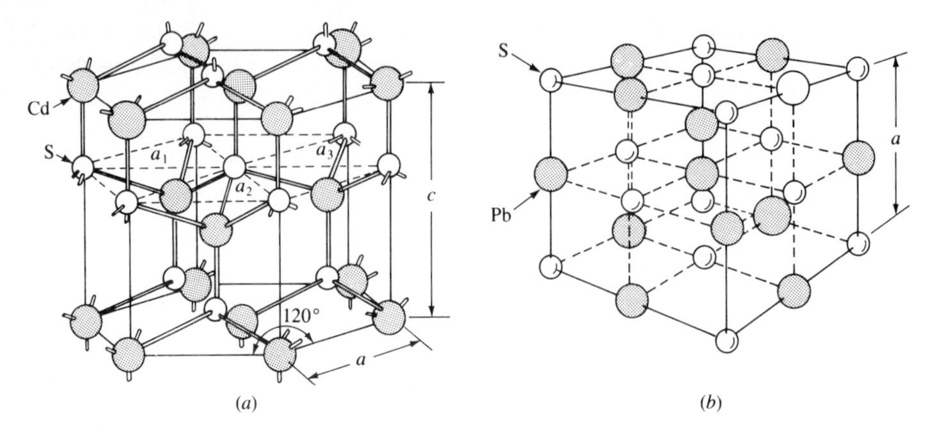

FIGURE 3-7
(*a*) Wurtzite and (*b*) halite crystal structures.

crystal structure. Crystals that can have more than one crystal structure are called allotropic or polymorphic. Crystalline SiC semiconductor is polymorphic.

Single-element semiconductors such as silicon and germanium have the same crystal structure as diamond. All III-V compounds semiconductors (one element from the group III in the periodic table and the other from group V) such as GaAs and InP have the zinc blende (or zinc sulfide) structure. II-VI compounds semiconductors such as ZnSe and CdTe have one of the following structures: either zinc blende, wurtzite, or halite (NaCl structure). The latter two structures are shown in Fig. 3-7.

The diamond structure is a special face-centered cubic structure, which is shown in Fig. 3-8. Elements (atoms) in the structure are bonded by four covalent bonds and produce tetrahedrons, which form one-eighth of the unit cell. The coordination number for each silicon atom is four. The atoms on the corners of the tetrahedral cubes provide atoms at each of the regular FCC lattice points.

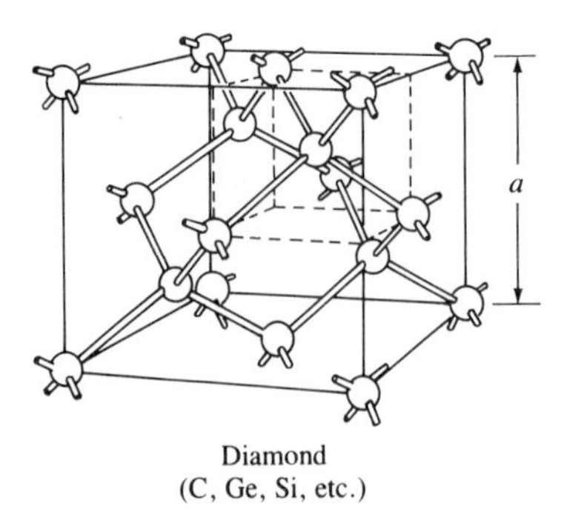

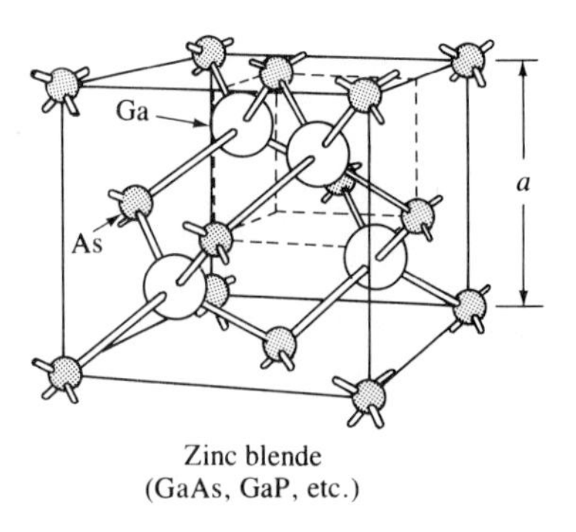

FIGURE 3-8
Diamond and zinc blende structures (Sze, 1981).

However, four additional atoms are present with the unit cell from the atoms in the center of the tetrahedral cubes. The diamond lattice can be viewed as an FCC lattice with two atoms associated with each lattice point. Therefore there are eight units per unit cell. The zinc blende structure results if the atoms in alternating layers are different from the atoms between the alternating layers, as shown in Fig. 3-8. In the figure, the dark circles represent one type of atom and the open circles the other type. For GaAs, for example, the dark circles represent As atoms and the open ones the Ga atoms. Each layer has the same kind of atom. Each arsenic atom (with five valence electrons) has four neighboring gallium atoms, each of which has three valence electrons. Each gallium atom has four neighboring arsenic atoms, each having five valence electrons. Together they form a primarily covalent-bonded semiconductor.

Processing characteristics are dependent on the crystallographic direction and plane. The $\{111\}$ planes have the highest density of atoms on the surface. Mechanical properties such as tensile strength are highest for $\langle 111 \rangle$ directions. Because of their interdependence, the choice of crystal orientation is usually fixed for specific applications and is determined from a number of considerations. For example, $\langle 111 \rangle$ oriented silicon is preferred for bipolar circuits and $\langle 100 \rangle$ silicon for MOS devices.

Any interruption in a perfectly periodic lattice may be called a defect. These defects are formed during crystal growth. The defects can be point defects, line defects (dislocations), or area and volume defects. Point defects are localized disruptions of the lattice involving one or possibly several atoms (Fig. 3-9). These include vacancy defects, interstitial defects, substitutional defects, Frenkel defects,

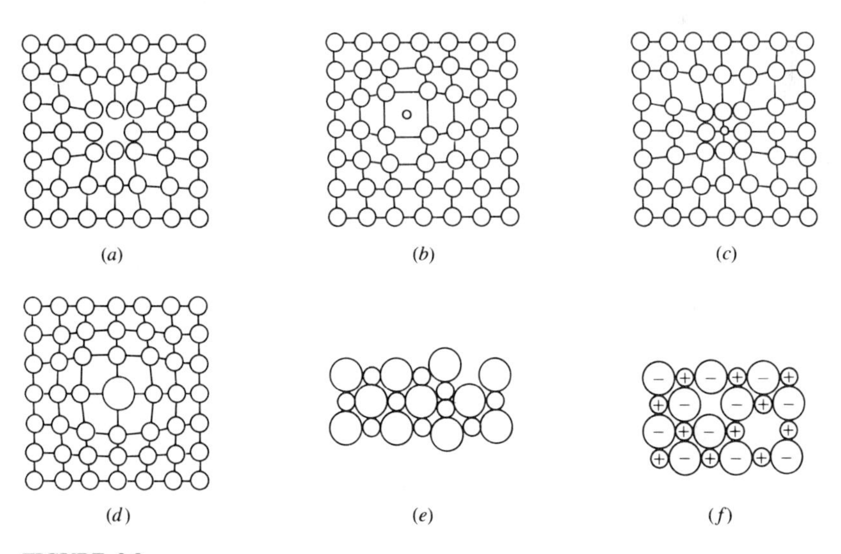

FIGURE 3-9
Point defects: (*a*) vacancy, (*b*) interstitial atom, (*c*) small substitutional atom, (*d*) large substitutional atom, (*e*) Frenkel defect, and (*f*) Schottky defect (Askeland, 1985).

and Schottky defects. A vacancy is produced when an atom is missing from a normal lattice point. An interstitial defect is formed when an extra atom is inserted into the lattice structure at a site that is not a normal lattice point. A substitutional defect results when a host atom is replaced by a foreign atom. Intentional impurities such as dopants normally occupy the interstitial or substitutional sites. It is important to recognize that usually substitutional impurities are electronically active whereas interstitial impurities are not. A Frenkel defect is a vacancy-interstitial pair formed when an ion jumps from a normal lattice point to an interstitial site, leaving behind a vacancy. A Schottky defect is a pair of vacancies in an ionically bonded material. Both an anion and a cation must be missing in order to maintain charge neutrality.

The equilibrium concentration of point defects can be obtained from probabilistic consideration (e.g., Brophy *et al.*, 1964). They are expressed as follows:

$$\frac{n}{N} = \exp\left(-\frac{E_d}{mkT}\right) \tag{3.1}$$

where n is the equilibrium concentration of point defects, N is the host atom concentration (e.g., 5×10^{22} atoms/cm^3 for silicon), E_d is the energy barrier against the defect formation, k is the Boltzmann constant, T is the temperature, and m is a constant depending on the type of point defect. For Frenkel defects in silicon, for example, $m = 2$ and $E_d = 1.1$ eV; for Schottky defects, $m = 1$ and $E_d = 2.3$ eV.

Dislocations are line imperfections in an otherwise perfect lattice. The two types are the screw dislocation and the edge dislocation. They are shown in Fig. 3-10. Screw dislocations can be illustrated by cutting partway along the plane *ABCD* in Fig. 3-10*a*, which is one of its regular lattice planes. When the

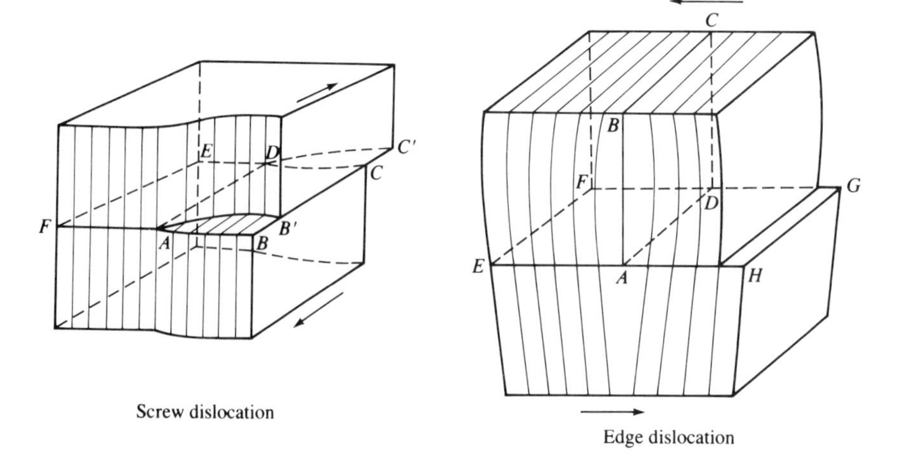

FIGURE 3-10
Screw and edge dislocations.

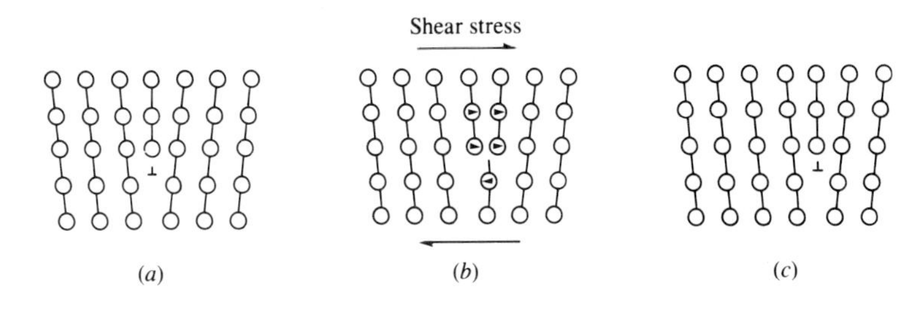

FIGURE 3-11
Propagation of dislocation (Askeland, 1985).

two halves of the crystal on either side of this plane are subjected to sufficiently large shearing forces, they separate by one atomic spacing and a screw dislocation occurs. The line of the screw dislocation so formed is *AD* in Fig. 3-10*a*, since this makes the boundary in the plane *ECBF* which divides the perfect crystal from the imperfect. An edge dislocation can be illustrated by inserting an extra half-plane of atoms, *ABCD*, in an otherwise regular lattice. An edge dislocation of this type is created when a shearing force is applied along the face of the crystal, parallel to a major crystallographic plane. The stress causing the dislocations in crystal growth is generated by the contraction that occurs during the cooling process.

When a shear force is applied to a crystal containing a dislocation, the dislocation can move by breaking the bonds between the atoms in one plane. The cut plane is shifted slightly to establish bonds with the original partial plane of atoms. This shift causes the dislocation to move one atom spacing to the side, as shown in Fig. 3-11. If this process continues, the dislocation moves through the crystal until a step is produced on the exterior of the crystal (Fig 3-11*b*). The crystal has now been deformed. The process by which a dislocation moves and causes crystal to deform is called slip. The plane along which the slip takes place is called the slip plane, and the direction in which the dislocation line moves is called the slip direction. For silicon, the slip plane is $\{111\}$ and the slip direction is $\langle 110 \rangle$.

When using the Dash technique (1959) with the Czochralski method, all the $\{111\}$ slip planes are either oblique or perpendicular to either the crystal axis or pulling direction for $\langle 100 \rangle$ and $\langle 111 \rangle$ orientations. As a result, all dislocations that move on $\{111\}$ slip planes are conducted out of the crystal at some time, particularly if the crystal diameter is small. The necking process in the Dash technique accomplishes the conduction of all dislocations to the neck surface, leading to dislocation-free crystal growth. The side wall of the grown boule appears threaded due to the conduction of the dislocations to the outer surface.

Since slip occurs along the slip plane, the force applied to the crystal has to be translated into the force in the slip direction. The situation is shown in Fig. 3-12. For the cylindrical single crystal shown in the figure, the force projected onto the slip direction is $F \cos \lambda$, which is called the resolved shear force. In

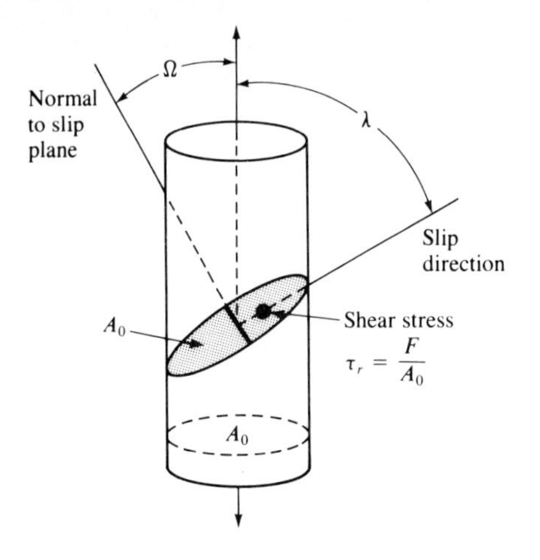

FIGURE 3-12
Resolved stress components (Askeland, 1985).

terms of shear stress (force divided by area), the resolved shear stress t_r is given by

$$t_r = \frac{F}{A_0} \cos \Omega \cos \lambda \tag{3.2}$$

since $A = A_0/\cos \Omega$. For slip to occur, this shear stress should exceed a critical value, $(t_r)_c$, i.e.,

$$t_r > (t_r)_c \tag{3.3}$$

Example 3.3. Calculate the resolved shear stress on the (111) $[\bar{1}01]$ slip system if a stress of 10,000 lb/in^2 is applied in the [001] direction of an FCC unit cell (see Fig. 3-13) (Askeland, 1985).

Solution. By inspection, $\lambda = 45°$ and $\cos \lambda = 0.707$. The normal to the (111) plane must be the [111] direction. Thus,

$$\cos \Omega = \tfrac{1}{3} = 0.577 \qquad \text{or} \qquad \Omega = 54.76°$$

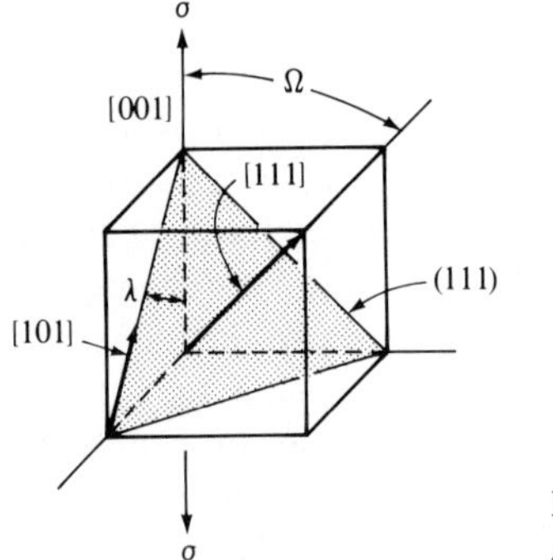

FIGURE 3-13
An FCC unit cell.

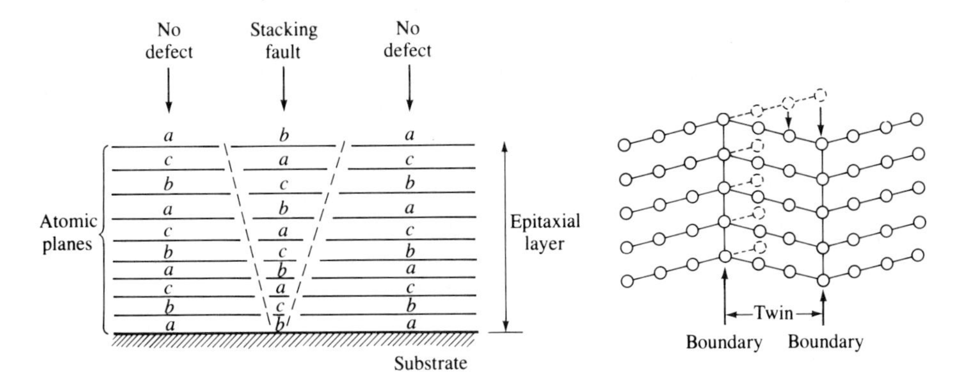

FIGURE 3-14
Stacking faults (Ghandhi, 1983) and twinning (Askeland, 1985).

$$\text{and} \qquad t_r = \frac{F}{A_0} \cos \Omega \cos \lambda = (10{,}000 \text{ lb/in}^2)(0.577)(0.709) = 4079 \text{ lb/in}^2$$

The gross defects that occur in a crystal are called either stacking faults or twinning (or twin boundaries). The former is usually present in single crystalline films grown by chemical vapor deposition. The latter is a form of dislocation and is more likely to be observed in polycrystalline structures. These defects are illustrated in Fig. 3-14. As shown in the figure, the stacking fault forms due to an improper stacking sequence of crystallographic planes.

Aggregates of oxygen precipitates are the most prevalent volume defect in silicon crystals. Since the electronic properties of crystals are dependent on the regular lattice structure of semiconductor crystal (more specifically, the lattice sites), it is natural that the presence of defects significantly alters the electrical properties of the semiconductor material. The effects of defects on electrical properties are treated in the next chapter.

3.3 CRYSTAL GROWTH AND IMPURITY DISTRIBUTION

Most bulk crystals are grown by the Czochralski technique. Therefore, Czochralski growth (Fig. 3-1*b*) is considered in detail here. As discussed in Sec. 3.1, the space above the melt is purged intensively with an inert gas (usually argon). This purging also removes undesired species in the growth atmosphere. In silicon crystal growth, for example, a large amount of silicon monoxide is produced by the reaction between the silicon melt and the silica crucible. The high strong evaporation rate of silicon monoxide can disturb the crystal pulling at relatively low pressure. The pressure in the gas phase has to be higher than 5 mbar to prevent this. The usual pressure lies between 5 and 1000 mbar (Zulehner, 1983). The purging also helps to reduce condensation of silicon monoxide and dissolution of oxygen to supersaturation levels in the melt. Another contaminant carried away by the purging gas is carbon monoxide. Carbon monoxide is formed from the hot graphite parts of the furnace by the chemical reduction of silicon monoxide at the hot graphite surface. If carbon monoxide comes into contact with the silicon melt, it is immediately absorbed, and the carbon content of the melt increases accordingly.

Incorporation of a desired impurity or dopant into the growing crystal is accomplished by adding a known amount of heavily doped material to the melt prior to crystal pulling. The saturation solubility of a solute in a liquid is different from that in a solid of the same material. Therefore, the amount of dopant incorporated into the crystal from the melt is different from the dopant concentration in the melt. The same is true for undesired impurities such as carbon, oxygen, and others. A distribution (equilibrium) coefficient is used to relate the impurity concentration in the melt to that in the crystal. The effective distribution (segregation) coefficient k_e is defined as follows:

$$k_e = \frac{C_s}{C_b} \tag{3.4}$$

where C_s is the concentration of the impurity in the crystal and C_b is that in the bulk melt. The (local) distribution coefficient k is defined in terms of the liquid phase solute concentration at the interface, C_l:

$$k = \frac{C_s}{C_l} \tag{3.5}$$

The distribution coefficients for various impurities in silicon are given in Table 3.1 and those for gallium arsenide are given in Table 3.2.

When an impurity has a distribution coefficient less than unity, the impurity is segregated at the interface. Therefore, the melt gets richer in the impurity content as the crystal is pulled. On the other hand, the melt is depleted of the impurity when the distribution coefficient is larger than unity. Almost all impurities including dopants have distribution coefficients less than unity, which leads to purification of the solid material in terms of undesired impurities. An exception is oxygen in silicon, which is preferentially incorporated into the growing crystal because its distribution coefficient is larger than unity (Table 3.1).

TABLE 3.1
Distribution coefficients of impurities in silicon (Pearce, 1983)

Impurity	Al	As	B	C	Cu	O*	P	Sb
k	0.002	0.3	0.8	0.07	4×10^{-4}	1.25	0.35	0.023

* The value for oxygen is still in dispute.

TABLE 3.2
Distribution coefficients for dopants in gallium arsenide (Milnes, 1973)

n-type dopant	S	Se	Te	Sn
k	0.3–0.5	0.1–0.3	0.059	0.08
n/*p*-type dopant	C	Ge		Si
k	0.2–0.8	0.01		0.14–2.0
p-type dopant	Be	Mg		Zn
k	3	0.1		0.4–1.0
Semi-insulating		Cr		Fe
k		5.7×10^{-4}		1×10^{-3}

A few comments are needed for Table 3.2. Unlike silicon, GaAs is a compound semiconductor. Since Ga is from the group III elements in the periodic table and As is from the group V elements, GaAs with a dopant out of the group IV such as C, Si, and Ge can be either n or p type depending on whether the group IV element is associated with the group III (Ga) or group V (As) element in its electronic configuration in GaAs. Therefore, depending on the processing conditions, the group IV element can be made an electron donor or acceptor. The need for semi-insulating GaAs arises from the difficulty in forming GaAs insulators for isolation purposes. Thus, isolation is accomplished by forming a mesa structure into a semi-insulating layer (refer to Fig. 1-14). As shown in Table 3.2, the dopants for semi-insulating GaAs are Cr and Fe.

A relationship between the effective distribution coefficient and the local distribution coefficient can be obtained under idealized conditions (Burton *et al.*, 1953). Although the meniscus of the interface is usually concave upward under typical growth conditions, an idealized case of a flat interface can be used with its axial coordinate x into the melt and origin at the interface. If one lets the interface velocity be V, a one-dimensional mass balance for an impurity can be written as follows:

$$-D \frac{d^2C}{dx^2} + V \frac{dC}{dx} = 0 \tag{3.6}$$

where C is the impurity concentration and D is the impurity diffusivity in the melt. The net flux at the interface or at $x = 0$ is the amount of impurity rejected at the interface minus the rate of incorporation into the crystal:

$$-D \frac{dC}{dx} = GC - GC_s \qquad \text{at } x = 0 \tag{3.7}$$

where G is the pull rate (growth rate) of the crystal. Another condition is that at $x = 0$, $C = C_l$. Utilizing the fact that the pulling is in the opposite direction to the positive x coordinate such that $V = -G$, Eq. (3.6) can be solved with the boundary conditions to yield

$$C = C_l(1 - k)e^{-Gx/D} + kC_l \tag{3.8}$$

where the definition of k [Eq. (3.5)] has been used for C_s. At some distance λ away from the interface, the impurity concentration in the melt reaches its bulk value due to the agitation caused by counterrotation of both the crystal being pulled and the crucible, i.e., $C = C_b$ at $x = \lambda$. Therefore, one has, from Eq. (3.8),

$$C_b = C_l(1 - k)e^{-G\lambda/D} + kC_l \tag{3.9}$$

Use of Eqs. (3.4) and (3.5) in Eq. (3.9) yields

$$k_e = \frac{k}{k + (1 - k)\exp(-G\lambda/D)} \tag{3.10}$$

The distance λ is often correlated to powers of kinematic viscosity μ_k, angular velocity ω, the diffusivity D, and the pull rate G. A typical form given by Liepold *et al.* (1980) is

$$\lambda = 1.8D^{1/3}G^{1/6}\omega^{-1/2} \tag{3.11}$$

The brunt of correction factors for nonideal conditions is usually borne by λ.

Example 3.4. A typical growth rate of 3-inch diameter silicon crystal by the Czochralski method is 2.0 mm/min. Assuming that the diffusivity is 5×10^{-5} cm^2/s and λ is 0.02 cm, calculate the ratio C_l/C_b for phosphorus and oxygen. Comment on the effect of the growth rate on the effective distribution coefficient.

Solution. From Eqs. (3.4) and (3.5), it follows that

$$\frac{C_l}{C_b} = \frac{k_e}{k} \tag{A}$$

From the given conditions,

$$\exp\left(-\frac{G\lambda}{D}\right) = \exp\left[-\frac{(0.20/60)0.02}{5 \times 10^{-5}}\right] = 0.264$$

Thus, Eq. (3.10) becomes

$$k_e = \frac{k}{k + 0.264(1 - k)} \tag{B}$$

From Table 3.1, one has

$$k_e = \frac{0.35}{0.35 + 0.264 \times 0.65} = 0.67 \text{ for P}$$

$$k_e = \frac{1.25}{1.25 + 0.264(-0.25)} = 1.056 \text{ for O}$$

From Eq. (A) and k values, one has

$$\frac{C_l}{C_b} = 1.91 \text{ for P}$$

$$\frac{C_l}{C_b} = 0.84 \text{ for O}$$

It is seen that $C_l/C_b > 1$ if $k < 1$ and $C_l/C_b < 1$ if $k > 1$. If the growth rate is very high, the exponential in Eq. (3.10) approaches zero. In this extreme, the effective distribution coefficient approaches unity regardless of the equilibrium distribution coefficient. In this extreme case, no enrichment of dopant in either phase occurs. In the other extreme case of very low growth rate, k_e approaches k.

One major assumption in deriving Eq. (3.10) is that the impurity concentration in the melt along the interface is uniform. This condition of uniform impurity concentration or dopant concentration, which is desired for radially uniform dopant distribution in the grown crystal (or more specifically wafers), is not easy to attain due to the convective motion in the melt. Thermal convection due to the temperature gradients in the melt is the source of nonuniform dopant concentrations. The thermal convection as affected by crystal/crucible rotation has been studied by Carruthers and Nassau (1968) and Carruthers (1977) and the results are shown in Fig. 3-15. When there is almost no rotation, the coldest area in the melt is the crystal-melt interface at the solidification temperature. Since the heat is provided laterally from the heater, the hottest point is in the middle of the melt near the crucible. This temperature gradient drives the melt to flow as shown in Fig. 3-15*a*. The motion of the flow increases the dopant concentration toward the center, resulting in a nonuniform concentration profile. When the rate of crystal rotation is increased (Zulehner, 1983), the flow pattern that results (Fig. 3-15*b*) leads to a melt flow from the crucible bottom to the crystal without evaporation of SiO, which causes a higher oxygen content in the crystal. Because of the extra eddy current below the melt surface as shown in the figure, the transport of SiO from the inner eddy current to the free melt surface is inhibited. Further, the radial melt flow from the center of the freezing interface to its edge reduces the enrichment of dopant in its center.

The usual industrial practice is to counterrotate the crystal and crucible for the maximum stirring effect with a low crucible rotation rate relative to the crystal rotation rate. The effects of the counterrotation are also shown in Fig. 3-15. In the case of low rotation rates for both crystal and crucible, two isolated eddy currents can exist (Fig. 3-15*c*). This arrangement should lead to a lower oxygen content in the crystal since the upper eddy current should facilitate the evaporation of SiO at the side of the crystal. In the case of high crystal rotation (Fig. 3-15*d*), a higher oxygen content but more uniform radial dopant distribution should result in the crystal. Although not shown in the figure, isorotation leads to flow patterns that are undesirable. This discussion on transport effects in the melt has been confined to the situation where the crucible is well within the

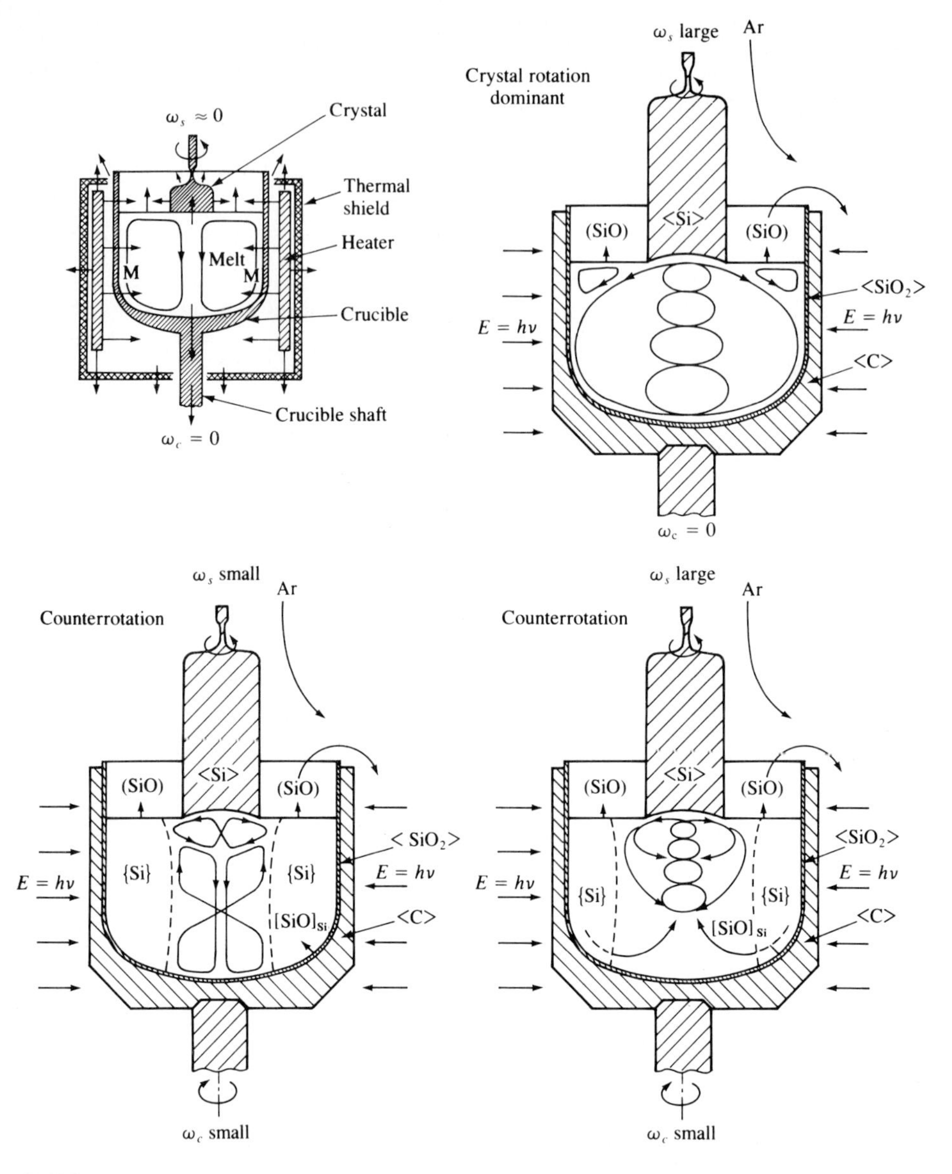

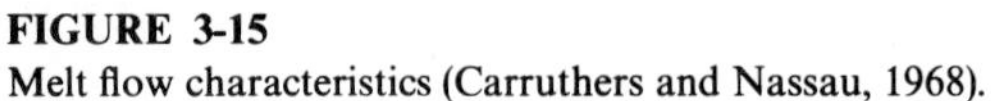

FIGURE 3-15
Melt flow characteristics (Carruthers and Nassau, 1968).

furnace. As the crystal is pulled, the whole crucible is also lifted upward and the characteristics of the heat transfer from the furnace to the crucible change, which in turn brings about changes in the flow pattern. Therefore, the flow pattern in the melt changes with the pulling of the crystal. Transport effects in melts in closed containers are detailed by Brown (1988).

The enrichment of dopant (also impurities) in the melt that takes place in the course of crystal pulling results in an increasing dopant concentration toward the butt (opposite to the seed end) when the distribution coefficient is less than one. The opposite is true for oxygen, since its distribution coefficient is larger than unity in silicon. A certain contributing factor is the decrease in surface areas available for crucible dissolution and SiO evaporation. The axial variation in the dopant concentration can be examined by writing a mass balance for any impurity as follows:

$$V_o C_o - VC = \int_0^{V_s} C_s(\alpha)\, d\alpha \tag{3.12}$$

where V_o is the original melt volume, V is the melt volume remaining in the crucible during pulling, V_s is the pulled crystal volume corresponding to V, C_o is the original impurity concentration in the melt, and C is the impurity concentration in the melt during pulling. The mass balance states that the amount of dopant removed from the melt [left-hand side of Eq. (3.12)] is the same as that incorporated into the crystal. Equation (3.12) can be differentiated with respect to x to give

$$-V\frac{dC}{dx} - C\frac{dV}{dx} = C_s\frac{dV_s}{dx} \tag{3.13}$$

where x is the axial coordinate with its origin at the seed end and normalized with respect to the boule length. If α is defined as the ratio of the melt density to the crystal density, one has

$$\begin{aligned} V &= V_o - \frac{V_s}{\alpha} \\ &= V_o - \frac{V_t x}{\alpha} \end{aligned} \tag{3.14}$$

Use of Eq. (3.14) in Eq. (3.13) yields

$$\frac{dC}{dx} = \left(\frac{V_t/\alpha - k_e V_t}{V_o - V_t x/\alpha}\right)C \tag{3.15}$$

If the $(1 - \beta)$ fraction of the original melt charge (V_o) remains in the crucible at the end of the crystal pulling, $V_t = \alpha\beta V_o$. Using this relationship in Eq. (3.15) and integrating the result yields

$$C = C_o(1 - \beta x)^{(\alpha k_e - 1)} \tag{3.16}$$

It then follows from Eq. (3.16) and the definition of k_e that

$$C_s = k_e C_o(1 - \beta x)^{(\alpha k_e - 1)} \tag{3.17}$$

Equation (3.17) gives the axial dopant distribution in the grown crystal and Eq. (3.16) provides the change in the dopant concentration in the melt as a function of the fraction of the boule pulled from the melt. The relationship of Eq. (3.17) for $\beta = 1$ (no melt at the end of the pulling) and $\alpha = 1$ has been used to describe the axial dopant distribution in a grown boule (Bridges *et al.*, 1958).

Example 3.5. Calculate the axial impurity distribution of boron and oxygen assuming that $k_e = k$ for the impurities in silicon. Use five equidistant points for the axial profiles. Calculate the concentration of B in the melt when the crystal is pulled three-fourths of the final boule length. Assume that the original impurity concentration is 10^{22} atoms/cm^3, $\beta = 1$, and $\alpha = 1.1$. Draw conclusions from the calculated profiles.

Solution. For $\beta = 1$ and $\alpha = 1.1$, Eq. (3.17) becomes

$$C_s(x) = k_e C_o(1 - x)^{(1.1k_e - 1)}$$

From Table 3.1, $k_e = k = 0.8$ for B and $k_e = k = 1.25$ for O. Thus, one has

$$\frac{C_s(x)}{C_o} = \begin{cases} 0.8(1 - x)^{0.12} & \text{for B} \\ 1.25(1 - x)^{0.375} & \text{for O} \end{cases}$$

The boron concentration one quarter-length away from the seed end, for instance, is given by

$$\frac{C_s(0.25)}{C_o} = 0.8(1 - 0.25)^{-0.12} = 0.828$$

Similar calculations lead to the following results:

x	**0**	**0.25**	**0.5**	**0.75**	**0.999**
C_s/C_o for B	0.8	0.828	0.869	0.945	1.833
C_s/C_o for O	1.25	1.12	0.96	0.74	0.09

As can be seen from the table, the impurity concentration increases rapidly toward the butt (opposite the seed end) when $\alpha k_e < 1$, as with boron. It decreases rapidly toward the butt when $\alpha k_e > 1$. When the crystal is pulled three-fourths of the final boule length, $x = 0.75$. The corresponding impurity concentration in the melt can be calculated from Eq. (3.16):

$$\begin{aligned} C &= C_o(1 - x)^{(1.1k_e - 1)} \\ &= 10^{22}(0.25)^{-0.12} \\ &= 1.18 \times 10^{22} \text{ atoms/cm}^3 \end{aligned}$$

Although the axial variation of dopant concentration is a problem, the radial distribution at a given axial position is much more of a problem since the grown boule is sliced into wafers and thus the wafers inherit the radial variation. Therefore, the performance of a circuit made on the center of a wafer will be different from that made on the other parts of the same wafer if the radial variation is significant. An approximate relationship for the radial profile at any axial position x (Lee, 1987a) is

$$C_s(r, x) = \frac{k_e C_o(1 - \beta x)^{(\alpha k_e - 1)}\{1 - (3r^2 - 2r^3)[1 - \mu(1 - \beta x)^b]\}}{0.3 + 0.7\mu(1 - \beta x)^b} \tag{3.18}$$

where

$$\mu = \frac{C_{os}\, k_{es}}{C_{oc}\, k_{ec}} \qquad b = k_{es} - k_{ec} \tag{3.19}$$

Here C_{oc} is the dopant concentration in the melt, k_{ec} the effective distribution coefficient at the center of the meniscus of the melt-crystal interface, C_{os} and k_{es} are the counterparts at the edge of the meniscus, and r is the normalized radius of the crystal with its origin at the crystal center. Undoubtedly, the mixing pattern, as dictated by the melt flow, determines the parameters μ and b for the radial distribution. Use of a magnetic field (Hoshikawa *et al.*, 1980) is one way of controlling the radial variation of dopant concentration. The dopant transport as affected by a magnetic field has been studied by Kim and Langlois (1986).

The axial dopant profile grown by the floating-zone method can be obtained in a similar manner as for Czochralski-grown crystals. The balance equation similar to Eq. (3.12) is

$$V_m C = V_m C_m + \int_0^{V_s} (C_m - C_s)\, dV_s \tag{3.20}$$

where V_m is the constant volume of the molten zone and C_m is the original dopant concentration in the polysilicon rod. The balance states that the amount of dopant in the molten zone of constant volume (which is typically 1.5 cm long) is equal to the amount of dopant originally in the zone plus the amount rejected at the interface from the current resolidified volume V_s. Noting that $V_s = V_t x$ and that $C = C_m$ at $x = 0$, one can follow similar procedures as for Eq. (3.16) to arrive at the following:

$$\frac{C_s(x)}{C_m} = 1 - (1 - k_e) \exp\left(-\frac{k_e V_t x}{V_m}\right) \tag{3.21}$$

Example 3.6. A polysilicon rod for floating-zone growth has been doped in such a way that after crystal growth the axial dopant distribution is uniform. Find the original dopant distribution profile that will yield the final uniform distribution. Let the desired final uniform dopant concentration be C_d and the original distribution in the rod be $f(x)$.

Solution. Eq. (3.20) can be rewritten for the problem as follows:

$$V_m C = V_m f(x) + \int_0^{V_s} (f - C_s)\, dV_s$$

Differentiating this with respect to x yields

$$V_m \frac{dC}{dx} = V_m \frac{df}{dx} + V_t(f - C_s) \tag{A}$$

For no axial variation, dC/dx should be zero. Further, $C_s = C_d$ if the constant (uniform) concentration is to be C_d. It follows from Eq. (A) that

$$V_m \frac{df}{dx} = -V_t(f - C_d)$$

Solving this for f yields

$$f = C_1 e^{-V_t x/V_m} + C_d$$

At $x = 0$, C_s is also at C_d and f should be that corresponding to the desired C_d, or $f(0) = C_d/k_e$ since $C_s = k_e C$ and $C(0) = f(0)$. Thus,

$$f(x) = C_d\left(\frac{1 - k_e}{k_e}\right) \exp\left(-\frac{V_t x}{V_m}\right) + C_d \tag{B}$$

If polysilicon is made having the dopant profile as specified by Eq. (B), then the axial dopant concentration profile in the crystal, grown by the floating-zone method, should be uniform at C_d.

Radial dopant concentration variations that are present in the form of spikes in intervals of 50 ~ 500 μm are typical of crystal grown by the floating-zone technique. These variations are attributed to the misalignment of the rotational axis of the crystal and the thermal axis of symmetry of the crystal growth apparatus. The variations can be virtually eliminated by neutron transmutation doping (Herrmann and Herzer, 1975).

In the Czochralski technique, there are two major obstacles to overcome before obtaining a very high pull rate. There is also a lower limit on the pull rate that is related to a remelting phenomenon. At the faster end, the pull rate is limited by the rate of heat removal for solidification. Even when the constraint from the cooling capacity is removed, the maximum pull rate is eventually limited by lowering the freezing point.

The maximum pull rate (growth rate) is constrained by the heat balance at the melt-crystal interface, given by

$$\Delta H_s \frac{dm}{dt} - \lambda_l \frac{dT_l}{dx_l} A = -\lambda_s \frac{dT_s}{dx_s} A \tag{3.22}$$

where A is the interface area, λ_l and λ_s are the thermal conductivities for liquid and solid, respectively, T_l and T_s are the corresponding temperatures along the axial coordinates x_l and x_s, dm/dt is the mass solidification rate, and ΔH_s is the latent heat of solidification (fusion). In terms of the linear growth rate G (length/time), Eq. (3.22) can be rewritten as follows:

$$\Delta H_s G \rho_s - \lambda_l \frac{dT_l}{dx_l} = -\lambda_s \frac{dT_s}{dx_s} \tag{3.23}$$

If one neglects heat conduction into the melt, then the usual result obtained for the maximum pull rate is

$$G_{\max} = \frac{-\lambda_s}{\Delta H_s \rho_s} \left(\frac{dT_s}{dx_s} \right)_{x_s=0} \tag{3.24}$$

where ρ_s is the crystal density. If a radially averaged temperature T is used, the heat balance for the crystal is

$$\frac{d}{dx}\left(\lambda_s \frac{dT}{dx}\right) - G\rho_s C_p \frac{dT}{dx} - \frac{2}{R} q_r = 0 \tag{3.25}$$

where the interface is at the origin of the axial coordinate x. Here R is the radius, C_p is the specific heat of the crystal, and q_r is the net rate of radiation heat transfer. Heat loss from the crystal is dominated by radiation. The heat loss by radiation is quite complicated, depending on the apparatus geometry and material, and the position x. Rea (1981) simulated Eq. (3.25) for the purpose of obtaining the gradient in Eq. (3.24) for the maximum pull rate. His results are given in Fig. 3-16 as the dashed line along with the experimental results by Digges and Shima (1980). It is seen that the theoretical maximum pull rate is higher than the maximum experimental growth rate, and that the pull rate decreases with increasing crystal diameter.

The presence of dopant in a melt can limit the rate at which a crystal can be grown without loss of crystallinity. The limitation applies to the dopants whose

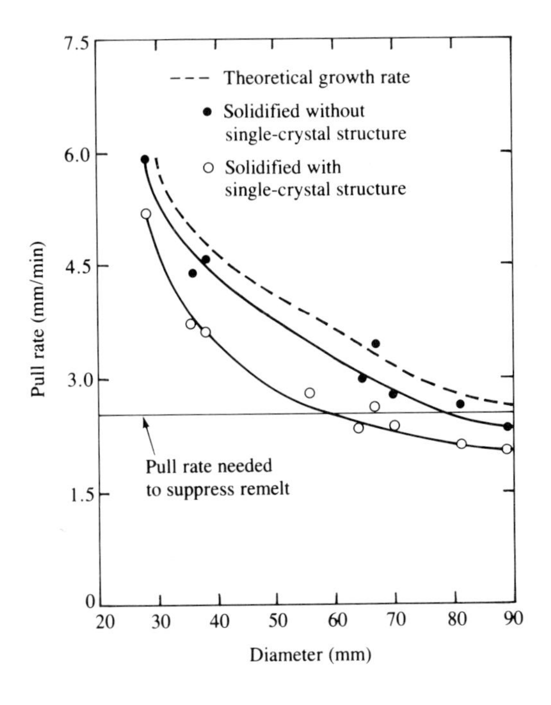

FIGURE 3-16
Theoretical and experimental pull rates for Czochralski-grown crystals. The dashed line is the theoretical maximum pull rate (Rea, 1981). The experimental results are by Digges and Shima (1980).

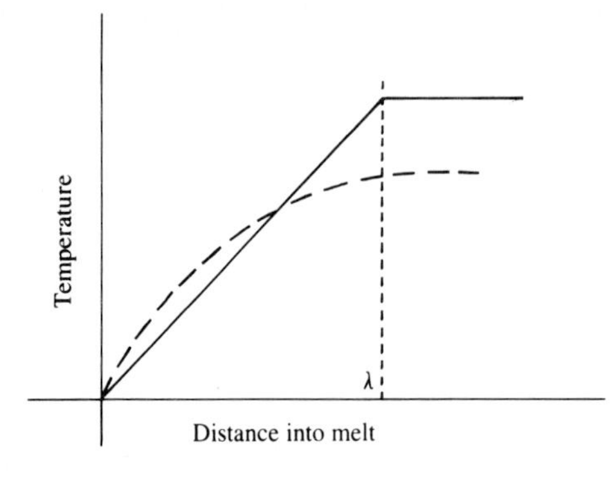

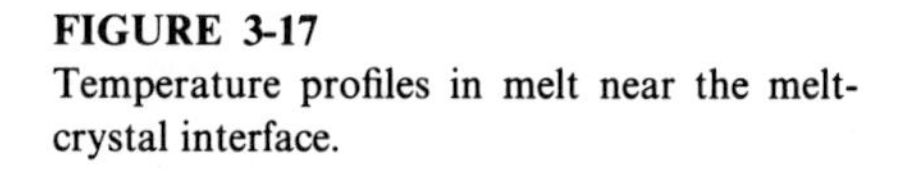

FIGURE 3-17
Temperature profiles in melt near the melt-crystal interface.

effective distribution coefficients are less than unity. When the coefficient is less than unity, the dopant rejected at the melt-crystal interface enriches the melt. The enrichment of solute in the melt leads to lowering of the melting point. This so-called 'constitutional supercooling' can cause spurious nucleation and polycrystalline growth. Thus, the growth rate is limited in the sense that polycrystalline growth must be avoided. In this light, consider the two temperature profiles near the interface shown in Fig. 3-17 with the axial coordinate x into the melt and the origin at the interface.

The dashed curve is for the freezing point. When the effective distribution coefficient is less than unity, the dopant concentration is the highest at the interface and decreases toward the bulk melt. Since freezing point depression is largest at the interface, the freezing temperature increases toward the bulk melt as shown in the figure. The actual melt temperature is lowest at the interface due to the solidification since the melt is heated beyond the melting point. When the actual melt temperature is lower than the freezing temperature, solidification takes place. In Fig. 3-17, solidification takes place in the melt near the interface since a portion of the actual melt temperature curve lies below the freezing curve. This spurious nucleation can be avoided (Hurle, 1961) if

$$\frac{dT_f}{dx} < \frac{dT}{dx} \qquad \text{at } x = 0 \tag{3.26}$$

If the condition is satisfied, the two curves do not intersect each other and the actual melt temperature T will be higher than the freezing temperature T_f everywhere. Therefore, no spurious nucleation will take place, at least from the viewpoint of supercooling. The maximum rate of growth, that is to be ultimately limited by supercooling when the heat removal problem is completely solved, is given (Lee, 1987*b*) by

$$\frac{G_u}{G_m} = \frac{1}{1-k}\{1 + \beta_T[k + (1-k)e^{-G_u/G_m}](1-u)^v\} \tag{3.27}$$

where G_u is the ultimate rate and

$$G_m = \frac{D}{\lambda} \tag{3.28}$$

$$\beta_T = \frac{\Delta H_s(T_o - T_{fo})}{R_g T_{fo}^2 Y_o} \tag{3.29}$$

$$v = \frac{(1-k)\exp(-G_u/G_m)}{k + (1-k)\exp(-G_u/G_m)} \tag{3.30}$$

Here $(T_o - T_{fo})$ is the initial lowering of the freezing temperature, T_o is the bulk melt temperature, R_g is the gas constant, Y_o is the initial dopant mole fraction in the melt, and u is the fraction of the crystal as measured from the seed end.

Example 3.7. For silicon, G_m is of the order of 10^{-3} to 10^{-4} cm/s and the value of β_T is $0.106/Y_o$ for bulk melt temperature of 1470 °C. Calculate the ultimate rate G_u for a silicon melt doped with 10^{20} atoms/cm^3 of antimony for $u = 0.8$ (80 percent crystallized). The concentration of silicon in the melt can be taken approximately as 5×10^{22} atoms/cm^3, for which $Y_o = 0.002$. Assume $G_m = 0.003$ cm/s. Compare the calculated value with the heat-removal limited maximum rate given in Fig. 3-16 for a 120-mm diameter crystal, which is about 13 cm/h.

Solution. From the problem statement, $u = 0.8$ and $k = 0.023$ from Table 3.1. As $\beta_T = 0.106/Y_o$ and $Y_o = 0.002$, $\beta_T = 52.8$. As apparent from Eq. (3.30), v approaches unity when $k \ll 1$. Thus, Eq. (3.27) can be simplified to

$$\frac{G_u}{G_m} = 1 + \beta_T e^{-G_u/G_m}(1-u)$$

Using the given values, the above equation can be rewritten as

$$\frac{G_u}{G_m} = 1 + 52.8e^{-G_u/G_m}(0.2)$$

Solving this by trial and error for G_u/G_m yields a value of 2.19. For G_m of 0.003 cm/s, $G_u = 0.00657$ cm/s = 24 cm/h. This compares with the heat-removal limited rate of 13 cm/h.

In addition to the upper limits on the growth rate, first from heat removal and then from constitutional supercooling, there is also a lower limit on the growth rate. Remelting and solidifying can take place simultaneously at the interface if the pull rate (growth rate) is below a certain critical value. This occurs when microfluctuations in temperature at the interface dominate the growth process at relatively small pull rates. Based on a local sinusoidal fluctuation in the growth rate, van Run (1981) gives the following lower limit:

$$G > \frac{G_T}{2} \tag{3.31}$$

where G_T is the maximum rate as dictated by the heat-removal rate, which is given by Eq. (3.24).

3.4 OXYGEN PRECIPITATION

The most prevalent, unintentional impurity in Czochralski-grown silicon crystal is oxygen. Oxygen in silicon arises mostly from the dissolution of the crucible during growth. Typical values range from 10^{17} to 10^{18} atoms/cm^3 (Zulehner, 1983). The flow characteristics of a melt dominate the resulting content and distribution of oxygen in the crystal. In general, the concentration has a local radial maximum at the center and increases axially toward the seed end. A magnetic field is an effective means of achieving uniform radial distribution (Suzuki *et al.*, 1981), as was the case with dopant distribution.

As an impurity, oxygen has three effects (Patel, 1977): donor formation, yield strength improvement, and defect generation by oxygen precipitation. In a grown crystal, over 95 percent of the oxygen atoms occupy interstitial lattice sites, each oxygen bridging two host lattice atoms. The remainder of the oxygen polymerizes into complexes, such as SiO_4. The oxygen complexes in general act as a donor, thus causing distortion of the intended resistivity obtained by intentional doping. The formation of the donor complexes is most favored in the 400 to 500 °C temperature range, and the rate of formation is proportional to the fourth power of oxygen concentration (Kaiser, 1957). However, the complexes are unstable above 500 °C. Therefore, annealing of grown crystal is carried out between 600 and 700 °C to dissolve the complexes. The annealed crystal is quenched to room temperature to avoid reforming the complexes. Wafer annealing is preferred to ingot annealing because of the difficulty in quenching a large-diameter crystal (usually by blow-drying). Complex formation can also occur during device processing. Therefore, the trend toward low-temperature processing poses a dilemma (Pearce, 1983) since low temperature promotes the formation of the complexes.

The usual oxygen content in grown crystal is so high that the crystal is supersaturated with oxygen even at high temperatures. Thus, precipitation of oxygen should take place in principle. However, the formation of the first precipitation nuclei does not take place readily in the case of homogeneous nucleation. In the crystal grown by the Czochralski method with the usual oxygen content, this first precipitation is only 1 to 5 percent of the total oxygen present. In crystals with oxygen content less than 6×10^{17} atoms/cm^3, the amount of precipitation is too small to measure. For oxygen contents higher than 10^{18} atoms/cm^3, however, the amount can be substantial, and numerous visible oxygen precipitates exist in most cases (Zulehner, 1983).

The precipitation of oxygen has beneficial effects if the amount of precipitates is moderate. Along with the interstitial oxygen, the precipitates act as a hardening additive through the mechanism of solution hardening and thus increase the yield strength of silicon (Sumino *et al.*, 1980). The precipitates also act as centers of gettering undesired impurities and allow for the formation of a

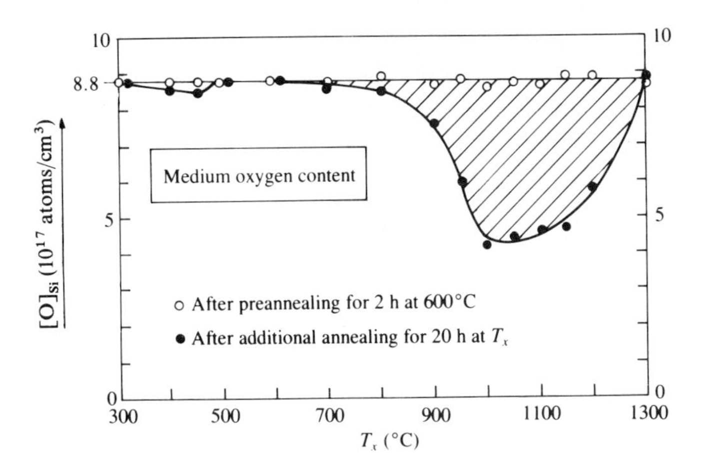

FIGURE 3-18
Typical procedures of dealing with oxygen in silicon (Zulehner, 1983).

denuded, defect-free zone at the wafer surface where junctions are formed. The precipitates are in reality an SiO_2 phase rather than a pure oxygen phase. The volume mismatch, which occurs as the precipitates grow in size, causes a compressive strain on the lattice, leading to dislocation formation. The precipitates also lead to other types of defects. Therefore, oxygen precipitation has to be controlled carefully so that the beneficial effects can be realized and the harmful effects can be avoided or minimized.

The usual procedure in dealing with oxygen is to anneal at around 600 °C to dissolve (mainly) donor type oxygen complexes, and then anneal at a high temperature to obtain the desired quantity of oxygen precipitates. Such a procedure is shown in Fig. 3-18. Note that the amount of precipitated oxygen is the difference between the open circles and the corresponding dark circles in the shaded area.

Another unintended impurity that is present in relatively large amounts is carbon. It is mainly due to carbon in the graphite parts of the furnace that are transported to the melt. Carbon occupies substitutional lattice sites and does not precipitate. It is also electrically inactive. However, it does have the undesirable effect of aiding in the formation of defects (Matsushita *et al.*, 1980).

Because of the importance of oxygen precipitation, its behavior is of interest in terms of the size and the change in the quantity of precipitates. The change in the amount of interstitial oxygen (and thus the amount of precipitates) is often described by

$$\frac{C - C^*}{C_o - C^*} = \exp\left(-\frac{t}{t_r}\right) \tag{3.32}$$

$$t_r = \left(\frac{3}{4\pi N}\right)^{2/3} (3D)^{-1}[v_m(C_o - C^*)]^{-1/3} \tag{3.33}$$

which is an approximation of the result of Ham's theory (1958) at a sufficiently long time (Hu, 1981), based on the assumption that there is a fixed number of precipitates during annealing. Here C is the concentration of the interstitial oxygen, C^* is the saturation concentration, N is the number of precipitates per unit volume, v_m is the molecular volume of the precipitate, D is the diffusivity of oxygen, and C_o is the initial value of C. The diffusivity is rather well established and is given by

$$D_s = D_o \exp\left(-\frac{E}{kT}\right) \tag{3.34}$$

where

$$D_o = 0.02 \sim 0.07 \text{ cm}^2/\text{s}$$

$$E = 2.42 \sim 2.44 \text{ eV}$$

The growth of the precipitates is often presented in the following empirical form (Wada and Inoue, 1985):

$$r = bt^n \qquad \text{for } n = 0.5 \sim 0.9 \tag{3.35}$$

where r is the precipitate size, b is a constant, and t is time.

NOTATION

A, A_o	Area (L^2)
b	Constant in Eq. (3.35)
C	Impurity concentration in melt (mol/L^3); oxygen concentration in Eq. (3.22) (mol/L^3)
C_b	Impurity concentration in bulk melt
C_l	Impurity concentration in the melt side of melt-crystal interface
C_m	Constant original impurity concentration in polysilicon rod
C_{oh}	Initial impurity concentration in melt; initial oxygen concentration in Eq. (3.32)
C_s	Impurity concentration in crystal
C^*	Saturation concentration of oxygen in crystal
D	Impurity diffusivity in melt (L^2/t)
D_{oh}	Preexponential factor for diffusivity (L^2/t)
D_s	Interstitial oxygen diffusivity in solid (L^2/t)
E	Activation energy
F	Force in Eq. (3.2) (ML/t^2)
G	Growth (pull) rate (L/t)
G_m	D/λ (L/t)
G_T	Maximum G dictated by heat-removal rate
G_u	Ultimate pull rate (L/t)
ΔH_s	Latent heat of fusion (E/mol or E/M)
k	Local distribution coefficient (dimensionless); Boltzmann constant
k_e	Effective distribution coefficient (dimensionless)

m	Mass (M)
q_r	Rate of radiation heat exchange (E/tL^2)
r	Size of oxygen precipitate (L)
R	Crystal radius (L)
R_g	Gas law constant
t	Time (t)
t_r	Resolved shear stress given by Eq. (3.2) (M/t^2L); quantity given by Eq. (3.33) (t)
T	Temperature (T); crystal temperature (T)
T_f	Freezing temperature in Eq. (3.26)
T_l	Melt temperature
T_{oh}	Bulk melt temperature
T_{foh}	T_f in the absence of impurity
u	Fractional crystal length as measured from seed end
V	Melt volume remaining after pulling (L^3)
V_m	Volume of molten zone in floating-zone technique
V_s	Boule (crystal) volume
V_t	Total boule volume
x	Normalized axial coordinate for melt
x_s	Normalized axial coordinate for crystal
Y_o	Initial mole fraction of dopant

Greek letters

α	ρ_l/ρ_s
β	Fraction of initial melt that is crystallized
β_T	Quantity defined by Eq. (3.29)
λ	Angle in Fig. 3-12; "boundary-layer" thickness in Eq. (3.9) (L)
ρ_l	Melt density (M/L^3)
ρ_s	Solid (crystal) density
ω	Angular velocity (rad/t)
Ω	Angle in Fig. 3-12

Subscripts

c	Center of the meniscus at melt-crystal interface
l	Liquid or melt
s	Edge of the meniscus at the melt-crystal interface

Units

E	Energy (ML^2/t_2)
L	Length
M	Mass
t	Time
T	Temperature

PROBLEMS

3.1. Calculate the equilibrium concentrations of Frenkel and Schottky defects in silicon using the information given in the text. Also determine the ratio of Frenkel to Schottky defects. Determine the number of Frenkel point defects on the face of a silicon wafer assuming the point defects are uniformly distributed.

3.2. Consider (111) silicon crystal growth by the Czochralski technique. Noting that the direction perpendicular to the (111) plane is [111], the pulling force is perpendicular to the growing plane. On the other hand, the major stress due to the temperature gradient is the radial stress, which is in the direction of [110]. Determine the resolved shear stress, which is in the direction of [110]. Determine the resolved shear stress in the [111] and [110] directions. Explain why the angular stress determines wafer deformation.

3.3. Suppose that the pull rate for As-doped silicon crystal is 2.0 nm/min. Determine the pull rate for P-doped silicon crystal that will yield the same axial dependence of the dopant profile as for the As-doped crystal. Assume that the "boundary-layer" thickness λ is that given by Eq. (3.11) and that all the other operating conditions except for the pull rate remain the same. Assume also that the diffusivities in the melt are the same at 5×10^{-5} cm^2/s and that λ for As-doped silicon is 0.02 cm.

3.4. Equation (3.10) can be rewritten as follows:

$$k_e = \frac{1}{1 + [(1-k)/k]\exp(-G\lambda/D)} \tag{A}$$

It follows then that the effective distribution coefficient is almost unity if

$$\frac{1-k}{k}\exp\left(-\frac{G\lambda}{D}\right) < \beta \tag{B}$$

where β is much less than unity. The only operating condition that can be used to satisfy the condition of Eq. (B) is the pull rate G. Give the physical reasoning for the effective distribution coefficient being almost unity regardless of the intrinsic distribution coefficient when the pull rate is sufficiently high. The fact that k_e is unity does not necessarily mean uniform axial distribution of dopant according to Eq. (3.19). Explain why the distribution is not necessarily uniform.

3.5. Resistivity is usually inversely proportional to dopant concentration. Suppose that undoped polysilicon is first charged and a small amount of heavily doped silicon (resistivity of 0.01 Ω·cm) is then added. For one run, the resistivity of the grown crystal at the midsection is 1 Ω·cm. The amount of heavily doped silicon is to be changed in the next run to obtain a resistivity of 0.5 Ω·cm at the midsection. Determine the increase in the amount of heavily doped silicon that is needed. Assume all operating conditions remain the same including the amount of the initial charge of undoped silicon. Discuss whether the answer is dependent on the axial position of the crystal. Neglect the volume change due to the small amount of heavily doped polysilicon.

3.6. A simple and yet useful relationship for a sustainable growth rate can be obtained (Pearce, 1983) by considering a simple heat balance, accounting only for the major heat source and sink. The major heat source is the latent heat of solidification; the dominant heat sink is the radiation loss. Thus one can write the following heat balance:

$$\Delta H_s G \rho_s = \frac{2}{R} q_r = \frac{2}{R} Fs(T_s^4 - T_e^4)$$

where F is the view factor, s is the Stefan-Boltzmann constant [1.355×10^{-12} cal/(s·cm^2·K^4)], T_s is the crystal temperature, and T_e is the effective temperature of the surrounding. In the Czochralski growth, the view factor F changes. Since the dimensionless view factor is less than unity, one way of expressing the maximum growth rate is as follows:

$$G_{\max} = \frac{2s(T_s^4 - T_e^4)}{R \Delta H_s \rho_s} \tag{A}$$

Given $\Delta H_s = 264$ cal/g and $\rho_s = 2.33$ g/cm^3 for silicon, calculate the growth rate as a function of crystal diameter and compare with the experimental data given in Fig. 3-16. Assume $T_s = 1470\,°\text{C}$ and $T_e = 1425\,°\text{C}$. Here T_e was choosen to fit the maximum growth rate (0.45/60 cm/s) in Fig. 3-16 for a 3-cm diameter crystal.

3.7. In comparing the annealing behavior given by Eq. (3.32) for oxygen precipitation with the experimental data in Fig. 3-18, care should be taken that redistribution of precipitates by diffusion will occur at high temperatures and Eq. (3.32) cannot be valid there. Assuming that Eq. (3.32) is valid up to 1000 °C at which the amount of precipitated oxygen is maximum, determine t_r at 900 °C. Explain how one can obtain t_r values at other temperatures than 900 °C.

3.8. Derive Eq. (3.21) from Eq. (3.20).

3.9. Holmes (1963) treated solute separation (segregation) at the melt-crystal interface as a phenomenon of solute adsorption/desorption on the growing solid surface at the interface. The coverage of the surface by the solute (adsorbate) can be written as follows:

$$\theta = \frac{C_s}{C_t}$$

where C_s is the solute concentration in the crystal and C_t is the concentration of the total sites for the adsorption. A balance at the interface can be written as follows:

$$k_a C_l(1 - \theta) - k_d \theta = \frac{GC_s}{C_t} = G\theta \tag{A}$$

where k_a and k_d are the adsorption and desorption rate constants, respectively, C_l is the solute concentration in the melt at the interface, and G is the growth rate. The balance states that the net rate of adsorption is the same as the rate at which the solute is incorporated into the crystal. Noting that the distribution coefficient k is defined as $k = C_s/C_l$, derive an expression for the distribution coefficient k. Determine the conditions under which k becomes a constant. Discuss the implication.

3.10. An expression for the maximum growth rate $G_{\max}$ can be obtained by combining Eq. (3.25) with the following heat balance for the inert gas flowing countercurrently with respect to the direction of pulling in the Czochralski apparatus:

$$v \rho_g C_{pg} \frac{dT_f}{dg} = -\frac{2}{R} q_r \tag{A}$$

where v is the gas velocity, T_f is the gas temperature, and the subscript g is for the fluid. Combining Eq. (A) with Eq. (3.25) yields

$$\frac{d}{dz}\left(\lambda_s \frac{dT}{dz}\right) - G\rho C_p \frac{dT}{dz} = -v\rho_g C_{pg} \frac{dT_f}{dz} \quad \text{(B)}$$

$$T = T_m \qquad \text{for } z = 0 \quad \text{(C)}$$

$$\lambda \frac{dT}{dz} = -Fs(T^4 - T_w^4) \quad \text{(D)}$$

where T_m is the solidification temperature, z is the axial coordinate with the origin at the melt-crystal interface and H is the height to the seed end, and T_w is the effective temperature of the surrounding at $z = H$. In the boundary condition of Eq. (D), the gas-solid film heat transfer was neglected in comparison to the heat transfer. Based on Eqs. (B), (C), (D), and Eq. (3.24), find an expression for G_{max} in terms of measured temperatures.

REFERENCES

Askeland, D. R.: *The Science and Engineering of Materials*, PWS Publishers, Boston, Mass., 1985.
Bonora, A. C.: in *Semiconductor Silicon 1977*, p. 154, Electrochemical Society, Pennington, N.J., 1977.
Bridges, H. E., *et al.*: *Transistor Technology*, vol. 1, Van Nostrand, New York, 1958.
Brophy, J. H., R. M. Rose, and J. Wulff: *The Structure and Properties of Materials*, vol. II, Wiley, New York, 1964.
Brown, R. A.: *Am. Inst. Chem. Engrs. J.*, vol. 34, p. 881, 1988.
Burton, J.A., R. C. Prim, and P. Slichter: *J. Chem. Phys.*, vol. 21, p. 1987, 1953.
Carruthers, J. R.: *J. Appl. Phys.*, vol. 30, p. 459, 1959.
———: *J. Crystal Growth*, vol. 42, p. 379, 1977.
——— and K. Nassau: *J. Appl. Phys.*, vol. 39, p. 5205, 1968.
Dash, W. C.: *J. Appl. Phys.*, vol. 30, p. 459, 1959.
Digges, Jr., T. G., and R. Shima: *J. Crystal Growth*, vol. 50, p. 865, 1980.
Ghandhi, S. K.: *VLSI Fabrication Principles*, Wiley, New York, 1983.
Goodman, C. H. L. (ed.): *Crystal Growth: Theory and Techniques*, vol. 1, Plenum, New York, 1974.
Ham, F. S.: *J. Phys. Chem. Solids*, vol. 6, p. 335, 1958.
Hannay, N. B. (ed.): *Semiconductors*, Reinhold, New York, 1959.
Herrmann, H. A., and H. Herzer: *J. Electrochem. Soc.*, vol. 122, p. 1568, 1975.
Holmes, P. J.: *J. Phys. Chem. Solids*, vol. 24, p. 1239, 1963.
Hoshikawa, I., H. Kohda, H. Nakanishi, and K. Ikuta: *Jap. J. Appl. Phys.*, vol. 19, p. 133, 1980.
Hu, S. M.: *J. Appl. Phys.*, vol. 52, p. 3974, 1981.
Hurle, D. T. J.: *Solid State Electron.*, vol. 3, p. 37, 1961.
Kaiser, W.: *Phys. Rev.*, vol. 105, p. 1751, 1957.
Kern, W.: *RCA Rev.*, vol. 39, p. 278, 1978.
Kim, K. M., and W. E. Langlois: *J. Electrochem. Soc.*, vol. 133, p. 2590, 1986.
Kittel, C.: *Introduction to Solid State Physics*, 4th ed., Wiley, New York, 1971.
Lee, H. H.: *J. Crystal Growth*, vol. 83, p. 1610, 1987*a*.
———: *J. Electrochem. Soc.*, vol. 134, p. 971, 1987*b*.
Liepold, M. H., T. P. O'Donnel, and M. A. Hagan: *J. Crystal Growth*, vol. 40, p. 366, 1980.
Matsushita, Y., S. Kishino, and M. Kanamori: *Jap. J. Appl. Phys.*, vol. 19, p. L101, 1980.
Milnes, A. G.: *Deep Impurities in Semiconductors*, Wiley, New York, 1973.
Mullin, J. B., B. W. Straughan, and W. S. Brickell: *J. Phys. Chem. Solids*, vol. 26, p. 782, 1965.
Pamplin, B. R.: *Crystal Growth*, Pergamon, New York, 1980.
Patel, J. R.: in *Semiconductor Silicon 1977*, p. 521, Electrochemical Society, Pennington, N.J., 1977.

Pearce, C. W.: in S. M. Sze (ed.), *VLSI Technology*, McGraw-Hill, New York, 1983.
——— and V. J. Zaleckas: *J. Electrochem. Soc.*, vol. 126, p. 1436, 1979.
Pfann, W. G.: *Zone Melting*, Wiley, New York, 1958.
Rea, S. N.: *J. Crystal Growth*, vol. 54, p. 267, 1981.
Robbins, H., and B. Schwartz: *J. Electrochem. Soc.*, vol. 123, p. 1909, 1976.
Small, M. B., and I. Crossley: *J. Crystal Growth*, vol. 27, p. 35, 1974.
Sumino, K., *et al.: Jap. J. Appl. Phys.*, vol. 19, p. L49, 1980.
Suzuki, T., N. Isawa, Y. Okubo, and K. Hoshi: in *Semiconductor Silicon 1981*, p. 90, Electrochemical Society, Pennington, N.J., 1981.
Sze, S. M.: *Physics of Semiconductor Devices*, Wiley, New York, 1981.
Teal, G. K., and J. B. Little: *Phys. Rev.*, vol. 78, p. 647, 1950.
Van Run, A. M. J. G.: *J. Crystal Growth*, vol. 53, p. 441, 1981.
Wada, K., and N. Inoue: *J. Crystal Growth*, vol. 71, p. 111, 1985.
Zulehner, W., *J. Crystal Growth*, vol. 65, p. 189, 1983.

CHAPTER
4

ELECTRICAL CHARACTERISTICS OF PROCESSED MATERIALS

4.1 INTRODUCTION

The ultimate goal of microelectronics processing is to relate the processing conditions with the desired properties of processed materials, the final products of which are various types of microelectronic devices. Furthermore, it is desirable to obtain quantitative relationships between these properties and conditions, so that the process can be controlled precisely to give usable products. Such a lofty goal has not been attained at present, although some aspects of the overall processing could be argued to be amenable to such a treatment. In many cases, the qualitative picture of which process variable causes which processing phenomenon is not clear. Two reasons are the large number of processing steps involved in producing the final products and the complicated interdependence of the steps.

Even when one focuses only on electrical properties, there is a question as to how one should relate the processing conditions to them. The approach taken in this chapter is that there are essentially four different groups of materials involved in almost all integrated circuits and that there are major problems related to electrical properties that result from the processing of these materials. The first group is the materials used for contacts and interconnects. Next are dielectrics that are used either in a capacitive mode or for isolation. Substrate materials are the third group from which junctions are formed. Packaging materials make up the last group. Major electrical characteristics and problems are treated for each group. These are then related to processing conditions. Packaging materials and associated problems are discussed in Chapter 11.

4.2 METAL CONTACTS AND INTERCONNECTS

In integrated circuits, metallization films, in the form of metal contacts and interconnects, provide access between the semiconductor device and the outside world. They are deposited either by physical means, such as thermal evaporation or sputtering, or by chemical means, such as chemical vapor deposition.

Metal contacts can be ohmic or rectifying. In general, ohmic contacts are desired for metal contacts in integrated circuit (IC) applications so that current flows with minimum resistance. A rectifying contact is very similar to a *pn* junction. Current flows in the "forward" direction but almost no current flows when reverse-biased. The major difference between the metal-semiconductor contact and a *pn* junction is that majority carriers are responsible for current in the former. Thus, the rectifying contacts can provide better performance than typical *pn* junctions in terms of switching speed and high-frequency properties. The concepts involved in rectifying contacts carry over directly to ohmic contacts.

Consider the energy bands shown in Fig. 4-1. A metal has its own work function Ω_m, which is the energy required to remove an electron in the Fermi level to the vacuum outside the metal. A semiconductor also has its own work function as shown in Fig. 4-1. The electron affinity qX is measured from the vacuum level to the semiconductor conduction band edge. As a metal and a semiconductor (*n* type in Fig. 4-1*b*) are brought into contact, an equilibrium is established such that the Fermi level for the metal aligns with that for the semiconductor and they become equal. The conduction and valence band edges (E_c and E_v) also move by the same amount, which is $E_{Fs} - E_{Fm}$, where the subscripts *s* and *m* are for semiconductor and metal in their respective Fermi level. This difference in Fermi level is also equal to $q(\Omega_m - \Omega_s)$ and is the static potential energy qV_0 for the Schottky barrier diode shown in Fig. 4-1*b*. Because of the static potential, a depletion region forms on the semiconductor side, in contrast with a *pn* junction where the depletion (transition) region extends into both *p* and *n* semiconductors. The potential barrier against electron injection from the metal into the semiconductor conduction band, which is denoted by Ω_{Bn} for the *n*-type semiconductor and is equal to $(\Omega_m - X)$, is the most important quantity for metal-semiconductor contacts. The potential barrier height for the ideal cases being considered is given by

$$\Omega_{Bn} = \Omega_m - X \tag{4.1}$$

$$q\Omega_{BP} = E_g - q(\Omega_m - X) \tag{4.2}$$

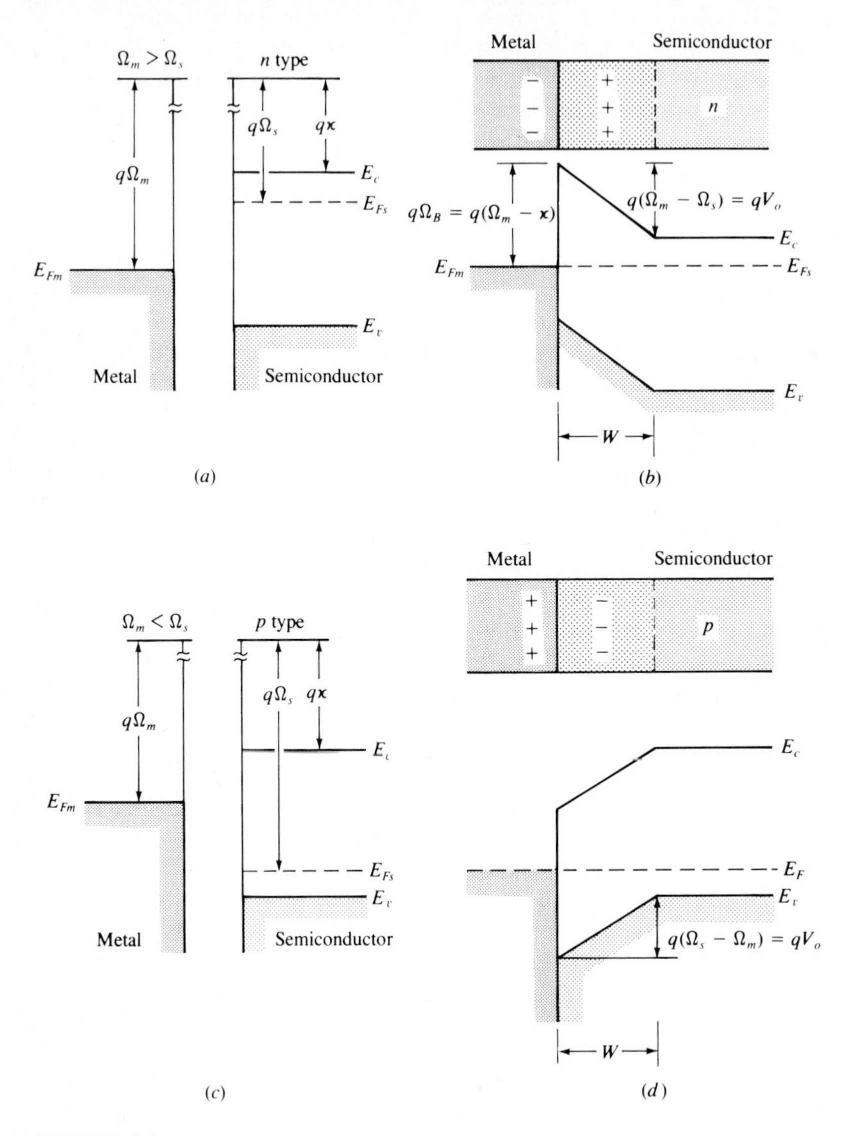

FIGURE 4-1
A Schottky barrier formed by contacting a semiconductor with a metal: (*a*) band diagrams before joining for the metal and *n*-type semiconductor and (*b*) equilibrium band diagram for the junction; (*c*) band diagrams before joining for the metal and *p*-type semiconductor and (*d*) band diagram for the junction (Streetman, 1980).

where E_g is the semiconductor bandgap. As shown in Fig. 4-1*c* and *d* for a Schottky barrier on a *p*-type semiconductor, aligning the Fermi levels at equilibrium requires a positive charge on the metal side and a negative charge on the semiconductor side, with the depletion region again confined to the semiconductor. In terms of work function potentials, ohmic and rectifying contacts can be categorized as follows:

$$\left.\begin{array}{l}\Omega_m > \Omega_s \ n \text{ type} \\ \Omega_m < \Omega_s \ p \text{ type}\end{array}\right\} \quad \text{for rectifying contacts} \tag{4.3}$$

$$\left.\begin{array}{l}\Omega_m < \Omega_s \ n \text{ type} \\ \Omega_m > \Omega_s \ p \text{ type}\end{array}\right\} \quad \text{for ohmic contacts} \tag{4.4}$$

Ideal metal-semiconductor contacts are ohmic when the charge induced in the semiconductor from aligning the Fermi levels is provided by majority carriers. As shown in Fig. 4-2*b* for an *n*-type semiconductor, the Fermi levels are aligned at equilibrium by transferring electrons from the metal to the semiconductor. This lowers the electrostatic potential relative to the metal at equilibrium (Fig. 4-2*b*). In this case the barrier to electron flow between the metal and the semiconductor is small and easily overcome by a small voltage. Similarly, $\Omega_m > \Omega_s$ for a *p*-type semiconductor (Fig. 4-2*c*) and holes flow easily across the junction. Unlike the rectifying contacts, no depletion region occurs in the semiconductor in these cases since the electrostatic potential difference required to align the Fermi levels at equilibrium calls for accumulation of majority carriers in the semiconductor (Streetman, 1980).

Although the barrier height for an *n*-type semiconductor is given by Eq. (4.1) for ideal metal-semiconductor contacts, there are two additional factors that have to be considered. One has to do with the lowering of the barrier height when an electric field E is applied. This Schottky barrier lowering $\Delta\Omega$ is given (Sze, 1981) by

$$\Delta\Omega = \left(\frac{qE}{4\pi\alpha_s}\right)^{1/2} \tag{4.5}$$

where α_s is the semiconductor permittivity in contact with a metal. The second has to do with the nonideality of the metal-semiconductor interface. A Schottky barrier junction, unlike a *pn* junction that occurs within a single crystal, includes a termination of the semiconductor crystal. This surface contains surface states (energy levels) due to incompleted covalent bonds and other effects such as impurity segregation or an oxide layer, which can lead to charges at the metal-semiconductor interface. These effects can be combined to give the following barrier height for typical metal-semiconductor contacts (Sze, 1981) for an *n*-type semiconductor:

$$\Omega_{Bn} = C_2(\Omega_m - X) + (1 - C_2)\left(\frac{E_g}{q} - \Omega_o\right) - \Delta\Omega = C_2\Omega_m + C_3 \tag{4.6}$$

where Ω_o is the energy level at the surface and C_2 is a constant for which $(1 - C_2)/C_2$ is proportional to the density of surface states in number of states per square centimeter per electronvolt. An ideal metal-semiconductor is one where there are no surface states such that $C_2 = 1$, in which case Eq. (4.6) reduces to Eq. (4.1) (except for the Schottky lowering term). In the other extreme case of

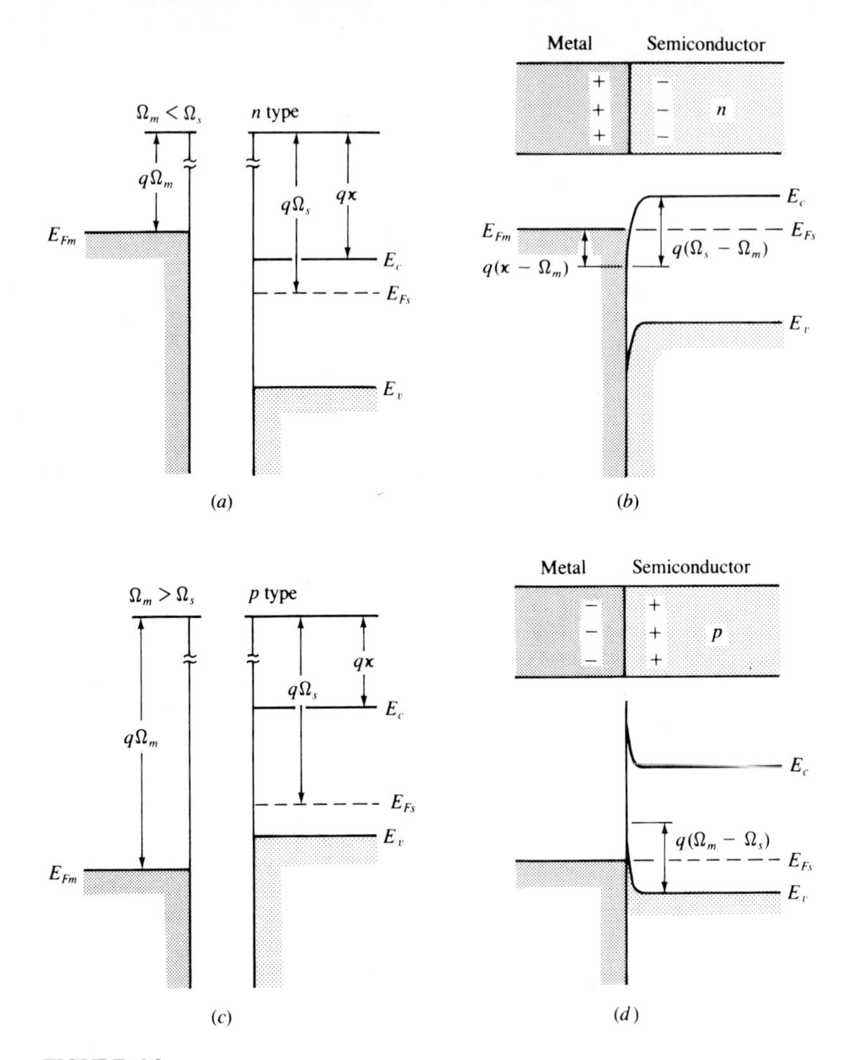

FIGURE 4-2
Ohmic metal-semiconductor contacts: (*a*) an *n*-type semiconductor and (*b*) the equilibrium band diagram for the junction; (*c*) a *p*-type semiconductor and (*d*) the junction at equilibrium (Streetman, 1980).

infinitely many surface states, C_2 approaches zero and Eq. (4.6) reduces to

$$q\Omega_{Bn} = (E_g - q\Omega_o) - q\,\Delta\Omega \tag{4.7}$$

It is seen that the barrier height is independent of the metal work function and is determined entirely by doping ($\Delta\Omega$ term) and the surface properties of the semiconductor. In this case, the Fermi level at the interface is "pinned" by the surface states at the value Ω_0 above the valence band. Experimental results for various metals fitted to Eq. (4.6) in the form of $C_2\,\Omega_m + C_3$ are given in Table 4.1. Typical

TABLE 4.1
Constants and quantities for barrier height (Sze, 1981)

Semiconductor	C_2, V	C_3, V	X, V	$q\Omega_o$, eV
Si	0.27 ± 0.05	-0.55 ± 0.22	4.05	0.30 ± 0.36
GaP	0.27 ± 0.03	-0.01 ± 0.13	4.0	0.66 ± 0.2
GaAs	0.07 ± 0.05	$+0.49 \pm 0.24$	4.07	0.53 ± 0.33
CdS	0.38 ± 0.16	-1.20 ± 0.77	4.8	1.5 ± 1.5

barrier heights for various combinations of metals and semiconductors are given in Table 4.2. The metal work function is given in Fig. 4-3. Although the barrier heights for GaP and GaAs are represented in the form of Eq. (4.6) in Table 4.1, the barrier heights for III-V compounds are typically represented by Eq. (4.7). As can be seen from Fig. 4-4, the III-V compound Fermi levels are independent of the metal (Fermi-level pinning), which is, for instance, approximately 0.8 eV for n-type GaAs.

The current-voltage relationship for metal-semiconductor (n-type) contacts (Sze, 1981) is given by

$$J = A^*T^2 \exp\left(-\frac{q\Omega_{Bn}}{kT}\right)(e^{qV/kT} - 1) \tag{4.8}$$

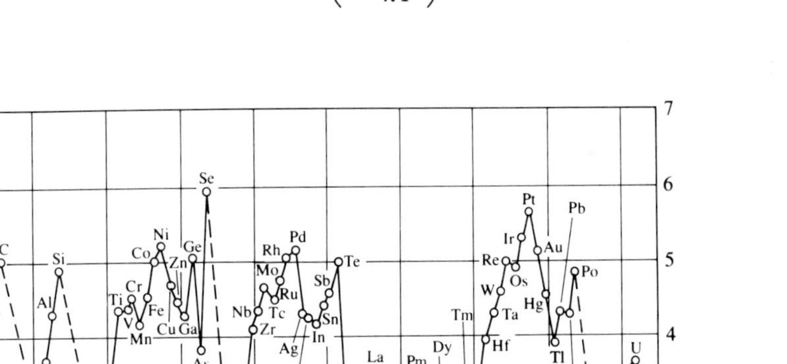

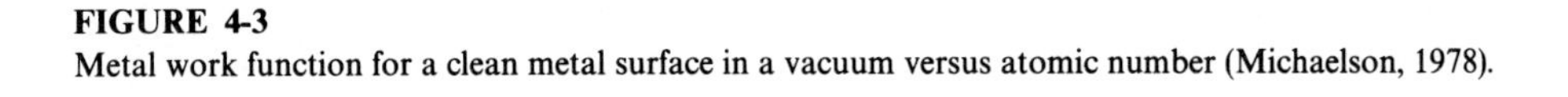
FIGURE 4-3
Metal work function for a clean metal surface in a vacuum versus atomic number (Michaelson, 1978).

TABLE 4.2
Measured Schottky-barrier heights in volts at 300 K (Sze, 1981)

Semi-conductor	Type	E_x, eV	Ag	Al	Au	Cr	Cu	Hf	In	Mg	Mo	Ni	Pb	Pd	Pt	Ta	Ti	W
Diamond	*p*	5.47			1.71													
Ge	*n*	0.66	0.54	0.48	0.59		0.52		0.64			0.49	0.38					0.48
Ge	*p*		0.50		0.30				0.55									
Si	*n*	1.12	0.78	0.72	0.80	0.61	0.58	0.58		0.40	0.68	0.61		0.81	0.90		0.50	0.67
Si	*p*		0.54	0.58	0.34	0.50	0.46				0.42	0.51	0.55				0.61	0.45
SiC	*n*	3.00		2.00	1.95													
AlAs	*n*	2.16			1.20										1.00			
AlSb	*p*	1.63			0.55													
BN	*p*	7.50			3.10													
BP	*p*	6.00			0.87													
GaAb	*n*	0.67			0.60													
GaAs	*n*	1.42	0.88	0.80	0.90		0.82	0.72							0.84	0.85		0.80
GaAs	*p*		0.63		0.42			0.68										
GaP	*n*	2.24	1.20	1.07	1.30	1.06	1.20	1.84		1.04	1.13	1.27			1.45		1.12	
GaP	*p*				0.72													
InSb	*n*	0.16	0.18†		0.17†													
InAs	*p*	0.33			0.47†													
InP	*n*	1.29	0.54		0.52													
InP	*p*				0.76													
CdS	*n*	2.43	0.56	Ohmic	0.78		0.50					0.45	0.59	0.62	1.10		0.84	
CdSe	*n*	1.70	0.43		0.49		0.33								0.37			
CdTe	*n*		0.81	0.76	0.71										0.76			
ZnO	*n*	3.20		0.68	0.65		0.45		0.30					0.68	0.75	0.30		
ZnS	*n*	3.60	1.65	0.80	2.00		1.75		1.50	0.82				1.87	1.84	1.10		
ZnSe	*n*		1.21	0.76	1.36		1.10		0.91				1.16		1.40			
PbO	*n*		0.95						0.93			0.96	0.95					

† 77 K.

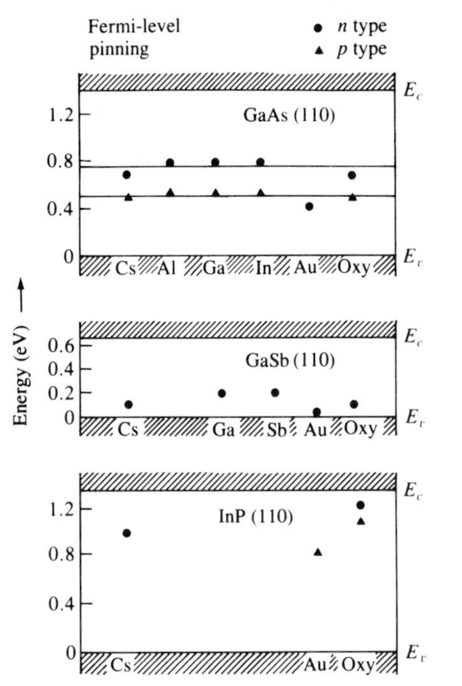

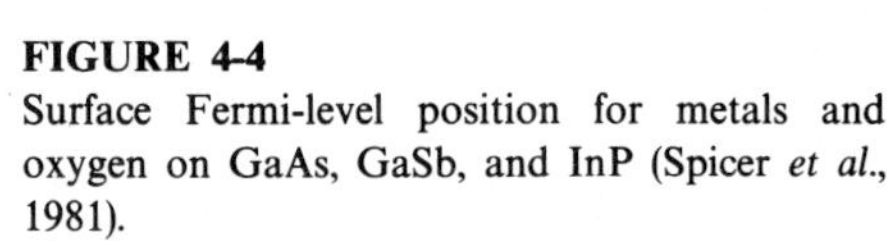

FIGURE 4-4
Surface Fermi-level position for metals and oxygen on GaAs, GaSb, and InP (Spicer *et al.*, 1981).

where J is the current density (flux), T is temperature, V is the applied potential, and A^* is the effective Richardson constant. The effective Richardson constant is a weak function of temperature, doping concentration, and electric field. It is shown in Fig. 4-5 for the conditions given there. When current transport is dominated by the tunneling effect, as is the case for heavily doped semiconductors or low-temperature operation, the current-voltage relationship (Padovani and Stratton, 1966) is given by

$$J \sim \exp\left(-\frac{q\Omega_{Bn}}{E_o}\right) \tag{4.9}$$

where
$$E_o = \frac{qh}{4\pi}\left(\frac{N_D}{\alpha_s m_e}\right)^{1/2}$$

where N_D is the donor concentration and m_e is the effective electron mass. The current-voltage relationship for typical contacts, considering both diffusion and tunneling currents, can be represented (Chang and Sze, 1970) by

$$J = A^*T^2 \exp\left(-\frac{q\Omega_{Bn}}{kT}\right)(e^{qV/nkT} - 1) \tag{4.10}$$

where n is the ideality factor defined as

$$n = \frac{q}{kT}\frac{\partial V}{\partial(\ln J)} \tag{4.11}$$

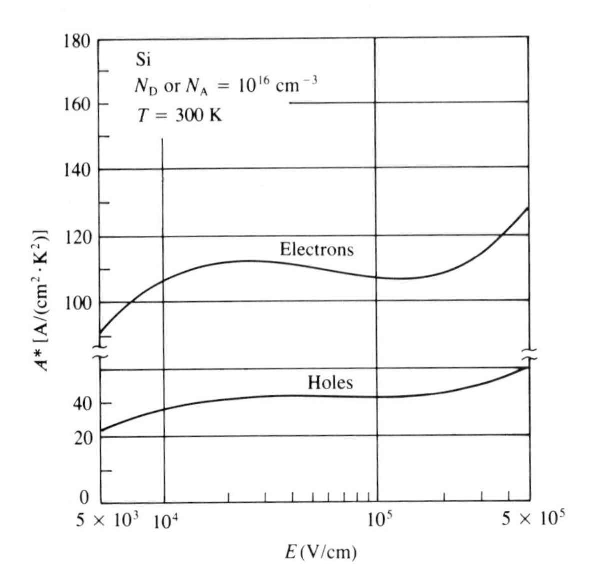

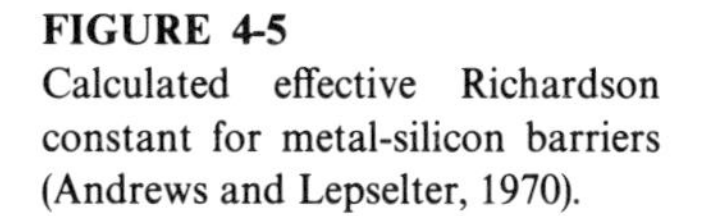

FIGURE 4-5
Calculated effective Richardson constant for metal-silicon barriers (Andrews and Lepselter, 1970).

An ohmic contact is one that has a negligible contact resistance relative to the bulk. Thus, it should supply the required current with a voltage drop that is sufficiently small compared with the drop across the active regions of the device. An important figure of merit for ohmic contacts is the specific contact resistance R_c (Chang *et al.*, 1966), defined as follows:

$$R_c = \left(\frac{\partial J}{\partial V}\right)^{-1}\Bigg|_{V=0} \tag{4.12}$$

For metal-semiconductor contacts with lower doping concentrations (say, $N_D < 10^{17}$ atoms/cm^3), Eq. (4.8) applies and the use of the equation in Eq. (4.12) yields

$$R_c = \frac{k}{qA^*T} \exp\left(\frac{q\Omega_{Bn}}{kT}\right) \tag{4.13}$$

For contacts with higher doping levels (say, $N_D > 10^{19}$), the tunneling process dominates and Eqs. (4.9) and (4.12) yield

$$R_c \sim \exp\left(\frac{q\Omega_{Bn}}{E_o}\right) \sim \exp\left(\frac{\Omega_{Bn}}{N_D^{1/2}}\right) \tag{4.14}$$

In both cases the contact resistance is exponentially proportional to the barrier height Ω_{Bn}, but in the tunneling range the specific contact resistance also varies exponentially with $N_D^{-1/2}$. Both Eqs. (4.13) and (4.14) show that high doping concentration, low barrier height, or both, must be used to obtain low values of R_c. These are exactly the approaches used for all ohmic contacts. For wide-gap semiconductors including silicon, it is not easy to make ohmic contacts, since a metal does not generally exist with a low enough work function to yield a low barrier.

In such cases the general technique for making an ohmic contact is to establish a heavily doped surface layer in the contact region such as a metal-n^+-n or a metal-p^+-p contact. With the heavily doped layer, the depletion width is small enough to allow carriers to tunnel through the barrier. For example, Au containing a small percentage of Sb can be alloyed to n-type Si, forming an n^+ layer at the semiconductor surface for an excellent ohmic contact. Also, Al makes a good ohmic contact to p-type Si, since the required p^+ surface layer is formed during a brief heat treatment of the contact following Al deposition. For n-type Si, however, n^+ diffusion or ion implantation treatment is required before aluminum contacting.

The lowering of the barrier height for an n-type semiconductor (or raising of the barrier height for a p-type semiconductor) by introducing a thin heavily doped layer (10 nm or less) can be analyzed with the idealized profiles given in Fig. 4-6. From Poisson's equation, the electric field distribution is obtained as follows:

$$E = \begin{cases} -E_m + \dfrac{qn_1 x}{\alpha_s} & \text{for } 0 < x < a \\[2ex] -\dfrac{qn_2(W - x)}{\alpha_s} & \text{for } a < x < W \end{cases} \tag{4.15}$$

where the symbols are given in Fig. 4-6. The maximum electric field at the metal-semiconductor E_m is given by

$$E_m = \frac{q[n_1 a + n_2(W - a)]}{\alpha_s} \tag{4.16}$$

If the dopant concentration in the heavily doped layer n_1 is much larger than that in the underlying layer n_2, or $n_1 a \gg n_2(W - 1)$, Eqs. (4.5) and (4.16) yield

$$\Delta\Omega = \frac{q}{\alpha_s}\left(\frac{n_1 a}{4\pi}\right)^{1/2} \tag{4.17}$$

Another more significant effect than the barrier lowering by heavily doping a thin layer is the tunneling effect. As indicated by Eq. (4.14), the resistance is lowered by a factor of exp $(\Omega_{Bn}/N_D^{1/2})$ due to the tunneling.

Example 4.1. Calculate the barrier lowering $\Delta\Omega$ for $n_1 a = 10^{12}$ and 10^{13} cm^{-2} for an Au-Si contact. Assume that $n_1 a \gg n_2(W - a)$. For an Au-Si contact, the semiconductor permittivity α_s is $12\alpha_0$, where α_0 is the free-space permittivity.

Solution. $\Delta\Omega$ is given by Eq. (4.17). The values of q and α_s are

$$q = 1.602 \times 10^{-19} \text{ C (coulomb)}$$

$$\alpha_s = 12\alpha_0 = 12(8.854 \times 10^{-14}) \text{ F/cm}$$

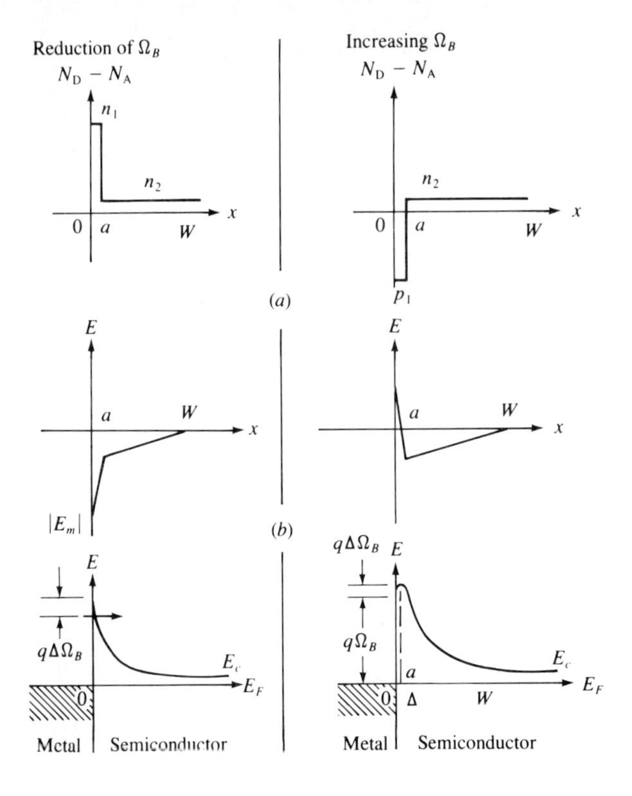

FIGURE 4-6
Idealized controlled barrier contacts with a thin n^+ or p^+ layer on an n-type substrate for barrier reduction (n^+ layer) or barrier raising (p^+ layer) (Sze, 1981).

Thus,

$$\Delta\Omega(\text{V}) = \frac{1.602 \times 10^{-19}}{12(8.854 \times 10^{-14})} \left(\frac{n_1 a}{4 \times 3.14} \right)^{1/2}$$

$$= 4.25 \times 10^{-8} \text{ (V·cm)}(n_1 a)^{1/2}$$

For $n_1 a = 10^{12}$ and 10^{13} cm^{-2}, the above equation yields:

$$\Delta\Omega = 0.0425 \text{ V} \qquad \text{for } n_1 a = 10^{12}$$

$$\Delta\Omega = 0.134 \text{ V} \qquad \text{for } n_1 a = 10^{13}$$

Example 4.2. One way of determining the barrier height experimentally is to obtain a current-voltage curve from the metal-semiconductor contact (diode) and then extrapolate the curve to zero voltage (bias). The current thus obtained is called the saturation current (at zero bias). Obtain a relationship from Eq. (4.10) that can be used to determine the barrier height based on the saturation current. For an Au-Si contact, the saturation current is relatively independent of doping up to 10^{17} cm^{-3}

for an n-type silicon, but increases exponentially with doping thereafter (Sze, 1981). The saturation current at $N_D = 10^{19}$ cm^{-3} and 300 K is approximately 2×10^{-4} A/cm^2. Determine the barrier height for the Au-Si contact ($N_D = 10^{19}$ cm^{-3} and $a = 10$ nm). The effective Richardson constant A^* is about 110 A/(cm^2·K^2) for electrons. Determine the fraction of the total barrier lowering due to the tunneling effect.

Solution. Equation (4.10) is a general relationship for I-V (or J-V) characteristics of metal-semiconductor contacts. Since the saturation current density (or equivalently the current) J_s is that corresponding to zero bias ($V = 0$), Eq. (4.10) can be arranged to give

$$J_s = A^*T^2 \exp\left(-\frac{q\Omega_{Bn}}{kT}\right) \tag{A}$$

Solving this for $q\Omega_{Bn}$ yields

$$q\Omega_{Bn} = kT \ln\left(\frac{A^*T^2}{J_s}\right) \tag{B}$$

which is the desired relationship. Using Eq. (B), one has

$$\begin{aligned} q\Omega_{Bn} &= kT \ln \frac{110 \times 300^2}{2 \times 10^{-4}} \\ &= 8.62 \times 10^{-5} \times 300 \text{ (eV)} \ln (1.65 \times 10^{10}) \\ &= 0.61 \text{ eV} \end{aligned}$$

According to Table 4.2, the barrier height for an Au-Si contact without an n^+ thin layer, which is 10 nm thick in this example, is 0.8 eV. Thus, the total reduction in the barrier height is $0.8 - 0.61 = 0.19$ eV. For $n_1 = N_D = 10^{19}$ and $a = 10$ nm, $n_1 a = 10^{13}$ cm^{-2}. Example 4.1 results gave $\Delta\Omega$ of 0.134 V for the $n_1 a = 10^{13}$ cm^{-2}. Assigning the balance of the total reduction to tunneling, the fraction of the total due to the tunneling is $(0.19 - 0.134)/0.19 = 0.295$. It is seen in this example that approximately 30 percent of the barrier reduction is due to the tunneling effect.

The simple theories considered so far are for ideal situations from the processing point of view. For ohmic contacts, for example, minimization of the barrier height is the prime processing objective from the standpoint of good electrical properties. To achieve this objective, however, the processing has to cope with the realities of corrosion and interdiffusion of atoms across the metal-semiconductor interface, in addition to the interfacial states already taken into consideration in modeling the I-V characteristics.

A brief discussion of a phase diagram is useful before proceeding further. Equilibrium phase diagrams of Au-Si and Al-Si are shown in Fig. 4-7. A liquidus line is one below which liquid and solid coexist. Below a solidus line only solid phases exist. The relative concentration at any temperature on the liquidus and solidus lines refers to alloys having the lowest free energy at solid and liquid states, respectively. The temperature at which the free energy curves for liquid

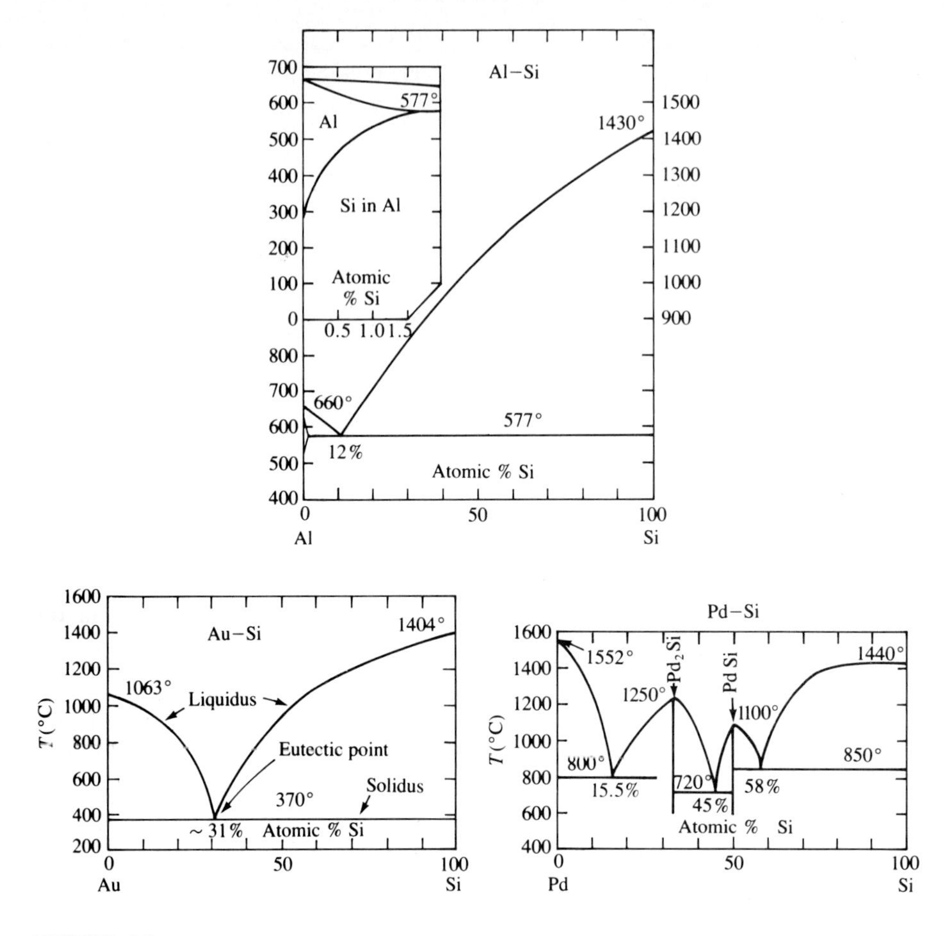

FIGURE 4-7
Phase diagrams (Hansen, 1958).

and solid states coincide at their extrema along a common horizontal tangent (along an isotherm) is called the congruent point, where freezing occurs without any composition change. In a eutectic system a given alloy composition remains in a liquid state down to the lowest possible temperature, called the eutectic temperature, where it solidifies. Any deviations from the eutectic composition are accommodated by a rejection of the surplus species into the solid phase (in a region enclosed by the liquidus and solidus) of the excess pure metals (Au-Si) or dilute alloys (Al-Si). The eutectic point is a special case of congruent point. When interacting elements are rather dissimilar in terms of crystal structure, atomic volume, and electronegativity, different intermetallic compounds can occur in a phase diagram, as shown in Fig. 4-7 for Pd-Si, which themselves act as end phases in eutectic systems. More details on phase diagrams can be found in Reed-Hill (1986) and other works.

When two dissimilar solids are allowed to make a contact involving interatomic distances, they are rarely separated by an abrupt interface. In the absence of interfacial oxides or contaminations they interdiffuse intimately, even at room temperature, to form a very thin (2 to 50 monolayers thick) interfacial layer having a composition almost equal to the lowest melting point eutectic (if it exists) in this phase diagram and a structure almost totally disordered and, therefore, metastable. In the case of covalently bonded elemental semiconductors such as Si, deposited Au forms an interfacial layer with a composition of the Au-Si eutectic with $T_m = 370\,°\text{C}$ (Fig. 4-7). In doing so, the covalent bonds of Si are broken and the Si atom behaves like a metal atom with much lower bond strengths to other Si or Au atoms. The metallic Si atoms move quickly through the Au overlayer to reach the surface and oxidize, even at room temperature. Au also diffuses interstitially into crystalline Si and moves away from this alloy layer (Mukherjee, 1981).

The interdiffusion between metal and semiconductor can cause a penetration of a *pn* junction interface by the diffusing metal, a phenomenon referred to as junction spiking. This junction spiking caused by the local dissolution of Si may be solved by depositing a metal such as Al with Si added. The amount of Si required should be determined by the maximum process temperature and the Al-Si phase diagram.

Example 4.3. When Al is used as the contact for Si, junction spiking may occur. The problem is compounded by the fact that the junction depth is small, say around 0.3 μm. Using the phase diagram of Fig. 4-7 for Al-Si, determine the amount of Si needed if annealing is carried out at 450 °C. Discuss the consequences of adding more Si than is dictated by the phase diagram for both p^+ and n^+ Si substrates.

Solution. According to the inset of the phase diagram for Al-Si in Fig. 4-7, 0.5 wt % Si is required at 450 °C. At that temperature, 0.5% Si in Al is in equilibrium with Al. If the Al contacts are all to p^+ Si, having a Si content higher than is dictated by the phase diagram is acceptable for solving the junction spiking problem. Because of the excess Si present in the Al, some precipitation of Si occurs in the contact window. If n^+ Si is used, the precipitation leads to a nonohmic contact to n^+ Si, since the recrystallized Si precipitate contains Al, which is a *p* dopant (Fraser, 1983).

In addition to the interdiffusion problem, metallization for practical metal contacts should take into consideration the factors associated with both the adhesion of the deposited film and the potential for corrosion. To solve these problems, multilayer metallization is often used, involving three different layers in the case of metal contacts. The first is the adhesion layer (*A*), typically Ti, Cr, Ta, etc. These elements are capable of reducing SiO_2 or Ga_2O_3 locally (Baglin and Poate, 1978) to form a strong adhesive bond with the semiconductor surface and the surrounding oxide. In the middle is the barrier layer (*B*) whose task is to inhibit solid-state interactions and diffusion, as well as preventing the loss of Ti or Cr into the top conducting or contact layer (*C*) (Mukherjee, 1981). Ideal relationships for the three layers should be as follows (Nicolet, 1978):

1. A and C should not diffuse into B but the A-B and B-C interface bonding should be strong.
2. B should not diffuse into or form compounds with A and C.
3. The couples A-B and B-C should not form galvanic cells (Chapter 11) with relatively large electrochemical potential to prevent corrosion.
4. The combinations A-B, B-C, and C-A should not have a low-temperature eutectic alloy composition.
5. A, B, and C should have similar thermal expansion coefficients, small thermal and electrical contact impedances at the interfaces, and B should be chemically inert and pinhole free.

In practice not all the conditions are satisfied simultaneously and some compromises have to be made. Intermetallics are often used as barriers, because of their relatively high stability (lower free energy). An intermetallic barrier having a free energy lower than other possible compounds formed with A and C ensures its own stability. Foreign atoms can also be introduced into the grain boundaries of B to hinder or block the diffusion of A and C. The crystal structure in metal contacts is polycrystalline, i.e., crystalline within each grain but surrounded by the boundaries of the adjacent grains or the grain boundaries. Diffusion along grain boundaries is energetically most favorable. Any interdiffusion takes place along grain boundaries first, since the boundaries offer the least-resistance paths for solid-state diffusion. If the annealing time is relatively short and the temperature is not too great, typically up to 450 °C, then the foreign atoms present in the grain boundaries can successfully hinder grain boundary diffusion (Gupta and Rosenburg, 1975).

While barrier height is the key feature for a metal contact's electrical properties, resistivity is the key to understanding metal interconnects. Electrical conduction in metals is due to electrons. The electrical resistivity results from the scattering of these electrons by the metal lattice. Therefore, any factor such as temperature, dissolved impurities and vacancies that leads to more scattering also leads to higher resistivity. In general, the resistivity increases linearly with temperature above room temperature. The temperature coefficient of resistance α is used to represent the temperature dependence of resistivity, ρ:

$$\alpha = \frac{1}{\rho}\frac{d\rho}{dT} \tag{4.18}$$

According to Matthiessen's rule (e.g., Wilson, 1958), the product $\alpha\rho$ is a constant for a given metal:

$$\alpha\rho = \text{constant} \tag{4.19}$$

Since α is positive, a decrease in α means an increase in the resistivity.

The resistivity or the conductivity, which is the inverse of the resistivity, of a thin-film metal is different from that of the bulk metal. According to Maissel

(1970), the film conductivity λ is given by

$$\frac{\lambda}{\lambda_0} = 1 - \frac{3}{2\beta}\int_1^{\infty}\left(\frac{1}{t^3} - \frac{1}{t^5}\right)(1 - e^{-\beta t})\,dt \tag{4.20}$$

where λ_0 is the bulk conductivity and β is the ratio of film thickness to electron mean free path. For extreme values of β and allowing for elastic electron scattering, Eq. (4.20) reduces to

$$\frac{\lambda}{\lambda_0} = \begin{cases} 1 - \dfrac{3(1-p)}{8\beta} & \text{for } \beta \gg 1 \\ \dfrac{3\beta}{4}(1 + 2p)\left(\ln\dfrac{1}{\beta} + 0.423\right) & \text{for } \beta \ll 1 \end{cases} \tag{4.21}$$

where p is the fraction of electrons elastically scattered. This is a somewhat artificial quantity. Equation (4.20) is for the case of zero p. Similar derivations give the temperature coefficient of resistance, which is the same as that given by Eq. (4.21) for $\beta \gg 1$ but is given for $\beta \ll 1$ by

$$\frac{\alpha}{\alpha_0} = \frac{1}{\ln(1/\beta) + 0.423} \qquad \text{for } \beta \ll 1 \tag{4.22}$$

Example 4.4. The mean free path of electrons in copper at room temperature is approximately 40 nm, and the bulk copper resistivity is 1.69 $\mu\Omega\cdot$cm. The mean free path of electrons in most metals lies between 10 and 60 nm at room temperature. Suppose that the copper is used for an interconnection, and its thickness is 1 μm. Calculate the conductivity of the copper film using the zero p value. Discuss the effect of film thickness on the conductivity.

Solution. The ratio β for the problem is

$$\beta = \frac{10^{-4}}{40 \times 10^{-7}} = \frac{1000}{40} = 25 \gg 1$$

Using the first part of Eq. (4.21) for the case $\beta \gg 1$, one has, for $p = 0$,

$$\frac{\lambda}{\lambda_0} = 1 - \frac{3}{8\beta} = 1 - \frac{3}{8 \times 25} = 0.985$$

Thus, the film conductivity is $0.985\lambda_0$ or 0.583 $(\mu\Omega\cdot\text{cm})^{-1}$ since conductivity is the inverse of resistivity. As seen in this example and from Eq. (4.21), the conductivity decreases with decreasing film thickness. Physically, this relates to collisions with the surface becoming a significant fraction of the total number of collisions with decreasing thickness.

The treatments up to this point are for ideal films: a completely continuous film with uniform thickness everywhere. Real thin films are anything but continuous during their early stages of growth and are made up of small islands which may or may not be physically connected to one another. As opposed to ideal films, it is not unusual for discontinuous films to have a negative temperature

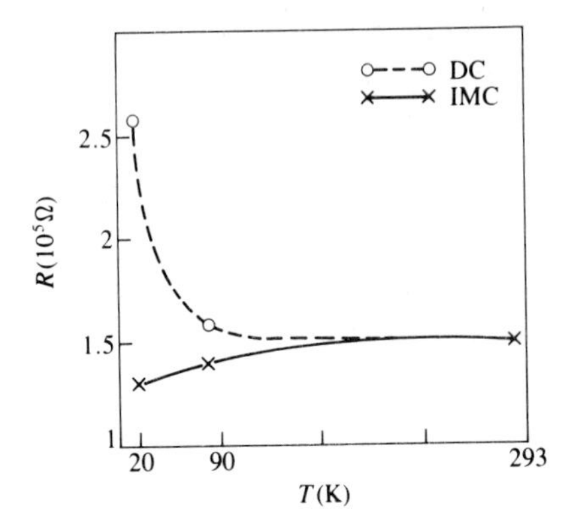

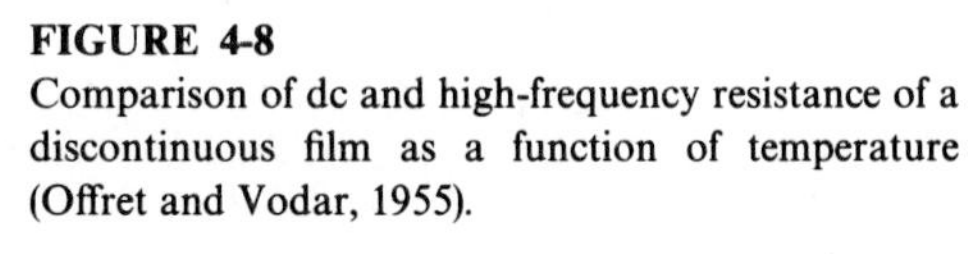

FIGURE 4-8
Comparison of dc and high-frequency resistance of a discontinuous film as a function of temperature (Offret and Vodar, 1955).

coefficient of resistance, i.e., the resistivity decreases with increasing temperature (Neugebauer and Wilson, 1966). In discontinuous films, the grains making up the film are separated by air spaces. The negative temperature coefficient arises because of the activation energy associated with electron transitions from grain to grain. A structure in which metallic grains are separated by small spaces is electrically equivalent to a series of capacitors, and consequently its ac impedance is less than the dc value. Furthermore, one would expect this difference between ac and dc conductance to become progressively larger as the temperature is reduced, since the dc resistance of discontinuous films increases rapidly with decreasing temperature (Maisscl, 1970). An illustration of this is shown in Fig. 4-8 for platinum on Pyrex. Theories of conduction in discontinuous films can be found in the above reference.

A host of structural defects in a metal thin film can be annealed out of the film by relatively mild heat treatment. This mild annealing, when carried out

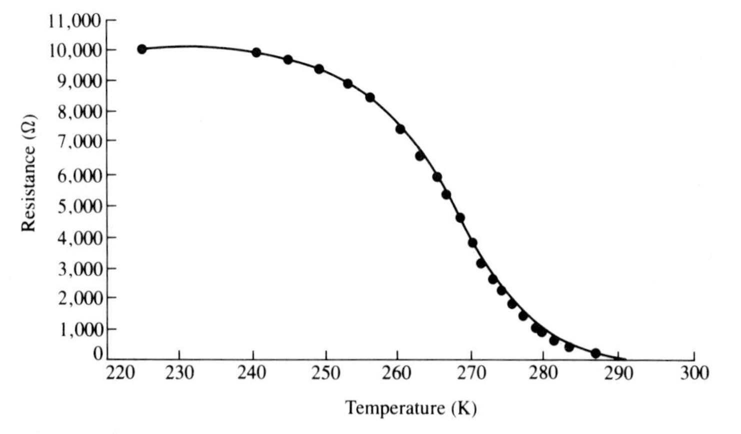

FIGURE 4-9
Resistance as a function of annealing temperature for an evaporated gold film (Wilkinson and Birks, 1949).

optimally, results in a significant reduction in the resistivity. An example is shown in Fig. 4-9. Films deposited at lower temperatures consist of small grains. At lower temperatures, the film particles have lower mobility and remain more or less where they first join to the substrate. Thus, a continuous film can not form. On the other hand, an island structure is likely to form when the deposition is carried out at high temperatures. Further, a continuous film deposited at lower temperatures can form an island structure if annealed at high temperatures since then small grains start agglomerating because of higher mobility. Thus, an optimal temperature and annealing time should be sought for a given material. Another problem that leads to a reduction in conductivity is oxidation of deposited metal. A suitable choice of a passivating layer can protect the underlying metal interconnects from oxidation (and corrosion) problems.

Silicides are often used as interconnects because of their low resistivity (Table 4.3) and relative stability during later processing. The silicides can also be used as rectifying contacts because of the wide range of barrier height that can be obtained. Figure 4-10 shows barrier height correlated with heats of formation of the silicides.

One major cause of interconnect failure is the phenomenon known as electromigration (or electron transport). Electromigration is a process in which atoms migrate in a direction not dictated by the concentration gradient but by the orientation of an electric field and/or the direction of electric current. Atoms migrate toward an electron current when they have a negatively charged state. Atoms migrate against the electron current if the atom has a positively charged state. This material transport of the conductive material is driven by momentum transfer between positive metal ions and electrons, which are moving under the

TABLE 4.3
Silicide resistivites at 300 K (Frazer, 1983)

Silicide	Starting form	Resistivity (μΩ·cm)
TiS_2	Metal/poly Si†	13–16
	Co-sputtered	25
$ZrSi_2$	Metal/poly Si	35–40
$TaSi_2$	Metal/poly Si	35–45
	Co-sputtered	50–55
WSi_2	Co-sputtered	70
$CoSi_2$	Metal/poly Si	17–20
	Co-sputtered	25
$NiSi_2$	Metal/poly Si	50
	Co-sputtered	50–60
$PtSi_2$	Metal/poly Si	28–35
Pd_2Si	Metal/poly Si	30–35

† Obtained by sintering a metal layer on polysilicon.

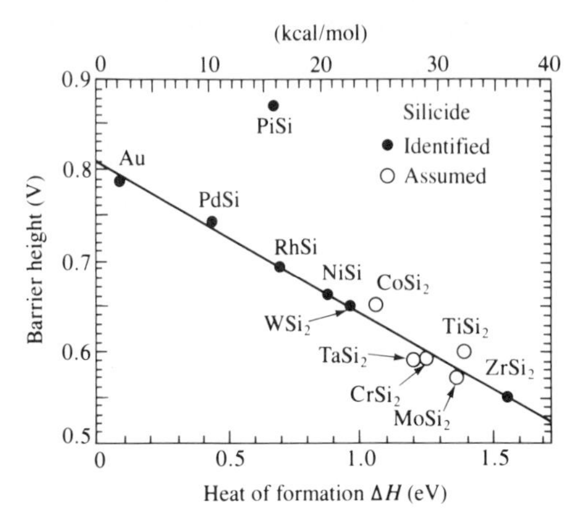

FIGURE 4-10
Barrier height of transition metal silicides as a function of heat of formation (Andrews, 1975).

influence of the electric field applied along the conductor. The eventual consequence is the physical opening of an interconnection, as shown in Fig. 4-11. This open-circuit-type failure is catastrophic. In some applications, even a small increase in the resistance of the interconnect stripe, due to local mass depletion, can cause a device malfunction. Since metal ions are generally positive, material accumulation occurs in the direction of the anode. The ion flux J_i is given (Huntington and Grone, 1961) by

$$J_i = \frac{ND_v}{kT} Zq\rho j \qquad \text{where } D_v = D_0 \exp\left(-\frac{E_v}{kT}\right) \tag{4.23}$$

where N is the metal atom density, D_v is the diffusivity of vacancies, ρ is the resistivity, j is the current density (flux), and Z is an "effective charge" of the metal atoms. As might be expected from Eq. (4.23), the mass flow is greater at higher temperatures. For Al, for instance, the opening in the metal line (Fig. 4-11)

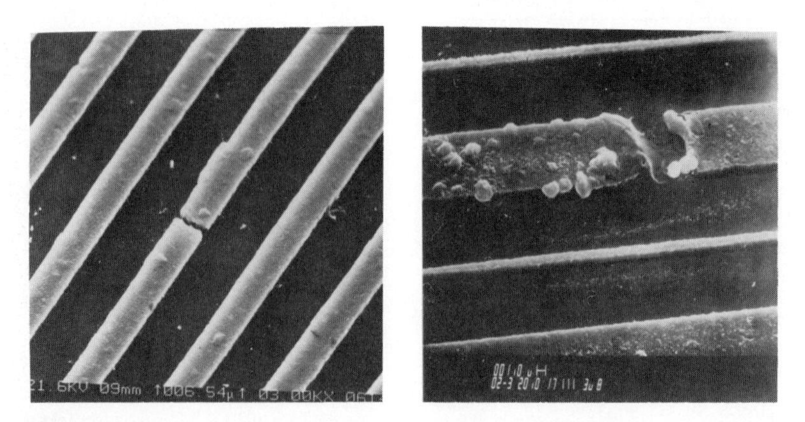

FIGURE 4-11
SEM micrographs of electromigration failure (Vaidya *et al.*, 1980).

forms whenever the electron flow is in the direction of increasing temperature (Bleck and Meieran, 1967).

Electromigration resistance of film conductors can be increased by several techniques. These include alloying with Cu, the incorporation of discrete layers such as Ti, encapsulating the conductor in a dielectric material, or incorporating oxygen during film deposition (Grangulee *et al.*, 1975). The mean time to failure (MTF, MTTF, or MTF t_{50} for time to 50 percent failure) of the conductor can be related to the current density j in the conductor as follows:

$$\text{MTF} = \frac{kV_c T}{j^n D_b} \qquad (h/w < 0.5) \tag{4.24}$$

where k is a constant, V_c is the conductor volume, h and w are the stripe height and width, and D_b is usually taken as the grain boundary diffusivity of the atoms. The exponent n can range from 1 to 15 (e.g., D'Heurle and Ho, 1978) but is usually around 2 or 3.

Example 4.5. Most electromigration studies are based on MTF experiments. Accelerated life testing experiments are usually carried out to plot failure time against cumulative percent failure on a lognormal plot. Such a plot is shown in Fig. 4-12 for TiW/Al conductors (Towner, 1985). From the plot, determine the lognormal relationship. As Eq. (4.24) implies, a plot of MTF plotted against the current density yields the value of n and a plot of MTF versus $1/T$ yields the activation energy for D_b. Towner obtained the following values from her MTF data:

$$n = 4$$

$$E_d = 0.52 \text{ eV}$$

where E_d is the activation energy for the diffusivity D_b. For the current density of 4×10^6 A/cm², the MTF is 0.3 h at 200 °C. Write an expression for the MTF.

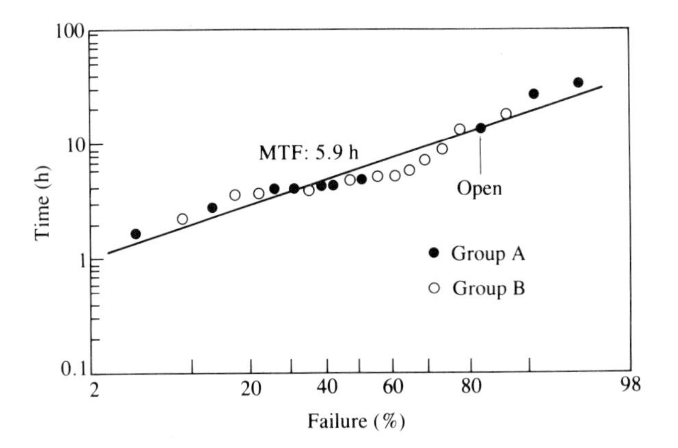

FIGURE 4-12
Lognormal probability plot of interlayer short-circuit failure times in TiW/Al conductors (Towner, 1985).

Solution. The lognormal relationship can be written as follows:

$$F = bt^a \tag{A}$$

where a and b are constants and F is the fractional cumulative failure. One can rewrite Eq. (A) as

$$\ln F = a \ln t + \ln b \tag{B}$$

From Fig. 4-12, one has the following pairs: $(F, t) = (0.02, 1)$; $(F, t) = (0.5, 5.9)$ where time is in hours. Using the two points in Eq. (B), one obtains

$$F = 0.02t^{1.8} \qquad (t \text{ in hours}) \tag{C}$$

Noting that $D_b = D_0 \exp(-E_d/kT)$, where k is Boltzmann's constant, one can rewrite Eq. (4.24) as follows:

$$\text{MTF} = K'Tj^{-n} \exp\left(\frac{E_d}{kT}\right) \qquad \text{where } K' = \frac{kV_c}{D_0} \tag{D}$$

since for a given interconnect material, all the constants can be absorbed into K'. From the data provided, one has, for Eq. (D),

$$0.3 = K'(473)(4 \times 10^6)^{-4} \exp\left(\frac{0.52}{8.62 \times 10^{-5} \times 473}\right)$$

Solving this for K' yields a value of 4.7×10^{17} h/[K·(A/cm^2)^4]. Thus, Eq. (4.24) can be rewritten as follows:

$$\text{MTF} = 4.7 \times 10^{17} j^{-4} \exp\left(\frac{0.52}{kT}\right)$$

In addition to electromigration, stress-induced void formation in the interconnect line is another major cause of metallization failures. This problem becomes acute as the line width narrows and the pattern structure becomes complex. Numerical simulation of stresses can be utilized to study these failures (Groothuis and Schroen, 1987).

4.3 DIELECTRICS

Dielectrics are used for insulation between conducting layers, for diffusion and ion implantation masks, for diffusion from doped oxides, for gettering impurities, and for passivation to protect devices from impurities, moisture, and scratches. Perhaps the most important application of dielectrics in ICs is their use as insulators. Dielectrics are deposited by physical and chemical means. Plasma deposition is also used. Some of the important dielectrics in microelectronics are silicon oxides, silicon nitrides, and alumina.

Electrical insulators are best considered as capacitor dielectrics. The capacitance C of a parallel-plate capacitor is given by

$$C = \frac{\alpha\alpha_0 A}{d} \tag{4.25}$$

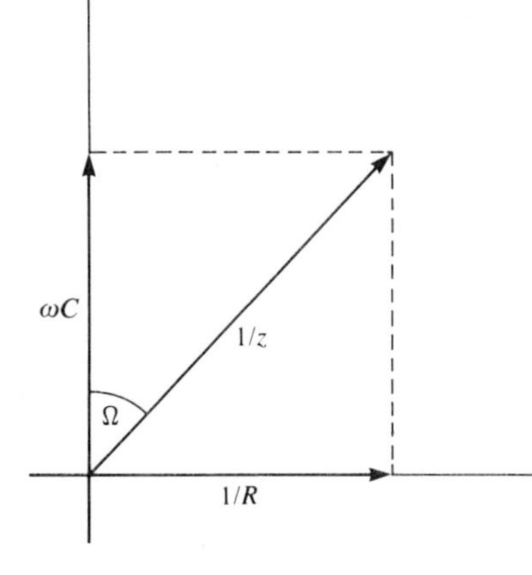

FIGURE 4-13
Vector diagram of impedance.

where α_0 is the permittivity of free space (8.85×10^{-14} F/cm), α is the relative permittivity, also known as the dielectric constant, A is the electrode area, and d is the dielectric thickness. Insulators always have a finite parallel resistance R, and the total impedance can be found from a vector diagram as in Fig. 4-13, where ω is 2π times the frequency and z is the impedance. The sine of the angle Ω is a measure of the energy absorbed in the insulator. Since $\tan \Omega$ is small for insulators, one may write

$$\sin \Omega \doteq \tan \Omega = \frac{1}{\omega RC} \tag{4.26}$$

Tan Ω is often referred to as the loss of the insulator. This represents capacitor loss during ac operation. In general, the loss decreases exponentially with increasing bandgap of the dielectric material.

Ideally, the conductance of an insulating film is zero. Real insulators, however, show carrier conduction when the electric field or temperature is sufficiently high. An estimate of the electric field in an insulator under biasing conditions can be made by

$$E_i = E_s\left(\frac{\alpha_s}{\alpha_i}\right) \tag{4.27}$$

where E_i and E_s are the electric fields in the insulator and the semiconductor, respectively, and α_i and α_s are the corresponding permittivities. When the dielectric layer is too thin or the electric field strength is too high, tunneling occurs and electrons pass from one electrode to the other by means of the tunneling effect. The currrent density J_t in such cases is given by

$$J_t = \frac{\alpha_1 E^2}{\Omega} \exp\left(-\frac{\alpha_2 \Omega^{3/2}}{E}\right) \tag{4.28}$$

where α_1 and α_2 are constants, Ω is the tunnel barrier height, and E is the electric field given by V/d, V being the applied voltage and d the insulator thickness.

Although tunneling is to be avoided in the nominal use of an insulator, it can be exploited to gain control over conductance. This is the case for memory circuits, where a thin insulator is inserted between two conductors. The leakage current of an MOS capacitor for SiO_2 can be represented by Eq. (4.28) as well. This equation is also known as the Fowler-Nordheim current density relationship. For Si_3N_4, there are five well-defined trap levels near the valence band edge of Si_3N_4 (Kapoor and Bibyk, 1980). For the nitride, the leakage current is often represented by the Poole-Frenkel current density relationship:

$$J = A_1 E \exp\left\{\frac{-q[\Omega - (qE/\pi\alpha_i)^{1/2}]}{kT}\right\} \tag{4.29}$$

where A_1 is a constant and E is again given by V/d. The leakage current should be as small as possible for a good insulator.

Another important quantity in insulators is the maximum dielectric strength, also referred to as the breakdown field strength. This dielectric breakdown occurs when the passage of a sufficient current density produces a localized meltdown of the dielectric. The breakdown can be thermal or electric. At low temperatures, the breakdown is believed to be due to the growth of the electron current from collision ionization at the point of electron injection. Forlani and Minnaja (1964) give the following breakdown field E_B for this case:

$$E_B = b_1 d^{-b_2} \tag{4.30}$$

where b_1 and b_2 are constants. The thermal breakdown at high temperatures (Klein and Levanson, 1969) is given by

$$E_B = \frac{1}{b} \ln \frac{\lambda}{a_1 q d A (E_B)^2} \tag{4.31}$$

where a_1 and b are constants and λ is the film thermal conductivity.

Example 4.6. The breakdown field for SiO_2 and Si_3N_4 is of the order of 10^7 V/cm. According to the data of Harari (1978), the breakdown field is 2.8×10^{11} V/cm for SiO_2 thickness (Al-SiO_2-Si) of 5 nm and 1.2×10^7 V/cm for the thickness of 30 nm. Assuming that the breakdown is electric in nature, estimate the breakdown field at 15 nm. Do the same at 150 nm.

Solution. Using the two data points in Eq. (4.30),

$$2.8 \times 10^7 = \frac{b_1}{(5)^{b_2}}$$

$$1.2 \times 10^7 = \frac{b_1}{(30)^{b_2}}$$

From these two equations, one gets

$$E_B(\text{V/cm}) = \frac{5.99 \times 10^7}{[d(\text{nm})]^{0.473}} \tag{A}$$

At $d = 15$ nm, Eq. (A) yields $E_B = 1.66 \times 10^7$ V/cm. At $d = 150$ nm, $E_B = 0.56 \times 10^7$ V/cm. While the calculated E_B at 15 nm compares well with experimental data, that at 150 nm compares with an experimental value of 0.8×10^7 V/cm, thus underestimating the breakdown field at large thicknesses.

Dielectric breakdown is catastrophic, and yet it is time-dependent (taking relatively a long time for it to occur) unless it involves a thermal runaway process. Time-dependent dielectric breakdown (TDDB) is a major cause of reliability failures in MOS circuits. The TDDB phenomenon is often studied based on time-to-failure experiments. As was the case in electromigration failure, lognormal plots are used for cumulative percent failure. The TDDB studies have not yet become as sophisticated as those for electromigration.

Example 4.7. Dielectric breakdown generally results in the growth of a conductive polyfilament through the holes created by the meltdown. This short-circuits the gate to the underlying silicon substrate. McPherson and Baglee (1985) used a thermodynamic argument to arrive at the following equation for TDDB:

$$\mathrm{TF}(f\%) = A_1 \exp\left(\frac{\Delta H}{kT}\right) \exp\left[\beta(T)S\right] \qquad \text{(A)}$$

where TF(f%) is the time to failure for f% breakdown of test circuits, ΔH is the enthalpy change required to activate the polyfilament growth at breakdown, A_1 is a constant, and β and S are given by

$$\beta(T) = A_2 + \frac{A_3}{T} \qquad \text{(B)}$$

$$S = E_B - E \qquad \text{(C)}$$

Here E is the applied electric field and A_2 and A_3 are constants. The quantity β is referred to as the electric field acceleration parameter. Show that β can be obtained from Eq. (A) as follows:

$$\beta = \left[\frac{\partial \ln (\mathrm{TF})}{\partial S}\right]_T \qquad \text{(D)}$$

where the subscript indicates a given constant temperature. They report the following data:

T, °C	E, MV/cm	$t_{50\%}$, h	Standard deviation
25	6	4×10^{11}	13.1
	7	1.1×10^6	6.6
	8	90	3.2
85	6	—	—
	7	1.7×10^4	5.1
	8	12	2.3
150	6	7×10^5	7.5
	7	2.7×10^2	3.1
	8	2.5	1.4

They obtained these data based on lognormal plots of cumulative percent failed versus time. Here $t_{50\%}$ is the TF(50%). Use the data to determine the temperature dependence of the electric field acceleration parameter.

Solution. Taking the logarithm of Eq. (A),

$$\ln(\text{TF}) = \ln A_1 + \frac{\Delta H}{kT} + \beta(T)S \tag{E}$$

Use of Eq. (E) in (D) yields

$$\beta = \left[\frac{\partial \ln(\text{TF})}{\partial S}\right]_T \tag{F}$$

Since $S = E_B - E$, Eq. (F) can be rewritten as

$$\beta = -\left[\frac{\partial \ln(\text{TF})}{\partial E}\right]_T \tag{G}$$

The value of β at a given temperature is the slope when ln (TF) is plotted against E according to Eq. (G). Such plots should yield the following:

T, °C	β, decades/(MV·cm)	$10^3/T$, K^{-1}
25	11.1	3.356
85	7.26	2.793
150	6.27	2.364

A plot of β versus $1/T$ yields the following:

$$\beta(T) = -5.4 + 4.7 \times \frac{10^3}{T}$$

The leakage current and the dielectric breakdown are closely related to processing conditions. These conditions affect the thickness uniformity of the deposited layer, pinholes that might be present in the insulator, the stress generated during processing, and the charge traps due to impurities and interface conditions. When a thickness distribution exists in the deposited layer, the leakage and breakdown problems are likely to occur at the thinnest part of the insulator. The presence of pinholes can easily lead to breakdown, as discussed in Example 4.6 for shorting. When stress is generated during thermal processing, stress-induced defects are generated, which in turn cause the formation of traps. This problem can become acute when the geometry surrounding the insulator is complex (e.g., Mitsuhashi *et al.*, 1987). Also, edge effects leading to avalanche breakdown (Rusu and Bulucea, 1979) can occur. Other than the pinhole and thickness uniformity problems, insulator failure is controlled by the charge traps in the insulator. The charge traps also lead to a flat-band voltage shift (Chap. 1), contributing device instability as well.

There are four types of charge traps in Si-SiO_2. The interface region of Si-SiO_2, for an oxide formed from native silicon, can be divided into several

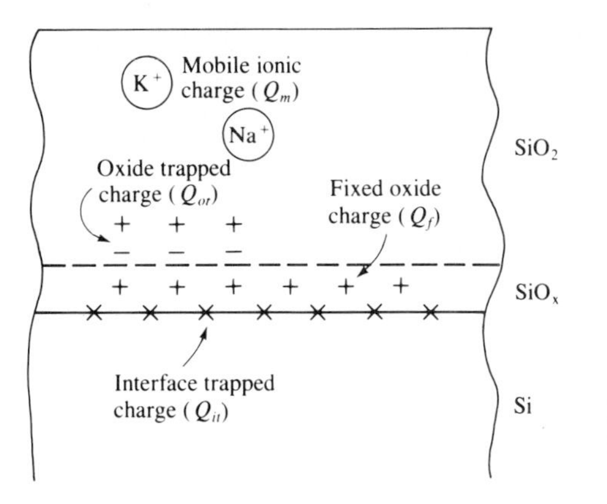

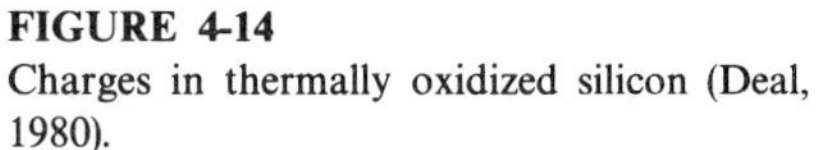

FIGURE 4-14
Charges in thermally oxidized silicon (Deal, 1980).

layers by chemical composition and structural differences. Adjacent to the single-crystal silicon is a layer of SiO_x $(1 < x < 2)$. Adjacent to the SiO_x is a strained region of SiO_2 roughly 1 to 4 nm thick. Finally, there is another bulk region of strain-free amorphous SiO_2 (Nicollian and Brews, 1982). The first trap type is for interface-trapped charges as shown in Fig. 4-14. The interface charges have energy states in the silicon forbidden bandgap and can interact electrically with the underlying silicon. A low-temperature hydrogen anneal (450 °C) effectively neutralizes the interface-trapped charges (Deal, 1980). The second trap type shown in the figure is from fixed oxide charges in the SiO_x layer. These cannot be charged or discharged. The number of charges for this type ranges from 10^{10} to 10^{12} charges/cm^2. According to Deal, rapid cooling from high oxidation temperatures leads to lower values of the charge. The third type of oxide trapped charge can be positive or negative depending on the charge carrier trapped in the bulk oxide, and is in general associated with defects in the oxide. The charge density for this type ranges from 10^9 to 10^{13} cm^{-2} and can be annealed out by a low-temperature treatment. The fourth type is due to ionic charges attributed to alkali ion impurities in the oxide, such as sodium, as well as negative ions and heavy metals. The charge density ranges from 10^{10} to 10^{12} cm^{-2}. Because of the mobile nature of the charges under bias conditions, particularly with alkali and lighter ions, it is important to minimize this type of charge trap. Common techniques to minimize this type of charge trap include cleaning the furnace tube in a chlorine envionment, gettering with phosphosilicate glass, and using masking layers such as silicon nitride (Katz, 1983). Both the interface-trapped and oxide-trapped charges must be annealed to ensure that they do not contribute to the mobile ionic charge.

As opposed to silicon, there are no satisfactory insulators for GaAs at present. This is the main reason why GaAs-based ICs are based on the MESFET structure rather than MIS. Isolation is by mesa formation (Chap. 1). The problem is due to the formation of segregated arsenic atoms at the insulator-semiconductor interface (e.g., Offsey *et al.*, 1986). This results in Fermi-level pinning at the middle of the bandgap.

It should be appreciated that different processing methods lead to different breakdown fields. Although the breakdown field depends on the insulator thickness, plasma-deposited SiO_2 gives a value of around 3 to 6 MV/cm whereas ($SiH_4 + O_2$)-deposited SiO_2 yields a value of around 8 MV/cm. The thermally grown SiO_2 value can be in excess of 10 MV/cm. For Si_3N_4, the value is around 10 MV/cm for LPCVD-deposited films, whereas it is around 5 MV/cm for plasma-deposited ones.

4.4 SUBSTRATES AND JUNCTIONS

Unmodified wafers as such or epitaxial films grown on wafers are used as substrates for ICs. In almost all cases, *pn* junctions are formed within the substrates by doping either thermally or by ion implantation. Epitaxial films are usually grown by chemical vapor deposition methods.

A major problem for junction is the breakdown that occurs when a sufficiently high electric field is applied to the IC. When breakdown occurs, a very large current is created. There are three different types of the breakdown that normally occur under reverse bias (Moll, 1964): thermal instability, the tunneling effect (or Zener effect), and avalanche breakdown (multiplication).

When the reverse current is high due to high reverse voltage, the junction temperature increases due to heat dissipation. This temperature increase in turn increases the reverse current in comparison with its value at lower voltages. The thermal runaway that eventually sets in destroys the junction. This thermal breakdown is important at room temperature for those junctions with large saturation currents. As discussed in Sec. 1.5, a *pn* junction biased in the reverse direction exhibits a small, essentially voltage-independent saturation current, given by Eq. (1.38). Thus, the thermal instability has to do with the temperature effect on the saturation current. Since the heat generaton is proportional to IV_R, where I is the current (or equivalently J_R, the reverse current density) and V_R is the reverse voltage, any heat generation curve lying above the saturation current at a given temperature will lead to thermal runaway (Strutt, 1966). This thermal runaway condition is shown in Fig. 4-15. Note that the maximum reverse voltage that can be allowed decreases with decreasing temperature. The maximum voltage for all possibilities is denoted by V_U in Fig. 4-15, which is called the turnover voltage.

Example 4.8. The temperature dependence of the two terms in Eq. (1.38) is similar (Sze, 1981). Thus, only one term need be used for the temperature dependence:

$$I_0 \sim \frac{D_p p_{ne}}{L_p} = \left(\frac{D_p}{t_p}\right)^{1/2} \frac{n_i^2}{N_D} \qquad \text{where } L_p = (D_p t_p)^{1/2} \tag{A}$$

The temperature dependence of n_i (e.g., Streetman, 1980) is given by

$$n_i = 2\left(\frac{2\pi kT}{h^2}\right)^{3/2} (m_n m_p)^{3/4} \exp\left(-\frac{E_g}{2kT}\right) \tag{B}$$

where h is Planck's constant, and m_n and m_p are the effective mass of an electron and a hole, respectively. If one assumes that (D_p/t_p) is proportional to β, it follows from

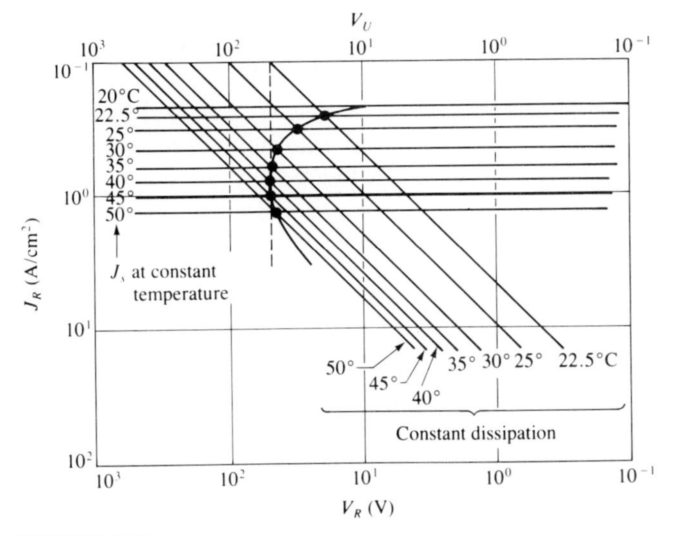

FIGURE 4-15
Reverse *I-V* characteristics of thermal breakdown (Strutt, 1966).

Eqs. (A) and (B) that

$$I_0 \sim J_R \sim T^{(3+\beta/2)} \exp\left(-\frac{E_g}{kT}\right) \tag{C}$$

Ordinarily, the exponential term dominates and one may write

$$J_R(J_s) \sim \exp\left(-\frac{E_g}{kT}\right) \tag{D}$$

Assuming that Fig. 4-15 is for a bandgap of 1.43 eV, calculate the reverse voltage that can cause thermal runaway at 45 °C for a semiconductor with a bandgap of 1.28 eV. Note that, at this temperature, the value of V_R causing the runaway is approximately 70 volts according to Fig. 4-15.

Solution. According to Eq. (D), the ratio of J_R for the two different semiconductors is

$$\begin{aligned}\frac{J_R(1.28)}{J_R(1.43)} &= \frac{\exp(-1.28/kT)}{\exp(-1.43/kT)}\\ &= \exp(0.15/kT)\\ &= \exp(0.15/8.62 \times 10^{-5} \times 318)\\ &= 237\end{aligned}$$

Therefore, J_s for the semiconductor with a bandgap of 1.28 eV is 237 A/cm^2 at 45 °C. The intersection of $J_R = 237$ A/cm^2 with the constant dissipation line at 45 °C is the upper limit for the semiconductor of 1.28 eV. J_R of 237 A/cm^2 is out of scale in Fig. 4-15 but it can be seen that V_R is much less than 1 volt since, even at J_R of 100 A/cm^2, V_R is less than 1 volt (when the 45 °C line is drawn to J_R of 100 A/cm^2). It is seen that the bandgap has a very significant effect on the thermal breakdown.

The upper limit on V_R is 70 volts for a semiconductor with a bandgap of 1.43 eV, whereas it is less than 1 volt for one with a bandgap of 1.28 eV.

The mechanism of breakdown at low applied fields is due to the tunneling effect whereas at high applied fields it is due to avalanche multiplication. Also, the breakdown for lightly doped junctions is usually due to the avalanche breakdown.

Avalanche multiplication (or impact ionization) is the most important mechanism in junction breakdown, since the avalanche breakdown voltage imposes an upper limit on the reverse bias for most diodes, on the collector voltage of bipolar transistors, and on the drain voltage of MESFETs and MOSFETs (Sze, 1981). The breakdown is caused by the impact ionization of host atoms by energetic carriers. Normal lattice-scattering events can result in the creation of electron-hole pairs if the carrier being scattered has sufficient energy. If the electric field in the transition region of the junction is large, an electron entering from the p side may be accelerated to high enough kinetic energy to cause an ionizing collision with the lattice. The original and generated electrons can have further ionizing collisions and therefore each incoming carrier can initiate the creation of large number of carriers.

Avalanche breakdown owes its name to this snowballing process. If one assigns a probability P to a carrier of either type having an ionizing collision with the lattice, the carrier multiplication M, which is the number of ionizing collisions per incident carrier, is given (Streetman, 1980) by

$$M = \frac{1}{1 - P} \tag{4.32}$$

The probability P is often correlated with $(V/V_B)^n$ where V_B is the breakdown voltage and n is a constant ranging from 3 to 6 depending on the type of material used for the junction. The breakdown voltage, as determined by doping concentration (Sze and Gibbons, 1966), is shown in Fig. 4-16 for one-sided abrupt (p^+n or n^+p) junctions and linearly graded ones (for which the dopant concentration changes linearly with the slope of a). M_B in the figure is the background dopant concentration. The dashed lines indicate the maximum doping beyond which the tunneling mechanism dominates the voltage breakdown characteristics. Approximate relationships for the breakdown voltage (Sze, 1981) are

$$V_B(\mathrm{V}) = \begin{cases} 60(E_g/1.1)^{3/2}(N_B/10^{16})^{-3/4} & \text{(abrupt)} \\ 60(E_g/1.1)^{6/5}(a/3 \times 10^{20})^{-2/5} & \text{(graded)} \end{cases} \tag{4.33}$$

For silicon, the maximum field E_m at V_B is given by

$$E_m(\mathrm{V/cm}) = \frac{4 \times 10^5}{1 - (1/3)\log(N_B/10^{16})} \tag{4.34}$$

For p^+nn^+ or p^+pn^+ diodes, a mechanism called punch-through can occur before avalanche breakdown takes place, if the width W of lightly doped n or p is not large enough. Since the transition (depletion) region width W increases with

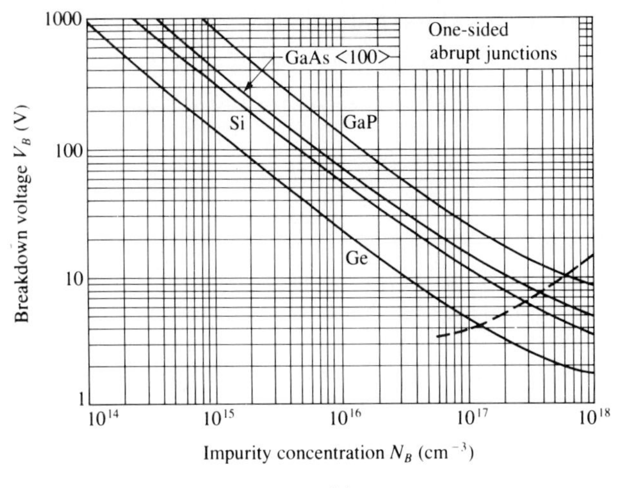

(*a*)

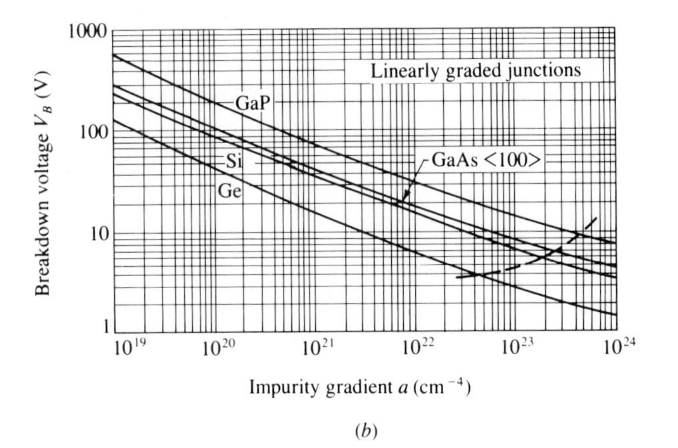

(*b*)

FIGURE 4-16

Avalanche breakdown voltage versus impurity concentration: (*a*) one-sided abrupt junction; (*b*) linearly graded junction. The dashed line indicates the maximum doping beyond which the tunneling mechanism dominates the breakdown (Sze and Gibbons, 1966).

reverse bias and extends primarily into the lightly doped region, it is possible for W to increase until it fills the entire length of this region. The result of this punch-through is a breakdown below the value expected from Eq. (4.33). This, however, can be avoided if the width W is made larger than the width at avalanche breakdown. The base region of a bipolar junction transistor can encounter a similar problem. In this case, the junction is in the form of p^+np^+. At high reverse bias, the transition region of p^+ can also fill the base region, which is also a punch-through. The same situation can also arise in MOS transistors in the form of a short channel effect.

For switching applications, the transition from forward bias to reverse bias must be nearly abrupt and the transient time short. The delay time in the output response is approximately proportional to carrier lifetime. Thus, a small lifetime is desired for a fast switch. The carrier lifetime can be substantially reduced by introducing impurities in the middle of the bandgap, which means deep-level impurities. An example is gold in silicon. These impurities act as recombination centers.

When electrons in the conduction band and holes in the valence band recombine, a band-to-band recombination is said to occur. Holes in the valence band can be excited by light, for instance, such that if the incident energy is higher than the bandgap energy, they jump to the conduction band and recombine with electrons there. Electrons in the conduction band may make the transition to the valence band and recombine with holes there. In the process, the electrons release energy on the order of the energy gap size in the form of light or heat. Although band-to-band recombination is important for optoelectronic materials such as gallium arsenide (involving direct transition), the detailed structure of the energy bands of silicon makes this process extremely unlikely there. Instead, the recombination process involves imperfections or impurities in the silicon, and the process takes place through these (impurity) intermediate recombination centers. Imperfections within the semiconductor disrupt the perfect periodicity of the crystal lattice and as a result can introduce new energy levels into the forbidden gap between the edges of valence and conduction bands in much the same way that donor and acceptor impurities do. These energy levels then act as intermediate states (stepping stones) for the transition of electrons and holes between the conduction and valence bands.

Shown in Figure 4-17 are recombination processes (Schockley and Reed, 1952; Sah *et al.*, 1957). Figure 4-17*a* represents band-to-band recombination. Single-level recombination is shown in Fig. 4-17*b* in which there is only one trapping energy level present in the bandgap; multilevel recombination involving more than one trapping energy level is shown in Fig. 4-17*c*. There are four distinct processes in the single-level recombination (Fig. 4-17*b*): electron capture, electron emission, hole capture and hole emission, all involving the trapping energy level. Each of these can be considered an elementary rate process. Thus, the rate of electron capture r_a should be proportional to the concentration of free electrons in the conduction band and also to the concentration of the trapping centers that are not occupied by electrons:

$$r_a = k_a n N_t(1 - f) \tag{4.35}$$

where n is the free conduction band electron concentration, N_t is the concentration of the trapping centers, f is the fraction of N_t occupied by electrons, and k_a is the rate constant. The probability f that a center is occupied by an electron follows the Fermi-Dirac (Chap. 1) distribution:

$$f = \frac{1}{1 + \exp\left[(E_t - E_F)/kT\right]} \tag{4.36}$$

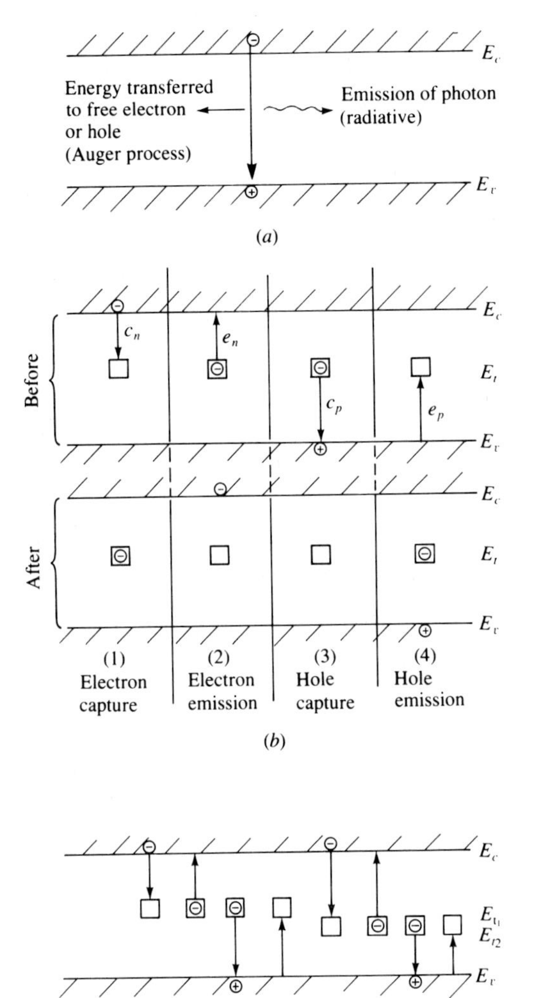

FIGURE 4-17
Recombination processes: (*a*) band-to-band recombination, (*b*) single-level recombination, and (c) multilevel recombination (Sah *et al.*, 1957).

where E_t is the trapping energy level. The rate constant is given by

$$k_a = v_{th}\rho_n \tag{4.37}$$

where v_{th} is the thermal velocity of the carrier given by ($3kT/m^* \sim 10^7$ cm/s at room temperature) and ρ_n is the capture cross section for an electron ($\sim 10^{-15}$ cm^2). The electron capture process is analogous to an adsorption process. The rate of electron emission r_b is simply proportional to the amount of trapped centers or occupied electrons, much the same way as in desorption:

$$r_b = k_{en}N_t f \tag{4.38}$$

where the rate constant k_{en} is given by

$$k_{en} = v_{th}\,\rho_n n_i \exp\left(\frac{E_t - E_i}{kT}\right) \tag{4.39}$$

The rate of hole capture r_c and that of hole emission r_d can be written in the same manner as for an electron:

$$r_c = v_{th}\,\rho_p p N_t f \tag{4.40}$$

$$r_d = k_{ep} N_t(1 - f) \qquad \text{where } k_{ep} = v_{th}\,\rho_p n_i \exp\left(\frac{E_i - E_t}{kT}\right) \tag{4.41}$$

where p is the hole concentration, ρ_p is the capture cross section for holes, and E_i is the intrinsic Fermi level. It should be noted that the capture of holes by a center corresponds to the transition of an electron from a center to the valence band and therefore the rate is proportional to the concentration of centers occupied by electrons, or $N_t f$.

Example 4.9. Derive the relationship given by Eq. (4.39). Utilize the fact that the rates of the two processes through which transition into and out of the conduction band takes place must be the same at equilibrium.

Solution. At equilibrium, $r_a = r_b$. Thus,

$$k_a n N_t(1 - f) = k_{en} N_t f$$

or

$$k_{en} = \frac{k_a n(1 - f)}{f} \tag{A}$$

From Chap. 1, one has

$$n = n_i \exp\left(\frac{E_K - E_i}{kT}\right) \tag{B}$$

Substituting Eqs. (B) and (4.36) into Eq. (A) yields Eq. (4.39) when Eq. (4.37) is used for k_a. The rate constant k_{ep} can be obtained in a similar manner.

At steady state but under non-equilibrium conditions, the net rate of recombination U is $(r_a - r_b)$, which is in turn equal to $(r_c - r_d)$. Thus,

$$U = r_a - r_b = r_c - r_d \tag{4.42}$$

With the expressions already available for all individual rates, the equality in Eq. (4.42) can be solved for the occupancy probability f, in much the same way as for the coverage in adsorption-desorption processes (except for the nonequilibrium nature of U), which yields

$$f = \frac{\rho_n n + \rho_n n_i e^{(E_i - E_t)/kT}}{\rho_n(n + n_i e^{(E_t - E_i)/kT}) + \rho_p(p + n_i e^{(E_i - E_t)/kT})} \tag{4.43}$$

Use of this expression in Eq. (4.42) yields the expression for the net rate of recombination U. The essential features of the net rate of recombination can best be examined for the case of $\rho_p = \rho_n = \rho$, in which case the expression for U becomes

$$U = \frac{\rho v_{th} N_t (pn - n_i^2)}{n + p + 2n_i \cosh\left[(E_t - E_i)/kT\right]} \tag{4.44}$$

As one might expect, at equilibrium $pn = n_i^2$, and the net rate of recombination is zero. Therefore, $(pn - n_i^2)$ can be viewed as the driving force for recombination. The recombination rate approaches a maximum as the energy level of the trapping (recombination) center approaches the midgap level, i.e., when E_t approaches E_i. Thus, the most effective recombination centers are those located near the middle of the bandgap. It can be shown under low injection cases (e.g., Sze, 1981) that the lifetime of minority carriers t_j is given by

$$t_j = (\rho_j v_{th} N_t)^{-1} \qquad \text{for } j = \begin{cases} p \text{ for } n\text{-type semiconductor} \\ n \text{ for } p\text{-type semiconductor} \end{cases} \tag{4.45}$$

It is immediately seen that the lifetime is inversely proportional to the trap concentration. The reduction in the lifetime brings about a decrease in the majority carrier concentration (carrier removal). Thus the resistivity of semiconductors increases with increasing trap concentration.

Traps with an energy level near the conduction or valence bands are called shallow-level traps. It should be pointed out that deep-level traps are often introduced intentionally, as with gold in silicon, to reduce the lifetime of the minority carrier such that the transient delay in switching is minimized, as discussed earlier in this section. Although impurities and/or crystal imperfections (defects) provide traps, those provided by structural defects are more damaging. They provide not only the traps but also conduits for diffusion along a dislocation, for example, of undesired materials or impurities. An example is shown in Fig. 4-18. As shown in the figure, diffusion of the emitter material along dislocations can lead to shorting of the circuit, if diffusion pipes or spikes are formed. These form due to more rapid diffusion along the dislocations. Therefore, the major concern in the processing of wafers or epitaxial films for use as substrates centers on the

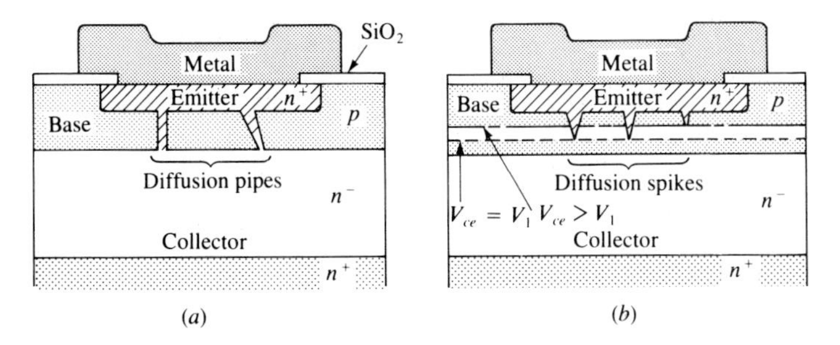

FIGURE 4-18
(*a*) Diffusion pipes and (*b*) diffusion spikes through the base along dislocations (Wang and Kakihana, 1974).

control of structural defects. The major source of structural defects is the stress generated during high-temperature processing, particularly during cooling of heated material. One major reason for the desire for low-temperature processing has to do with the reduction of the stress generated during processing and the corresponding decrease in the number of defects. Another source of structural defects is nucleation due to the presence of foreign material in the processing environment (e.g., the presence of oxygen). Still another source is the damage made by ion implantation.

High-temperature processing involves (thermal) diffusion and oxidation. For processing, wafers are usually stacked on a boat with narrow spacings between the individual wafers in a row. Such an arrangement gives rise to a geometric factor for radiation heat transfer, which is the dominant heat transfer process, such that the central region of the wafer heats up or cools down much more slowly than the outer region. Thus a radial temperature gradient develops, which during cooling produces a compressive stress in the central region and a dominant tensile tangential stress in the outer region. Slip due to the stresses occurs more severely in the outer region since the tangential stress is always much larger than the compressive stress at the center. When the temperature gradient is very large, slip also occurs in the central region. The pattern of thermal slip is shown in Fig. 4-19. The case of a larger temperature gradient, and thus higher stress, is shown in Fig. 4-19*a*; the case of moderate stress is shown in Fig. 4-19*b*. The slip occurs when the stress exceeds the critical stress required for dislocation movement. Another mode of slip is observed when an epitaxial film is grown on a wafer. The wafer is placed on a susceptor heated by an inductive heating method such as radio-frequency heating. In general, a hot spot develops in the middle of the susceptor (and thus wafer) due to the cold gas flow and conduction away from the center. The temperature nonuniformity becomes worse when the wafer(s) has an initial bow and is placed with the concave side up, because of poor contact of the wafer face with the susceptor. The result is a thermal slip predominant in the central region of the wafer. Use of a hollow dimple in the susceptor, as discussed in Chap. 6, and placing the wafer with the convex side up help to minimize the problem.

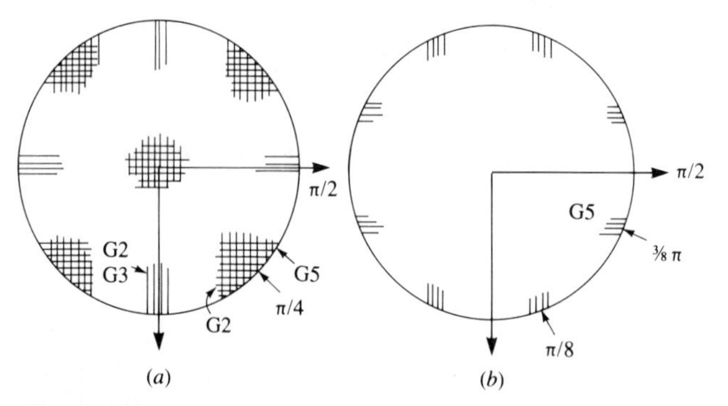

FIGURE 4-19
Patterns of slips due to stress (Hu, 1977).

One method of reducing the thermal stresses is to use temperature ramping (Hu, 1986). The temperature ramping can be accomplished by ramping the furnace temperature. This can also be accomplished by controlled insertion and withdrawal of wafers across graded temperature zones in the furnace. A drawback of the method is the complication of impurity redistribution and clustering during the temperature ramping (Lorettini and Nobili, 1984).

The severity of slip is dependent to a large extent on the state of material being processed such as surface damage, surface flatness, and the presence of dissolved or precipitated oxygen in the substrate. The effect of oxygen on thermal slip is very complex. While dissolved oxygen and small oxygen clusters can strengthen the silicon lattice and help resist thermal slip (Chiou *et al.*, 1984), larger "oxygen precipitates" can act as dislocation nucleation sites and aggravate thermal slip and wafer warpage (Shimizu *et al.*, 1985).

Stacking faults are another major source of defects. They can be caused by impurities at the substrate surface or by mismatch of the epitaxial film at the substrate surface in the initial phase of epitaxy. Stacking faults can also be induced by oxidation. They are formed by the condensation of excess silicon self-interstitials at the silicon–silicon oxide interface that are generated by thermal oxidation (Hu, 1974). The oxidation-induced stacking faults can be eliminated by removing nucleation centers such as impurities, mechanical damage, and the damage incurred by ion implantation and reactive ion etching.

Although low-temperature processing is desirable for reducing stress-induced defects, it is detrimental to oxide isolation involving complex geometries (Tamaki *et al.*, 1981). When one volume of silicon is oxidized to 2.25 volumes of silicon oxide in a laterally confined space, enormous stresses are generated. These stresses can be relaxed mainly through viscous flow of the oxide at sufficiently high temperatures. Since the viscosity decreases with increasing temperature, higher temperatures are required for the silicon dioxide to flow and wet the surrounding structure, thereby relieving the stress.

Although defects originate from processing of a material, the material itself determines the susceptibility to defect generation, and depends on the material's structure and the impurity content. The structure as determined by crystal orientation can have a significant effect on defect generation. For instance, the (111) surface of silicon is much more susceptible to pyramidal defects than the (100) surface, but the (111) surface misoriented by just a few degrees produces no discernible pyramidal defects (Mendelson, 1964). The presence of impurities can either prevent or enhance defect generation. The different effects of oxygen in silicon on defects discussed earlier is an example. Gettering with impurities is another example.

NOTATION

a	Thickness of heavily doped layer (L) in Fig. 4-6
a_1, b	Constants in Eq. (4.31)
A	Area (L^2)

A_1	Constant in Eq. (4.29)
A^*	Effective Richardson constant ($A/L^2/T^2$)
b_1, b_2	Constants in Eq. (4.30)
C	Capacitance (F)
C_1, C_2, C_3	Constants in Eq. (4.6)
d	Thickness
D_b	Boundary diffusivity (L^2/t)
D_p	Hole diffusion coefficient (L^2/t)
D_v	Diffusivity of vacancies (L^2/t)
E	Electric field (V/L)
E_B	Breakdown field (V/L)
E_F	Fermi level (E)
E_g	Bandgap energy (E)
E_i	Electric field in insulator (V/L)
E_m	Maximum field (V/L)
E_o	Quantity defined in Eq. (4.9)
E_s	Electric field in semiconductor (V/L)
E_t	Trap energy level (E)
f	Probability defined by Eq. (4.36)
h	Planck constant
ΔH	Heat of formation (E/mol)
j	Current density in Eq. (4.23) (A/L^2)
J	Current density (A/L^2)
J_i	Ion flux (ion/t/L^2)
J_R	Reverse current density
J_s	Saturation current density
J_t	Tunneling current density
k	Boltzmann constant
k_a	Rate constant given by Eq. (4.37)
k_{en}	Rate constant given by Eq. (4.39)
k_{ep}	Rate constant defined in Eq. (4.41)
K	Constant in Eq. (4.24)
m_e	Effective electron mass (M)
M	Carrier multiplication factor in Eq. (4.30)
MTF	Mean time to failure (t)
n	Electron density (L^{-3}); constant defined by Eq. (4.11)
n_1, n_2	Electron densities in Fig. 4-6
n_i	Intrinsic electron density (L^{-3})
N	Metal atom density (L^{-3})
N_B	Background (lightly doped) dopant concentration (L^{-3})
N_D	Donor concentration (atoms/L^3 or L^{-3})
N_t	Trap density (L^{-3})
p	Hole concentration (L^{-3})
p_{ne}	Hole density in n-type material at equilibrium (L^{-3})
P	Probability of a carrier having an ionizing collision with lattice

q	Elementary charge (C)
r_a	Rate of electron capture
r_b	Rate of electron emission
r_c	Rate of hole capture
r_d	Rate of hole emission
R	Resistance (Ω)
R_c	Specific contact resistance (ΩL^2)
S	$E_B - E$
t	Time
t_j	Minority carrier lifetime (t)
T	Temperature (T)
TF	Time to failure (t)
U	Net rate of recombination ($t^{-1}L^{-3}$)
v_{th}	Thermal velocity (L/t)
V	Applied potential
V_c	Film volume (L^3)
W	Junction width (L)
x	Axial distance (L)
X	Electron affinity potential (V)
Z	Effective charge

Greek letters

α	Temperature coefficient of resistance (T^{-1}); relative permittivity
α_0	Permittivity of free space (F/L)
α_i	Permittivity of insulator
α_s	Semiconductor permittivity (F/L)
α_1, α_2	Constants in Eq. (4.28)
β	Ratio of film thickness to electron mean free path; temperature-dependent constant in Example 4.7
λ	Film conductivity ($\Omega^{-1}L^{-1}$)
λ_0	Bulk conductivity
ρ	Resistivity (ΩL); capture cross section (L^2)
ρ_n, ρ_p	Capture cross section of electrons and holes, respectively (L^2)
ω	Angular frequency (t^{-1})
Ω	Work function potential (V); angle in Fig. 4-13
Ω_m	Metal work function potential (V)
Ω_o	Work function potential at surface (V)
Ω_s	Semiconductor work function potential (V)
Ω_{Bn}	Barrier height for n-type semiconductor (V)
Ω_{Bp}	Barrier height for p-type semiconductor (V)
$\Delta\Omega$	Barrier height reduction (V)

Units

A	Ampere
C	Charge

E	Energy
F	Farad
L	Length
M	Mass
t	Time
T	Temperature
V	Volt
Ω	Ohm

PROBLEMS

4.1. For ohmic contacts, one can choose a metal with a low energy barrier. One can also form a very thin, heavily doped layer on top of substrate before metallization. A combination of both is typically used for ohmic contacts. Calculate the maximum reduction in the barrier height that can be attained for *n*-type silicon. Assume that the thickness of the heavily doped layer is 10 nm and that the semiconductor permittivity is ten times the permittivity for free space.

4.2. The work function of intrinsic silicon is 4.85 eV. Using Fig. 4-3, give one metal each for *n*-type and *p*-type silicon for an ohmic contact. Note that the bandgap energy for silicon is 1.12 eV.

4.3. For ohmic contacts, the barrier height should be as small as possible. Based on the result of Prob. 4.2, find a most suitable metal for *n*- and *p*-type silicon.

4.4. The data obtained by Chang and Sze (1970) for Au-Si barriers are given below:

$J(\text{A/cm}^2)$	10^{-5}	10^{-3}
V(volts)	0.092	0.220

The data are in the region where a plot of ln J versus V yields a straight line. Write an expression for the J-V relationship based on the data including all numerical values of the constants involved. The doping level (*n* type) is 10^{16} atoms/cm^3 and temperature is 296 K. Also, calculate the effective Richardson constant. Compare your result with that given in Fig. 4-5. Use 0.8 V as the barrier height.

4.5. Consider the metal interconnect in Example 4.5. Because of the push for smaller device dimensions, the interconnect width is to be halved with the corresponding decrease in the height by the same factor. Find a constraint on the current density when the MTF is to remain the same.

4.6. For SiO_2, the leakage current at 298 K is 4×10^{-11} A/cm^2 at an electric field strength of 6×10^6 V/cm (Av-Ron *et al.*, 1978). According to Example 4.6, the breakdown field E_B is given by

$$E_B(\text{V/cm}) = \frac{5.99 \times 10^7}{[d(\text{nm})]^{0.473}} \tag{A}$$

Suppose that the leakage current is for silicon dioxide 20 nm thick. Estimate the percent change in the time to failure due to dielectric breakdown using the results of Example 4.7 when the thickness changes to 10 nm.

4.7. An accurate expression for the avalanche breakdown voltage (Sze, 1981) is given by

$$V_B = \frac{\alpha_s E_m^2}{2qN_B} \tag{A}$$

For silicon, an expression for the maximum field E_m for an abrupt junction is given by

$$E_m = \frac{4 \times 10^5}{1 - 0.333 \log_{10} (N_B/10^{16})} \quad \text{V/cm} \tag{B}$$

where the lightly doped background doping concentration N_B is in cm^{-3}. According to Fig. 4-16, the transition from avalanche breakdown to tunneling begins at an N_B of approximately 3×10^{17} cm^{-3} for silicon, and the corresponding breakdown voltage is approximately 6.3 volts. Check these values with Eqs. (A) and (B). Take the permittivity of silicon as 11.7 times the free-space permittivity.

4.8. As discussed in the text, the punch-through occurs when the metallurgical width for the lightly doped region of p^+nn^+ junctions, W_{mt}, is less than the depletion layer width at breakdown, W_m. W_m for abrupt junctions is given by

$$W_m = \frac{\alpha_s E_m}{qN_B} \tag{A}$$

The punch-through voltage V_{pt} for a given W_{mt} (Sze, 1981) is given by

$$\frac{V_{pt}}{V_B} = \left(\frac{W_{mt}}{W_m}\right)\left(2 - \frac{W_{mt}}{W_m}\right) \tag{B}$$

Calculate the punch-through voltage using the calculated value of V_B in Prob. 4.7 when W_{mt} is half of W_B. Determine the metallurgical width W_{mt} that is required to avoid punch-through before avalanche breakdown occurs.

4.9. Show that for n-type semiconductors, the probability f that a recombination center is occupied by an electron is almost unity while it is almost zero for p-type semiconductors. Assume that the trap energy level is not too close to the conduction or the valence band. Show also that Eq. (4.45) follows from Eq. (4.44).

4.10. When an interconnect consists of one metal, Eq. (4.23) gives the flux for electromigration. One way of preventing electromigration is to add another metal to the interconnect such that the flux due to electron current (electromigration) is counteracted by the diffusive flux arising from the concentration difference. An example is the addition of the Cu to Al interconnect, thereby increasing the lifetime by more than an order of magnitude (Blech, 1977). A pseudo-state balance for Cu in the interconnect is

$$J_i - D\frac{\partial N}{\partial X} = V_{\text{Cu}} N - D\frac{\partial N}{\partial X} = 0$$

$$V_{\text{Cu}} \equiv \frac{D_v}{kT} Z^* q\rho j$$

where N is the concentration of Cu in the interconnect. Explain how one can determine the effective charge Z^* of Cu from experimental data of N versus X. The value of Z^* determined by Blech at 450 °C is around 7 (dimensionless), although the value appears to decrease with increasing current flux j. Assume that the Cu atomic diffusivity D is the same as the vacancy diffusivity D_v.

4.11. The net rate of surface recombination U_s is exactly the same as that given by Eq. (4.44), provided that n and p are replaced by the surface concentrations n_s and p_s (Grove, 1967):

$$U_s = v_{th} N_{st} \frac{p_s n_s - n_i^2}{n_s + p_s + 2n_i \cosh[(E_t - E_i)/kT]} \tag{A}$$

where N_{st} is the surface trap center concentration. The flux of minority carriers reaching the surface must equal U_s. If the recombination rate within the surface space-charge region is not too high, this flux can be approximated by the flux of minority carriers reaching the edge of the surface space-charge region:

$$D_p \left. \frac{\partial p_n}{\partial x} \right|_{x=x_d} = s_0 \frac{p_s n_s - n_i^2}{n_s + p_s + 2n_i \cosh[(E_t - E_c)/kT]} \tag{B}$$

where

$$s_0 = v_{th} N_{st}$$

x_d is the position at the plane, and s_0 is the recombination velocity in the absence of a space-charge region. Even when equilibrium does not prevail, the pn product is still a constant, although it is different from n_i^2. Then

$$p_s n_s = p_n(x_d) n_n(x_d) \doteq p_n(x_d) N_D \tag{C}$$

Noting that $p_0 N_D \doteq n_i^2$, Eq. (B) can be rewritten as

$$D_p \left. \frac{\partial p_n}{\partial x} \right|_{x_d} = s[p_n(x_d) - p_0] \tag{D}$$

where

$$s = \frac{s_0 N_D}{n_s + p_s + 2n_i \cosh[(E_t - E_i)/kT]} \tag{E}$$

Equation (D) can be used as a boundary condition for describing the minority carrier distribution within the bulk semiconductor. Show that the maximum surface recombination velocity s_{max} is given by

$$s_{max} = \frac{s_0 N_D}{4n_i} \tag{F}$$

Calculate the bulk lifetime of minority carriers and the maximum surface recombination velocity for the following conditions:

$$v_{th} = 10^7 \text{ cm/s}$$

$$\rho = 10^{-15} \text{ cm}^2$$

$$n_i \text{ for Si} = 10^{10} \text{ cm}^{-3}$$

For the silicon wafer, the donor concentration is 10^{17} cm^{-3}, $N_{st} = 10^{10}$ cm^{-2}, and $N_t = 10^{10}$ cm^{-3}. Assume that the trap energy level is in the middle of the bandgap.

REFERENCES

Andrews, J. M.: "Extended abstracts," Electrochemical Society Spring Meeting, Abstract 191, p. 452, 1975.

——— and M. P. Lepselter: *Solid State Electron.*, vol. 13, p. 1011, 1970.

Av-Ron, M., M. Shatzkes, T. H. DiStefano, and I. B. Cadoff: in S. T. Pantelider (ed.), *The Physics of SiO_2 and Its Interfaces*, Pergamon, New York, 1978.

Baglin, J. E. E., and J. M. Poate: in J. M. Poate *et al.* (eds.), *Thin Films—Interdiffusion and Reactions*, chap. 9, Wiley, New York, 1978.

Blech, I. A., and E. S. Meieran: *Appl. Phys. Lett.*, vol. 11, p. 263, 1967.

———: *J. Appl. Phys.*, vol. 48, p. 473, 1977.

Chang, C. Y., Y. K. Fang, and S. M. Sze: *Solid State Electron.*, vol. 9, p. 695, 1966.

——— and S. M. Sze: *Solid State Electron.*, vol. 13, p. 727, 1970.

Chiou, H. D., J. Moody, R. Sandfort, and F. Shimura: in *VLSI Science Technology 1984*, p. 59, Electrochemical Society, Pennington, N.J., 1984.

Deal, B. E.: *IEEE Trans. Elect. Dev.*, vol. ED-27, p. 606, 1980.

D'Heurle, F. M., and P. S. Ho: in J. M. Poate *et al.* (eds.), *Thin Films—Interdiffusion and Reactions*, chap. 8, Wiley, New York, 1978.

Forlani, R., and N. Minnaja: *Phys. Stat. Solidi*, vol. 4, p. 311, 1964.

Fraser, D. B.: in S. M. Sze (ed.), *VLSI Technology*, chap. 9, McGraw-Hill, New York, 1983.

Grangulee, A., P. S. Ho, and K. N. Tu (eds.), *Low Temperature Diffusion and Application to Thin Films*, Elsevier, New York, 1975.

Groothius, S. K., and W. H. Schroen: *25th Annual Proceedings on Reliability Physics*, San Diego, Calif., p. 60, IEEE, 1987.

Grove, A. S.: *Physics and Technology of Semiconductor Devices*, Wiley, New York, 1967.

Gupta, D., and R. Rosenburg: *Thin Solid Films*, vol. 25, p. 171, 1975.

Hansen, M.: in K. Anderko (ed.), *Constitution of Binary Alloys*, 2d ed., McGraw-Hill, New York, 1958.

Harari, E.: *J. Appl. Phys.*, vol. 49, p. 2478, 1978.

Hu, S. M., *J. Appl. Phys.*, vol. 45, p. 1567, 1974.

———: *J. Vac. Sci. Tech.*, vol. 14, p. 17, 1977.

———: in *Semiconductor Silicon 1986*, p. 722, Electrochemical Society, Pennington, N.J., 1986.

Huntington, H. B., and A. R. Grone: *J. Phys. Chem. Solids*, vol. 20, p. 76, 1961.

Kapoor, V. J., and S. B. Bibyk: in J. Luchoveky (ed.), *The Physics of MOS Insulators*, p. 117, 1980.

Katz, L. E.: in S. M. Sze (ed.), *VLSI Technology*, chap. 4, McGraw-Hill, New York, 1983.

Klein, N., and N. Levanson: *J. Electrochem. Soc.*, vol. 116, p. 963, 1969.

Lorettini, L., and D. Nobili: *Mat. Chem. Phys.*, vol. 10, p. 21, 1984.

Maissel, L. I.: in L. I. Maissel and R. Glang (eds.), *Handbook of Thin Film Technology*, chap. 13, McGraw-Hill, New York, 1970.

McPherson, J. W., and D. A. Baglee: *23rd Annual Proceedings on Reliability Physics*, Orlando, Fla., p. 1, IEEE, 1985.

Mendelson, S.: *J. Appl. Phys.*, vol. 35, p. 1570, 1964.

Michaelson, H. B.: *IBM J. Res. Dev.*, vol. 22, p. 72, 1978.

Mitsuhashi, J., H. Mutoh, Y. Ohno, and T. Matukawa: *25th Annual Proceedings on Reliability Physics*, San Diego, Calif., p. 60, IEEE, 1987.

Moll, J. L.: *Physics of Semiconductors*, McGraw-Hill, New York, 1964.

Mukherjee, S. D.: in M. J. Howes and D. V. Morgan (eds.), *Reliability and Degradation*, chap. 1, Wiley, New York, 1981.

Neugebauer, C. A., and R. H. Wilson: *Basic Problems in Thin Film Physics*, p. 579, Vanderhoeck and Ruprecht, Goettengen, 1966.

Nicolet, M. A.: *Thin Solid Films*, vol. 52, p. 415, 1978.
Nicollian, E. H., and J. R. Brew: *MOS Physics and Technology*, Wiley, New York, 1982.
Offret, M., and M. D. Vodar: *J. Phys. Radium.*, vol. 17, p. 237, 1955.
Offsey, S. D., J. M. Woodall, A. C. Warren, P. D. Kirchnee, T. I. Chappell, and G. D. Pettit: *Appl. Phys. Lett.*, vol. 48, p. 475, 1986.
Padovani, F. A., and R. Stratton: *Solid State Electron.*, vol. 9, p. 695, 1966.
Reed-Hill, R. E.: *Physical Metallurgy Principles*, Van Nostrand, Princeton, N.J., 1966.
Rusu, A., and C. Bulucea: *IEEE Trans. Elect. Dev.*, vol. ED-26, p. 201, 1979.
Sah, C. T., R. N. Noyce, and W. Schockley: *Proc. IRE*, vol. 45, p. 1228, 1957.
Schockley, W., and W. T. Reed: *Phys. Rev.*, vol. 87, p. 835, 1952.
Shimizu, H., T. Watanabe, and Y. Kabui: *Jap. J. Appl. Phys.*, vol. 24, p. 815, 1985.
Spicer, W. E., I. Lindau, P. Skeath, C. Y. Su, and P. Chye: *Phys. Rev. Lett.*, vol. 24, p. 256, 1981.
Streetman, B. G.: *Solid State Electronic Devices*, 2d ed., Prentice-Hall, Englewood Cliffs, N.J., 1980.
Strutt, M. J. O.: *Semiconductor Devices*, 2d ed., Wiley, New York, 1966.
Sze, S. M.: *Physics of Semiconductor Devices*, 2d ed., Wiley, New York, 1981.
——— and G. Gibbons, *Appl. Phys. Lett.*, vol. 8, p. 111, 1966.
Tamaki, Y., S. Isomne, S. Mizuo, and H. Higuchi: *J. Electrochem. Soc.*, vol. 128, p. 644, 1981.
Towner, J. M.: *23rd Annual Proceedings on Reliability Physics*, Orlando, Fla., p. 81, IEEE, 1985.
Vaidya, S., D. B. Frazer, and A. K. Sinha: *18th Proceedings on Reliability Symposium*, New York, p. 165, IEEE, 1980.
Wang, A. C. M., and S. Kakihana: *IEEE Trans. Elect. Dev.*, vol. ED-21, p. 667, 1974.
Wilkinson, P. G., and L. S. Birks: *J. Appl. Phys.*, vol. 20, p. 1168, 1949.
Wilson, A. H.: *The Theory of Metals*, Cambridge University Press, New York, 1958.

CHAPTER 5

CHEMICAL RATE PROCESSES AND KINETICS

5.1 INTRODUCTION

The processing of wafers for fabrication of integrated circuits (ICs) involves various chemical and physical rate processes. Chemical rate processes are considered in this chapter. Chemical vapor deposition (CVD), which is one of the major means by which ICs are fabricated, is an example of a chemical rate process. Rate processes of a physical nature, such as deposition by condensation as well as physicochemical processes such as plasma processes, are considered in Chap. 9.

Many chemical rate processes are involved in the unit processing steps of Fig. 5-1. As indicated in the figure, each time a new mask is used for further processing, as prescribed by the composite circuit layout, certain unit processing steps are carried out according to the pattern being imprinted onto the wafer surface via lithography. This lithography-unit processing sequence is repeated until the intended IC is fabricated. Take as an example the manufacturing (fabrication) process for an NMOS silicon gate IC shown in Fig. 1-19. It is seen that four unit processing steps are required to complete the pattern specified by the first mask. These are ion implantation, two oxidation steps for field and gate oxides, and polysilicon deposition. For the second mask, the lithography-unit processing sequence involves diffusion of *n*-type dopant into the source-drain regions and deposition of silicon oxide. The third mask is used to deposit aluminum for contacts or metallization by condensation of evaporated metal. It is seen that the fourth mask defines metal contacts such that all the chip surface except

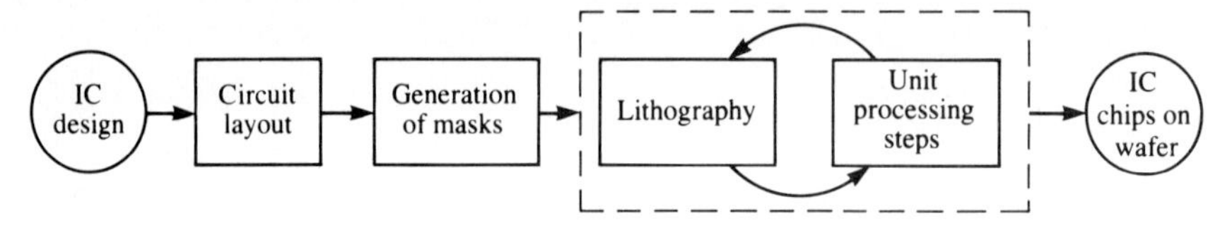

FIGURE 5-1
Device fabrication sequence.

for the contacts can be covered with a protective layer, i.e., protective passivation. Although not shown explicitly in Fig. 1-19, etching is almost always carried out at each mask level to define and delineate a specific pattern such as opening windows (refer to Fig. 1-18). For device structures other than MOSFET, epitaxial film growth is also involved, as in the BJT structure with a buried layer (refer to Fig. 1-12*b*).

Although some of the chemical rate processes involved in IC fabrication have been brought out in the example just considered, there are still more chemical rate processes: in fact, there are too many to list. The processing techniques based on chemical reactions are constantly evolving and changing, and yet there are still more to explore. These diverse reactions belong to certain classes according to the nature of the kinetics. Their importance, however, is derived from their ultimate purpose in the fabrication sequence. The kinetics of the reactions are considered based on their natures, either homogeneous, heterogeneous, gas-solid, or photochemical.

Many of the IC fabrication steps involve deposition of solid material onto a solid substrate from the gas phase through chemical reactions. This chemical vapor deposition may involve simple gas phase reactions that lead to the formation of a particular solid, which deposits simply onto the substrate, or gas phase reactions that only lead to another gas species, which upon adsorbing onto the substrate surface goes through a further chemical transformation leading to the formation of the deposited solid. The former is called homogeneous deposition and the latter is called heterogeneous deposition. For many cases, it is not always clear whether the deposition is homogeneous or heterogeneous, although most of the depositions in this chapter are heterogeneous.

It is important in any study of deposition to make a distinction between the observed rate and the rate that is intrinsic to the process. The observed rate is often the rate obscured by mass and heat transport effects. Unless one knows how the intrinsic rate is affected by transport effects, it is difficult to identify the exact nature of the observed rate. In this chapter, the intrinsic rate and kinetics are examined; transport effects are covered in the next chapter. The intrinsic kinetics of gas-solid reactions, such as the oxidation of a native silicon substrate, are often very simple and rapid. Diffusion effects tend to dominate their observed rates, because of the very low gas diffusivity in the solid. For this reason, the discussion of gas-solid reactions in this chapter includes transport effects.

Chemical processing often involves high-temperature reactions. The electrical properties of ICs are largely determined by the way the dopants are distributed. In the course of high-temperature processing, the dopants begin to redistribute themselves, and this can defeat the purpose of specifying the dopant profile for desired device performance. Another reason for the desire for low-temperature processing is the stress-induced defects that are generated upon cooling heated material. Therefore, it is always desirable to carry out a particular processing step at the lowest allowable temperature. One way of accomplishing this is to use photochemical reactions. Another principle of IC processing is to minimize the number of mask levels, which results in fewer processing steps and lower levels of undesired impurities that are introduced between transfers for the masking. One way of accomplishing this is to use selective reactions. These subjects are also treated in this chapter.

The chemical reactions that are perhaps unique to IC processing are those involved in depositing films with crystalline structure. The ability to deposit films of monocrystalline structure on a foreign substrate, for instance, can bring about a step change in the device performance and at the same time eliminate many processing steps. The subject of the growth processes used to produce films of a crystalline structure is treated first because of its importance in further advancing IC technology from a materials science perspective and because the general concepts central to most other types of deposition processes are involved.

5.2 GROWTH PROCESSES OF FILMS OF CRYSTALLINE STRUCTURE

Epitaxy is a term applied to the processes of growing a monocrystalline film on a substrate. In epitaxial processes the substrate serves as a seed crystal. While this aspect is similar to Czochralski growth, for example, it differs from the recrystallization process in that the crystal can be grown below the melting point. Typically, epitaxy is carried out by CVD techniques although molecular beam epitaxy (MBE) based on evaporation can also be used. Epitaxy is termed homoepitaxy when a crystal is grown epitaxially on a substrate of the same material, as in silicon film grown on silicon substrate. It is termed heteroepitaxy when a crystal is grown on a foreign substrate, as in a silicon film grown on sapphire or gallium arsenide grown on a silicon substrate.

There are certain advantages in using homoepitaxial film, in particular in silicon ICs (Pearce, 1983). The epitaxial films grown on silicon wafer offer the device designer a means of controlling the doping profile in a device structure beyond that available through diffusion or ion implantation of dopants. The buried layer in a BJT structure is a good example of homoepitaxy. The epitaxial films are also free from carbon and oxygen impurities compared to the melt-grown substrate. Although homoepitaxy offers some advantages, heteroepitaxy is the one that holds the greatest promise. One apparent advantage lies in the potential reduction of wafer cost, particularly for gallium arsenide. Note that only a very small fraction, say several micrometers, of the whole wafer is used for the actual IC fabrication. Therefore, a film of gallium arsenide, which is much more expensive than silicon, grown on a low-cost substrate, e.g. silicon, can result

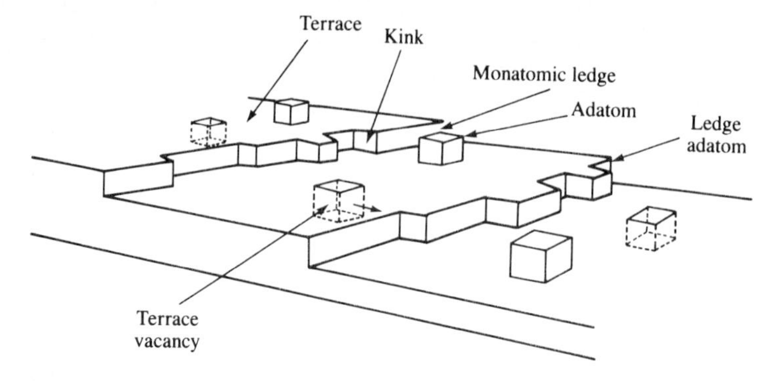

FIGURE 5-2
Surface of a growing film.

in a significant reduction in overall wafer cost. More importantly, heteroepitaxy can lead to new device fabrication technology and applications by offering increased material flexibility in device design.

It should be noted that film growth by CVD techniques does not always lead to the epitaxial (monocrystalline) film. It can also lead to polycrystalline or amorphous films. Consider the epitaxial crystal growth process. Epitaxial growth is an "ordered" process of adding atoms into a regular monocrystalline structure. In order for this ordered process to take place, certain gas phase species have to adsorb onto the substrate surface. On the atomic scale, the surface is not flat but rather has kinks, edges, and steps, as shown in Fig. 5-2. These adsorbed species, called adspecies or adatoms if they are atoms, then migrate on the surface until they find their new equilibrium sites. The thermal vibrations of the adatom at an adsorbed site causes its movement (in jumps) on the surface. This movement is termed surface diffusion. As surface diffusion by random jumps of adatoms takes place, some of them are close enough (one jump away) to the energetically favorable step sites, which are at a lower energy level, and the next jump into one of the step sites results in the incorporation of the adatom into the crystal structure, leading to the propagation of the steps and thus crystal growth. Stable clusters of adatoms can also form on the top layer, providing sink sites for additional adatoms. The growth by lateral propagation of steps of atomic-sized steps has been observed by electron microscopy on a properly cleaned silicon surface (Abbink *et al.*, 1968; Ogden *et al.*, 1974).

Example 5.1. Consider an adatom held on the surface by a small binding energy E_d. If the vibration frequency of the adatom is v_0, the probable number of times this atom jumps out of the energy well, or the jump frequency f, is given by

$$f = qv_0 \exp\left(-\frac{E_d}{kT}\right) \tag{A}$$

where k is Boltzmann's constant and q is the number of equivalent neighboring sites. For a one-dimensional random walk of an atom, the root mean square distance can be used as the net displacement. Then the expected value of X^2, $\langle X^2 \rangle$, is given by

$$\langle X^2 \rangle = n_j a^2 \tag{B}$$

where X is the net displacement, n_j is the number of jumps, and a is the distance of jump, which is the distance between two nearest neighboring sites. If t is the time to make n_j jumps, then n_j is equal to ft. Thus, $\langle X^2 \rangle = fta^2$. It is customary to define the diffusion coefficient as

$$D_s = \frac{\langle X \rangle^2}{2bt} = \frac{fa^2}{2b} \tag{C}$$

where D_s is the surface diffusivity and b is the number of coordinate directions in which the jumps may occur with equal probability. For diffusion equally probable in two directions ($b = 2$), use of Eq. (A) in Eq. (C) yields

$$D_s = \frac{qa^2 v_0}{4} \exp\left(-\frac{E_d}{kT}\right) \tag{D}$$

(*a*) Based on a replication technique for electron microscopy that can detect 0.3 nm steps on surfaces, Abbink *et al.* (1968) obtained an estimate of the surface diffusivity of silicon on a silicon surface, which is 10^{-3} cm^2/s at 800 °C. Assuming that $q = 6$, $a = 0.3$ nm, and using the usual v_0 value of 10^{13} s^{-1}, calculate the binding energy E_d.

(*b*) Silicon epitaxy is usually carried out at about 1000 °C and the corresponding film growth rate is about 1 μm/min. Calculate the distance an adatom would have moved while the crystal has grown by 1 μm. Assume that the adatom does not desorb.

(*c*) Part (*b*) is for the case where the adatom stays adsorbed for 1 minute. In reality, however, an adatom may also desorb. The mean residence time of an adatom t_r can be estimated from

$$t_r = \frac{1}{v_0} \exp\left(\frac{E_{\text{des}}}{kT}\right) \tag{E}$$

where E_{des} is the energy barrier for desorption. What is the probable distance an adatom can travel before it desorbs? Take the desorption energy as that for the Si—Si bond, which is 50 kcal/mol.

Solution

(*a*) For $q = 4$, $a = 3 \times 10^{-8}$ cm, and $v_0 = 10^{12}$ s^{-1}, Eq. (D) can be written as

$$D_s = \frac{6 \times (3 \times 10^{-8})^2 \times 10^{13}}{4} \exp\left(-\frac{E_d}{2 \times 1073}\right)$$

$$= 10^{-3} \text{ cm}^2/\text{s}$$

where $R = kN_A = 2$ cal/(mol·K)

Solving this for E_d yields

$$E_d = 5.6 \text{ kcal/mol}$$

(*b*) At 1000 °C,

$$D_s = 0.0135 \exp\left(-\frac{5600}{2 \times 1273}\right) = 1.5 \times 10^{-3} \text{ cm}^2/\text{s}$$

From Eq. (C), the net displacement is given by

$$\langle X^2 \rangle^{1/2} = (2bD_s t)^{1/2} = 2(1.5 \times 10^{-3} \times 60)^{1/2} = 0.6 \text{ cm}$$

(*c*) For the data given,

$$t_r = \frac{1}{10^{13}} \exp\left(\frac{50{,}000}{2 \times 1273}\right)$$

$$= 3.14 \times 10^{-5} \text{ s}$$

The probable distance is then given by

$$\langle X^2 \rangle^{1/2} = 2(D_s t)^{1/2} = 2 \times (0.0015 \times 3.4 \times 10^{-5})^{1/2} = 4.5\ \mu\text{m}$$

Therefore, the probability is that all adatoms can be incorporated into the crystal structure if the spacing between steps is smaller than 9 μm. This is two times the average distance of migration according to the problem statement.

The role of step propagation in epitaxial growth indicates that the epitaxial growth rate will be faster if the substrate surface is prepared to expose steps on the surface. This can be accomplished by slightly misorienting the substrate crystal orientation, which would result in the surface shown in Fig. 5-3. Ever since the demonstration by Tung (1965) of the effect of crystal misorientation, all silicon epitaxial films have been grown on slightly misoriented (3 to 7° off a major crystal axis) substrates. The same can be stated for heteroepitaxy, as discussed later. The essential questions regarding any epitaxy are what adspecies are present and whether or not the adspecies go through a chemical transformation at the time of incorporation into the crystal lattice or prior to the incorporation.

Although there are no definitive details available for any epitaxy, the epitaxial growth process does consist of three events in series. These are adsorption of gas phase species, followed by surface diffusion of the adspecies, which is then followed by the incorporation of the adspecies into the crystal structure. This epitaxial process can be represented as follows:

$$B(g) \longrightarrow A(g) + B_1$$

$$A(g) + S \rightleftharpoons A \cdot S \tag{5.1}$$

$$A \cdot S \longrightarrow A_1(c) + B_2(g) \tag{5.2}$$

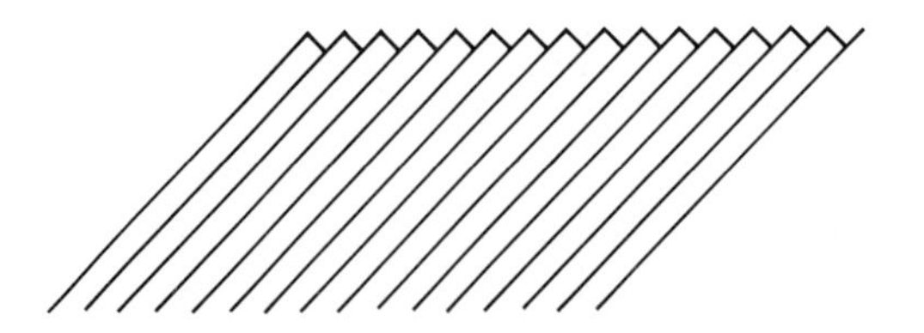

FIGURE 5-3
Shape of misoriented film surface.

The first step is for the formation of gas phase species by homogeneous reactions in the gas phase, which upon adsorption become adspecies. An adspecies is the adsorbed species that is actually responsible for film growth. The second step represents the adsorption of the gas phase species on vacant surface sites, S, to form adspecies, which can also desorb. In the course of surface migration (diffusion), some of the adspecies desorb. The last step is for incorporation of the adspecies into the crystal structure (lattice). As shown in Eq. (5.2), the adspecies $A \cdot S$ releases the other element of the adspecies, B_2, and the atom A_1 becomes a part of the crystal structure, $A_1(c)$. The last step may involve another gas phase species. The surface diffusion step is included in Eq. (5.1) in the sense that only those adspecies still remaining in the adsorbed state will participate in the last step of incorporation into the crystal lattice. Therefore, Eq. (5.1) represents the equilibrium adspecies concentration prevailing under constant growth conditions. If the adspecies were adatoms, one more step (Bloem and Claassen, 1980) may be involved:

$$A(g) + S \rightleftharpoons A \cdot S$$

$$A \cdot S \rightleftharpoons A_1 \cdot S + B_2(g)$$

$$A_1 \cdot S \longrightarrow A_1(c)$$

where the middle step represents the transformation of adspecies into adatoms $A_1 \cdot S$.

Kinetics of epitaxial film growth can be derived on the basis of the individual steps of Eqs. (5.1) and (5.2). It should be remembered that these represent only macroscopic steps that are consistent with epitaxial crystal growth and do not necessarily represent the detailed mechanism of the actual crystal growth. An alternative would be to determine the detailed mechanism and then derive the kinetics. The ample experience with catalytic reactions has revealed that the resulting kinetics is too cumbersome to use even when the mechanism is known, which is rare, and the same kinetic behavior can be described equally well by simplified kinetics. For instance, the two-step kinetic model proposed by Boubart (1972) is sufficient for many real catalytic reactions. Approximate kinetics based on Eqs. (5.1) and (5.2) may be sufficient to model many epitaxial crystal growth processes.

Since almost all types of deposition follow similar steps as those represented in Eqs. (5.1) and (5.2), it is appropriate here to describe the essential features for deriving a kinetic expression based on the growth steps. In deriving the kinetics, each step is taken as an elementary step. A rate process is elementary if the order of the rate is the same as the stoichiometry of the rate process. For the adsorption/desorption process represented by Eq. (5.1), for example, the net rate r_a is given by

$$r_a = k_f C_A C_v - k_r C_{A \cdot S} = k_f\left(C_A C_v - \frac{C_{A \cdot S}}{K_A}\right) \qquad \text{where } K_A = \frac{k_f}{k_r} \tag{5.3}$$

and where k_f is the rate constant for the forward step (adsorption), k_r is the same for the reverse step (desorption), C_A is the gas phase volumetric concentration of species A, $C_{A \cdot S}$ is the surface concentration of the adspecies $A \cdot S$, and C_v is the surface concentration of vacant sites. Because the steps are elementary, the order of the forward rate is the same as the stoichiometry of the forward step, which is unity. Thus, the forward rate is $k_f C_A C_v$. It should be recognized that the order cannot be fractional for elementary reactions since an integer number of molecules (atoms) and not a fraction of molecule interact when a reaction step occurs. For the second step [Eq. (5.2)], one has for the rate r_s:

$$r_s = k_s C_{A \cdot S} \tag{5.4}$$

where k_s is the rate constant.

If $r_a/k_f \ll 1$, it follows from Eq. (5.3) that

$$C_{A \cdot S} = K_A C_A C_v$$

If there are C_t number of sites per unit area, the total sites consist of vacant sites C_v (surface concentration of vacant sites) and those sites occupied by adspecies $C_{A \cdot S}$. Thus, the total site balance is

$$C_t = C_v + C_{A \cdot S}$$

The site balance and the expression for $C_{A \cdot S}$ lead to

$$\frac{C_v}{C_t} = \frac{1}{1 + K_A C_A}$$

The deposition rate r_D is the same as the rate at which the adspecies are incorporated into the crystal, or $r_D = r_s$. It follows from Eq. (5.4) that

$$\begin{aligned} r_D = r_s = k_s C_{A \cdot S} &= k_s K_A C_A C_v \\ &= \frac{C_t k_s K_A C_A}{1 + K_A C_A} \end{aligned} \tag{5.5}$$

where the expression for C_v/C_t has been used. A concept used in deriving the deposition kinetics is that the step of Eq. (5.2) is the rate-limiting (rate-controlling) step. When one elementary step is rate-limiting, the fact does not mean that the net rates for the other steps [in this case, the step of Eq. (5.1)] are zero. Rather, the ratio r_a/k_f is small compared to r_s/k_s. This is the basis for the pseudo steady-state assumption for non-rate-limiting elementary steps. The kinetics of the form of Eq. (5.5) are known as Langmuir-Hinshelwood kinetics in the field of catalytic reactions (e.g., Butt, 1980). Although there is a close analogy between the Langmuir-Hinshelwood kinetics and deposition kinetics, there are subtle differences. In catalytic reactions, a reactant adsorbs onto an active site(s), reacts in the adsorbed state to form a product(s), and then the adsorbed product desorbs and is released into the fluid phase. In deposition, a gas phase species (precursor) adsorbs onto a site forming an adspecies. This adspecies migrates on the surface to energetically favorable sites and then becomes a part of crystal structure. Thus,

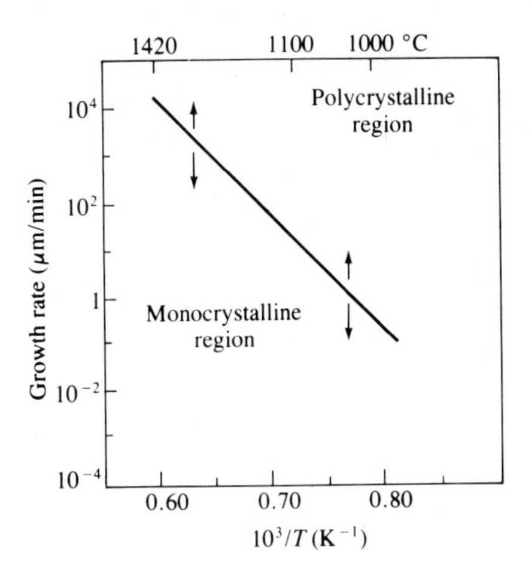

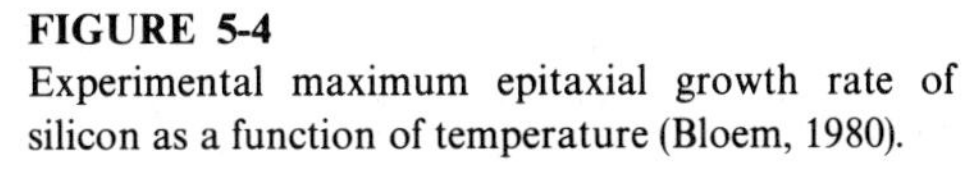
FIGURE 5-4
Experimental maximum epitaxial growth rate of silicon as a function of temperature (Bloem, 1980).

there is only the incorporation step and no desorption of product is involved except in the release of byproducts. This difference has to be recognized in deriving the kinetics of deposition, which is treated in detail in Sec. 5.4.

The conditions for monocrystalline or polycrystalline growth can be examined in the light of the epitaxial process just considered. A polycrystalline structure is a composite material made of grains of monocrystalline structure separated by grain boundaries. The polycrystalline structure results when the probability of adspecies getting together to form stable nuclei or islands is higher than the probability of the adspecies migrating toward steps for incorporation into the crystal lattice. Higher adspecies concentration and lower surface mobility increase the probability of the adspecies forming stable islands. Since higher temperatures favor desorption (and thus lower adspecies concentration) and higher mobility of adspecies, it is expected that the monocrystalline growth would be favored at higher temperatures. Shown in Fig. 5-4 is an estimated, experimental maximum growth rate of monocrystalline silicon obtained as a function of growth temperature (Bloem, 1980).

It is seen that the maximum epitaxial growth rate increases with increasing temperature. Since lower temperatures favor polycrystalline structures, the epitaxial growth at the lower temperatures is possible only when the partial pressure of the source gas, e.g., silane, is low. This, in turn, means a lower epitaxial growth rate. Note that low partial pressures lead to low adsorption rates and thus low adspecies concentrations.

Example 5.2. The rate at which gas phase molecules strike a solid surface per unit surface area, Z, can be obtained from simple kinetic theory of gases (e.g., Moore, 1955):

$$Z = \frac{p}{(2\pi mkT)^{1/2}}$$

where m is the mass of the molecule, p is the partial pressure of the molecule, k is the Boltzmann constant, and T is temperature. Not all molecules striking the surface adsorb. Only those with sufficient energy to overcome the energy barrier E_a can adsorb. Further, only the fraction of the surface still vacant is available for adsorption. Therefore the rate of adsorption r_a in molecules per time per unit area is given by

$$r_a = Z(1 - \alpha) \exp\left(-\frac{E_a}{kT}\right)$$

$$= \frac{p(1 - \alpha)}{(2\pi mkT)^{1/2}} \exp\left(-\frac{E_a}{kT}\right) \quad \text{(A)}$$

where α is the coverage defined by the number of adsorbed molecules (sites), N_a, divided by the total number of sites, N_t. The rate of desorption, r_d, in molecules per time per unit area can be expressed as the probable vibrational frequency, v_0, of the adsorbed molecule overcoming the desorption energy barrier, E_D, times the number of adsorbed molecules:

$$r_d = v_0 N_a e^{-E_D/kT} \quad \text{(B)}$$

(a) Assuming that adsorption and desorption are in equilibrium and that only a small fraction of the equilibrium concentration of the adspecies is involved in crystal growth, obtain an expression for the equilibrium adspecies concentration. For silicon epitaxy based on SiH_4, the mole fraction of SiH_4 in H_2 carrier gas is 0.001 at atmospheric pressure at 1273 K. Calculate the equilibrium concentration for the given conditions. As an approximation, take N_t as the number of silicon atoms per unit surface area and use an atomic height of 3 angstroms. The molecular weight of silicon is 28 g/mol and the density is 2.3 g/cm^3. Use SiH_2 as the adspecies and assume complete conversion of SiH_4 to SiH_2 in the gas phase. Take v_0 as 10^{13} s^{-1}. Assume that the heat of adsorption $(E_D - E_a)$ is 40 kcal/mol.

(b) Make conclusions on the effect of temperature and pressure on the equilibrium concentration.

Solution

(a) At equilibrium, $r_a = r_d$. Thus,

$$\frac{p(1 - N_a/N_t)}{(2\pi mkT)^{1/2}} \exp\left(-\frac{E_a}{kT}\right) = v_0 N_a \exp\left(-\frac{E_D}{kT}\right)$$

Solving this for the equilibrium concentration of the adsorbed species $(N_a)_{eq}$ yields

$$(N_a)_{eq} = \frac{1}{1/N_t + [(2\pi mkT)^{1/2} v_0 \exp(-Q/kT)]/p} \quad \text{where } Q = E_D - E_a \quad \text{(C)}$$

Now $k = 1.38 \times 10^{-16}$ g·cm^2/(s^2·molecule K)

$m = 30/(6 \times 10^{23}) = 5 \times 10^{-23}$ g/molecule

$p = (0.001)(1.013 \times 10^6)$ g/(cm·s^2)

$$N_t = \frac{\rho_s h N_A}{M}$$

$$= \frac{2.3(3 \times 10^{-8})}{28}(6 \times 10^{23}) = 1.48 \times 10^{15} \text{ sites/cm}^2$$

where ρ_s = density of silicon
h = atomic height
M = molecular weight of silicon
N_A = Avogadro's number

$$\frac{(2\pi mkT)^{1/2}v_0}{p}e^{-Q/kT} = \frac{[6.28(5 \times 10^{-23})(1.38 \times 10^{-16})(1273)]^{1/2} \times 10^{13}}{1.013 \times 10^3} \times \exp\left(-\frac{40{,}000}{2 \times 1273}\right)$$

$$= 1.4 \times 10^{-14} \text{ cm}^2/\text{molecule}$$

Therefore, from Eq. (C),

$$(N_a)_{eq} = \frac{1}{1/(1.48 \times 10^{-15}) + 1.4 \times 10^{-14}} = 0.068 \times 10^{15} \text{ molecules/cm}^2$$

(*b*) When $\alpha(=N_a/N_t)$ is small, as in part (*a*), Eq. (C) can be approximated by

$$(N_a)_{eq} = \frac{p}{(2\pi mkT)^{1/2}v_0}\exp\left(\frac{Q}{kT}\right) \tag{D}$$

It is seen that the equilibrium concentration of the adspecies decreases exponentially with temperature and increases linearly with partial pressure. The same conclusion can be drawn from Eq. (C).

As one might expect from the discussion of the epitaxial process, the growth rate depends on the substrate crystal orientation. Since (111) planes have the highest density of atoms on the surface, the film grows most easily on these planes for silicon. Silicon epitaxy is usually carried out on (111) or (100) substrates. Although the (111) planes of gallium arsenide also have the highest density of atoms, epitaxial gallium arsenide is usually grown on (100) substrates because of difficulties in growing the film in the [111] direction, particularly on As-exposed (111) substrate (Laporte *et al.*, 1980). The major reason for the use of (100) substrates is the ease with which the crystal can be cleaved along the edge (110) planes of (100) substrate. The difference here is that unlike silicon epitaxy, gallium arsenide epitaxy involves two different atoms, gallium and arsenic. As detailed in Chap. 3 for the zinc blende structure, both (111) and (100) gallium arsenide crystals consist of alternating layers of gallium and arsenic. The (111) Ga face, i.e. the structure ending with gallium atoms at the surface, has gallium atoms with no free electrons, since all their valence electrons are covalently bonded to the arsenic atoms below the exposed gallium layer. The (111) As face has arsenic atoms, each with two free electrons. On the other hand, the (100) Ga face has one free electron and the (100) As face has three free electrons. In terms of covalent bonds that can form with the other type of atom, i.e., Ga with As face or As with Ga face, the (100) face can have two "dangling bonds" since they can form two covalent bonds, regardless of the type of the exposed surface atoms. On the other hand, the (111) Ga face has one dangling bond, but the (111) As face has two dangling bonds. These facts are important in gallium arsenide heteroepitaxy.

Example 5.3. Explain why the (100) face of gallium arsenide can have two dangling bonds regardless of the type of surface atom [when the (100) Ga face has only one free electron] whereas the (100) As face has three free electrons.

Solution. All bulk atoms in crystalline gallium arsenide are tetrahedrally coordinated. This means that one covalent bond is formed entirely by two electrons from arsenic atoms and the other three are formed by sharing electrons from both gallium and arsenic since the number of valence electrons for gallium is three whereas that for arsenic is five. For the (100) Ga face, one free electron can be shared by one arsenic atom above the surface for one bond and the other bond can form by the donation of two electrons from another arsenic atom above the surface. For the As (100) face, two out of the three free electrons are donated to one bond for one gallium atom and the third electron is shared with another gallium atom. The donation of two electrons by an arsenic atom is the reason why the (111) Ga face can have one dangling bond even though the gallium atom does not have any free electrons.

The general epitaxial process considered earlier also applies for gallium arsenide epitaxy. The difficulty of growing epitaxial gallium arsenide on a [(111)As] substrate compared to the growth on a [(111)Ga] substrate can be explained in terms of the difficulty of the migration of adspecies to the steps (Laporte *et al.*, 1980), as evident from the number of free surface electrons. In view of the fact that the Ga face is more inert than the As face, whether or not the substrate is (111) or (100) crystal, one would expect the adspecies to be different for different face elements.

The source gas for silicon epitaxy is usually one of the chlorosilanes (SiH_xCl_{4-x}; $x = 0$ to 4) and the carrier gas is usually hydrogen, although nitrogen can be used and hydrochloric acid gas is also used with SiH_4 ($x = 4$) to prevent homogeneous nucleation. For gallium arsenide epitaxy, there are three different processes depending on the gallium and arsenic source gases. The first is called the halide process (Knight *et al.*, 1965) since $AsCl_3$ is used as the arsenic source. The chloride along with hydrogen is passed over a liquid gallium boat and then over the substrate for the process. The second is called the hydride process (Tietjen *et al.*, 1966) since the arsenic source is in the form of AsH_3. Hydrogen chloride in hydrogen carrier gas is passed over a liquid gallium boat, the products of which are combined with the arsenic hydride in hydrogen. Then the mixture is passed over the substrate. Metal organic chemical vapor deposition (MOCVD) is based on the use of metal organics for the gallium source, typically $Ga(CH_3)_3$, along with AsH_3 and H_2. Unlike the halide and hydride processes, the gallium source can be introduced independently by bubbling hydrogen gas through liquid trimethyl gallium maintained at a low temperature, say 0 °C. MOCVD has emerged as the dominant CVD process for compound semiconductors because of the flexibility MOCVD offers in growing epitaxial films of three- and four-component semiconductors and heteroepitaxial films.

Although many experimental results are reported in the literature for epitaxy carried out in the absence of a dopant, simultaneous doping is always carried out in practice by introducing a dopant source gas along with the semiconductor source gas. The purpose of silicon epitaxy in the fabrication of the bipolar junction structure, for instance, is to introduce a layer of highly doped silicon on the substrate, which is later covered by other material. The effect of dopant on the growth rate is well known (Farrow and Filby, 1971; Nakayama *et al.*, 1986). For silicon, acceptor atoms accelerate the growth rate whereas donor atoms retard it. As the device sizes get smaller and shallower, this effect will become more significant. For such devices, it is important that the dopant distribution at the growing surface be uniform during growth or an uneven surface will result due to the difference in the growth rate due to the dopant content.

The desire to grow an epitaxial film on an insulator, which is usually amorphous, has led to a method known as epitaxial lateral overgrowth (ELO). For silicon, it is called silicon over insulator (SOI) technology. As the name implies, this epitaxy relies on the lateral growth of an epitaxial film over the insulator, as illustrated in Fig. 5-5 (Jastrzebski, 1983, 1984). The motivation is to synthesize dielectrically isolated silicon islands, so that flexibility can be provided in the fabrication of a variety of integrated circuits. As shown in Fig. 5-5 for SOI, the epitaxial silicon is grown on the windows between silicon dioxide strips (Fig. 5-5*a*). When the growing silicon reaches the height of the silicon dioxide strip, the epitaxial layer grows laterally over the insulator (Fig. 5-5*b*) and eventually meets with other layers to form the epitaxial film on the dioxide strip (Fig. 5-5*c*). For the ELO to be effective, the lateral growth rate should be larger than the vertical growth rate so that the epitaxial layer spreads out over the insulator. At the same time nucleation on the insulator should be prevented, since the epitaxial layer cannot form over that part of the insulator where nucleation has already taken place. The first of these requirements can be satisfied if a silicon source gas is used which yields a much higher lateral growth rate than vertical growth rate. For instance, it has been shown (Rathman *et al.*, 1982) that for short times the selective growth with a high lateral to vertical growth ratio (40:1) can be realized with SiH_4. The fact that the lateral-vertical growth ratio depends on the silicon source gas is an indication that the adspecies is different for different

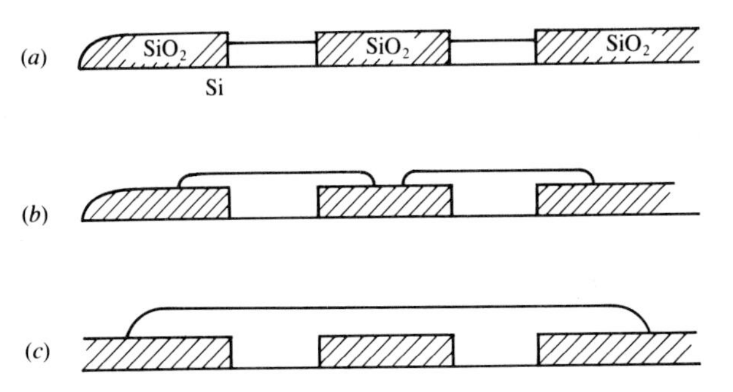

FIGURE 5-5
Growth of epitaxial silicon over silicon dioxide, SOI (Jastrzebski, 1983).

source gases, for this ratio is mainly dependent on the rate of surface diffusion. The other requirement for ELO can be met by carrying out the ELO in cycles of growth and etching. In general, the nuclei formed on an insulator are unstable compared to the epitaxial structure and therefore the amount of epitaxial silicon etched by passing gaseous HCl is much smaller than the silicon nuclei removed from the insulator. Furthermore, it takes a certain amount of time for the nucleation to take place, which is called the incubation period. Therefore, the growth period is usually of the order of or longer than the incubation period, which is about 1 to 60 seconds for SiH_4. The growth period decreases with increasing temperature and supersaturation (Claassen and Bloem, 1980). Recently, a process has been reported (Furumura *et al.*, 1986) that provides perfectly selective silicon epitaxy for the filling of windows between SiO_2 islands when $SiHCl_3$ in H_2 is used in the temperature range of 900 to 1000 °C at 100 Pa and high fluid velocities of 400 cm/s. Perfect selective growth of polycrystalline silicon has been made possible by introducing trichloroethylene.

The epitaxial growth of gallium arsenide over an oxide or metal mask is relatively easy, because nucleation of GaAs on the oxide does not readily take place (Tausch and Lapierre, 1965) and the ratio of lateral to vertical growth rate is larger for GaAs than for Si. However, ELO based on MOCVD is much more difficult since a chlorine compound cannot be used. Thus any unwanted nuclei formed (Asai and Ando, 1985) cannot be readily removed (etched).

Selective epitaxy, as implied by the lateral epitaxial overgrowth, was initially proposed by Joyce and Baldrey (1962). It is very attractive in widening device applications (Bozler and Alley, 1980) and device fabrication flexibility. The basic phenomenon central to selective epitaxy is nucleation behavior. The critical radius of a stable cluster r^* below which the cluster cannot exist is often estimated from thermodynamics:

$$r^* = \frac{2\Omega v}{kT \ln (p/p_e)} \tag{5.6}$$

where Ω is the surface free energy, v is the atom volume, p_e is the equilibrium vapor pressure of the source gas, and p is the partial pressure of the gas such that p/p_e is the degree of supersaturation. Although Eq. (5.6) is often used, the supersaturation under typical growth conditions is so high that the typical critical radius approaches molecular dimensions and thus the above relationship based on classical thermodynamics, which was derived for ensembles of at least a thousand molecules, is invalid.

Example 5.4. Calculate the critical radius of a silicon cluster for supersaturations of 5 and 10^3 at 1000 °C. For Si, $v = 2 \times 10^{-21}$ cm³, $\Omega = 10^{-4}$ J/cm², and SiH_4 equilibrium vapor pressure is about 10^{-6} atm (Claassen and Bloem, 1980).

Solution. For $p/p_e = 5$, from Eq. (5.6),

$$r^* = \frac{(2 \times 10^{-4})(10^7)(2 \times 10^{-21})}{(1.38 \times 10^{-16})(1273)(\ln 5)} = 140 \text{ nm}$$

For $p/p_e = 10^3$,

$$r^* = \frac{(2 \times 10^{-4})(10^7)(2 \times 10^{-21})}{(1.38 \times 10^{-16})(1273)(\ln 5)} = 33 \text{ nm}$$

The kinetic approach by Walton (1962) may be used for the cluster of interest here consisting of a small number of atoms. By considering the formation of a cluster consisting of n atoms from individual adatoms, Walton arrived at the following conclusion on the critical radius:

$$n^* \propto r_a^{n^*+1} \tag{5.7}$$

where n^* is the number of atoms for a stable nucleus. Equation (5.7) indicates that the critical size decreases with an increasing rate of adsorption. Experimental results on nucleation based on electron microscopy indicate that stable nuclei contain at least five atoms and most contain many more (Lewis and Campbell, 1967).

The other quantity of interest in selective epitaxy, which can also be measured, is the saturation concentration. The saturation concentration is the maximum surface concentration of nuclei beyond which no new nuclei form and the clusters merely increase in size. After a certain period of time, however, clusters can start coalescing. The constant cluster concentration before coalescence is referred to as the saturation concentration. For example, the saturation concentration in Fig. 5-6 is approximately 8×10^3 cm^{-2}. It can be seen that the coalescence starts at about 25 seconds of exposure time. The incubation period for nucleation was reported to be 3 seconds. Two separate cases have to be considered for the saturation concentration (Frankel and Venables, 1970): initially complete condensation, in which the nucleation is completed within a short time interval, and incomplete condensation. The case of incomplete condensation is less of a problem than that of complete condensation since it takes a relatively

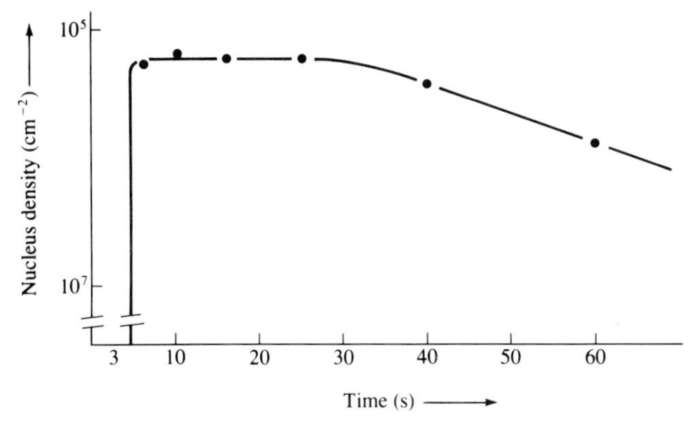

FIGURE 5-6
Nucleation behavior (Claassen and Bloem, 1980).

longer time for nucleation to take place. In the case of complete condensation, the size distribution is almost uniform (Frankel and Venables, 1970). The saturation concentration N_s has been derived by Lewis and Campbell (1967) and their results are:

$$N_s = \begin{cases} N_t \exp\left(-\dfrac{E_D - E_d}{kT}\right) & \text{incomplete condensation} \quad (5.8a) \\ \left(\dfrac{N_t r_a}{v_0}\right)^{1/2} \exp\left(\dfrac{E_d}{2kT}\right) & \text{complete condensation} \quad (5.8b) \end{cases}$$

where N_t is the surface concentration of total adsorption sites and r_a is the rate of adspecies adsorption given (Example 5.2) by

$$r_a = \frac{p(1 - N_a/N_t)}{(2\pi mkT)^{1/2}} \exp\left(-\frac{E_a}{kT}\right) \tag{5.9}$$

When the coverage is small, this can be rewritten as

$$r_a = \frac{p}{(2\pi mkT)^{1/2}} \exp\left(-\frac{E_a}{kT}\right) \tag{5.10}$$

Depending on the temperature, the same system can show complete condensation behavior at low temperatures and incomplete condensation behavior as the nucleation temperature increases. The transition temperature T_0 is that corresponding to the break point shown in Fig. 5-7. The complete condensation case, which is more relevant to selective epitaxy, is based on the idealized model of rapid nucleation and uniform growth giving uniform sizes at any stage of growth. In reality, however, variations in size are expected. Therefore, Eq. (5.8*b*) for

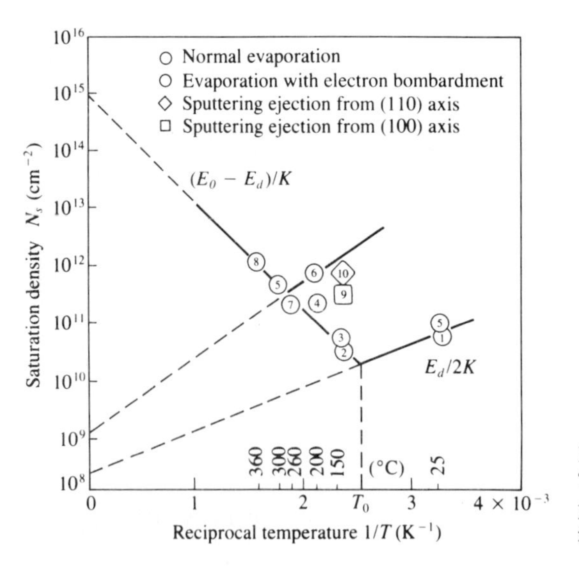

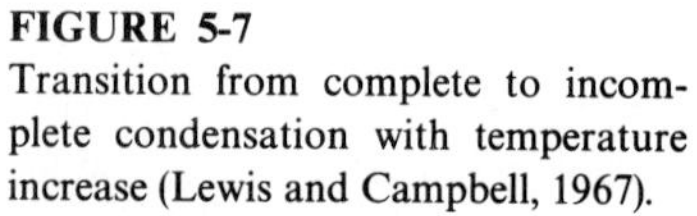

FIGURE 5-7
Transition from complete to incomplete condensation with temperature increase (Lewis and Campbell, 1967).

narrow size distributions may be written as

$$N_s = \left(\frac{N_t r_a}{v_0}\right)^n \exp\left(\frac{E_d}{2kT}\right) \tag{5.11}$$

where n is a constant.

Heteroepitaxy is much more difficult to carry out in practice than homoepitaxy. Heteroepitaxy involves growing a film of crystalline structure different from the substrate. Therefore, there is a lattice mismatch at the interface between the two crystalline structures. The two different materials also have different thermal expansion coefficients. As the grown film goes through heating and cooling cycles in the fabrication, a bending of the film can result. This bending can cause severe problems in lithography and other later processing steps. Another major problem, which is usually identified only after the fact, has to do with electrical properties that are affected by the defects at the interface or those propagated through the grown film.

The lattice mismatch, μ, is usually expressed as

$$\mu = \frac{a_s - a_f}{a_s} \tag{5.12}$$

where a_s and a_f are the lattice constants of substrate and film, respectively. To be more precise, these are the lattice constants along the major crystallographic directions since the constants depend on the direction unless the crystal is simple cubic. In essence, the thermal expansion incompatibility, if any, is also included in Eq. (5.12), since for monocrystalline material, the thermal expansion is that of the unit cell. Therefore, the mismatch at different temperatures is also a measure of the thermal expansion incompatibility. In order to go through the transition from the substrate lattice to film lattice, there has to be a buffer layer (or transition layer) whose lattice constant changes from the substrate to the film. This can be accomplished by introducing intentional impurities or growing intermediate layers of different lattice constants for the buffer layer. The lattice constant resulting from the introduction of the intentional impurities can be calculated using the Vegard coefficient w_i, which is the ratio of the radius of impurity atom to that of host atom in the solid solution. The coefficient is determined from x-ray diffraction data from the interstitial solid solution (Aleksandrov, 1975). The relationship is

$$a_c = a_o \prod_i^n (1 + w_i x_i) \tag{5.13}$$

where a_c is the lattice constant for the solid solution, a_o is the same for the pure solvent, and x_i is the fraction of impurity atom i in the solid solution. Note that the relationship holds only for small values of x_i. When there is only one impurity, Eq. (5.13) may be replaced by

$$a_c = a_o[(1 - x) + wx] \tag{5.14}$$

Example 5.5

(*a*) For the heteroepitaxy of GaAs on Si substrate, calculate the lattice mismatch. Suppose that GaP and $GaAs_{0.5}P_{0.5}$ layers of 10 nm each are grown on Si and then the GaAs is grown so that there are two intermediate layers for the buffer layer.

(*b*) Calculate the lattice mismatch for each interface. Use the following data and assume, for approximate calculations, that the atomic radii ratio does not change with the interstitial substitution:

$$a_{\text{Si}} = 5.43 \text{ Å}$$

$$a_{\text{GaP}} = 5.45 \text{ Å}$$

$$a_{\text{GaAs}} = 5.65 \text{ Å}$$

Solution

(*a*) According to Eq. (5.12),

$$\mu = \frac{5.43 - 5.65}{5.43} = -0.041$$

It is seen that the lattice mismatch is approximately 4 percent.

(*b*) For the Si-GaP interface,

$$\mu_{\text{Si-GaP}} = \frac{5.43 - 5.45}{5.43} = -0.0037$$

For the GaP-$GaAs_{0.5}P_{0.5}$ interface, one has to calculate the lattice contant of $GaAs_{0.5}P_{0.5}$. Since the As-P atom ratio is 0.5, it can be calculated on the basis of 50% GaAs and 50% GaP. Thus,

$$a_{\text{GaAs}_{0.5}\text{P}_{0.5}} = \frac{a_{\text{GaAs}} + a_{\text{GaP}}}{2} = 5.55$$

$$\mu_{\text{GaP-GaAsP}} = \frac{5.45 - 5.55}{5.45} = -0.018$$

For the $GaAs_{0.5}P_{0.5}$-GaAs interface,

$$\mu = \frac{5.55 - 5.65}{5.55} = -0.018$$

It is seen that by introducing two intermediate layers, the maximum lattice mismatch is reduced to 1.2 percent compared to the 4 percent for the case of no buffer layers.

In theory, one can solve the lattice mismatch and thermal expansion problems as a single problem by combining Eqs. (5.12) and (5.13) and insisting that the lattice mismatch be zero at the maximum processing temperature. Combining the equations leads to

$$\mu = \frac{a_s(T) - (a_f)_o \prod [1 + w_i(T)x_i]}{a_s(T)} \tag{5.15}$$

where it has been assumed that intentional impurities are only in the film, which would be the typical case. Theoretically, therefore, one can satisfy the following condition to solve the lattice mismatch and thermal expansion problems:

$$\mu(T_m) = 0 \tag{5.16}$$

where T_m is the maximum growth temperature. In practice, it is very difficult to satisfy Eq. (5.16). Nevertheless, it provides a direction to pursue. The linear thermal expansion coefficient, β, which is a function of temperature, is defined by

$$\frac{dl}{dT} = \beta l \qquad \text{or} \qquad \frac{da}{dT} = \beta a \tag{5.17}$$

where l is the length. For monocrystalline material along a given direction, l is the same as the lattice constant, which is reflected in Eq. (5.17). In general, the thermal expansion coefficient is inversely proportional to bond strength.

Example 5.6. It is proposed to have a graded silicon content in a buffer layer between a silicon wafer of 5 cm diameter and a GaAs film. For the maximum growth temperature of 750 °C, plot the lattice mismatch as a function of silicon content changing linearly from 100 percent at the wafer-buffer interface to 0 percent at the GaAs film in a linear manner. Assume the following thermal expansion coefficients at 300 K to be constants. The coefficients are

$$\beta_{\text{Si}} = 2.5 \times 10^{-6} \text{ cm/(cm·°C)}$$

$$\beta_{\text{GaAs}} = 6.5 \times 10^{-6} \text{ cm/(cm·°C)}$$

$$a_{\text{Si}} = 5.43 \text{ Å}$$

$$a_{\text{GaAs}} = 5.65 \text{ Å}$$

Solution. Using Eq. (5.14) in Eq. (5.12) leads to

$$\mu(T) = \frac{a_s(T) - (a_f)_o[(1 - x) + wx]}{a_s(T)} \tag{A}$$

Solving Eq. (5.17) for constant β,

$$a = a_i \exp [\beta(T - T_i)]$$

where T_i = initial temperature (=25 °C)
a_i = value of a at T_i

Since β is quite small, the above can be well approximated by

$$a = a_i[1 + \beta(T - T_i)]$$

$$a(750\,°\text{C}) = a_i[1 + 725\beta] \tag{B}$$

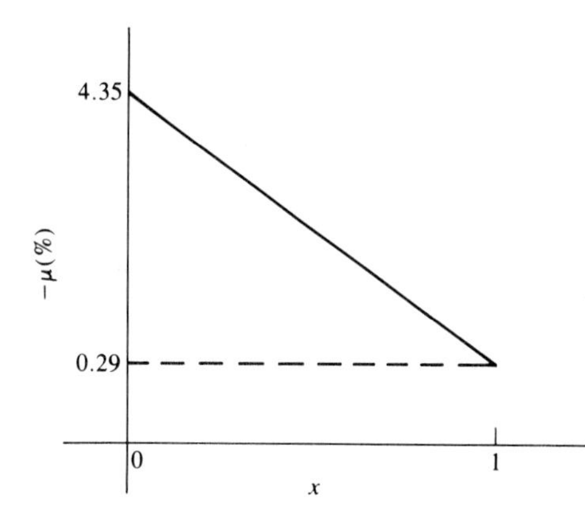

FIGURE 5-8
Change of lattice mismatch with impurity content.

At $T = T_m = 750\,^\circ\mathrm{C}$, Eq. (A) can be written as

$$\mu(T_m = 750\,^\circ\mathrm{C}) = \frac{a_{\mathrm{Si}}(1 + 725 \times 2.5 \times 10^{-6}) - a_{\mathrm{GaAs}}(1 + 725 \times 6.5 \times 10^{-6})[(1 - x) + 5.43x/5.65]}{a_{\mathrm{Si}}(1 + 725 \times 2.5 \times 10^{-6})}$$

$$= \frac{5.43(1 + 725 \times 2.5 \times 10^{-6}) - 5.65(1 + 725 \times 6.5 \times 10^{-6})(1 - 0.0389x)}{5.43(1 + 725 \times 2.5 \times 10^{-6})}$$

$$= 1 - 1.04353(1 - 0.0389x)$$

$$= -0.04351(1 - 0.9329x)$$

where x is the fraction of Si in GaAs film. Note that w in this case was assumed to be the lattice constant ratio at room temperature. It is seen that the lattice mismatch changes linearly with the silicon content, as shown in Fig. 5-8. The lattice mismatch shown in the figure represents the worst possible case since the change in length, $\Delta l/l$, consists of not only the thermal expansion contribution but also the elastic strain. Note also that the change in the entire buffer zone is large but the change between adjacent layers is small.

Another critical factor in heteroepitaxy is the crystal orientation of the surfaces involved. More specifically, the substrate surface orientation should be compatible with the film orientation. For instance, the heteroepitaxy of GaAs on Si substrate by MOCVD has been found to yield good results (Saki *et al.*, 1986) only when the Si(100) plane misoriented 2° toward [011] was cleaved *in situ* at a high temperature to expose the steps of the plane. The atomic steps along the Si [011] direction coincide with the [011] direction of the grown GaAs.

As one might surmise, the growth of ternary III-V compounds, mostly used for optoelectronic applications, on GaAs substrates is relatively easy because a very close lattice match can be made (e.g., Wagner *et al.*, 1981). The flexibility offered by MOCVD makes it the dominant chemical epitaxial process for compound semiconductors.

5.3 HOMOGENEOUS REACTIONS

Homogeneous gas phase reactions by themselves are not of interest in semiconductor processing. They are, however, relevant to the extent that they produce the species actually responsible for film growth or the precursors for deposition, and are also the main source of undesired impurities. The gas phase reactions are generally very fast at typical growth/deposition temperatures. Source gases for semiconductor processing are mostly chlorosilanes (SiH_xCl_{4-x}; $x = 0$ to 4), metal organics, halides, and hydrides. Carrier gases are usually either hydrogen or nitrogen.

Despite the fact that homogeneous reactions are responsible for producing the deposition precursors, no detailed relevant mechanistic studies comparable to the well-known hydrogen bromide reaction network are available. The handful of homogeneous reaction kinetics available in the literature indicate that most reactions of interest could be well represented by nth-order kinetics and, in particular, pseudo first-order kinetics:

$$r_A = kC_A^n \qquad \text{where } k = k_0 \exp\left(-\frac{E}{RT}\right) \tag{5.18}$$

and where r_A is the rate of reaction per unit volume, C_A is the concentration of the species of interest A, n is the order of the reaction, and k is the rate constant, which can be represented in Arrhenius form. In the Arrhenius expression, k_0 is the preexponential factor, E is the reaction activation energy, and R is the gas constant, which is the product of Avogadro's number and the Boltzmann constant. Most reactions of interest are decomposition reactions, and therefore pseudo first-order kinetics are often adequate for describing them. The usual analysis methods (e.g., Levenspiel, 1972) can be used to determine the order and the rate constant, given kinetic data.

The gas phase pyrolysis of silane has been reported (Newman *et al.*, 1979) to be approximately unimolecular, involving molecular hydrogen elimination:

$$SiH_4 \longrightarrow SiH_2 + H_2 \tag{5.19}$$

where the activation energy is 52.7 kcal/mol and the preexponential factor is 2.14×10^{13} s^{-1}. When silane is used for silicon epitaxy with a hydrogen carrier gas, a small amount of HCl is always added to prevent gas phase polymerization and homogeneous nucleation of silicon. Dichlorosilane (SiH_2Cl_2) readily decomposes to $SiCl_2$ and H_2 at the typical growth temperatures (around 1000 °C) (Sedgwick and Smith, 1976). It has been found (Aoyama *et al.*, 1983) that $SiHCl_3$, $SiCl_2$, SiH_2Cl_2, $SiCl_4$, and HCl are present in the gas phase when $SiCl_4$ in H_2 is used for silicon epitaxy.

The rate of consumption of silane in excess oxygen (more than two times the stoichiometric amount) for vapor phase deposition of SiO_2 on silicon wafers (typically up to 450 °C) has been shown to follow first-order kinetics (Strater, 1968). The overall reaction is

$$SiH_4 + O_2 \longrightarrow SiO_2 + 2H_2 \tag{5.20}$$

It was later found (Tobin *et al.*, 1980) that, at pressures near 0.5 torr, a secondary reaction also takes place:

$$SiH_4 + 2O_2 \longrightarrow SiO_2 + 2H_2O \tag{5.21}$$

although it only contributes a maximum of 10 percent relative to the reaction of Eq. (5.21). The reactions appear to be both homogeneous and heterogeneous in that direct precipitation of SiO_2 from the gas phase onto the wafer appears to take place.

Example 5.7. Suppose more than one reactant is involved in a reaction such as Eq. (5.21). In these cases, the *n*th-order kinetics of Eq. (5.18) can be written as

$$r_A = kC_A^{n_1}C_B^{n_2}\cdots \tag{A}$$

Show that one can rewrite the above equation, in terms of the concentration of species A only, in the following form:

$$r_A = kf(C_A) \tag{B}$$

where $f(C_A)$ is a function of C_A.

Solution. A single reaction can be written as

$$a\text{A} + b\text{B} + \cdots \longrightarrow d\text{D} + e\text{E} + \cdots$$

where *a*, *b*, etc., are the stoichiometric coefficients and A, B, etc., are the species. If A is taken as the main species, the stoichiometry allows one to write:

$$\frac{(C_A)_i - C_A}{a} = \frac{(C_B)_i - C_B}{b} = \frac{C_D - (C_D)_i}{d} = \frac{C_E - (C_E)_i}{e} \tag{C}$$

where the subscript *i* is for the initial or reactor inlet concentration. Equation (C) allows one to write the concentrations of the other species in terms of the concentration of species A and the initial (or reaction inlet) concentrations, which are known, i.e.,

$$C_B = (C_B)_i - \frac{b}{a}[(C_A)_i - C_A]$$

$$C_D = (C_D)_i + \frac{d}{a}[(C_A)_i - C_A]$$

etc.

Substitution of Eq. (C) into Eq. (A) yields

$$r_A = kC_A^{n_1}\left\{(C_B)_i - \frac{b}{a}[(C_A)_i - C_A]\right\}^{n_2}\cdots$$

$$= kf(C_A)$$

Example 5.8. Strater (1968) carried out isothermal experiments with a tubular reactor packed with quartz spheres with a constant residence time of 4.3 s for the reaction of Eq. (5.20). His reduced experimental data are summarized below. Determine the kinetics of the reaction. Note that no detectable reaction took place at temperatures less than approximately 190 °C.

T (°C)	223	257	320	400
% SiH_4 reacted	11.0	39.7	72.6	92.5

Solution. Since the reaction follows first-order kinetics, the problem is that of determining the activation energy and preexponential factor. Assuming plug flow for the tubular reactor, the mass balance for SiH_4 is

$$v \frac{dC_A}{dz} = -r_A = -kC_A \tag{A}$$

where z is the axial coordinate, C_A is the concentration of SiH_4, and v is the superficial fluid velocity. Defining $t_r = L/v$, $y = z/L$, and $x = (C_{A,\,in} - C_A)/C_{A,\,in}$, where L is the reactor length, t_r is the residence time, and x is the conversion of SiH_4, Eq. (A) can be rewritten as

$$\frac{dx}{dy} = kt_r(1 - x); \qquad x = 0 \text{ at } y = 0 \tag{B}$$

Integration of Eq. (B) from the reactor inlet ($y = 0$) to the outlet ($y = 1$) yields

$$x_{out} = 1 - \exp(-kt_r) \tag{C}$$

where x_{out} is the conversion at the outlet. Rearranging and taking the natural logarithm of the result gives

$$-kt_r = \ln(1 - x_{out}) \tag{D}$$

Since $k = k_0 \exp(-E/RT)$, one can take the logarithm of both sides of Eq. (D) to obtain

$$\ln k_0 + \frac{-E}{RT} = \ln\left[-\frac{1}{t_r}\ln(1 - x_{out})\right] = Y$$

A linear plot of Y versus $1/T$ can be made to obtain the activation energy and preexponential factor:

$1/T$	0.002016	0.001887	0.001686	0.001486
Y	−3.608	−2.140	−1.200	−0.507

A best fit of the data gives a slope ($-E/R$) of −4625 K. Therefore, the activation energy is

$$E = 4625 \times 2[\text{cal/(mol·K)}] = 9.25 \text{ kcal/mol}$$

The apparent preexponential factor determined from the line is 650 s^{-1}.

The extraction of halogens from halides (typically chlorides) by hydrogen often occurs in semiconductor processing. Examples are the reaction between $SiCl_4$ and H_2 in silicon epitaxy and that between BCl_3 and H_2 when the chloride is used for doping. The extent to which these reactions proceed is limited by thermodynamic equilibrium. The reactions are often thought to involve successive extraction of chlorine for chlorides, for instance, yielding HCl each step. Thus, one would have the following sequence for $SiCl_4$:

$$SiCl_4 + H_2 \longrightarrow SiHCl_3 + HCl$$

$$SiHCl_3 + H_2 \longrightarrow SiH_2Cl_2 + HCl$$

etc.

Reactions of BCl_3 have been reported (Carlton *et al.*, 1970) to consist of the following steps:

$$BCl_3 + H_2 \longrightarrow BHCl_2 + HCl$$

$$BHCl_2 \longrightarrow BCl + HCl$$

In view of the fact that the reactions can often proceed to the equilibrium composition at high temperatures, consider the equilibrium composition. For a single reaction, the equilibrium constant K is related to the standard free energy change ΔG° by

$$K = \exp\left(-\frac{\Delta G^\circ}{RT}\right) \tag{5.22}$$

The equilibrium constant K is defined in terms of the equilibrium activities a_i of the reactants and products. For a general reaction

$$a\mathrm{A} + b\mathrm{B} \rightleftharpoons c\mathrm{C} + d\mathrm{D} \tag{5.23}$$

the equilibrium constant is

$$K = \frac{a_\mathrm{C}^c\, a_\mathrm{D}^d}{a_\mathrm{A}^a\, a_\mathrm{B}^b} \tag{5.24}$$

For the reaction, $\Delta G^\circ = cG_\mathrm{C}^\circ + dG_\mathrm{D}^\circ - aG_\mathrm{A}^\circ - bG_\mathrm{B}^\circ$. The activities refer to equilibrium conditions in the reaction mixture and are defined as the ratio of the fugacity in the equilibrium mixture f_i to that in the standard state, f_i^0:

$$a_i = \frac{f_i}{f_i^0} \tag{5.25}$$

Standard states at given temperatures are commonly chosen as the pure component at one atmosphere for gases and solids, pure liquid at its vapor pressure for liquids, and solutes in one molar solution or dilute concentration such that the activity is unity for the solutes in the solution. For reactions with a standard state of unit fugacity, the expression for the equilibrium constant becomes

$$K = \frac{f_C^c f_D^d}{f_A^a f_B^b} \tag{5.26}$$

In general, the fugacity of the pure component is usually known. However, the fugacity of the component in the equilibrium mixture or its dependence on the composition is not known so that it is necessary to make assumptions about the behavior of the reaction mixture. The simplest and most common assumption is that the mixture behaves as an ideal solution. Then the fugacity at equilibrium f is related to that of the pure compound f' at the same pressure and temperature by

$$f_i = f' y_i \tag{5.27}$$

where y_i is the mole fraction. Use of Eq. (5.27) in Eq. (5.26) yields

$$K = \frac{(f'_C)^c (f'_D)^d}{(f'_A)^a (f'_B)^b} K_y \qquad \text{where } K_y = \frac{y_C^c y_D^d}{y_A^a y_B^b} \tag{5.28}$$

For ideal gases, the fugacity is equal to the partial pressure so that the equilibrium constant can be written as

$$K = \frac{p_C^c p_D^d}{p_A^a p_B^b} = K_y P^{(c+d)-(a+b)} \tag{5.29}$$

where P is the total pressure. For a solid component taking part in a reaction, fugacity variations with pressure are small and can usually be ignored. Hence,

$$(f/f_0)_{\text{solid component}} = 1 \tag{5.30}$$

The equilibrium composition, as governed by the equilibrium constant, changes with temperature. From thermodynamics the rate of this change is given by the van't Hoff equation:

$$\frac{d(\ln K)}{dT} = \frac{\Delta H^\circ}{RT^2} \tag{5.31}$$

where ΔH° is the change in the heat of formation due to the reaction. This can be integrated to give

$$\ln \frac{K_2}{K_1} = \frac{1}{R} \int_{T_1}^{T_2} \frac{\Delta H^\circ(T)}{T^2}\, dT \qquad \text{where } K_i = K(T_i) \tag{5.32}$$

When the heat of reaction can be considered to be nearly constant over the temperature interval, Eq. (5.32) reduces to

$$\ln \frac{K_2}{K_1} = -\frac{\Delta H^\circ}{R}\left(\frac{1}{T_2} - \frac{1}{T_1}\right) \tag{5.33}$$

It is useful to recognize that the equilibrium constant for an overall reaction consisting of a series of individual reaction steps is the product of the individual equilibrium constants K_i:

$$K = \prod_i K_i \tag{5.34}$$

Example 5.9. For the following reactions:

$$SiCl_4 + H_2 \underset{}{\overset{K_1}{\rightleftharpoons}} SiHCl_3 + HCl \tag{A}$$

$$SiHCl_3 + H_2 \underset{}{\overset{K_2}{\rightleftharpoons}} SiH_2Cl_2 + HCl \tag{B}$$

for which the overall reaction is

$$SiCl_4 + 2H_2 \underset{}{\overset{K}{\rightleftharpoons}} SiH_2Cl_2 + 2HCl \tag{C}$$

assume that the following are constant:

Species	ΔG°_{298}, kcal/mol	ΔH°_{298}, kcal/mol
$SiHCl_3$	−104.5	−113.0
$SiCl_4$	−136.2	−145.7
HCl	−22.8	−22.1

(*a*) Show that Eq. (5.34) holds.

(*b*) Explain why the forward reaction of Eq. (A) is not favored.

(*c*) Determine the equilibrium conversion of $SiCl_4$ and equilibrium composition at 1000 °C, assuming that only the reaction of Eq. (A) takes place. Assume also that the feed concentration of $SiCl_4$ is 1 mol % in H_2 at 1 atm.

Solution

(*a*) Assuming ideal gas,

$$K_1 = \frac{p_{SiHCl_3} p_{HCl}}{p_{SiCl_4} p_{H_2}}$$

$$K_2 = \frac{p_{SiH_2Cl_2} p_{HCl}}{p_{SiHCl_3} p_{H_2}}$$

The product of K_1 and K_2 is

$$K_1 K_2 = \frac{p_{SiH_2Cl_2} p^2_{HCl}}{p_{SiCl_4} p^2_{H_2}}$$

From Eq. (C),

$$K = \frac{p_{SiH_2Cl_2} p^2_{HCl}}{p_{SiCl_4} p^2_{H_2}}$$

It is seen that $K = K_1 K_2$.

(*b*) For the reaction of Eq. (A),

$$\Delta G^{\circ}_{298} = -104.5 - 22.8 - (-136.2) = 8.9 \text{ kcal}$$

$$\Delta H^{\circ}_{298} = -113 - 22.1 - (-145.7) = 10.6 \text{ kcal}$$

From Eq. (5.22),

$$(K_1)_{298} = \exp\left(-\frac{8900}{2 \times 298}\right) = 3.27 \times 10^{-7}$$

Since ΔH° is constant, Eq. (5.33) can be used for K_1 at 1273 K:

$$\ln \frac{(K_1)_{1273}}{(K_1)_{298}} = \frac{-10{,}600}{2}\left(\frac{1}{1273} - \frac{1}{298}\right) = 13.62$$

Thus,

$$(K_1)_{1273} = 0.269$$

The equilibrium concentrations (partial pressures) at 25 °C are related to

$$(K_1)_{298} = 3.27 \times 10^{-7} = \left(\frac{p_{SiHCl_3}\, p_{HCl}}{p_{SiCl_4}\, p_{H_2}}\right)_{eqb} \ll 1$$

This means that $p_{SiCl_4} p_{H_2} \gg p_{SiHCl_3} p_{HCl}$, which in turn means that the reaction does not proceed readily to the right of Eq. (A). At 1000 °C, one has

$$(K_1)_{1273} = 0.269 = \left(\frac{P_{SiHCl_3}\, P_{HCl}}{P_{SiCl_4}\, P_{H_2}}\right)_{eqb} < 1$$

Even at 1000 °C, K_1 is less than unity, indicating that the equilibrium product concentrations (or the maximum conversion to the products) are small relative to the reactant concentrations.

(*c*) Let x be the equilibrium conversion of $SiCl_4$. Then

$$\underset{C_s(1-x)}{SiCl_4} + \underset{(C_h - C_s x)}{H_2} \rightleftharpoons \underset{C_s x}{SiHCl_3} + \underset{C_s x}{HCl}$$

Assuming ideal gas, the partial pressure is equivalent to the concentration ($P_i = C_i RT$). Therefore,

$$(K_1)_{1273} = 0.269 = \frac{C_s^2 x^2}{C_s(1-x)(C_h - C_s x)}$$

$$= \frac{x^2}{(1-x)(C_h/C_s - x)}$$

$$= \frac{x^2}{(1-x)(99-x)}$$

or

$$x = 0.96$$

Therefore, the equilibrium concentrations in terms of the mole fractions are

$SiCl_4$:	0.04 mol %	$SiHCl_3$:	0.96 mol %
HCl:	0.96 mol %	H_2:	98.04 mol %

It is seen from parts (*b*) and (*c*) that the reaction will proceed to the right only when the temperature is very high.

Metal organics used in semiconductor processing are usually metal alkyls such as trimethyl gallium, $Ga(CH_3)_3$, and triethyl indium, $In(C_2H_5)_3$. In GaAs epitaxy, for instance, $Ga(CH_3)_3$ and AsH_3 are typically used. A study of the gas phase reactions of trimethyl and triethyl gallium (TMG and TEG) in H_2 and N_2 (Yoshida and Wantanabe, 1985) has revealed that the reaction products of TMG are methane plus a small amount of ethane in H_2, but that they are methane, ethane, and propane in N_2; the reaction products of TEG are mainly ethylene in both N_2 and H_2. These results, along with the work of Leys and Veenvliet (1981), indicate that TMG in H_2 follows the reaction:

$$2Ga(CH_3)_3 + 3H_2 \longrightarrow 2Ga + 6CH_4 \tag{5.35}$$

Although no carbide formation has been reported in the case of TMG, trimethyl aluminum (TMA) leads to Al_4C_3 formation, presumably through the formation of dimeric TMA followed by pyrolysis (Tromson-Carli *et al.*, 1981). Undoubtedly, the gas phase reactions of metal alkyls do involve alkyl radicals (e.g., Squire *et al.*, 1985). The presence of hydrogen in great excess, which is the usual practice, minimizes the problems that can be caused by radicals. Although the reactions of metal alkyls are not simple, they have been reported to be well represented by first-order kinetics, at least for TMG (Yoshida and Wantanabe, 1985) and trimethyl indium (Larsen and Stringfellow, 1986).

The decomposition reactions of metal alkyls, MR_n, where M is the metal, R is the alkyl, and n is the number of the alkyl group, are often postulated (e.g., Jacko and Price, 1964) to proceed according to the following elementary steps:

$$MR_n \longrightarrow MR_{n-1} + R$$

$$MR_{n-1} \longrightarrow MR_{n-2} + R$$

$$\vdots$$

$$MR \longrightarrow M + R$$

Propagation and recombination of R

This postulated mechanism will always lead to first-order kinetics for the decomposition rate of the metal alkyl. However, the kinetics for the formation of various products can be quite complicated, the complexity depending on the extent of the propagation and recombination/disproportionation of the alkyl radicals (see Probs. 5.8 and 5.9).

At low temperatures, metal alkyls and hydrides can form adduct compounds. Examples for II-V and III-V compound semiconductor adducts are

$$(CH_3)_2Zn + SeH_2 \longrightarrow (CH_3)_2ZnSeH_2$$

$$(CH_3)_3Ga + AsH_3 \longrightarrow (CH_3)_3GaAsH_3$$

The formation of these compounds, often occurring in the feed line at room temperature, can lead to undesirable polymer formation (Mullin *et al.*, 1981).

Although the formation of the adduct compounds appears to give problems in the current practice of MOCVD based on hydrides for group V metals, the adducts between III-V compounds, such as R_3MNR_3 [e.g., $(CH_3)_3GaP(CH_3)_3$] where M and N are III and V metals and R is the alkyl group, do offer alternatives for the precursors for MOCVD (Bradley *et al.*, 1986). These adducts are liquid at room temperature and are reported to be stable at the growth temperature (ca. 650 °C) in the presence of group V metal hydrides.

Most homogeneous reactions in semiconductor processing involve decomposition reactions. Some of the reactions are believed to be unimolecular as in the silane decomposition of Eq. (5.19). Because a low-pressure process can offer better thickness uniformity of grown film, many films are grown in a low-pressure environment, typically in the range of 0.1 to 10 torr. In this low-pressure range and lower, the kinetics obtained at atmospheric pressure are not directly applicable to the low-pressure range if the reaction is unimolecular or involves a third body (a gas molecule not involved in the reaction stoichiometry). Recognition of this fact is important in order to avoid making an incorrect conclusion with regard to the growth precursor. An example would be the epitaxial silicon growth in ultra-high vacuum where the extent of the reaction of Eq. (5.19) is negligible because of the decrease of the rate constant with decreasing pressure (Meyerson *et al.*, 1986). This means that the film growth precursors can be different depending on the total pressure.

Although the detailed account of this pressure effect is complicated, the essence of the pressure effect is not. The basic mechanism of a unimolecular reaction, A → products, is due to Lindemann, from which the more complicated theories such as RRKM theory (e.g., Gardiner, 1972) evolved:

$$A + M \underset{k_{de}}{\overset{k_e}{\rightleftharpoons}} A^* + M \tag{5.36}$$

$$A^* \xrightarrow{k_{uni}} \text{products} \tag{5.37}$$

where A* represents an A molecule that is energized as a result of the bimolecular encounter (collision) and M represents any molecule (including other A molecules) that A molecules may encounter in the reaction system. If A has been diluted with a large amount of diluent gas, then M can be assumed to approximately represent diluent gas molecules. This is typical in semiconductor processing where the mole fraction of the source gas is very low. For example, mole fractions of silane are usually less than 0.01 with the rest being either hydrogen or nitrogen. Making a pseudo steady-state assumption (see Sec. 5.6) on the concentration of A*, C_{A*}, leads to the following rate law:

$$r = \frac{k_e C_A C_M}{1 + k_{de} C_M / k_{uni}} \tag{5.38}$$

where r is the rate of formation of the products. When the total pressure is high such that $k_{de} C_M \gg k_{uni}$, Eq. (5.38) can be written as

$$r_{high} = \frac{k_{uni} k_e}{k_{de}} C_A = k_\infty C_A \qquad \text{where } k_\infty = \frac{k_{uni} k_e}{k_{de}} \tag{5.39}$$

The value of k_∞ is the usual first-order rate constant. When the pressure is very low, Eq. (5.39) reduces to

$$r_{low} = k_e C_A C_M = k' C_A \qquad \text{where } k' = k_e C_M \tag{5.40}$$

Two items are of interest in the low-pressure range. One is that the first-order rate constant k' changes in proportion to the total pressure if the mole fraction of the reactant (A in this case) is very small, since C_M is then proportional to the total pressure. The other is that the activation energy in the low-pressure range, which is that corresponding to k_e, is different from the activation energy in the high-pressure range, which is that corresponding to $(k_{uni} k_e / k_{de})$. The first-order rate constant in the low-pressure range is

$$k = \frac{k_e C_M}{1 + k_{de} C_M / k_{uni}} \tag{5.41}$$

which follows from Eq. (5.38). The concentration C_M can be taken as that corresponding to the total pressure. An example of the effect of total pressure on the first-order rate constant is shown in Fig. 5-9 for the reaction of cyclopropane to propylene at 765 K (Gardiner, 1972). Note that at very low pressures, the rate constant again becomes constant. This is due to the fact that activation occurs in collisions with the vessel wall rather than with gas molecules when the mean free path of the molecules is comparable to the dimension of the experimental vessel.

5.4 HETEROGENEOUS REACTIONS AND DEPOSITION KINETICS

Film deposition in semiconductor processing can be monocrystalline, polycrystalline, or amorphous. As discussed in detail in Sec. 5.2, monocrystalline film growth is an orderly process, as is polycrystalline film growth. Although no definitive work on the nature of adspecies is available, it is clear that some adspecies are always involved in crystalline film growth. It is not always clear, however, in the case of amorphous deposition that adspecies are involved. Nevertheless, it is evident in many cases that the substrate catalyzes the reactions responsible for the amorphous deposition.

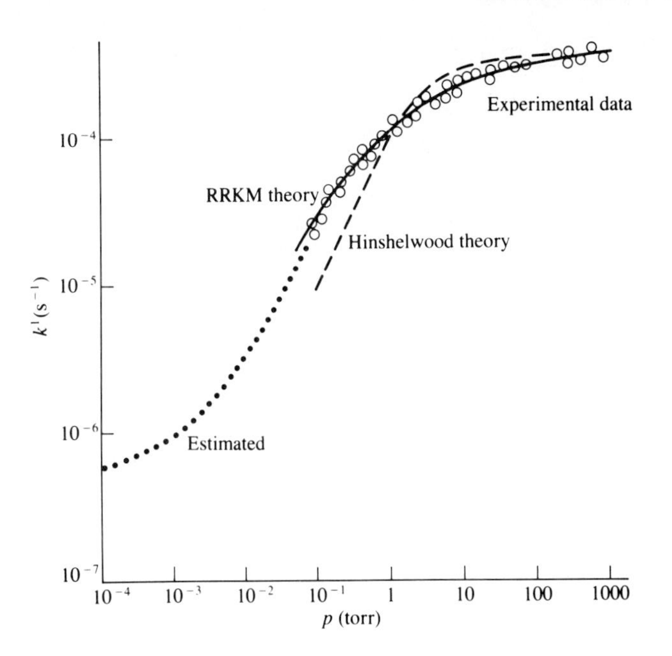

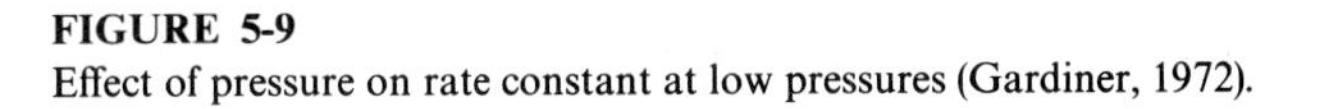
FIGURE 5-9
Effect of pressure on rate constant at low pressures (Gardiner, 1972).

The kinetics of crystalline film growth can be derived in a general form on the basis of the growth sequence described in Sec. 5.2. The gas phase reaction leading to the formation of the growth precursor is often equilibrium limited. Further, it is essential for uniform deposition that the gas phase composition be uniform. Thus, the deposition is typically carried out in the region where the composition is uniform, which can be realized under either equilibrium conditions or complete conversion to adspecies precursor. Therefore, it is often assumed that the gas phase reaction is at equilibrium. Under that assumption, Eqs. (5.1) through (5.3) are written as follows:

$$\mathrm{B}(g) \underset{}{\overset{K}{\rightleftharpoons}} \mathrm{A}(g) + \mathrm{B}_1(g) \tag{5.42}$$

$$\mathrm{A}(g) + \mathrm{S} \overset{K_a}{\rightleftharpoons} \mathrm{A}\cdot\mathrm{S} \tag{5.43}$$

$$\mathrm{A}\cdot\mathrm{S} \xrightarrow{k_s} \mathrm{A}_1(c) + \mathrm{B}_2(g) \tag{5.44}$$

It follows from Eq. (5.42) that

$$K = \frac{p_{\mathrm{A}}\, p_{\mathrm{B}_1}}{p_{\mathrm{B}}} = \frac{C_{\mathrm{B}_0} x_e^2 (RT)}{1 - x_e} \tag{5.45}$$

where x_e and C_{B_0} are the equilibrium conversion and the feed concentration of species B, respectively. The ideal gas law has been used for the partial pressures in Eq. (5.29). The equilibrium step of Eq. (5.43) can be written as

$$K_a C_A C_v = C_{A \cdot S} \tag{5.46}$$

where the adsorption equilibrium constant K_a is the ratio of the adsorption to the desorption rate constants, C_A is the gas phase (volume) concentration of the adspecies precursor, $C_{A \cdot S}$ is the surface concentration of the adspecies and C_v is the surface concentration of vacant adsorption sites. Since the rate of film growth is identical to the rate at which the adspecies is incorporated into the crystal lattice, denoted by $A_1(c)$, the rate of growth can be written using Eq. (5.44) as follows:

$$r_G = k_s C_{A \cdot S} \tag{5.47}$$

The site balance yields

$$C_t = C_v + C_{A \cdot S} \tag{5.48}$$

This balance along with Eq. (5.46) gives

$$\frac{C_v}{C_t} = \frac{1}{1 + K_a C_A} \tag{5.49}$$

The growth rate can be obtained by using Eqs. (5.46) and (5.49) in Eq. (5.47):

$$r_G = \frac{k_s C_t K_a C_A}{1 + K_a C_A} \tag{5.50}$$

The concentration of the adspecies precursor is usually not known. Therefore, it is necessary to relate C_A to C_{B0}. Noting that $C_A = C_{B0} x_e$, the growth rate becomes

$$r_G = \frac{k x_e C_{B0}}{1 + K_a x_e C_{B0}} \qquad \text{where } k = k_s C_t K_A \tag{5.51}$$

and x_e is given by Eq. (5.45).

A few comments are in order. If the equilibrium conversion of the source species is complete, $x_e = 1$ and the rate of growth becomes

$$r_G = \frac{k C_{B0}}{1 + K_a C_{B0}} \tag{5.52}$$

If there are adsorbed species other than the adspecies, the rate balance should include these species, in which case Eq. (5.48) becomes

$$C_t = C_v + \sum_i C_{i \cdot s} \tag{5.53}$$

where the $C_{i \cdot s}$ are the concentration of all the adsorbed species, including the adspecies. The equilibrium relationships also need to be written for all adsorbed species. Finally, the rate of growth given by Eq. (5.50) is in moles per time per surface area. The usual measurement of the growth rate is in length per time. The conversion of r_G to the linear growth rate G is

$$r_G = \frac{G\rho}{M_w} \tag{5.54}$$

where ρ and M_w are the density of the film material and its molecular weight, respectively.

It should be understood that the steps given by Eq. (5.42) through (5.44) only represent a macroscopic growth process that is consistent with the present understanding of the growth process. Microscopic growth steps are more complicated although no definitive details are available, even for silicon epitaxy. Even when detailed steps are known, the kinetics based on the macroscopic growth process should be sufficient for the purpose of describing the deposition processes.

Example 5.10. Polycrystalline silicon (polysilicon) is usually deposited in an N_2 carrier gas at 600 to 650 °C. Assuming that SiH_2 is the adspecies precursor, derive a rate expression for the polysilicon growth.

Solution. The gas phase reaction for the adspecies precursor is that given by Eq. (5.19):

$$SiH_4 \longrightarrow SiH_2 + H_2 \tag{A}$$

The sequence of adsorption followed by incorporation into the polycrystalline structure Si(p) is

$$SiH_2 + S \xrightleftharpoons{K_a} SiH_2 \cdot S \tag{B}$$

$$SiH_2 \cdot S \xrightarrow{k_s} Si(p) + H_2 \tag{C}$$

The sequence of Eqs. (B) and (C) is the same as that represented by Eqs. (5.43) and (5.44). Therefore, one has, from Eq. (5.50),

$$r_G = \frac{k_s C_t K_a C_{SiH_2}}{1 + K_a C_{SiH_2}} \tag{D}$$

where C_{SiH_2} is the concentration of SiH_2. The rate constant for the reaction of Eq. (A) at 600 °C can be calculated from the activation energy given in the text [below Eq. (5.19)]:

$$k = 2.14 \times 10^{13} \exp\left(-\frac{52{,}700}{2 \times 873}\right)$$

$$= 1.67 \ s^{-1}$$

If the residence time of the feed gas is of the order of a few seconds, the conversion would be complete since the conversion x is given by

$$x = 1 - \exp(-kt_r)$$

In such a case, $C_{\mathrm{SiH_2}} = C^{\circ}_{\mathrm{SiH_4}}$ where the superscript is for the feed composition. Then

$$r_G = \frac{kC^{\circ}_{\mathrm{SiH_4}}}{1 + K_a C^{\circ}_{\mathrm{SiH_4}}} \tag{E}$$

Otherwise, $C_{\mathrm{SiH_2}} = C^{\circ}_{\mathrm{SiH_4}} x$. For polysilicon deposition, the value of $K_a C_{\mathrm{SiH_4}}$ is sufficiently small compared to unity in Eq. (E) unless the temperature is low enough.

The growth sequence of Eqs. (5.42) through (5.44) does not necessarily lead to growth kinetics of the form of Eq. (5.50), as in the above example, when the kinetics are expressed in terms of the feed concentration. Consider the case where the formation of the adspecies precursor is severely limited by equilibrium so that conversion is quite small. The equilibrium relationship of Eq. (5.45) can then be rearranged to approximate x_e:

$$x_e = \left(\frac{K}{RTC_{\mathrm{B_0}}}\right)^{1/2} \tag{5.55}$$

Use of this in Eq. (5.51) leads to

$$r_G = \frac{k'C_{\mathrm{B_0}}^{1/2}}{1 + K'_a C_{\mathrm{B_0}}} \qquad \text{where } k' = k\left(\frac{K}{RT}\right)^{1/2},\ K'_a = K_a\left(\frac{K}{RT}\right)^{1/2} \tag{5.56}$$

It is seen that the growth rate becomes a function of $C_{\mathrm{B_0}}^{1/2}$ rather than $C_{\mathrm{B_0}}$. It is noted that low-pressure CVD, which tends to yield grown films of better thickness uniformity, shifts the equilibrium to the right of a reversible reaction (forward reaction) when the reaction results in a net increase in moles, as in Eq. (5.42). This is one form of the Le Chatelier's principle, which may be stated as follows: when a system is in equilibrium and one of the factors that determine the equilibrium point is altered, the system always reacts so as to tend to counteract the original alteration. The total pressure effect being considered can readily be verified from Eq. (5.29).

The deposition kinetics of an amorphous film, for which crystalline growth is always possible by either raising the temperature or lowering the precursor concentration as in silicon or gallium arsenide, also follows from the discussion in Sec. 5.2; i.e., an amorphous film results when the rate of adsorption of the adspecies precursor is much higher than the rate of surface diffusion of the adspecies. The macroscopic amorphous deposition can be represented by

$$\mathrm{B}(g) \underset{}{\overset{K}{\rightleftharpoons}} \mathrm{A}(g) + \mathrm{B_1}(g) \tag{5.42}$$

$$\mathrm{A}(g) + \mathrm{S} \xrightarrow{k_a} \mathrm{A \cdot S} \tag{5.57}$$

$$\mathrm{A \cdot S} \xrightarrow{k_s} \mathrm{A}(a) + \mathrm{B_2}(g) \tag{5.58}$$

where A(a) is the amorphous atom deposited. Equation (5.57) reflects the domination of adsorption over desorption, i.e., the desorption rate is negligible compared to the adsorption rate. Since the two surface rates in sequence should be the same at a constant rate, one has, from Eqs. (5.57) and (5.58),

$$r_G = k_a C_A C_v = k_s C_{A \cdot S} \tag{5.59}$$

Use of the site balance [Eq. (5.48)] and the equality in Eq. (5.59) yields

$$r_G = \frac{k_s k_a C_t C_A}{k_s + k_a C_A} \tag{5.60}$$

Unlike crystalline film growth, the adspecies does not migrate very much, if at all, because of the high surface concentration of adspecies and amorphous nuclei. It is likely, as reported by Scott *et al.* (1981), that the adspecies simply crosslink in the adsorbed state with neighboring adspecies for the formation of amorphous film. Therefore, k_s should be much larger than $k_a C_A$, in which case the growth rate becomes

$$r_G = k_a C_t C_A = k'_a C_A \qquad \text{where } k'_a = k_a C_t \tag{5.61}$$

If the precursor formation reaction [Eq. (5.42) above] is not limited by chemical equilibrium, one has, from Eqs. (5.42) and (5.61),

$$r_G = k'_a x C_{B0} \tag{5.62}$$

where x is the conversion of species B. In amorphous silicon deposition, the reaction of Eq. (5.19) is often considered the precursor reaction (Scott *et al.*, 1981; Beers and Bloem, 1982). Since the activation energy for adsorption is small, the activation energy for the amorphous growth would be close to that for the precursor reaction. In fact, the activation energy for the growth of amorphous silicon from SiH_4 has been reported to be 54 and 51 kcal/mol, respectively, in the above references. These compare with the activation energy for the precursor reaction [Eq. (5.19)] of 57 kcal/mol. When equilibrium prevails for the precursor reaction of Eq. (5.42), the growth rate is that given by Eq. (5.62) with x replaced by the equilibrium conversion.

Example 5.11. Show that the activation energy for amorphous silicon growth is approximately equal to the activation energy of the precursor reaction when conversion to the precursor is small.

Solution. The precursor reaction of Eq. (5.19) is of first order. Therefore, the conversion in a growth apparatus is given by

$$x = 1 - \exp(-kt_r)$$

where t_r = residence time for plug glow or batch reactor.

When the conversion is small, x can be approximated by

$$x = kt_r$$

Thus, the growth rate of Eq. (5.56) becomes

$$r_G = k'_a k t_r C_{B0}$$

Taking the logarithm,

$$\ln r_G = \ln t_r C_{B_0} + \ln (k'_a k)_0 - \frac{E_a + E}{RT}$$

where E_a is the activation energy for adsorption and E is the activation energy for reaction. If E_a is small compared to E, the activation energy for the growth is approximately E, which is 57 kcal/mol for the reaction of Eq. (5.19).

Any structure that is not crystalline is called amorphous. The amorphous structure can cover a wide range of "randomness." It can still have a certain structure(s), but the structure is random and differs from point to point. As the deposition temperature is increased the structure becomes more orderly. Therefore, the nature of high-temperature amorphous deposition, made possible by relatively high source gas pressure, is different from that of low-temperature amorphous deposition. Because of the presence of some solid-state order at high temperatures, the adspecies are likely to migrate again and the full rate form of Eq. (5.60) needs to be used in place of Eq. (5.61).

Consider the kinetics of epitaxial silicon film growth. The kinetics are largely dependent on the source gas, type of carrier gas, total pressure, and the concentration of HCl gas since these conditions change the adspecies precursor. It is generally believed that SiH_2 and $SiCl_2$ are the adspecies precursors (Bloem, 1980) for relatively high total pressure (say, higher than 0.1 torr around 1000 °C or 10 torr around 800 °C). In addition to etching caused by the reaction between HCl and substrate, the concentration of HCl determines the extent to which the two adspecies contribute to the growth rate (Lee, 1984). For silicon epitaxy, based on silane in hydrogen carrier gas with a small amount of hydrochloric acid, the gas phase precursor reactions can be written as follows:

$$SiH_4 \longrightarrow SiH_2 + H_2 \tag{5.63}$$

$$SiH_2 + 2HCl \rightleftharpoons SiCl_2 + 2H_2 \tag{5.64}$$

if SiH_2 and $SiCl_2$ are the adspecies precursors. Epitaxy is usually carried out at temperatures higher than 900 °C. At these temperatures, the conversion of SiH_4 to SiH_2 is practically complete. The conversion, in fact, is 99 percent at 800 °C, even for a residence time of only 0.01 s at atmospheric pressure. Since conversion is complete, the equilibrium relationship for Eq. (5.64) can be written as

$$K = \frac{p_0 x (p^0_{H_2} + 2xp_0 + p_0)^2}{p_0 (1 - x)(p^0_{HCl} - 2xp_0)^2} \tag{5.65}$$

where p_0 is the feed silane partial pressure, x is the conversion of SiH_2 to $SiCl_2$, and the superscript denotes the feed value. Since the feed composition is usually less than 1 mol % and that of hydrogen is greater than 99 percent, one can approximate Eq. (5.65) to write

$$K' = \frac{x}{(1 - x)(p^0_{HCl} - 2xp_0)^2} \qquad \text{where } K' = \frac{K}{(p^0_{H_2})^2} \tag{5.66}$$

The growth sequence by the two adspecies can be written as follows:

$$\begin{aligned} SiH_2 + S &\underset{}{\overset{K_A'}{\rightleftharpoons}} SiH_2 \cdot S \\ SiH_2 \cdot S &\overset{k_1'}{\longrightarrow} Si(c) + H_2 \\ SiCl_2 + S &\overset{K_B'}{\rightleftharpoons} SiCl_2 \cdot S \\ SiCl_2 \cdot S + H_2 &\underset{k_r'}{\overset{k_2'}{\rightleftharpoons}} Si(c) + 2HCl \end{aligned} \tag{5.67}$$

Note that the reverse path of the last step represents etching by HCl. The procedures followed earlier yield

$$r_G = \frac{k_1(1-x)p_O + k_2 x p_O - k_r(p_{HCl}^0 - 2xp_O)^2}{1 + K_A(1-x)p_O + K_B x p_O} \tag{5.68}$$

where the definitions of the rate constants can be found in the Notation and x is given by Eq. (5.66). It is seen that the growth rate consists of three parts: growth due to the adspecies, SiH_2 [first term in Eq. (5.68)], growth due to the adspecies, $SiCl_2$ (second term), and etching by HCl (third term). The relative growth rates due to SiH_2 and $SiCl_2$ play an important role in the epitaxial lateral overgrowth discussed in Sec. 5.2. The kinetics based on the two adspecies have been shown to describe the growth well, regardless of the silicon source material, and also the effects of adding HCl and using different mixtures of carrier gas (Stassinos *et al.*, 1985).

Epitaxy, as well as low-pressure polycrystalline deposition, particularly at pressures lower than 1 torr, could involve different adspecies than those at atmospheric pressure. Meyerson *et al.* (1986) have found that the rate constant for silane pyrolysis at low pressures (as discussed in Sec. 5.3) is so small that the reaction of Eq. (5.63) should be negligible. It is believed that dissociatively adsorbed silane is the growth precursor at low pressures and low temperatures where the reaction of Eq. (5.63) cannot be appreciable.

The mechanism for epitaxial growth by MOCVD based on gallium alkyls and arsenic hydride or chloride is still too unclear to write a macroscopic growth sequence. Perhaps what is understood best is that gallium atoms could be the adspecies. For the arsenic part, however, it is not clear in the case of arsenic hydride whether the hydride itself, As_2, or As_4 is the adspecies. It is known, however, that the gas phase reaction of Eq. (5.35) is complete at the typical growth temperature but that the conversion of arsenic hydride is not (Leys and Veenvliet, 1981). In addition to the uncertainty regarding the adspecies, it is not clear whether the growth takes place layer by layer, i.e., As layer followed by Ga layer, or in layers of GaAs. Another factor that has to be considered is whether the substrate surface is primarily Ga- or As-exposed.

Example 5.12. Assuming that Ga atoms and As_2 are the adspecies precursors and that the growth is in layers of GaAs for epitaxial MOCVD based on TMG and AsH_3, write a macroscopic sequence for the growth and then derive the growth kinetics. Assume that As_2 adsorbs dissociatively.

Solution. The adsorption/desorption of the precursors is

$$\mathrm{Ga} + \mathrm{S} \underset{}{\overset{K_G}{\rightleftharpoons}} \mathrm{Ga \cdot S}$$

$$\mathrm{As_2} + 2\mathrm{S} \underset{}{\overset{K_A}{\rightleftharpoons}} 2\mathrm{As \cdot S}$$

The incorporation into the lattice can be written as

$$\mathrm{Ga \cdot S} + \mathrm{As \cdot S} \xrightarrow{k_s} \mathrm{GaAs}(c)$$

For the adsorption/desorption equilibrium, one has

$$K_G C_G C_v = C_{G\cdot S} \tag{A}$$

$$K_A C_A C_v^2 = C_{A\cdot S}^2 \tag{B}$$

where C_G and C_A, respectively, are the concentrations of Ga and As_2, and $C_{G\cdot S}$ and $C_{A\cdot S}$, respectively, are the surface concentrations of Ga · S and As · S. One also has

$$C_t = C_v + C_{G\cdot S} + C_{A\cdot S}$$

$$= C_v[1 + K_G C_G + (K_A C_A)^{1/2}] \tag{C}$$

which follows from Eqs. (A) and (B) when they are used in Eq. (C). The rate of growth is

$$r_G = k_s C_{G\cdot S} C_{A\cdot S}$$

$$= k_s C_v^2 K_G C_G (K_A C_A)^{1/2} \tag{D}$$

which follows from Eqs. (A) and (B) for $C_{G\cdot S}$ and $C_{A\cdot S}$. Use of Eq. (C) in (D) yields

$$r_G = \frac{k C_G C_A^{1/2}}{[1 + K_G C_G + (K_A C_A)^{1/2}]^2} \quad \text{where } k = k_s C_t^2 K_G K_A^{1/2} \tag{E}$$

One can proceed further, if desired, to express the rate in terms of the feed concentration. Since the reaction of Eq. (5.35) is complete at the typical growth temperature, $C_G = C_{TMG}^\circ$, where the superscript denotes the feed composition. Further, the concentration C_A is $xC_{AsH_3}^\circ$, where x is the conversion of AsH_3 to As_2. Therefore, one has, from Eq. (E),

$$r_G = \frac{k C_{TMG}^\circ (x C_{AsH_3}^\circ)^{1/2}}{[1 + K_G C_{TMG}^\circ + (K_A x C_{AsH_3}^\circ)^{1/2}]^2}$$

As discussed in the introduction, there can be more than one proposed mechanism that matches the observed trends in behavior. This fact should be kept in mind when writing or using rate expressions. On the other hand, direct experimental evidence for or against a certain step should not be ignored.

Example 5.13. Green *et al.* (1984) carried out a detailed study of polycrystalline Al deposition by MOCVD. Aluminum is a typical metal contact for IC circuits. The metal deposition for the contact and device interconnection is commonly referred to as "metallization." Their conclusions are:

1. The triisobutyl aluminum used for the metallization decomposes reversibly in the gas phase in the presence of isobutylene:

$$\underset{\text{(CA)}}{[(CH_3)_2CHCH_2]_3Al} \overset{K}{\rightleftharpoons} \underset{\text{(DB)}}{[(CH_3)_2CHCH_2]_2AlH} + (CH_3)_2C{=}CH_2 \quad \text{(A)}$$

2. The dibutyl aluminum hydride decomposes catalytically on the substrate surface for aluminum deposition and the overall reaction is

$$[(CH_3)_2CHCH_2]_2AlH \longrightarrow Al + 2(CH_3)_2C{=}CH_2 + \tfrac{3}{2}H_2 \quad \text{(B)}$$

Assuming that catalytic decomposition takes place on the surface before surface migration, such that the adspecies is AlH_x, derive the deposition kinetics.

Solution. The catalytic surface reaction can be written as

$$[(CH_3)_2CHCH_2]_2AlH(a) \xrightarrow{k_A} AlH_x(a) + 2(CH_3)_2C{=}CH_2 + (x-1)H_2/2 \quad \text{for } x > 1 \quad \text{(C)}$$

where (a) indicates the adsorbed state. The deposition can be represented by

$$AlH_x(a)(=AlH_x \cdot S) \xrightarrow{k_s} Al(c) + xH_2/2 \quad \text{(D)}$$

Thus, the deposition rate is

$$r_G = k_s(C_{AlH_x} \cdot S) \quad \text{(E)}$$

At a constant deposition rate, the rate of formation of adspecies is equal to the deposition rate. The catalytic reaction leading to the formation of the adspecies is

$$[(CH_3)_3CHCH_2]_2AlH + S \overset{K_a}{\rightleftharpoons} [(CH_3)_2CHCH_2]_2AlH \cdot S \quad \text{(F)}$$

$$[(CH_3)_2CHCH_2]_2AlH \cdot S \xrightarrow{k_A} AlH_x \cdot S + 2(CH_3)_2CHCH + xH_2/2 \quad \text{(G)}$$

From Eqs. (F) and (G),

$$K_a C_{DB} C_v = C_{DB \cdot S} \quad \text{(H)}$$

$$r_c = k_A C_{DB \cdot S} \quad \text{(I)}$$

where r_c is the rate of the adspecies formation. The site balance is

$$C_t = C_v + C_{DB \cdot S} + C_{AlH \cdot S}$$

or

$$C_t - C_{AlH_x} \cdot S = C_v(1 + K_a C_{DB}) \quad \text{(J)}$$

Thus, from Eqs. (I) and (J),

$$r_c = \frac{k_A K_a C_{DB}(C_t - C_{AlH_x} \cdot S)}{1 + K_a C_{DB}}$$

Since $r_G = r_c$, this identity can be used to obtain

$$C_{AlH_x} \cdot S = \frac{k_A K_a C_{DB}}{k_s(1 + K_a C_{DB}) + k_A K_a C_{DB}}$$

Use of Eq. (I) in Eq. (E) yields

$$r_G = \frac{k_s k_A K_a C_{DB}}{k_s(1 + K_a C_{DB}) + k_A K_a C_{DB}} \tag{K}$$

For the concentration, C_{DB}, one can use the equilibrium of Eq. (A):

$$K_e = \frac{P_0 RT x_e^2}{1 - x_e} \quad \text{and} \quad C_{DB} = (C_{CA})x_e = (P_0 RT)x_e \tag{L}$$

where x_e is the equilibrium conversion of CA [see Eq. (A)] and P_0 is the partial pressure of triisobutyl aluminum in the feed. Equation (L), when used in Eq. (K), then gives the deposition rate.

The example just considered should reveal that the common practice of plotting the logarithm of growth/deposition rate against the inverse of temperature for determining the activation energy can lead to erroneous conclusions. For instance, the growth rate given in Eq. (5.13) does not depend on only one rate constant and therefore the plot may consist of segments of straight lines for various ranges of temperature. Interpretation of the result for a small temperature range or an average straight line would lead to erroneous conclusions about the activation energy and thermal behavior of the system.

Silicon dioxide films are frequently used in ICs as insulators between metal layers and as passivating layers over devices. They are also used as an ion implantation and diffusion mask, and also as a capping layer over doped regions to prevent outdiffusion during heat cycles. One method of silicon dioxide deposition is to use tetraethylorthosilane (TEOS) in the 700 to 800 °C region. Adams and Capio (1979) reported that the following rate expression represented their deposition rate well:

$$r_{SiO_2} = \frac{k p_{TEOS}^{1/2}}{1 + K p_{TEOS}^{1/2}} \tag{5.69}$$

They found that oxygen, at concentrations of 0 to 20 percent by volume, did not have any effect on the deposition rate. The growth of silicon dioxide has largely been superseded by deposition involving oxidation of silane at around 400 °C. Maeda and Nakamura (1981) correlated their data in the following form:

$$r_{SiO_2} = k p_{SiH_4}^n p_{O_2}^m \tag{5.70}$$

where the individual orders of the reaction, n and m, depend on whether the deposition rate increases ($n = m = \frac{1}{2}$; Baliga and Ghandhi, 1973) or decreases ($n = 2$, $m = -1$) with increasing oxygen partial pressure. At temperatures lower than 500 °C, the following reaction is favored:

$$SiH_4 + O_2 \longrightarrow SiO_2 + 2H_2 \tag{5.71}$$

whereas at temperatures higher than 500 °C, the following reaction is favored:

$$SiH_4 + 2O_2 \longrightarrow SiO_2 + 2H_2O \tag{5.72}$$

Silicon nitride (Si_3N_4) films can also be used as insulating or passivating films. It is the passivating film of choice for silicon devices because it serves as a very good barrier to the diffusion of both water and sodium, which can cause devices to corrode. Ammonia is the usual nitridation reactant, and the silicon source is either silane (700 to 900 °C) or dichlorosilane (650 to 750 °C):

$$3SiH_4 + 4NH_3 \longrightarrow Si_3N_4 + 12H_2 \tag{5.73}$$

$$3SiH_2Cl_2 + 4NH_3 \longrightarrow Si_3N_4 + 6HCl + 6H_2 \tag{5.74}$$

Although no mention of doping has been made so far, almost all film deposition, except for insulating and passivating films, involves codeposition of a dopant. In a typical film deposition, therefore, the dopant species deposits as well as the host species (e.g., silicon). It is necessary to understand how the presence of adsorbed dopant on the surface of the growing film affects the overall growth kinetics. Because of the depletive chemisorption phenomenon (Clark, 1970), further adsorption of dopant species around the adsorbed dopant becomes depletive. This means that the adsorption sites around the adsorbed dopant, i.e., circle of influence, are not favorable for further adsorption of dopant species, although the host species can adsorb. In this circle of influence, the adsorption of host species is promoted if the dopant atom is electron-deficient compared to the host atom but is retarded if it is electron-excessive. In general, therefore, acceptor atoms promote film growth while donor atoms retard growth. When the population of the adsorbed dopant species is such that the circles of influence cover the whole surface, the growth rate approaches an asymptotic value and no further effects of dopant can be detected. This asymptotic growth rate, r_{asymp}, and the growth rate in the absence of dopant, r', can be used to describe the growth as affected by dopant (Stassinos and Lee, 1986), r_G, as follows:

$$r_G = \frac{(1 + \alpha)yr_{\text{asymp}} + r'}{1 + y + \alpha y} \tag{5.75}$$

where α is the number of adsorption sites in the circle of influence and y is defined as

$$y = \frac{K_d C_d}{1 + KC} \tag{5.76}$$

Here C_d and C, respectively, are the dopant and gas phase concentrations, and K_d and K are the corresponding adsorption equilibrium constants. As discussed in Sec. 5.2, the dopant effect will play a more important role as devices become smaller and shallower for relatively low doping levels (say, less than 10^{17} cm^{-3}).

Example 5.14. Everstyn and Put (1973) reported the growth rate of polysilicon film at 700 °C as a function of dopant concentration. Noting that, at this high temperature, the adsorption equilibrium constant K_a [K in Eq. (5.70)] in Eq. (E) of Example 5.10 is small such that $r' = kC^{\circ}_{SiH_4}$, show that a linear plot in the form of $y = ax + b$ can be made from Eq. (5.75) for the determination of the parameters. Note that r' and r_{asymp} are known *a priori* from the data.

Solution. Since $K = 0$, $y = K_d C_d$ from Eq. (5.76). Rearranging Eq. (5.75) yields

$$r_G + y(1 + \alpha)r_G = y(1 + \alpha)r_{\text{asymp}} + r'$$

or

$$r_G = y(1 + \alpha)(r_{\text{asymp}} - r_G) + r'$$

Since $y = K_d C_d$, one has

$$r_G = K_d(1 + \alpha)C_d(r_{\text{asymp}} - r_G) + r' \tag{A}$$

Equation (A) is in the form of $y = ax + b$, where x is $C_d(r_{\text{asymp}} - r_G)$ and b is r'. The slope of the linear plot based on the data yields the value of $K_d(1 + \alpha)$. The plot should go through the point corresponding to the value of r', which is known at $C_d = 0$.

5.5 GAS-SOLID REACTIONS

The gas phase reaction of interest in semiconductor processing is almost exclusively oxidation of the native silicon substrate. Oxidation of native gallium arsenide does not produce a good insulator. As the current interest in nitridation of silicon (silane) indicates, use of other gas phase reactions will appear as the need arises. Although CVD silicon dioxide can also be used for the same purpose as thermally grown native silicon dioxide, thermal oxidation has been the preferred technique when a low charge density level is desired at the interface between the oxide and silicon.

In general, a gas-solid reaction can be represented as follows:

$$\text{A}(g) + b\text{B}(s) \longrightarrow m\text{M}(s) + q\text{Q}(g) \tag{5.77}$$

Here a gaseous (g) species A reacts with a solid (s) species B to form a solid product M and a gaseous product Q. The lower case letters are stoichiometric coefficients. For silicon oxidation, Eq. (5.77) is

$$O_2(g) + \text{Si}(s) \longrightarrow \text{SiO}_2(s)$$

The thickness of the oxide layer is usually small, and the substrate is flat so that a one-dimensional, flat-plate geometry can be assumed for the oxidation. The transport of the gaseous reactant from gas phase to the interface between the solid product layer and the solid reactant can be visualized as shown in Fig. 5-10. The concentration decreases because of the mass transfer resistance at the gas phase interface. The transport within the product layer is by diffusion and no reaction takes place. Therefore, the concentration in the product layer decreases linearly with the distance into the solid. A condition necessary for the shell progressive model (SPM; e.g., Levenspiel, 1972) shown in Fig. 5-10, in which the completely reacted outer shell of product layer moves progressively inward with time, is that the rate of diffusion in the product layer be much smaller than the potential rate of the reaction. In other words, the reaction is limited by the rate at which the reactant diffuses, which is usually valid for semiconductor material. For the SPM, the one-dimensional mass balance for the gaseous reactant in the

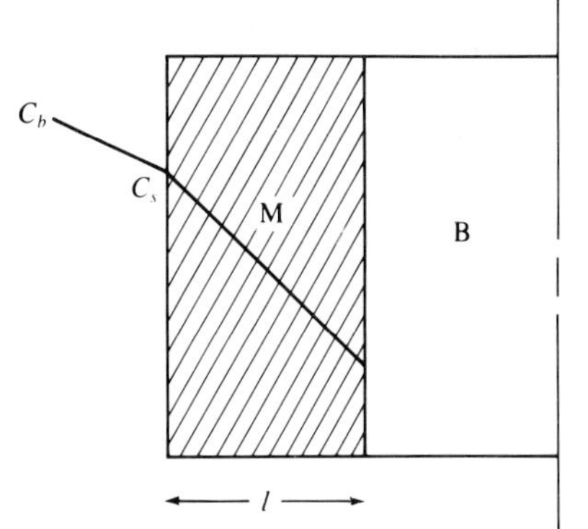

FIGURE 5-10
Concentration gradients for the gas-solid reaction.

product shell is

$$\beta \frac{dC}{dt} = \frac{d}{dx}\left(D \frac{dC}{dx}\right) \tag{5.78}$$

where β is the porosity of the product shell, D is the diffusivity of the reactant in the product shell, C is the reactant concentration, t is time, and x is the axial coordinate. If one makes a pseudo steady-state assumption, Eq. (5.78) becomes

$$\frac{d}{dx}\left(D \frac{dC}{dx}\right) = 0 \tag{5.79}$$

The relative error involved in making this assumption is $C_b/(6b\rho_s)$ (Bischoff, 1963), where C_b is the bulk gas concentration and ρ_s is the density of the product shell. Since C_b/ρ_s is approximately 10^{-3} for gas-solid reactions, one can readily make the pseudo steady-state assumption.

Under constant conditions, the rate of mass transfer, r_m, from the gas phase to the solid surface should be equal to the rate at which the gaseous reactant diffuses through the product shell, r_d, which in turn should be equal to the rate of reaction at the solid product–reactant interface, r_c (rate based on interfacial area):

$$r_m = r_d = r_c \tag{5.80}$$

If one uses the film mass transfer coefficient k_m for external mass transfer, one has, for r_m,

$$r_m = k_m(C_b - C_s) \qquad \text{(mol/area time)} \tag{5.81}$$

where C_s is the concentration of the gaseous reactant at the solid surface. Because of the pseudo steady-state assumption, $dC/dx = 0$ [Eq. (5.79)], which in turn means that the concentration changes linearly with x. Thus, r_d is given by

$$r_d = -D \frac{dC}{dx} = D\left(\frac{C_s - C_i}{l}\right) \tag{5.82}$$

where l is the thickness of the product shell, as shown in Fig. 5-10. The reaction takes place at the concentration at the solid product–reactant interface, C_i, and therefore one has

$$r_c = kf(C_i) \qquad (= kC_i \text{ for first-order reaction}) \tag{5.83}$$

where $f(C_i)$ is the concentration dependence of the rate of reaction. The concentration readily measurable is the bulk gas concentration C_b. The two identities in Eq. (5.80) can be used to express C_s and C_i in terms of C_b. Explicit expressions for these are possible for first-order kinetics, and are

$$\frac{C_i}{C_b} = \frac{1}{1 + k/k_m + kl/D} \tag{5.84}$$

$$\frac{C_s}{C_b} = \frac{1 + kl/D}{1 + k/k_m + kl/D} \tag{5.85}$$

From the definition of r_c (rate based on the gas-solid interfacial area A_i), one has

$$\left(\frac{m}{A_i}\right)\frac{dN_{\text{M}}}{dt} = \left(\frac{1}{A_i}\right)\frac{dN_{\text{A}}}{dt} = r_c \tag{5.86}$$

where N_{A} and N_{M}, respectively, are the number of moles A and M in Eq. (5.83). Note that the first identity in Eq. (5.86) follows from the reaction stoichiometry in Eq. (5.83). The moles of M created are

$$N_{\text{M}} = \frac{\rho_{\text{M}} A_i l}{M_{\text{M}}} = \rho'_{\text{M}} A_i l \qquad \text{where } \rho'_{\text{M}} = \frac{\rho_{\text{M}}}{M_{\text{M}}} \tag{5.87}$$

Use of Eq. (5.87) for dM_{M}/dt and Eqs. (5.83) and (5.84) for r_c in Eq. (5.86) yields

$$m\rho'_{\text{M}} \frac{dl}{dt} = \frac{kC_b}{1 + k/k_m + kl/D} \tag{5.88}$$

This can be solved with the condition $l = 0$ at $t = 0$ to give

$$l^2 + Al = Bt \tag{5.89}$$

where

$$A = 2D\left(\frac{1}{k} + \frac{1}{k_m}\right) \tag{5.90}$$

$$B = 2D\frac{C_b}{m\rho'_{\text{M}}} \tag{5.91}$$

Equation (5.89) describes how the product layer thickness l increases with time.

In silicon oxidation, a slightly modified form of Eq. (5.89) is known as the Deal-Grove model (Deal and Grove, 1965). In their model, the equilibrium concentration of oxidant in the silicon oxide, which is expressed as Hp (where H is Henry's constant and p is the partial pressure of the oxidant), was used in place of C_b. In the literature, the constant B is referred to as the parabolic rate constant and B/A as the linear rate constant. Experimental data of the oxide layer thickness, l, obtained as a function of time at a fixed temperature can be analyzed for the determination of A and B, with Eq. (5.89) rewritten in the following form:

$$l(t) = \frac{Bt}{l} - A \tag{5.92}$$

Example 5.15. The silicon oxide growth data at 920 °C for oxidation in wet oxygen that Deal and Grove obtained are given below:

t, h	0.11	0.30	0.40	0.50	0.60
l, μm	0.041	0.100	0.128	0.153	0.177

(*a*) Determine the linear and parabolic rate constants. The values of the linear and parabolic constants that they determined for various oxidation temperatures for wet oxidation are given below:

T, °C	B, μm^2/h	B/A, μm/h
920	0.203	0.406
1000	0.287	1.27
1100	0.510	4.64
1200	0.720	14.40

Suppose the activation energies are obtained based on the plots of ln B versus $1/T$ and ln B/A versus $1/T$. Explain what these activation energies represent.

(*b*) Determine the activation energies for the rate constants.

Solution

(*a*) The first set of data can be reduced as follows for Eq. (5.92):

l, μm	0.041	0.100	0.128	0.153	0.177
t/l, h/μm	2.683	3.000	3.125	3.268	3.390

Since Eq. (5.92) is in the form of $y = ax + b$, a plot of l versus t/l can be made and a straight line drawn. The slope of the line should yield

$$B = 0.2\ \mu\text{m}^2/\text{h}$$

The intercept of the line should yield

$$A = 0.50\ \mu\text{m}$$

Therefore, the linear rate constant $B/A = 0.40$ μm/h and the parabolic rate constant $B = 0.20$ μm^2/h.

(*b*) The parabolic constant [Eq. (5.91)] is

$$B = \frac{2DC_b}{M\rho'_M} \tag{A}$$

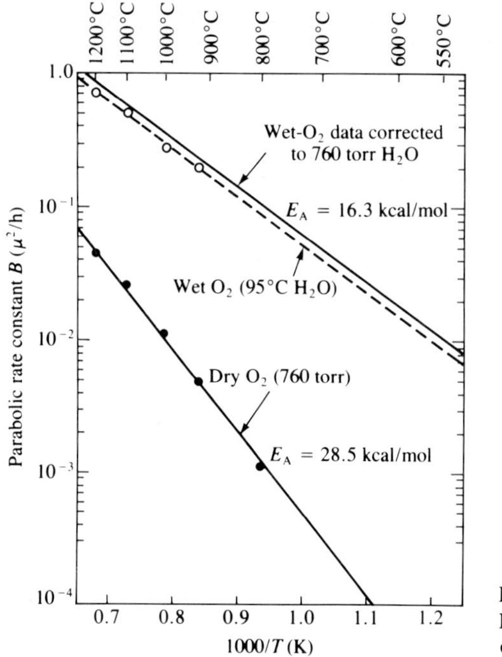

FIGURE 5-11
Linear plot for the parabolic rate constant.

Since the diffusivity is represented by $D = D_0 \exp(-E_D/RT)$, one can rewrite Eq. (A) as

$$\ln B = \ln\left(\frac{2D_0 C_b}{M\rho'_M}\right) - \frac{E_D}{RT} \tag{B}$$

Thus, a plot of ln B versus $1/T$ should yield the activation energy for diffusion of the oxidant in solid silicon dioxide. This plot based on the data in the second table of the problem is shown in Fig. 5-11. The activation energy determined from the slope is 16.3 kcal/mol, which is close to the activation energy of 18.3 kcal/mol for the diffusion of water in fused silica.

The linear rate constant B/A is

$$\frac{B}{A} = \left(\frac{C_b}{M\rho'_M}\right)\left(\frac{k}{1 + k/k_m}\right) \tag{C}$$

which follows from Eqs. (5.90) and (5.91). The mass transfer coefficient is a weak function of temperature and k/k_m is quite small. Therefore, Eq. (C) can be approximated by

$$\frac{B}{A} = \frac{C_b k}{M\rho'_M} = \frac{C_b k_0}{M\rho'_M} \exp\left(-\frac{E}{RT}\right) \tag{D}$$

where E is the activation energy for the reaction. Thus, the slope of a plot of ln (B/A) versus $1/T$ should yield the value of E. This plot is shown in Fig. 5-12. The activation energy from the plot is 45.3 kcal/mol.

The oxidants used for silicon oxidation are oxygen, steam, and a mixture of steam and oxygen. While the Deal-Grove model in the form of Eq. (5.89) works well for all oxide thickness ranges for oxidation in steam or mixtures of steam

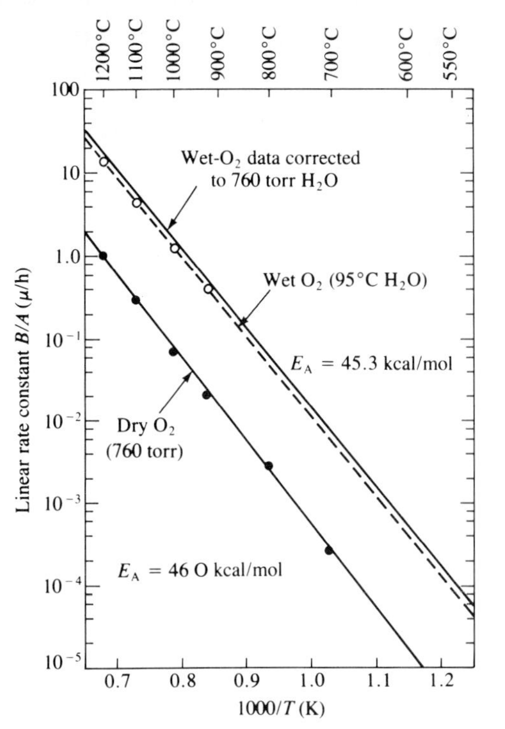

FIGURE 5-12
Linear plot for the linear rate constant.

and oxygen, oxidation in dry oxygen (say, less than 5 ppm H_2O) behaves as if there were an initial oxide layer before oxidation. Solving Eq. (5.88) with the initial condition, $l = l_i$ at $t = 0$, leads to the following form known as the Deal-Grove model:

$$l^2 + \mathrm{A}l = B(t + t_r); \qquad t_r = (l_i^2 + \mathrm{A}l_i)/B \tag{5.93}$$

The activation energies obtained using the correction for dry oxidation are: 28.5 kcal/mol for diffusion, which compares with 27 kcal/mol for the diffusivity in fused silica, and 46 kcal/mol for the reaction. The Deal-Grove model applies over wide ranges of temperature, total pressure and partial pressure of the oxidant. Initial rapid growth (compared to the rate predicted by the Deal-Grove model) in dry oxygen, that typically occurs for thicknesses less than 30 nm, has been the subject of active research recently because of ever-decreasing device sizes. For this thin oxide growth, Massoud *et al.* (1982) proposed the following empirical expression:

$$\frac{dl}{dt} = \frac{kC_b}{M\rho'_{\mathrm{M}}(1 + k/k_m + kl/D)} + C_2 \exp\left(-\frac{l}{L_2}\right) \tag{5.94}$$

where the last term is the correction term and C_2 and L_2 are empirical parameters. Various mechanisms have been proposed such as space charge effects (Deal and Grove, 1965), oxide structure effects (Revesz and Evans, 1969), oxide stress effects (Irene, 1983), oxygen solubility in the oxide, and a surface layer with a distribution of additional sites of oxidation (Massoud *et al.*, 1985). There does not yet appear to be any definitive proof for any of the proposed mechanisms (Blanc, 1986). The evidence based on TEM that the initial Si-SiO_2 interface is nonuniform (irregular curved surface), but that the interface becomes flatter as the oxidation progresses (Carim and Sinclair, 1986), may have a direct bearing on the behavior of thin oxide growth.

The results obtained by Deal and Grove were for lightly doped silicon (around 10^{16} cm^{-3}). For highly doped silicon (say, larger than 10^{19} cm^{-3}), the oxide growth rate in general increases with doping concentration. As the growth takes place, segregation of dopant at the Si-SiO_2 interface occurs, similar to the segregation observed in bulk crystal growth. If the dopant segregates into the oxide and remains, e.g., boron, the bond structure in the silica weakens, which permits both increased incorporation and diffusivity of the oxidizing species through the oxide (Katz, 1983). Therefore, the oxidation rate increases. Impurities that segregate into the silicon such as phosphorus also enhance the growth rate. Halogen species such as HCl, which is introduced to enhance dielectric breakdown strength, tend to increase the rate, at least in a dry oxygen environment (Hess and Deal, 1977). As might be expected, the growth rate depends on crystallographic orientation. In general, a (111) substrate gives a higher growth rate than a (100) substrate because more silicon-silicon bonds are available (Ligenza, 1961) to the oxidant. Polysilicon oxidation kinetics should follow a similar trend as the monocrystalline silicon oxidation in the context of gas-solid reaction kinetics. However, grain boundaries and grain orientation should play a major role when detailed kinetic behavior is examined.

Although silicon oxidation, in general, follows first-order intrinsic kinetics, not all gas-solid reactions are of first order. Typically, the intrinsic kinetics are represented in the following form:

$$r_c = kC^n \tag{5.95}$$

where f in Eq. (5.83) is C^n. For this more general case, the thickness of the product layer can be related to time (Hu, 1983) as follows:

$$l = \frac{w^{-n} - w^{1-n}}{a} \qquad \text{where } w = \frac{C_i}{C_b} \tag{5.96}$$

and w is related to time as follows:

$$t = \begin{cases} \dfrac{1}{b}\left[\dfrac{w^{-2n} - 1}{2} + \dfrac{(1-n)(w^{1-2n} - 1)}{2n - 1}\right] & \text{for } n \neq \frac{1}{2} \\ \dfrac{1}{2b}\left(\dfrac{1}{w} - 1 - \ln w\right) & \text{for } n = \frac{1}{2} \end{cases} \tag{5.97}$$

where

$$a = \frac{kC_b^{n-1}}{D} \qquad b = \frac{akC_b^n}{m\rho'_M} \tag{5.98}$$

These relationships were obtained by neglecting the external mass transfer resistance (large k_m) such that $C_b = C_s$. Note that a and b in these equations are similar to A and B in Eq. (5.89). Since $0 < w < 1$, a value of w, when used in Eq. (5.97), yields a value of t. The same value of w is used in Eq. (5.96) to obtain the value of l corresponding to the calculated value of t. This can be repeated for the whole range of w to generate l as a function of t.

Example 5.16. Suppose for a gas solid reaction that $a = 2\ \mu\text{m}^{-1}$, $b/a = 0.4\ \mu\text{m/h}$, and $n = 0.3$. Generate l versus t corresponding to $w = 1$, $w = 0.8$, and $w = 0.5$.

Solution. For the values given, Eqs. (5.96) and (5.97) become

$$l = 0.5(w^{-0.3} - w^{0.7}) \qquad \text{in } \mu\text{m} \qquad \text{(A)}$$

$$t = \frac{1}{0.8}\left[\frac{w^{-0.6} - 1}{2} - \frac{0.7(w^{0.4} - 1)}{0.4}\right] \qquad \text{in h} \qquad \text{(B)}$$

When $w = 1$, Eq. (A) yields $l = 0$ and Eq. (B) yields $t = 0$, which is expected since $l = 0$ and $w = 1$ at $t = 0$, i.e., $C_i = C_s$ at $t = 0$. For $w = 0.8$, Eq. (A) becomes

$$l = 0.5(0.8^{-0.3} - 0.8^{0.7}) = 0.214\ \mu\text{m}$$

and Eq. (B) becomes

$$t = \frac{1}{0.8}\left[\frac{0.8^{-0.6} - 1}{2} - \frac{0.7(0.8^{0.4} - 1)}{0.4}\right] = 0.221\ \text{h}$$

Similarly, one can obtain l versus t for other w values. These values are listed below:

t, h	0 ($w = 0$)	0.221 ($w = 0.8$)	0.852 ($w = 0.5$)
l, μm	0	0.214	0.308

Since $0 < w < \infty$, where $w = 0$ corresponds to $t = \infty$, one can generate l versus t for the whole time range of interest.

As the device dimension decreases, a need arises for thin insulator films (of the order of 10 nm). The top part of such an oxide is nitrided with NH_3 through rapid thermal nitridation (Ito *et al.*, 1980; Moslehi *et al.*, 1986) to enhance the dielectric strength of oxide insulator. The general analytical framework of gas-solid reactions should apply here also.

5.6 PHOTOCHEMICAL REACTIONS

The desire for low-temperature processing to minimize undesirable dopant redistribution has brought about an extensive use of photochemical vapor deposition. Other means of achieving low-temperature processing, involving plasma and ion-assisted chemical processes, are treated in Chap. 9.

Photochemical reactions are distinguished from thermal reactions by the presence of relatively large concentrations of highly excited species. These excited species may react faster than ground state species. The changes in chemical reactivity that may accompany the new electronic configuration of the species following the absorption of light energy are perhaps even more important. Thermal reactions involve excitation of a ground state molecule for which the excitation is distributed about all modes of translational, rotational, and vibrational excitation as well as electronic excitation. Photochemical reactions involve excitation of the electrons of the molecule. Therefore, reactive intermediates such as free radicals and atoms are produced at temperatures below those at which they are ordinarily encountered in thermal processes. The wavelength range of interest for photochemical reactions is less than 700 nm.

A molecule that has absorbed a quantum of radiation becomes energy-rich or excited in the absorption process. The energy E_p supplied by a photon is given by the Planck relation:

$$E_p = hf \tag{5.99}$$

where h is the Planck constant and f is the frequency. According to the Stark-Einstein law, one particle (molecule) is excited for each quantum of radiation energy absorbed. The energy absorbed is given by Eq. (5.99) if a species absorbs radiation at a frequency, f. Therefore, the excitation energy per mole E is obtained by multiplying this molecular excitation energy by Avogadro's number N_A:

$$E = N_A hf = \frac{N_A hC}{\lambda} = \frac{28{,}570}{\lambda(\text{nm})}\ \text{kcal/mol} \tag{5.100}$$

where C is the velocity of light. Upon absorbing the energy, an excited molecule can follow a number of routes. Of these, only four are of major interest in photochemical reactions. One is dissociation of the absorbing molecule into reactive fragments. An example is

$$NO_2^* \longrightarrow NO + O \tag{5.101}$$

where NO_2^* is the excited state of NO_2. Another route is direct reaction. An example is

$$O_2^* + O_3 \longrightarrow 2O_2 + O \tag{5.102}$$

The third route is deactivation (quenching) of the excited molecule. An example is

$$O_2^* + M \longrightarrow O_2 + M \tag{5.103}$$

where M is a third body which might be a reactor wall. The last is direct dissociation not involving an excited molecule:

$$O_3 + (hf)_{red} \longrightarrow O_2 + O \tag{5.104}$$

The process of creating an excited molecule by light absorption is called predissociation. For the first example above, the predissociation can be written as

$$NO_2 + hf \longrightarrow NO_2^* \tag{5.105}$$

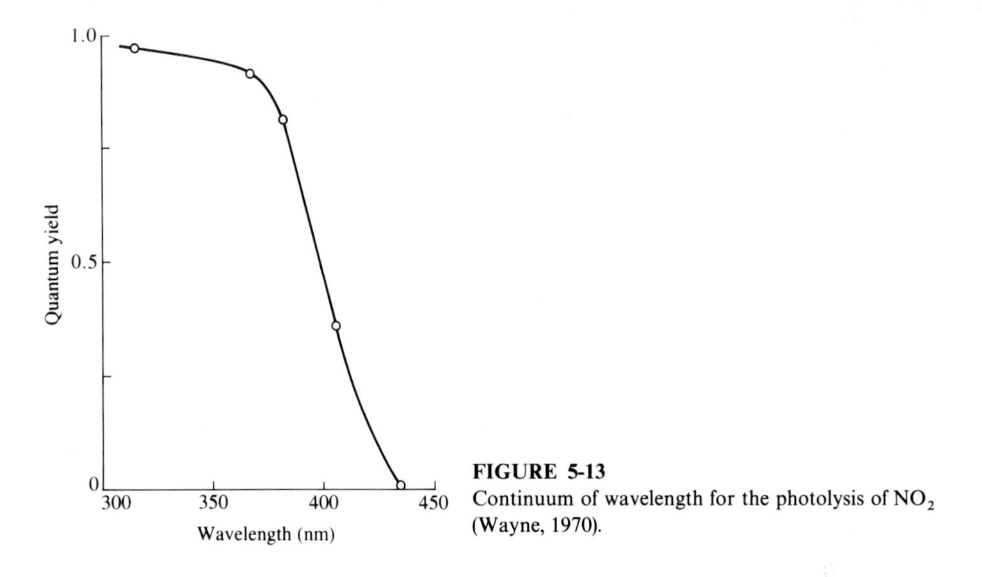

FIGURE 5-13
Continuum of wavelength for the photolysis of NO_2 (Wayne, 1970).

In order for a photochemical reaction to occur, a certain minimum energy is required. This means that the reactive system must be illuminated by light of the corresponding wavelength, less than or equal to a certain maximum. This threshold or maximum wavelength is often denoted by λ_{max}. The threshold may coincide with the beginning of a continuum or of a predissociation region. Shown in Fig. 5-13 is the continuum of wavelengths for the photolysis of NO_2; λ_{max} in this case is approximately 440 nm. The maximum wavelength that will produce dissociation is usually somewhat lower than that required for splitting the molecule. Thus, the 440-nm wavelength could be that corresponding to the predissociation limit.

Example 5.17. The dissociation energy of NO_2 to the ground state products is at least 71.8 kcal/mol (Pitts *et al.*, 1964). From Fig. 5-13, (*a*) calculate the energy that needs to be supplied by a laser beam or mercury lamp for the excitation of NO_2, assuming that the corresponding λ is 440 nm. (*b*) Calculate the energy required for a quantum yield of unity, assuming that the corresponding λ is 300 nm.

Solution

(*a*) From Eq. (5.100),

$$E_{\text{predissoc}} = \frac{28{,}570}{440} = 64.9 \text{ kcal/mol}$$

(*b*)

$$E_{\text{dissoc}} = \frac{28{,}570}{300} = 95.2 \text{ kcal/mol}$$

In photochemical reactions, the primary dissociation routes in Eqs. (5.101), (5.102), and (5.103) are followed by secondary reactions that involve free atoms

TABLE 5.1
Primary dissociative processes†

Species	Products	Wavelength,‡ nm	Continuum,§ nm	Quantum yield
H_2O	H + OH	<242		Around 1
H_2O_2	2OH	253.7		0.85 ± 2
H_2S	H + SH	200–255		Around 1
NH_3	NH_2 + H	<217		96% of reaction at λ = 184.9 nm
	N_2 + NH	Around 200		
HI	H + I	<327		Around 1
O_2	2O	Around 245.4		
N_2O	N_2 + O	Around 180		Around 1
	N + NO			12% at 123.6 nm
NO	N + O	<183		
NO_2	NO + O	<400		
$Pb(CH_3)_4$			<280	
$Pb_2(CH_5)_4$		Around 200	<350	
$Pb(C_6H_5)_4$		Around 255	<280	
$Zn(CH_3)_2$		Around 230	<265	
$Fe(CO)_5$		<410		
$Ni(CO)_5$		<395		

† Compiled from Calvert and Pitts (1966) and Rollefson and Burton (1939).
‡ λ_{max}.
§ λ at which the continuum begins.

and radicals, which often lead to chain reactions. The quantum yield defined as the number of molecules of reactant consumed for each photon of light absorbed, which should be not more than unity according to the Stark-Einstein law, can become larger than unity because of the secondary reactions. Since the law is for the primary dissociation, one can distinguish this primary quantum yield, Ω_p, from the measurable (experimentally) overall quantum yield, Ω_o (Wayne, 1970). If the light intensity absorbed per unit time per unit volume is I_{abs} and the rate of disappearance of the species of interest is r, then the overall quantum yield is

$$\Omega_o = \frac{-r}{I_{abs}} \tag{5.106}$$

The light intensity is usually expressed as quanta: one quantum corresponds to the energy given by hf.

Example 5.18. Thermopiles, photocells, and chemical actinometers (use of a reaction for which the quantum yield is accurately known) can be used to determine the light intensity (Wayne, 1970). Suppose that an ultraviolet laser beam of 300 nm results in an absorbed energy of 1 cal/min. Suppose the gas volume exposed to the light is 1 cm^3 and the rate of decomposition of the species is 2×10^{-6} mol/(min·cm^3). Determine the overall quantum yield.

Solution. From Eq. (5.100), one quantum for $\lambda = 300$ nm is

$$E_p = \frac{25{,}890 \times 10^3}{\lambda(6 \times 10^{23})} = 1.44 \times 10^{-19} \text{ cal} \qquad \text{where } N_A = 6 \times 10^{23} \text{ molecules/mol}$$

The quanta for the absorbed energy of 1 cal/min is

$$\frac{1}{1.44 \times 10^{-19}} = 6.94 \times 10^{18} \text{ quanta/min}$$

For the volume of 10 cm^3, I_{abs} is

$$I_{abs} = \frac{6.94 \times 10^{18}}{10} = 6.94 \times 10^{17} \text{ quanta/(min·cm}^3\text{)}$$

From Eq. (5.106), the overall quantum yield is

$$\Omega_o = \frac{r_p}{I_{abs}} = \frac{(2 \times 10^{-6})(6 \times 10^{23}) \text{ molecules/(mol·cm}^3\text{)}}{6.94 \times 10^{17} \text{ quanta/(min·cm}^3\text{)}} = 1.73$$

The values of the threshold wavelength and primary quantum yield (where possible) for the compounds of interest in semiconductor processing are given in Table 5.1. Recent results for photochemical reactions are summarized in Table 5.2.

It is sometimes desired to produce a photochemical reaction in a substance that is incapable of absorbing the available light. It is then necessary to introduce into the system a substance capable of absorbing that light and conveying the absorbed energy to the reactants. If this energy is sufficient to cause reaction and if the absorbing substance is not permanently transformed thereby, the process is known as photosensitization. As an example, consider dissociation of hydrogen. The continuum of the absorption spectrum of hydrogen or the dissociation of hydrogen begins at 85 nm. It thus follows that any wavelength larger than that value is incapable of disrupting the hydrogen molecule. It has been found that decomposition of hydrogen occurs when a trace of xenon is introduced into the

TABLE 5.2
Wavelength at which reactions initiated (Ehrlich and Tsao, 1983)

Wavelength, nm	Species
193	$Zn(CH_3)_2/NO_2$
193–257	SiH_4/NH_3, SiH_4/H_2O, Si_4, GeH_4
257	$Fe(CO)_5$, $W(CO)_5$, $Cr(CO)_6$, $Cd(CH_3)_2$, $Zn(CH_3)_2$, $Al_2(CH_3)_6$, $Ga(CH_3)_3$, CH_3Br, CH_3I, $Pb(CH_3)_4$
337.4–356.4	$Mn_2(CO)_{10}$
457–514	Cl_2
488	Br_2

hydrogen and the mixture is illuminated at the xenon resonance line of 147 nm. This process can be represented as follows:

$$\begin{aligned} \mathrm{Xe} + hf &\longrightarrow \mathrm{Xe}^* \\ \mathrm{Xe}^* + \mathrm{H}_2 &\longrightarrow \mathrm{Xe} + 2\mathrm{H} \end{aligned} \tag{5.107}$$

Another term often used is induced predissociation, which is significant only in the presence of some perturbation. The perturbation can be a magnetic field or collisions with other molecules.

The predissociation step can always be represented as follows:

$$\mathrm{A} + hf \xrightarrow{\Omega_p} \text{products} \tag{5.108}$$

Since one quantum results in dissociation of one molecule of species A, the rate of predissociation r_p is

$$r_p = \Omega_p I_{\mathrm{abs}} \tag{5.109}$$

Consider the kinetics of a simple photochemical reaction involving the photolysis of ozone-oxygen mixtures by red light where no predissociation is involved (Wayne, 1970) as in Eq. (5.104). The elementary reaction steps might be expected to proceed via the following mechanism:

$$\mathrm{O}_3 + (hf)_{\mathrm{red}} \xrightarrow{\Omega_1} \mathrm{O}_2 + \mathrm{O} \tag{5.110}$$

$$\mathrm{O} + \mathrm{O}_3 \xrightarrow{k_2} 2\mathrm{O}_2 \tag{5.111}$$

$$\mathrm{O} + \mathrm{O}_2 + \mathrm{M} \xrightarrow{k_3} \mathrm{O}_3 + \mathrm{M} \tag{5.112}$$

Assuming that these three equations are elementary reaction steps, one has

$$r_1 = \Omega_1 I_{\mathrm{abs}} \tag{5.113}$$

$$r_2 = k_2 C_{\mathrm{O}} C_{\mathrm{O}_3} \tag{5.114}$$

$$r_3 = k_3 C_{\mathrm{O}} C_{\mathrm{O}_2} C_{\mathrm{M}} \tag{5.115}$$

where C_{O}, C_{O_2}, C_{O_3}, and C_{M}, respectively, are the concentrations of atomic oxygen, oxygen, ozone, and a third body. The rate of disappearance of ozone $-r_{\mathrm{O}_3}$ and that of formation of atomic oxygen r_{O} follow from Eqs. (5.110) through (5.112):

$$-r_{\mathrm{O}_3} = r_1 + r_2 - r_3 \tag{5.116}$$

$$r_{\mathrm{O}} = r_1 - r_2 - r_3 \tag{5.117}$$

Atomic oxygen is a highly reactive intermediate, and it is possible to apply the pseudo steady-state hypothesis with respect to atomic oxygen. With this hypothesis, one has, from Eq. (5.117) ($r_{\mathrm{O}} = 0$),

$$r_1 = r_2 + r_3 \tag{5.118}$$

Use of Eq. (5.118) in Eq. (5.116) with the aid of Eq. (5.114) yields

$$-r_{O_3} = 2r_2 = 2k_2 C_O C_{O_3} \tag{5.119}$$

Equation (5.118) can now be solved with the aid of Eqs. (5.113) through (5.115) for the atomic oxygen concentration, which yields

$$C_O = \frac{\Omega_1 I_{abs}}{k_2 C_{O_3} + k_3 C_{O_2} C_M} \tag{5.120}$$

When this is combined with Eq. (5.119), there results

$$\Omega_o = \frac{-r_{O_3}}{I_{abs}} = \frac{2\Omega_1 k_2 C_{O_3}}{k_2 C_{O_3} + k_3 C_{O_2} C_M} \tag{5.121}$$

where the definition of the overall quantum yield [Eq. (5.106)] has been used.

Example 5.19. Castellano and Schumacher (1962) gave the following data for the photochemical decomposition of ozone by red light:

Ω_o	1.52	0.89	0.78	0.48	0.40
$C_{O_2} C_M/C_{O_3}$, mol/l	0.005	0.020	0.025	0.050	0.064

Determine the primary quantum yield Ω_1 and k_3/k_2 from the data.

Solution. Inversion of Eq. (5.121) yields

$$\frac{1}{\Omega_o} = \frac{1}{2\Omega_1} + \left(\frac{k_3}{2\Omega_1 k_2}\right)\left(\frac{C_{O_2} C_M}{C_{O_3}}\right) \tag{A}$$

Use of the data for $1/\Omega_o$ versus $C_{O_2} C_M/C_{O_3}$ yields

$1/\Omega_o$	0.658	1.124	1.282	2.083	2.5
$C_{O_2} C_M/C_{O_3}$, mol/l	0.005	0.020	0.025	0.050	0.064

A plot of $1/\Omega_o$ against $C_{O_2} C_M/C_{O_3}$ should yield the following slope and intercept:

$$\text{Slope} = \frac{k_3}{2\Omega_1 k_2} = 31.2 \text{ l/mol}$$

$$\text{Intercept} = \frac{1}{2\Omega_1} = 0.5$$

Therefore,

$$\Omega_1 = 1$$

$$\frac{k_3}{k_2} = 62.4 \text{ l/mol}$$

Consider the pseudo steady-state hypothesis used in the photolysis of ozone. The rigorous treatment, if a constant-volume batch reactor is used for the kinetic data, is to solve the following equation:

$$\frac{dC_{\mathrm{O}}}{dt} = r_{\mathrm{O}} = \Omega_1 I_{\mathrm{abs}} - k_2 C_{\mathrm{O}} C_{\mathrm{O}_3} - k_3 C_{\mathrm{O}} C_{\mathrm{O}_2} C_{\mathrm{M}} \tag{5.122}$$

along with the corresponding differential equations for C_{O_2} and C_{O_3}. If the rate of loss of active intermediate approaches the rate of formation in a time short compared with both the time over which reactant concentrations are effectively constant and the time over which the kinetics of reaction are studied, then the pseudo steady-state hypothesis can be applied. The applicability of the hypothesis can be tested directly, if desired, by solving the differential equations for the active intermediate concentration to find how closely the value approaches the pseudo steady-state value. In general, the hypothesis can be applied if the intermediate is highly reactive. The pseudo steady-state hypothesis is not arbitrary. In fact, the lifetime of the active intermediate can be determined experimentally (Rollefson and Burton, 1939; Wayne, 1970) and thus the hypothesis can also be verified experimentally.

Example 5.20. The photolysis of phosphine might be expected to proceed via the following mechanism:

$$\mathrm{PH_3} + hf \xrightarrow{\Omega_p} \mathrm{PH_2} + \mathrm{H}: \quad r = r_1 \tag{A}$$

$$2\mathrm{PH_2} \xrightarrow{k_2} \mathrm{P_2} + 2\mathrm{H_2}: \quad r = r_2 \tag{B}$$

$$\mathrm{PH_2} + \mathrm{H} \xrightarrow{k_3} \mathrm{PH_3}: \quad r = r_3 \tag{C}$$

$$2\mathrm{H} \xrightarrow{k_4} \mathrm{H_2}: \quad r = r_4 \tag{D}$$

Derive an expression for the rate of formation of P_2.

Solution. For the elementary steps, one has

$$r_1 = \Omega_p I_{\mathrm{abs}} \tag{E}$$

$$r_2 = k_2 C_{\mathrm{PH_2}}^2 \tag{F}$$

$$r_3 = k_3 C_{\mathrm{PH_2}} C_{\mathrm{H}} \tag{G}$$

$$r_4 = k_4 C_{\mathrm{H}}^2 \tag{H}$$

The rate of formation of P_2, $r_{\mathrm{P_2}}$, is given by

$$r_{\mathrm{P_2}} = r_2 = k_2 C_{\mathrm{PH_2}}^2 \tag{I}$$

If the pseudo steady-state hypothesis is applied to PH_2 and H, one has

$$0 = r_1 - r_2 - r_3 \tag{J}$$

$$0 = r_1 - r_3 - r_4 \tag{K}$$

With the aid of Eqs. (E) through (K), one can derive the following:

$$r_2 = k_2 C_{PH_2}^2 = \frac{k_4 \Omega_p I}{k_3(k_4/k_2)^{1/2} + k_4} \tag{L}$$

It is seen that the rate of formation of P_2 is proportional to the absorbed light intensity I_{abs}.

The fraction of light transmitted through an absorbing system is given by the Beer-Lambert law:

$$\frac{I_t}{I_0} = \exp(-uCd) \tag{5.123}$$

where I_t and I_0 are the transmitted and incident light intensities, C is the concentration of absorber, d is the depth of absorber through which the light beam has passed, and u is a constant of proportionality known as the extinction coefficient. The coefficient is dependent on the wavelength of radiation. The intensity of radiation absorbed is $I_0 - I_t$, so that

$$\frac{I_{abs}}{I_0} = 1 - \exp(-uCd)$$

$$= uCd \qquad \text{when } uCd \ll 1 \tag{5.124}$$

The product term uC is often referred to as the absorption coefficient.

Example 5.21. Sabin (1986) carried out a laser-activated CVD of silicon dioxide and nitride. The experimental values of the fractional light transmitted in a 1:1 mixture of nitrous oxide and silane for a 193-nm laser beam at 200 °C are given below:

p_{N_2O}, torr	0	15	20	30
Transmission	1.0	0.32	0.21	0.098

Assuming that the beam travels 4.36 cm through the gas before detection, calculate the absorption coefficient. Note that silane does not absorb light at 193 nm.

Solution. According to the Beer-Lambert law,

$$\frac{I_t}{I_0} = \exp(-uCd)$$

or

$$-\ln \frac{I_t}{I_0} = (ud)C$$

Thus, a plot of $-\ln(I_t/I_0)$ versus C should yield the value of ud:

$-\ln(I_t/I_0)$	0	1.139	1.548	2.322
$C_i = p_i/RT$, mol/cm³	0	5.09×10^{-7}	6.78×10^{-7}	10.17×10^{-7}

From the slope, one can find ud:

$$ud = 2.283 \times 10^6 \text{ cm}^3/\text{mol}$$

$$u = 2.283 \times 10^6/4.36 = 5.24 \times 10^5 \text{ cm}^2/\text{mol}$$

The absorption coefficient is uC. Thus $\alpha(\text{cm}^{-1}) = uC = 5.24 \times 10^5 C$ (mol/cm^3). At 15 torr, for example, the absorption coefficient is $\alpha_{15} = 0.267 \text{ cm}^{-1}$.

Photochemical film deposition involves not only the gas phase photolysis but also the photolysis of gas species adsorbed on the substrate. In the simplest case, the photochemical pathway for the adsorbed phase mimics that for the corresponding homogeneous phase. In more complicated cases, however, the pathway is different and new surface-catalyzed reactions appear (Ehlich and Tsao, 1983). Photochemical deposition has been attempted since the late 1970s for almost all phases of semiconductor processing (Chuang, 1983; Osgood, 1983) including epitaxy, nonepitaxial film deposition, oxidation, doping, metallization, and etching. The bulk of the work, however, has been on metallization based on metal organics and on etching based on alkyl halides.

For the metal organics, the photolysis is often considered to be initiated in a sequential manner as follows:

$$\text{ML}_n + hf \longrightarrow [\text{ML}_n{}^*] \longrightarrow \text{ML}_{n-x} + x\text{L}$$

$$\text{ML}_{n-x} \longrightarrow \text{M} + (n - x)\text{L}$$

or

$$\text{ML}_n + hf \longrightarrow \text{ML}_n{}^*$$

$$\text{ML}_n{}^* \longrightarrow \text{ML}_{n-1} + \text{L}$$

$$\text{ML}_{n-1} \longrightarrow \text{ML}_{n-2} + \text{L}$$

$$\vdots$$

where the ligands L can be alkyls or carbonyls and M is the metal. For $Cd(CH_3)_2$, for example, both will lead to the following sequence:

$$\text{Cd(CH}_3)_2 + hf \longrightarrow \text{Cd(CH}_3)_2{}^*$$

$$\text{Cd(CH}_3)_2{}^* \longrightarrow \text{CdCH}_3 + \text{CH}_3$$

$$\text{CdCH}_3 \longrightarrow \text{Cd} + \text{CH}_3$$

The photochemical reactions involving adsorbed species should depend on the nature of adsorption. Most of the work reported has dealt with physically adsorbed species. It has been found (Tsao *et al.*, 1983) that there appears to be a catalytic effect for freshly deposited film. Oxidation of gallium arsenide by a photochemical reaction with N_2O near room temperature (Bertrand, 1985) has also been reported to involve physisorbed N_2O.

Photochemical reactions are attractive because of the low-temperature processing they can provide, typically at or slightly above room temperature. A very ordered structure as in epitaxial films is not expected to result at these low temperatures but amorphous and polycrystalline deposition is possible. This is perhaps the reason why photochemical reactions are primarily used for metallization and etching. These reactions, however, can be combined with the usual chemical (thermally excited) reactions to deposit films of ordered structure at temperatures much lower than the temperature associated with the pure thermal reactions.

5.7 SELECTIVE DEPOSITION

Deposition on only the desired part of a surface consisting of different substances may be called selective deposition. The deposition of epitaxial silicon only on the silicon substrate, but not on the silicon oxide surface discussed earlier in Sec. 5.4, is an example. Deposition of tungsten metal and its selectivity with respect to silicon, which is to be considered here, is another example. Selective deposition does not require masking by its nature. It is conceivable that the whole IC fabrication sequence could be made resist-free if selective deposition could be perfected.

Chemical adsorption is the key to selective deposition of monocrystalline or polycrystalline silicon. If a species adsorbs chemically, only on a particular part of a surface, and it either strongly attracts all available surface electrons or physically blocks the adsorption sites, then no other species can adsorb onto the surface. That part of the surface then belongs only to the chemisorbing species. Indirect evidence for such effects is the accelerating or retarding effects of donor and acceptor impurities on the growth of silicon, considered in Sec. 5.5. Direct evidence has been reported (Meyerson and Yu, 1984) that complete monolayer coverage of an Si (100) surface by phosphine (adsorbed phosphine for $T < 450\,^{\circ}\mathrm{C}$ and phosphorus for $T < 550\,^{\circ}\mathrm{C}$) completely inhibits the adsorption of silane. This has been attributed to physical blocking of active sites.

Tungsten deposition based on WF_6 is known to result in selective deposition on silicon but not on silicon oxide (Shaw and Amick, 1970). Tungsten deposition can be carried out by starting with an inert gas environment and introducing hydrogen, or by having hydrogen present from the start. In both cases the film growth consists of two distinct steps. First the WF_6 is reduced by silicon and then the WF_6 is reduced by hydrogen. The overall reactions are

$$2WF_6 + 3Si \longrightarrow 2W + 3SiF_4 \tag{5.125}$$

$$WF_6 + 3H_2 \longrightarrow W + 6HF \tag{5.126}$$

The linear growth rate G (length/time) has been found to follow the kinetics given (Holman and Huegel, 1967; McConica and Krishnamani, 1986) by

$$G = kp_{H_2}^{1/2} \tag{5.127}$$

where p_{H_2} is the hydrogen partial pressure. The selective deposition of tungsten also occurs on silicides (Broadbent *et al.*, 1986).

Although selective deposition is one type of heterogenous reaction, a separate treatment is given here to draw attention to the significant opportunities the technique presents, not only for deposition but also other surface treatments of semiconductor surfaces. An example would be stabilization of segregated arsenic atoms on the surface of GaAs (Offsey *et al.*, 1986) by the technique, which can create opportunities for wider applications of semiconductor materials.

NOTATION

a, b, c, d, e	Stoichiometric coefficients
a	Distance between neighboring sites (L); lattice constant (L)
a_c	Lattice constant corrected for impurity atoms (L)
a_f	Film lattice constant (L)
a_i	Activity of species i
a_o	Lattice constant for pure crystal (L)
a_s	Substrate lattice constant (L)
A	Constant defined by Eq. (5.90) (L)
A_i	Interfacial area (L^2)
$A \cdot S$	Adsorbed species A
B	Parabolic rate constant defined by Eq. (5.91) (L^2/t)
b	Number of coordinate direction in which the jumps can occur with equal probability
C	Concentration (mol/L^3)
C_A	Concentration of species A
$C_{A S}$	Surface concentration of adspecies $A \cdot S$ (mol/L^2)
C_b	Bulk gas concentration
C_i	Concentration of species i; interfacial concentration
$C_{i, S}$	Surface concentration of adsorbed species $i \cdot S$
C_s	Concentration at solid surface (mol/L^3)
C_t	Surface concentration of total adsorption sites (mol/L^2)
C_v	Vacant site concentration (mol/L^2)
d	Distance (L)
D	Gas diffusivity in solid (L^2/t)
D_s	Surface diffusivity (L^2/t)
E	Activation energy for a reaction (E/mol)
E_a	Activation energy for adsorption
E_d	Activation energy for surface diffusion
E_D	Activation energy for desorption
E_p	Quantum energy
f	Fugacity (P); function of t; frequency (t^{-1})
f_i	Fugacity of species i
f^0	Fugacity at standard state
f'	Pure component fugacity
G	Linear growth rate (L/t)
ΔG	Change in the standard free energy due to a reaction (E)
h	Atomic height (L); Planck's constant ($= 4.14 \times 10^{-34}$ J·s)
H	Henry's constant (mol/L^3P)

ΔH	Change in the heat of formation due to a reaction (E)
I_0	Incident beam intensity (quanta/L^3t)
I_t	Transmitted beam intensity
I_{abs}	Absorbed beam intensity
k	Boltzmann constant ($= 1.38 \times 10^{-23}$ J/K); rate constant
k_a	Adsorption rate constant
k_m	Mass transfer coefficient (L/t)
k_o	Preexponential factor
k_r	$C_t^2 k_r'/RT$ in Eq. (5.68)
k_s	Surface reaction rate constant
k_1	$C_t k_1' K_a'/RT$ in Eq. (5.68)
k_2	$C_t k_2' C_{H_2}^{\circ} K_B'/RT$ in Eq. (5.68)
k_e, k_{de}, k_{uni}	Rate constants in Eqs. (5.42) and (5.43)
K	Equilibrium constant for a reaction (units dependent on reaction stoichiometry)
K_a	Adsorption equilibrium constant
K_A	K_A'/RT in Eq. (5.68)
K_B	K_B'/RT in Eq. (5.68)
K_i	Individual equilibrium constant
K_y	K based on mole fraction
l	Length (L); oxide layer thickness (L)
L	Reactor length (L)
m	Mass of a molecule; a constant
M	Molecular weight of silicon (M)
M_w	Molecular weight (M)
MR_n	Metal alkyl consisting of metal M and n number of alkyl group R
n	Order of reaction; a constant
n_j	Number of jumps
n^*	Number of atoms in a stable nuclei
N_a	Number of adsorption sites
$(N_a)_{eq}$	Equilibrium value of N_a
N_A	Avogadro's number ($=6.02 \times 10^{23}$ molecules/mol); number of moles of species A
N_M	Number of moles of species M
N_s	Saturation surface concentration of nuclei (number/L^2)
N_t	Number of total adsorption sites
p	Partial pressure (P)
p_e	Equilibrium partial pressure
p_0	Partial pressure in feed
P	Pressure; total pressure
q	Number of equivalent neighboring sites
Q	$E_D - E_a$ (E)
r	Molar growth rate in the absence of dopant (mol/t)
r_a	Rate of adsorption (mol/L^2t)
r_A	Rate of formation of species A (mol/L^3t)

r_G	Molar growth rate (mol/t)
r_d	Rate of desorption (mol/L^2t)
r^*	Critical nucleus size (L)
R	Gas constant [= 1.987 kcal/(mol·K), 8.314 J/(mol·K)]
S	Active site
t	Time (t)
t_r	Residence time
T	Temperature (T)
T_m	Maximum temperature
u	Extinction coefficient
v	Superficial fluid velocity (L/t); atomic volume (L^3)
v_0	Vibrational frequency of adspecies ($1/t$)
w_i	Ratio of ith impurity atom radius to host atom radius
x	Conversion
x_e	Equilibrium conversion
x_i	Fraction of the ith atom in a solid solution
X	Net displacement (L)
y	Mol fraction; normalized reactor length; quantity defined by Eq. (5.76)
y_i	Mole fraction of species i
z	Collision frequency (number/t)

Greek letters

α	Number of adsorption sites affected by an absorbed dopant; surface coverage ($= N_a/N_t$)
β	Linear thermal expansion coefficient ($1/T$); porosity
λ	Wavelength (L)
μ	Lattice mismatch
ρ	Density (M/L^3)
ρ_s	Silicon density
ρ'_M	Molar density of species M (mol/L^3)
Ω	Surface energy (E/mol L^2)
Ω_o	Overall quantum yield
Ω_p	Primary quantum yield

Superscript

*	Activated
0	Feed or reactor inlet

Units

E	Energy (ML^2/t^2)
L	Length
M	Mass
P	Pressure (M/Lt^2)
t	Time
T	Temperature

PROBLEMS

5.1. Assuming that SiH_2 and $SiCl_2$ are the adspecies for silicon epitaxy and that the ratio of the lateral to vertical growth rate in the epitaxial lateral overgrowth is entirely due to the difference in surface diffusivity, calculate the difference in the activation energy for diffusion between $SiCl_2$ and SiH_2 at 1000 °C for the ratio $R_{SiH_2} = 10R_{SiCl_2}$.

5.2. The typical epitaxial film growth rate for silicon is 1 μm/min at 1000 °C.

(*a*) Determine the rate of adspecies supply needed to maintain the growth at that rate. Compare the value with the maximum possible adsorption rate at that temperature. Take 5×10^{22} atoms/cm^3 as the silicon atom density and the mass of adsorption species as 5×10^{-23} g/molecule. Assume that the mole fraction of the silicon source gas is 0.001 at atmospheric pressure.

(*b*) Suppose the sticking coefficient is 0.1 and the activation energy for the adsorption is 4 kcal/mol. Determine the minimum partial pressure of the silicon source gas to maintain the growth rate.

(*c*) Suppose a low-pressure chemical vapor deposition (LPCVD) is to be used at a total pressure of 1 torr. Determine the minimum mole fraction of the silicon source gas.

5.3. Claassen and Bloem (1980) gave the following data from a silicon nucleation experiment on an SiO_2 substrate at 1000 °C with hydrogen as the carrier gas:

p_{SiH_4} (atm) $\times 10^5$	9	22	44	88	170
N_s (number/cm^2) $\times 10^8$	6.6	14	50	65	80

Determine the exponent n in Eq. (5.11). Note that they observed almost uniform size distribution. Use Eq. (5.11) for r_a.

5.4. Thin epilayers of GaAsAl-GaAs-GaAsAl are used as semiconductor lasers. Consider the growth of $GaAs_{1-x}Al_x$, where x is less than unity (say, 0.2), on GaAs substrate. Calculate the lattice mismatch as a function of x. The following can be used for the problem:

Substrate	GaAs	AlAs
Lattice constant, nm	0.565	0.566

Element	Al	As	Ga
Atomic radius, nm	0.143	0.125	0.135

5.5. The linear thermal expansion coefficients of silicon and silicon dioxide at 27 °C are as follows:

$$\beta_{Si} = 2.5 \times 10^{-6}(1/°C), \qquad \beta_{SiO_2} = 0.5 \times 10^{-6}\ (1/°C)$$

Suppose that SiO_2 is deposited by a CVD method on Si substrate of 2 cm diameter at 800 °C. Suppose it is subsequently cooled to 27 °C and that the film is free to move at the interface. Determine the difference in the diameter of the Si and SiO_2.

5.6. The decomposition reaction of silane at relatively high pressure has been reported to follow the following kinetics (*op. cit.*):

$$-r_{SiH_4} = 10^{13.3} \exp\left(-\frac{52{,}700}{RT}\right) C_{SiH_4} \quad \text{where } k \text{ is in s}^{-1} \tag{A}$$

According to the correction with the RRKM theory (Meyerson *et al.*, 1986), which is a more rigorous version of Lindemann's theory, the rate constant for pressures less than 10 torr at 800 °C decreases linearly with pressure in the following manner:

$$k\ (\text{s}^{-1}) = 1.5P\ (\text{torr}) \tag{B}$$

where P is the total pressure. In addition, the rate constant at 1000 torr is 1000 s^{-1}. Determine the rate constant at 800 °C in the following form of Eq. (5.47):

$$k = \frac{k_2 P}{1 + k_1 P} \tag{C}$$

The mean free path of a gas molecule λ is given by

$$\lambda = \frac{kT}{2^{1/2} P \pi d^2} \tag{D}$$

where d is the diameter of the molecule. As discussed in the text, the rate constant becomes independent of total pressure when the mean free path is comparable to the vessel dimension. Assuming that the shortest vessel dimension is 10 cm, calculate the pressure at which the rate constant becomes independent of total pressure and estimate the rate constant. Use the diameter of hydrogen for d, which may be taken as 0.2 nm.

5.7. The gas decomposition reaction (Larsen and Stringfellow, 1986) of PH_3 was carried out in a tube packed with thin quartz capillaries at a flow rate of 100 cm^3/min with a 10% PH_3/90% H_2 feed gas. The length of the reaction zone was 47 cm. The data read from the smooth curve drawn through the data points are given below:

% PH_3 decomposition	32	40	52	80
T, °C	600	700	800	900

Obtain the kinetics of the decomposition in the nth-order form given by Eq. (5.18). Assume constant hydrogen concentration and plug flow. Neglect the volume change due to the reaction.

5.8. The decomposition reaction of TMG in H_2 may be represented as follows:

$$Ga(CH_3)_3 \longrightarrow Ga(CH_3)_2 + CH_3$$

$$Ga(CH_3)_2 \longrightarrow Ga(CH_3) + CH_3$$

$$Ga(CH_3) \longrightarrow Ga + CH_3$$

$$CH_3 + H_2 \longrightarrow CH_4 + H$$

$$2H \longrightarrow H_2$$

Show that the kinetics of TMG decomposition is first order.

5.9. According to Yoshida and Wantanabe (1985), the main product of TMG decomposition in H_2 is methane with a small amount of ethane, whereas the main products are methane, ethane, and propane in N_2. For the decomposition in H_2, the hypothesis often made for the elementary steps would be those in Prob. 5.8, if the small amount of ethane is neglected. Write a series of elementary steps for the decomposition in N_2.

5.10. For the tungsten metallization based on WF_6, the overall reactions are

$$WF_6 + 3H_2 \longrightarrow W + 6HF$$

$$WF_6 + 3Si \longrightarrow 2W + 3SiF_4$$

As discussed in Sec. 5.7, these two reactions can take place simultaneously when bare silicon surface is exposed. From equilibrium considerations, determine which reaction will dominate. Take the following heats of formation as the free energy change and assume them to be constant:

Substance	HF	SiF_4
$-\Delta H$, kcal/mol	64.2	373

5.11. Trimethylaluminum (TMA) has been used for the epitaxial growth of AlSb. Two gas phase reactions of interest involving TMA are

$$Al_2(CH_3)_6 \xrightarrow{k_1} \text{Al adspecies precursors: } r_1$$

$$2Al_2(CH_3)_6 \xrightarrow{k_2} Al_4C_3 + 9CH_4\text{: } r_2$$

The second reaction causes an impurity problem because of Al_4C_3 deposited on growing AlSb. The nature of the reaction is not known. Suppose that the TMA reaction leading to the formation of the adspecies precursor follows first-order kinetics while the second follows second-order kinetics. How would you choose pressure and temperature so as to minimize the formation of Al_4C_3?

5.12. In ultra-high vacuum, almost no decomposition of SiH_4 occurs. The silicon growth may go through successive hydrogen extraction:

$$SiH_4 + S \rightleftharpoons SiH_4 \cdot S$$

$$SiH_4 \cdot S \rightleftharpoons SiH_3 \cdot S + H$$

$$SiH_3 \cdot S \rightleftharpoons SiH_2 \cdot S + H$$

For this problem, however, assume that SiH_4 is the adspecies mainly responsible for the growth so that the following growth steps are obeyed:

$$SiH_4 + S \underset{}{\overset{K_A}{\rightleftharpoons}} SiH_4 \cdot S$$

$$SiH_4 \cdot S \underset{}{\overset{K_s}{\rightleftharpoons}} Si(c) + 2H_2$$

Derive the kinetics of the silicon growth.

5.13. The kinetics of SiO_2 film growth from SiH_4 and O_2 have been correlated as follows:

$$r_{SiO_2} = kP_{SiH_4}^n P_{O_2}^m$$

Determine the composition for the highest rate of decomposition for $n = m = \frac{1}{2}$ for a given temperature and pressure. Assume that no other species are present in a stoichiometric feed and that the temperature is less than 500 °C. If the feed concentration of SiH_4 is C_{A_0} and that of O_2 is C_{B_0}, find the conversion of SiH_4 at which the rate is the maximum.

5.14. In thermal oxidation of native silicon, the growth of the silicon oxide layer is kinetically controlled if the oxide layer thickness is thin enough. Neglecting the film mass transfer resistance, a criterion for a kinetically controlled oxide growth rate can be obtained from Eq. (5.88):

$$\frac{kl}{D} \ll 1$$

Suppose that it is sufficient that $kl/D < 0.01$. Determine the thickness at which diffusion becomes important at 900 °C. At this temperature, $k = 1.2 \times 10^4$ $\mu m/h$ and D is 1.8×10^2 $\mu m^2/h$ for dry oxygen. Make conclusions from your result. Obtain an expression for the linear growth of the SiO_2 layer for the kinetically controlled growth with the condition $x = x_d$ at $t = 0$ where x is the oxide layer thickness.

5.15. Irene *et al.* (1986) reported the following data for the oxidation of native silicon at 800 °C for (100) and (111) orientations:

t, min	0	20	40	60	100	140
Thickness, nm; (100)	12	30	44	58	80	100
Thickness, nm; (111)	12	40	70	91	—	—

Determine the linear and parabolic rate constants for both orientations.

5.16. A doubled Ar laser can deliver 10 mW (milliwatts) power at 256 nm. Calculate the number of quanta imparted to a gas by the laser for 10 min illumination. Assume the gas completely absorbs the energy. One milliwatt-minute is equal to 0.0143 cal. Note that the energy of one quantum, E_p, is given by

$$E_p = \frac{4.315 \times 10^{-17} \text{ (cal)}}{\lambda \text{ (nm)}}$$

Determine the number of molecules activated and the energy imparted per mole.

5.17. For the same conditions in Prob. 5.16, suppose that aluminum deposition is carried out photochemically using an organic molecule of aluminum on a diameter of 1 μm. Calculate the thickness of the metal deposited, assuming that the overall quantum yield is unity. The density of aluminum is 2.7 g/cm^3. Calculate the time required for the film to reach a thickness of 2 μm. The beam diameter is 1 μm.

5.18. The photochemical decomposition of germanium is considered to take place as follows:

$$GeH_4 + hf \longrightarrow GeH_3 + H$$

$$GeH_4 + H \longrightarrow GeH_3 + H_2$$

$$2GeH_3 \longrightarrow 2Ge + 3H_2$$

Derive the kinetics of germanium formation.

5.19. For the photochemical decomposition of $Cd(CH_3)_2$ given in the text, show that the rate of decomposition is given by $\Omega_1 I_{abs}$. Based on this, calculate Ω_1 from the data in Fig. P5-19.

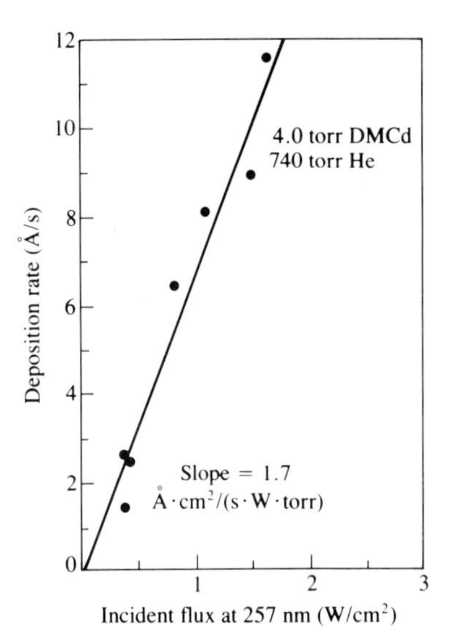

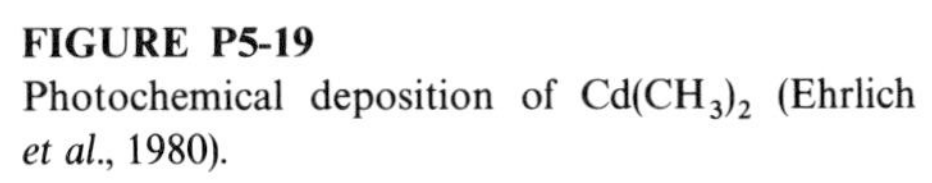

FIGURE P5-19
Photochemical deposition of $Cd(CH_3)_2$ (Ehrlich *et al.*, 1980).

5.20. McConica and Krishnamani (1986) assumed for tungsten deposition that adsorbed WF_6 undergoes consecutive reduction by adsorbed hydrogen:

$$WF_6 + S \rightleftharpoons WF_6 \cdot S$$

$$H_2 + 2S \rightleftharpoons 2H \cdot S$$

$$\vdots$$

$$WF_x \cdot S + H \cdot S \rightleftharpoons WF_{x-1} \cdot S + HF \cdot S$$

Suppose that all $(WF_x \cdot S)$ concentrations and $(H \cdot F)$ surface concentration are constant. Explain what this means. Determine the conditions under which the rate of tungsten deposition is simply proportional to $(C_{H_2})^{1/2}$.

5.21. Include the effect of an external mass transfer resistance for a relationship similar to Eq. (5.96).

REFERENCES

Abbink, H. C., R. M. Broudly, and G. P. McCarty: *J. Appl. Phys.*, vol. 39, p. 4693, 1968.
Adams, A. C., and C. D. Capio: *J. Electrochem. Soc.*, vol. 126, p. 1042, 1979.
Aleksandrov, L. N.: *J. Crystal Growth*, vol. 31, p. 103, 1975.
Aoyama, T., Y. Inoue, and T. Suzuki: *J. Electrochem. Soc.*, vol. 130, p. 203, 1983.
Asai, H., and S. Ando: *J. Electrochem. Soc.*, vol. 132, p. 2445, 1985.
Baliga, B. J., and S. K. Ghandhi: *J. Appl. Phys.*, vol. 44, p. 1990, 1973.
Beers, A. M., and J. Bloem: *Appl. Phys. Lett.*, vol. 41, p. 153, 1982.
Bertrand, P. A.: *J. Electrochem. Soc.*, vol. 132, p. 973, 1985.

Bischoff, K. B.: *Chem. Eng. Sci.*, vol. 18, p. 711, 1963.
Blanc, J.: *J. Electrochem. Soc.*, vol. 133, p. 1981, 1986.
Bloem, J.: *J. Crystal Growth*, vol. 50, p. 581, 1980.
Boudart, M.: *Am. Inst. Chem. Engrs J.*, vol. 18, p. 465, 1972.
Bozler, C. O., and G. D. Alley: *IEEE Trans. Elect. Dev.*, vol. ED-17, p. 1128, 1980.
Bradley, D. C., M. M. Faktor, and E. A. D. White: *J. Crystal Growth*, vol. 75, p. 101, 1986.
Broadbent, E. K., A. E. Morgan, J. M. DeBlasi, P. van der Putte, B. Coulman, B. J. Burrow, and D. K. Sadana: *J. Electrochem. Soc.*, vol. 133, p. 1715, 1986.
Butt, J. B.: *Reactor Kinetics and Reactor Design*, Prentice-Hall, Englewood Cliffs, N.J., 1980.
Calvert, J. G., and J. N. Pitts, Jr.: *Photochemistry*, Wiley, New York, 1966.
Carim, A. H., and R. Sinclair: in H. R. Huff, T. Abe, and B. Kolbesen (eds.), *Semiconductor Silicon 1986*, p. 458, The Electrochemical Society, Pennington, N.J., 1986.
Carlton, H. E., J. H. Oxley, E. H. Hall, and J. M. Blocher, Jr.: in J. M. Blocher and A. Withers (eds.), *Chemical Vapor Deposition, 2d International Conference*, The Electrochemical Society, Pennington, N.J., 1970.
Castellano, E., and H. J. Schumacher: *Z. Phys. Chem. Frankf.*, vol. 34, p. 198, 1962.
Chuang, T. J.: *Mat. Res. Soc. Symp. Proc.*, vol. 17, p. 45, 1983.
Claassen, W. A. P., and J. Bloem: *J. Electrochem. Soc.*, vol. 127, p. 194, 1980.
Clark, A.: *The Theory of Adsorption and Catalysis*, Academic Press, New York, 1970.
Deal, B. E., and A. S. Grove: *J. Appl. Phys.*, vol. 36, p. 3770, 1965.
Ehrlich, D. J., and J. Y. Tsao: *J. Vac. Sci. Technol.*, vol. B1, p. 969, 1983.
———, R. M. Osgood, Jr., and T. F. Deutsch: *IEEE J. Quantum Electr.*, vol. 16(11), p. 1233, 1980.
Everstyn, F. C., and B. H. Put: *J. Electrochem. Soc.*, vol. 120, p. 106, 1973.
Farrow, R. F. C., and J. D. Filby: *J. Electrochem. Soc.*, vol. 118, p. 149, 1971.
Frankel, D. R., and J. A. Venables: *Adv. Phys.*, vol. 19, p. 409, 1970.
Furumura, Y., F. Mieno, T. Nishizawa, and M. Maseda: *J. Electrochem. Soc.*, vol. 133, p. 379, 1986.
Gardiner, W. C., Jr.: *Rates and Mechanisms of Chemical Reactions*, 2d printing, chap. 5, W. A. Benjamin, Menlo Park, Calif., 1972.
Green, M. L., R. A. Levy, R. G. Nuzzo, and E. Coleman: *Thin Solid Films*, vol. 114, p. 367, 1984.
Hess, D. W., and B. E. Deal: *J. Electrochem. Soc.*, vol. 124, p. 735, 1977.
Holman, W. R., and F. J. Huegel: The First International Conference on *Chemical Vapor Deposition*, Gatlinburg, Tenn., p. 127, 1967.
Hu, S. M.: *Appl. Phys. Lett.*, vol. 42, p. 872, 1983.
Irene, E. A., *J. Appl. Phys.*, vol. 54, p. 5416, 1983.
———, H. Z. Massoud, and E. Tierney, *J. Electrochem. Soc.*, vol. 133, p. 1253, 1986.
Ito, T., T. Nozaki, and H. Ishikawa: *J. Electrochem. Soc.*, vol. 127, p. 2053, 1980.
Jacko, M. G., and S. J. W. Price: *Can. J. Chem.*, vol. 42, p. 1198, 1964.
Jastrzebski, L.: *J. Crystal Growth*, vol. 63, p. 493, 1983.
———: *J. Crystal Growth*, vol. 64, p. 253, 1984.
Joyce, B. D., and J. A. Baldrey: *Nature*, vol. 195, p. 458, 1962.
Katz, L. E.: in S. M. Sze (ed.), *VLSI Technology*, chap. 4, McGraw-Hill, New York, 1983.
Knight, J. R., D. Effer, and P. R. Evans: *Solid State Electron.*, vol. 8, p. 178, 1965.
Laporte, J. L., M. Cadoret, and R. Cadoret: *J. Crystal Growth*, vol. 50, p. 663, 1980.
Larsen, C. A., and G. B. Stringfellow: *J. Crystal Growth*, vol. 75, p. 247, 1986.
Lee, H. H.: *J. Crystal Growth*, vol. 69, p. 82, 1984.
Levenspiel, O.: *Chemical Reaction Engineering*, 2d ed., Wiley, New York, 1972.
Lewis, B., and D. S. Campbell: *J. Vac. Sci. Tech.*, vol. 4, p. 209, 1967.
Leys, M. R., and H. Veenvliet: *J. Crystal Growth*, vol. 55, p. 145, 1981.
Ligenza, J. R.: *Phys. Chem.*, vol. 65, p. 2011, 1961.
McConica, C. M., and K. Krishnamani: *J. Electrochem. Soc.*, vol. 133, p. 2543, 1986.
Maeda, M., and H. Nakamura: *J. Appl. Phys.*, vol. 52, p. 6651, 1981.
Manasevit, H. M., and W. I. Simpson: *J. Electrochem. Soc.*, vol. 166, p. 1725, 1969.
Massoud, H. Z., C. P. Ho, and J. D. Plummer: in J. D. Plummer *et al.* (eds.), Stanford University Technical Report TR DXG 501-82, July 1982.

———, J. D. Plummer, and E. A. Irene: *J. Electrochem. Soc.*, vol. 132, p. 2693, 1985.
Meyerson, B. S., and M. L. Yu: *J. Electrochem. Soc.*, vol. 131, p. 1366, 1984.
———, E. Ganin, D. A. Smith, and T. N. Nguyen: *J. Electrochem Soc.*, vol. 133, p. 1232, 1986.
Moore, W. J.: *Physical Chemistry*, 2d ed., Prentice-Hall, Englewood Cliffs, N.J., 1955.
Moslehi, M. M., S. C. Shates, and K. C. Saraswat: in H. R. Huff, T. Abe, and B. Kolbesen (eds.), *Semiconductor Silicon 1986*, p. 379, The Electrochemical Society, Pennington, N.J., 1986.
Mullin, J. B. S., J. C. Irvine, and D. J. Ashen: *J. Crystal Growth*, vol. 55, p. 92, 1981.
Nakayama, S., I. Kawashima, and J. Murota: *J. Electrochem. Soc.*, vol. 133, p. 1721, 1986.
Newman, C. G., H. E. O'Neal, M. A. Ring, F. Leska, and N. Shipley: *Int. J. Chem. Kinematics*, vol. 11, p. 1167, 1979.
Offsey, S. D., J. M. Woodall, A. C. Warren, P. D. Kirchner, T. J. Chappell, and G. D. Pettit: *Appl. Phys. Lett.*, vol. 48, p. 475, 1986.
Ogden, R., R. R. Bradley, and B. E. Watts: *Phys. Status Solidi*, vol. 26, p. 135, 1974.
Osgood, R. M., Jr.: *Ann. Rev. Phys. Chem.*, vol. 34, p. 77, 1983.
Pearce, C. W.: in S. M. Sze (ed.), *VLSI Technology*, chap. 1, McGraw-Hill, New York, 1983.
Pitts, J. N., Jr., J. H. Sharp, and S. I. Chan: *J. Chem. Phys.*, vol. 40, p. 3655, 1964.
Rathman, D. D., D. J. Silversmith, and J. A. Burns: *J. Electrochem. Soc.*, vol. 129, p. 2303, 1982.
Revesz, A. G., and R. J. Evans: *J. Phys. Chem. Solids*, vol. 30, 551, 1969.
Rollefson, G. K., and M. Burton: *Photochemistry*, Prentice-Hall, Englewood Cliffs, N.J., 1939.
Sabin, E. W.: in H. R. Huff, T. Abe, and B. Kolbesen (eds.), *Semiconductor Silicon 1986*, p. 284, The Electrochemical Society, Pennington, N.J., 1986.
Sakai, S., T. Soga, M. Takeyasu, and M. Umeno: MRS Symposium, Anaheim, Calif., Spring 1986.
Scott, B. A., R. M. Plecenik, and E. E. Simonyi: *Appl. Phys. Lett.*, vol. 39, p. 73, 1981.
Sedgwick, T. O., and J. E. Smith, Jr.: *J. Electrochem. Soc.*, vol. 123, p. 254, 1976.
Shaw, J. M., and J. A. Amick: *RCA Rev.*, vol. 31, p. 306, 1970.
Stassinos, E. C., and H. H. Lee: *J. Appl. Phys.*, vol. 60, p. 3906, 1986.
———, T. J. Anderson, and H. H. Lee: *J. Crystal Growth*, vol. 73, p. 21, 1985.
Strater, K.: *RCA Rev.*, vol. 30, p. 618, 1968.
Tausch, F. W., Jr., and A. G. Lapierre III: *J. Electrochem. Soc.*, vol. 112, p. 706, 1965.
Tietjen, J. J., and J. A. Amick: *J. Electrochem. Soc.*, vol. 133, p. 724, 1966.
Tobin, P. J., J. B. Price, and L. M. Campbell: *J. Electrochem. Soc.*, vol. 127, p. 2222, 1980.
Tromson-Carli, A., P. Gibart, and C. Bernard: *J. Crystal Growth*, vol. 55, p. 125, 1981.
Tsao, J. Y., R. A. Becker, D. J. Ehrlich, and F. J. Leonberger: *Appl. Phys. Lett.*, vol. 42, p. 559, 1983.
Tung, S. K., *J. Electrochem. Soc.*, vol. 112, p. 436, 1965.
Wagner, E. E., G. Hom, and G. B. Stringfellow: *J. Electron. Mater.*, vol. 10, p. 329, 1981.
Walton, D.: *J. Chem. Phys.*, vol. 37, p. 2182, 1962.
Wayne, R. P.: *Photochemistry*, American Elsevier, New York, 1970.
Yoshida, M., and H. Wantanabe: *J. Electrochem. Soc.*, vol. 132, p. 677, 1985.

CHAPTER 6

CHEMICAL VAPOR DEPOSITION REACTORS

6.1 CHEMICAL VAPOR DEPOSITION (CVD) REACTORS

The conventional chemical vapor deposition (CVD) reactors for epitaxial films are shown in Fig. 6-1. These are the horizontal, barrel, and pancake (vertical) reactors. Shown in Fig. 6-2 are the CVD reactors typically used for nonepitaxial film deposition. Empirical methods have mostly been used for optimal design and operation of these reactors. It is only recently that quantitative analytical approaches have been taken.

The horizontal reactor in Fig. 6-1 is perhaps the oldest and is usually operated with an inclined susceptor, made of graphite that is coated with SiC. The reactor is heated by a radio-frequency (rf) heater that heats only the susceptor. Therefore, the reactor wall, typically made of quartz, is "cold" and this mode of operation is referred to as cold-wall operation. The semiconductor source gas in hydrogen or other inert carrier gases is fed to the reactor, deposition occurs on the wafers as the gas mixture passes over the susceptor, and the spent gas exits.

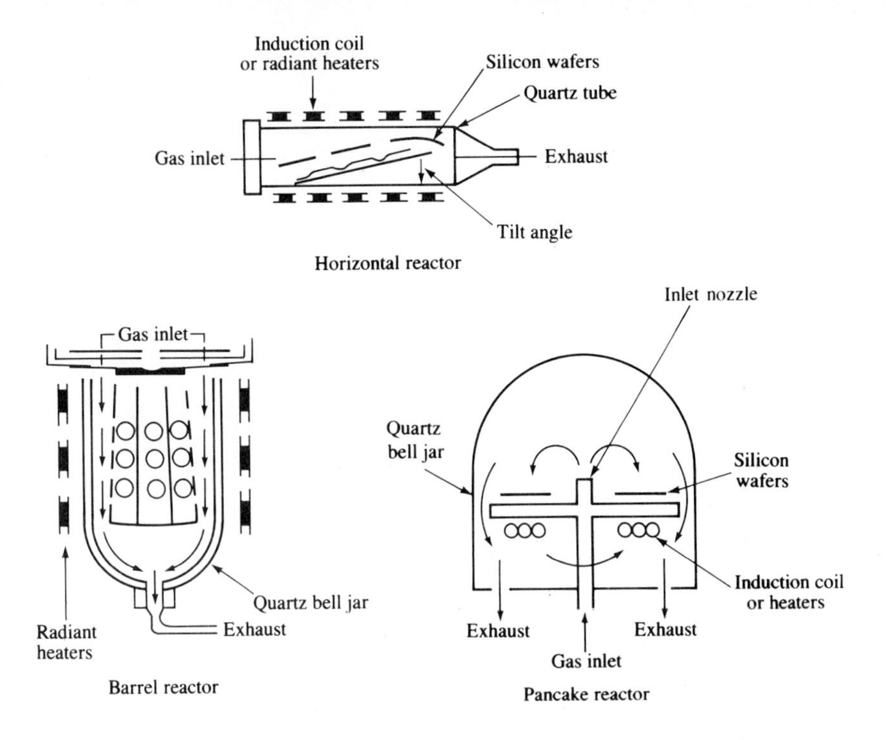

FIGURE 6-1
CVD reactors used mainly for epitaxial film growth.

Because of rf heating, homogeneous reactions are confined to the immediate vicinity above the hot susceptor. The susceptor is inclined so as to enhance the mass transfer above the susceptor in the direction of flow.

The barrel reactor is similar to the horizontal reactor in that one face of the barrel is subjected to the similar flow pattern as the susceptor in the horizontal reactor. An important difference, however, is that the flow direction in the barrel reactor is aligned with the force of gravity, unlike the horizontal reactor. This difference has an important bearing on free convection effects discussed later. Either radiant heating based on halogen lamps or rf heating is used for the reactor.

The pancake reactor is also called the vertical reactor because the flow is perpendicular to the susceptor. The primary objective of the reactor configuration is good mixing of the fluid above the susceptor. Both barrel and pancake susceptors are usually rotated during operation to enhance uniformity of the growing film's thickness. Although a material such as graphite, that will couple to the rf field, has to be used for induction-heated reactors, polysilicon or quartz susceptors are alternatives for radiantly heated reactors. CVD silicon nitride is coated onto the susceptor to prevent a gradual erosion of the susceptor in the presence of HCl.

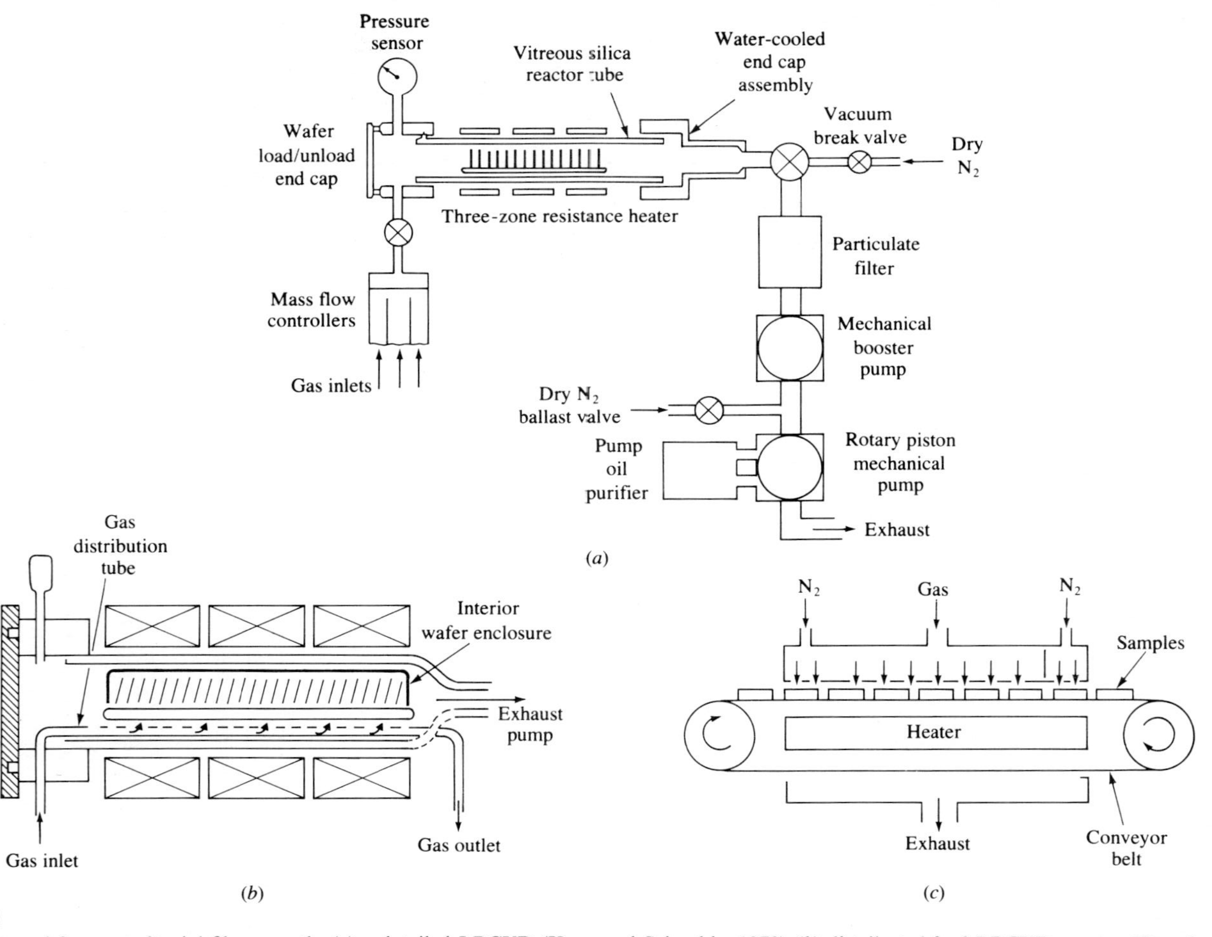

FIGURE 6-2
CVD reactors used for nonepitaxial film growth: (*a*) a detailed LPCVD (Kern and Schnable, 1979), (*b*) distributed-feed LPCVD reactor (Douglas, 1980), (*c*) continuous atmospheric reactor (Adams, 1983).

The performance of these reactors has been compared (Hammond, 1978). Horizontal reactors offer high capacity and throughput, but the control of the deposition over the entire susceptor is a problem. Pancake reactors give very uniform deposition but suffer from mechanical complexity. Barrel reactors are also capable of uniform deposition but are not suitable for extended operation at temperatures above 1200 °C (Pearce, 1983).

Nonepitaxial film deposition is usually carried out using external heating, and these are called hot-wall reactors. Because of the hot wall, deposition also occurs on the wall. A detailed version of a low-pressure CVD (LPCVD) reactor (Kern and Schnable, 1979) is shown in Fig. 6-2*a*. The total pressure in the reaction (heated) zone can range from 0.1 to 10 torr and the temperature from 300 to 900 °C. The source gas along with a carrier gas is introduced in one end and pumped out of the other. The fluid flow is perpendicular to the wafers mounted on a rack. The flowing fluid in the annulus between the wafers and the tube wall supplies the source gas. The mass transport of gas between two adjacent wafers is primarily by diffusion although the streamlines of the flow in the annulus can intrude into the space between wafers. Note that the source-gas supply is from both sides of the annulus, which may not be suggested by the schematics in the figure. As shown in Fig. 6-2, the gas feed can be distributed over the entire length of the boat, if desired. A continuous, atmospheric pressure reactor (Adams, 1983) is also shown in the figure. In this case, the wafers are carried through the reactor on a conveyor belt. The source gases flowing through the center of the reactor are contained by gas curtains formed by a very fast flow of nitrogen. The wafers are heated by convection. It should be noted that the LPCVD reactor can also be operated in the given configuration at atmospheric pressure. These reactors can accommodate large-diameter wafers and are suitable for processing large numbers of wafers. LPCVD reactors, in general, yield better thickness uniformity than reactors operated at atmospheric pressure at the cost of a lower rate of film growth.

There are two major differences between the CVD reactors and the usual chemical reactors. In CVD reactors, no recycle is used, partly because of the impurities that such a recycle stream can introduce and partly because of strict requirements on the controllability of the desired gas phase composition. Second, the flow regimes in CVD reactors can be diverse, depending on the pressure, temperature, and flow rate. Chemical reactors usually operate in the turbulent regime. In cold-wall CVD reactors, where a large temperature gradient exists between tube wall and hot susceptor, thermally driven free convection can play a major role in heat and mass transport.

The film deposition rate that is observed in a CVD reactor is not at its intrinsic rate given by the kinetics that have been considered in the previous chapter. Rather, it is the rate as affected by mass and heat transport, which in turn is largely dictated by fluid mechanics. Therefore, analysis and design of a CVD reactor is more complex than that of a chemical reactor. The deposition rate is close to its intrinsic rate only when the (potential) rate of mass transfer to the growing surface from the gas phase is much larger than the rate of growth.

Note that the potential rates determine the extent of mass transfer effects here since at steady state the rate of mass transfer is equal to the rate of growth. An equivalent way to visualize this is to compare bulk fluid properties with those at the surface. If the differences are small, rates approach the intrinsic rate. Otherwise the observed rates are lower.

Epitaxial film growth is usually carried out in a cold-wall reactor to prevent homogeneous nucleation in the gas phase, although some epitaxial reactors have hot walls. For nonepitaxial film growth, however, either a hot-wall or a cold-wall reactor can be used. If the overall reaction for the deposition is exothermic, a hot-wall reactor would be preferred. According to the van't Hoff equation [Eq. (5.31)], the reaction equilibrium constant decreases with increasing temperature for exothermic deposition, resulting in less deposition at higher temperatures at equilibrium. Thus, relatively less deposition occurs on the hot wall than on the substrate (susceptor), which is at a lower temperature. By the same reasoning, a cold-wall reactor would be preferred if the overall reaction for the deposition is endothermic since then deposition at a higher temperature is favored for the heated substrate.

The total pressure in a deposition reactor has profound effects on the deposition process. The most obvious is the associated change in concentration and thus the deposition rate due to the change in the total pressure. A subtle pressure effect is the increase in diffusivity accompanied by a decrease in total pressure. An even more subtle effect is the change in flow regime as the pressure is lowered. As the pressure is lowered, collisions of molecules with the reactor wall can become more dominant than collisions among molecules. When the mean free path of molecules is larger than the reactor radius, the laws of viscous flow, which assume that the gas behaves as a coherent medium, break down. In such cases, gas flow is characterized by molecules traveling independently of each other with random motions superimposed upon the direction of gas flow. This flow regime is called (free) molecular flow or Knudson flow as opposed to the familiar viscous flow. Molecular flow can dominate at as high a pressure as 0.5 torr. Note that the range of typical LPCVD total pressures is 0.1 to 10 torr. Note also that because of certain advantages of low pressure, the LPCVD is used not only for nonepitaxial films but also for epitaxial films.

An LPCVD process, when compared with the atmospheric CVD, yields a better thickness uniformity but at the cost of a lower deposition rate. This uniformity that the LPCVD offers is one of the major reasons for the wide use of the process. Another reason is a reduction in undesired impurities. The fluid velocity in the LPCVD reactor is higher by an order of magnitude than the velocity in the CVD reactor at atmospheric pressure. On the other hand, higher pressures than atmospheric pressure can also be used to obtain other advantages. Gas-solid reactions carried out at high pressures, as in the oxidation of native silicon (Hyrayama *et al.*, 1982), are an example although they cannot be classified as CVD.

The material and electrical properties of the grown film are the major considerations in any deposition. The material properties are thermal expansion (stress), density, thermal stability, and refractive index. Electrical properties depend on the use of the grown film. In epitaxial films, the main concern is the uniformity of resistivity as affected by defects such as dislocations and stacking faults. In insulating films, the breakdown voltage or dielectric strength could be of major concern. For metal contacts, the resistivity and lifetime are of interest. In addition to these properties, the thickness uniformity and conformity of grown film are important. Lithography cannot be carried out without uniformity. For nonepitaxial films, pinholes are often present in the grown film. Coping with poor adhesion of a grown film on the substrate and particulate contaminants are very real problems.

The type of material to be deposited, choice of chemicals, substrate preparation prior to the deposition, and temperature determine the material and electrical properties of grown films. For epitaxial films, higher temperature and lower growth rate tend to lead to better material and electrical properties. For instance, the defects' density decreases exponentially with both increasing temperature and decreasing growth rate (Chang and Baliga, 1984). This is not surprising in view of the discussion in the previous chapter on film structure, i.e., monocrystalline, polycrystalline, or amorphous. The same trends hold for nonepitaxial films. For a given deposition, therefore, the material and electrical properties define the broad boundary within which a CVD reactor can be operated. Once the feasible operating region is defined in terms of chemicals to be used, the type of material, and temperature, one is left with the design and operation of the reactor for thickness uniformity and conformity of windows into which the desired material is deposited.

This chapter is mainly concerned with uniformity and conformity for which the design and analysis of CVD reactors can have a direct bearing. Current photolithographic imaging techniques can demand a flatness tolerance of less than 2 μm (Benzing, 1984). This requirement will become more stringent as device dimensions continue to decrease. The uniformity problem, in terms of step coverage or conformity, is shown in Fig. 6-3. The conformity problem exists whenever deposition is onto an open window. This in turn causes the nonuniformity problem.

Crystallographic slip in epitaxial films that can be generated by temperature gradients within the wafer is another problem that the design and analysis of a CVD reactor can constrain. A high thermal flux normal to the wafer surface produces a concave-up bow in the wafer (Robinson *et al.*, 1982). Although this bow by itself is not sufficient to cause the slip, it causes the wafer to be in better thermal contact near its center than its edge and eventually the slip results. Therefore, uniformity of surface temperature is a desired requirement for CVD design and operation, in particular with respect to the heating source such as radiant heaters and their arrangement.

High wafer throughput, the uniformity and conformity, and uniform surface temperature can then be considered the design objectives of a CVD reactor. The pertinent fluid flow regimes are considered next, as the first step of the reactor design and analysis problem.

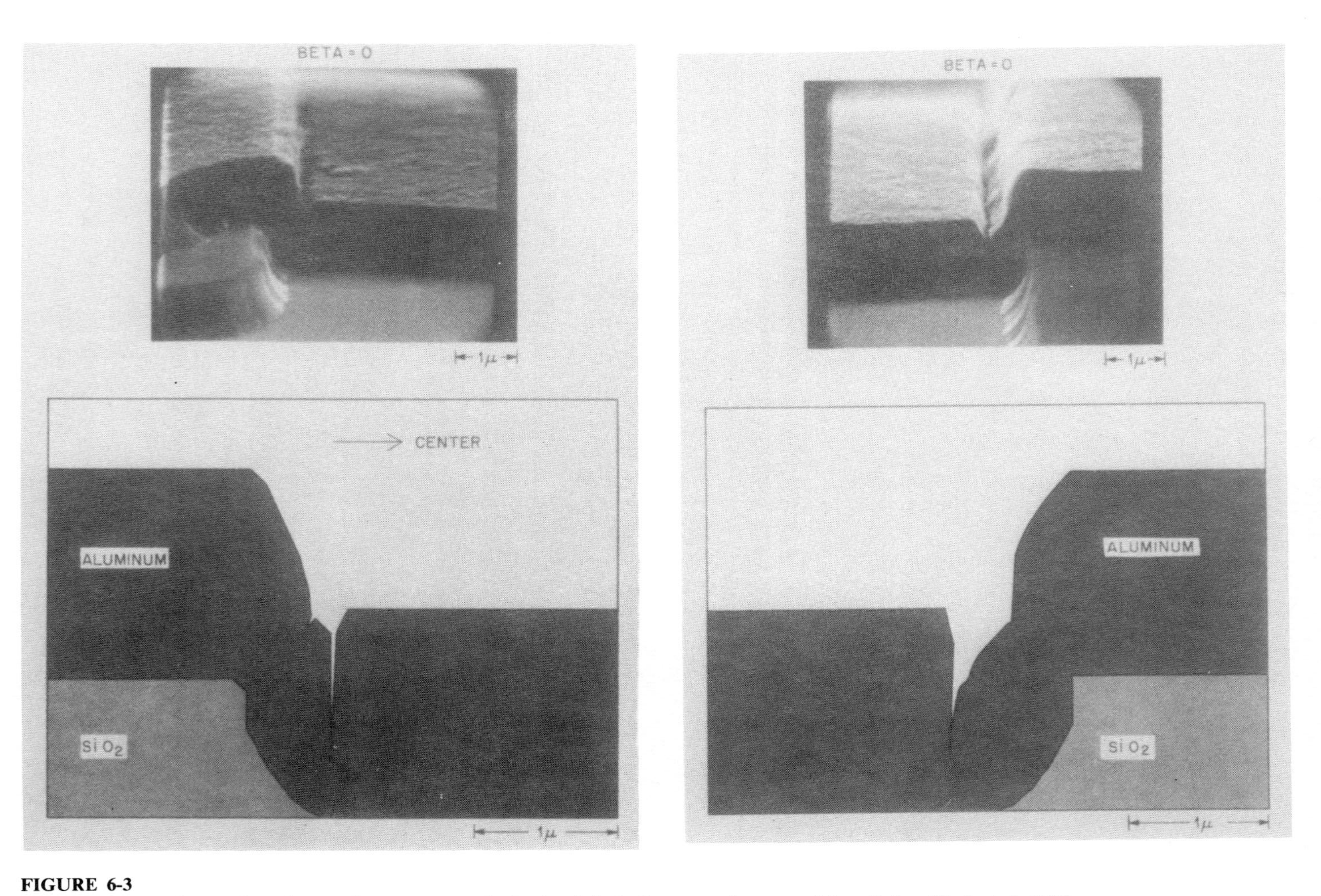

FIGURE 6-3
Step coverage: top figures show the actual step coverage obtained and the bottom ones show a model prediction (Blech *et al.*, 1978).

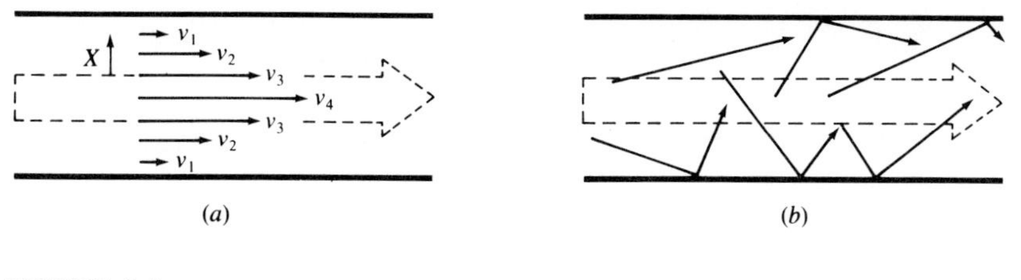

FIGURE 6-4
Viscous and molecular flows.

6.2 REGIMES OF FLUID FLOW

For flows in a channel or cylindrical tube, the transition from laminar to turbulent flow occurs at a Reynolds number (Re $= ud\rho/\mu$) of 2300, where u is the linear velocity, d is the tube diameter [hydraulic equivalent diameter for irregular shapes, which can be calculated as (4 × cross-sectional area)/(wetted perimeter)], and μ and ρ are the viscosity and density of the fluid, respectively. For typical CVD operation, the Reynolds number is around 100 or less when hydrogen is the carrier gas (Pearce, 1983). Thus, the flow regime in CVD is laminar unless buoyancy due to temperature gradients causes eddies.

The flow in a channel or cylindrical tube has a parabolic velocity profile, as indicated in Fig. 6-4(a). The volumetric flow rate Q in a cylinder follows from the Hagen-Poiseuille equation (e.g., Schlichting, 1960):

$$Q = \frac{\pi R_t^4}{8\mu l}(P_1 - P_2)$$

where l is the tube length, R_t is the tube radius, and P_1 and P_2, respectively, are the inlet and outlet pressures. In anticipation of the molecular flow results, a pressure flow rate Q_v can be defined as follows:

$$Q_v = QP' = \frac{\pi R_t^4 P'}{8\mu l}(P_1 - P_2) \tag{6.1}$$

where P' is the average of P_1 and P_2 and the subscript v is for viscous flow. At very low pressures, wall collisions are more frequent than intermolecular collisions. The viscous flow based on frequent intermolecular collisions, therefore, is not valid at these low pressures. The gas flow at very low particle densities is characterized by molecules traveling independently of each other with random motion superimposed upon the direction of gas transport. This molecular flow is shown in Fig. 6-4(b). The flow rate equivalent to Eq. (6.1) was derived for long cylindrical tubes by Knudsen:

$$Q_m = \frac{2\pi r^3 C}{3l}(P_2 - P_1) \tag{6.2}$$

where the subscript m is for molecular flow and the mean velocity C is given by

$$C = \left(\frac{8kT}{\pi m}\right)^{1/2} = 14{,}551\left(\frac{T}{M}\right)^{1/2} \text{ cm/s} \tag{6.3}$$

Noting that the mean free path of a molecule λ is given by

$$\lambda = \frac{kT}{2^{1/2}P\pi\alpha^2} \qquad \left(\lambda = \frac{5 \times 10^{-3}}{P(\text{torr})} \text{ cm at 300 K}\right) \tag{6.4}$$

where α is the molecular diameter, and using the simple kinetic theory (e.g., Bird *et al.*, 1960) for the viscosity, the ratio of Q_v to Q_m is

$$\frac{Q_v}{Q_m} = \frac{0.22r}{\lambda} \tag{6.5}$$

A more rigorous calculation gives a factor of 0.147 instead of 0.22 for Eq. (6.5). It can be seen from Eq. (6.5) that the molecular flow will dominate if the mean free path is much larger than the tube radius. The transition from the molecular flow to the viscous flow is gradual and the formulation by Knudsen for the transition flow Q_{tr} gives

$$Q_{tr} = Q_v + ZQ_m$$

or (6.6)

$$\frac{Q_{tr}}{Q_m} = \frac{Q_v}{Q_m} + Z = \frac{0.147r}{\lambda} + Z$$

where

$$Z = \frac{1 + 2.507r/\lambda}{1 + 3.095r/\lambda} \tag{6.7}$$

Example 6.1. Industrial LPCVD reactors operate at 0.05 torr and 910 °C for Si_3N_4 deposition based on SiH_2Cl_2/NH_3 and at 1 torr and 910 °C for SiO_2 based on SiH_2Cl_2/N_2O. Also, epitaxial silicon growth at 10^{-3} torr and 800 °C based on SiH_4/H_2 has also been reported in the literature (Meyerson *et al.*, 1986). Atmospheric silicon epitaxy can be carried out at 1000 °C based on SiH_2Cl_2/H_2. Assuming the effective diameter of the molecules in the mixture to be 0.5 nm for the first three cases and the reactor radius to be 8 cm, calculate the fraction of molecular flow for the four cases. For the epitaxial reactors, assume the reactor radius to be 2.5 cm. Redo the problem for a cavity for which $r = 2$ μm.

Solution. According to Eq. (6.4), the mean free path is given by

$$\lambda = \frac{1.38 \times 10^{-16}T}{1.333 \times 10^3 P(\text{torr}) \times 2^{1/2}\pi(5 \times 10^{-8})^2} = 9.333 \times 10^{-6}\,\frac{T(\text{K})}{P(\text{torr})} \text{ cm}$$

For H_2, the constant in the above equation is 30.84×10^{-6} instead of 9.33×10^{-6}. Therefore, one has

Cases	Si_3N_4	SiO_2	Epi Si	Atmospheric epi Si
λ, cm	0.22	0.011	10	5.16×10^{-5}

Equation (6.7) gives

$$Z = \frac{1 + 20.06/\lambda}{1 + 24.76/\lambda}$$

Use of the above and the calculated values of λ in Eq. (6.6) yields:

Cases	**Si_3N_4**	**SiO_2**	**Epi Si**	**Atmospheric epi Si**
Q_v/Q_m	5.35	107	0.037	2.35×10^4
Z	0.81	0.81	0.865	0.81
F, %	13	0.75	96	0

The fraction of the total flow due to the molecular flow F also follows from Eq. (6.6):

$$F = \frac{Z}{Q_v/Q_m + Z}$$

The F values are also given in the above table. The same procedures can be followed for a cavity with r of 2 μm. The results are:

Cases	**Si_3N_4**	**SiO_2**	**Epi Si**	**Atmospheric epi Si**
Q_v/Q_m	1.34×10^{-4}	2.67×10^{-3}	2.94×10^{-6}	0.57
Z	1	0.97	1	0.825
F, %	100	100	100	59

It is seen from the example that molecular flow plays a role in LPCVD reactions, its contribution to the total flow ranging from approximately 1 percent for SiO_2 deposition to 96 percent for epitaxial silicon deposition at very low pressure. The results for a 2-μm cavity show that the flow is entirely molecular, particularly for the film depositions of interest. Insulator, passivation film, and metal depositions are into windows of micrometer size. Therefore, deposition into these cavities is governed entirely by molecular flow. This is also true to a lesser extent for atmospheric processes. In considering uniform deposition, therefore, one has to distinguish between the uniformity over the entire wafer, which is the case, for example, for the epitaxial film growth of III-V compounds, and the uniformity for local deposition, which is the case for almost all processing steps on a wafer.

The stability of bulk flow, i.e., the absence of "rolls" (small rotating streamlines) and eddies, is of vital interest in achieving uniform deposition. The criterion for flow stability depends on whether the flow is fully developed before it reaches the susceptor. According to Schlichting (1960), the flow is fully developed at a distance from a flat entrance given by

$$l_v = 0.04H \text{ Re} \tag{6.8}$$

where H is the height of a channel and Re is based on the channel width. The thermal entrance length l_h for a fully developed radial profile, however, is seven times the velocity entrance length (Hwang and Chang, 1973):

$$l_h = 0.28H \text{ Re} \tag{6.9}$$

If the Schmidt number is the same as the Prandtl number, the concentration entrance length would be the same as the thermal length. For CVD gases, the two numbers are within one order of magnitude. The fully developed velocity and temperature profiles are shown in Fig. 6-5. The temperature profile, which is ultimately linear with the channel height, is for top and bottom planes maintained at two different temperatures. The velocity profile is given by

$$u(y) = \frac{P_1 - P_2}{2\mu l}\left[(0.5H)^2 - y^2\right] \tag{6.10}$$

where y is the coordinate for the channel height with the origin at the midpoint. For a cylindrical geometry, the constant in the denominator is 4 and H is the diameter.

Because of the wide ranges of pressure and temperature used in CVD, it is often of interest to know the pressure and temperature dependence of physical properties (see Table 6.1). Most of the dependencies are based on the hard sphere model. These are

$$\mu \propto T^{0.66} \qquad (P \leq 1 \text{ atm}) \tag{6.11}$$

$$D_{AB} \propto \frac{T^{1.75}}{P} \tag{6.12}$$

$$k \propto T^{0.69} \tag{6.13}$$

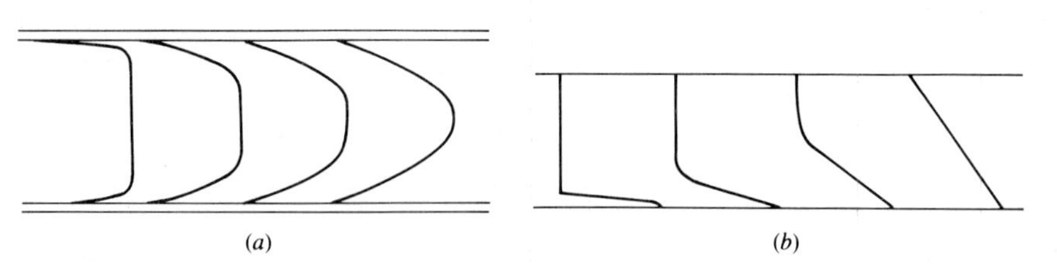

FIGURE 6-5
Development of (*a*) velocity and (*b*) temperature profiles.

TABLE 6.1
Relationships for physical properties (Bird *et al.*, 1960) and dimensionless groups

$$\mu[\text{g/(cm·s)}] = 2.6693 \times 10^{-5} \frac{(MT)^{1/2}}{\beta^2 \Omega_1} \qquad \beta \text{ in Å, } T \text{ in K}$$

$$D_{AB}(\text{cm}^2/\text{s}) = 0.0018583 \frac{[T^3(1/M_A + 1/M_B)]^{1/2}}{P\beta_{AB}^2 \Omega_2} \qquad P \text{ in atm, } \beta \text{ in Å}$$

$$k[\text{cal/(cm·s·K)}] = \left(C_p + \frac{5R}{4M}\right)\mu \qquad \text{where } C_p = \text{specific heat [cal/(g·K)]}$$

The Lennard-Jones parameters β, Ω_1, and Ω_2 for many gases can be found from the above reference.

Schmidt number	$\text{Sc} = \dfrac{\mu}{\rho D_{AB}}$
Prandtl number	$\text{Pr} = \dfrac{C_p \mu}{k}$
Reynolds number	$\text{Re} = \dfrac{\rho u d}{\mu}$
Grashof number	$\text{Gr} = \dfrac{\rho^2 g H^3 \Delta T}{\mu^2 T}$ for heat
	$= \dfrac{\rho^2 g H^3 \Delta x}{\mu^2 x}$ for mass (x = mole fraction)
Rayleigh number	$\text{Ra} = \text{Pr Gr}$
Froude number	$\text{Fr} = \dfrac{u^2}{g D_{AB}}$ where g = gravity
Nusselt number	$\text{Nu} = \dfrac{hL}{k}$ for heat where L = characteristic length
	$= \dfrac{k_m L}{D_{AB}}$ (=Sherwood number) for mass
Thermal diffusivity α (cm^2/s)	$= \dfrac{k}{\rho C_p}$
Momentum diffusivity (kinematic viscosity) (cm^2/s)	$= \dfrac{\mu}{\rho}$

Example 6.2. Silicon epitaxy in practice is carried out around a Reynolds number of 100 or less at atmospheric pressure. As the LPCVD pressure is lowered to 1 torr, the fluid velocity increases by a factor of between 10 and 100. For the LPCVD at 1 torr, calculate the entrance lengths, l_v and l_h, relative to the atmospheric process for the two velocity factors of 10 and 100. Assume the temperature to be the same for both atmospheric and low-pressure reactors. Calculate the entrance lengths for the LPCVD corresponding to the factor of 10 and H of 15 cm.

Solution. At the same temperature but different pressures, the Reynolds number ratio is

$$\frac{(\mathrm{Re})_l}{(\mathrm{Re})_a} = \frac{(\rho u)_l}{(\rho u)_a} \tag{A}$$

where the subscripts l and a, respectively, are for low and atmospheric pressure.

Equation (6.11) and the ideal gas law allow one to rewrite Eq. (A) as

$$\frac{(\mathrm{Re})_l}{(\mathrm{Re})_a} = \frac{(\rho u)_l}{(\rho u)_a} = \frac{1}{760}\left(\frac{u_l}{u_a}\right)$$

Therefore, the Reynolds number for the factor of 10 is 1.3 and it is 13 for the factor of 100. Thus, the entrance length l relative to the atmospheric process is

$$\frac{l_l}{l_a} = \frac{(\mathrm{Re})_l}{(\mathrm{Re})_a} = \frac{(\mathrm{Re})_l}{100} = \begin{cases} 0.013 \text{ for the factor of } 10 \\ 0.13 \text{ for the factor of } 100 \end{cases}$$

The Reynolds number for the factor of 10 is 1.3. Therefore, one has, from Eqs. (6.8) and (6.9),

$$l_v = 0.04H\ \mathrm{Re} = 0.04 \times 15 \times 1.3 = 0.78 \text{ cm}$$

$$l_h = 7 \times 0.78 = 5.46 \text{ cm}$$

It is clear from the example that the Reynolds number in a typical LPCVD reactor is much smaller than that for atmospheric CVD and that the thermal entrance length for the fully developed profile is not unreasonably long. However, the thermal entrance length for a horizontal reactor at atmospheric pressure with a height of 3 cm is 84 cm at a Reynolds number of 100, which is much longer than that for LPCVD reactors.

It is helpful to examine some of the commonly used assumptions of fluid flow. The flow of interest here is viscous. As opposed to the viscous flow, a perfect (frictionless and incompressible) fluid exerts no tangential forces (shearing stresses) on the surfaces in contact with it but acts on other fluid elements with normal forces (pressures) only. The essential differences between a real (viscous) and a perfect fluid are the existence of tangential stresses and the condition of no slip near solid walls for the real fluid. The assumption of a perfect fluid, which leads to potential (inviscid) flow theory, is reasonable for flows away from solid walls, and it leads to far-reaching simplifications of the equations of motion. Boundary-layer theory (Schlichting, 1960) is based on a combination of the two types of flow. Viscous flow exists in a very thin layer in the immediate neighborhood of the body in which the velocity gradient normal to the wall is very large (boundary layer), and potential flow dominates the remaining region where no such large velocity gradients occur and the influence of viscosity is unimportant. An assumption often made is that the fluid is incompressible. According to Schlichting, the density change due to compression $\Delta\rho/\rho$ is approximately given by

$$\frac{\Delta\rho}{\rho} = \frac{M^2}{2} \tag{6.14}$$

where the Mach number M is the ratio of the flow velocity to the sound velocity. Thus, the assumption of incompressible fluid is very reasonable when the flow velocity is much less than the sound velocity ($\sim 3.35 \times 10^4$ cm/s), which is more than satisfied for CVD reactors in view of the typical fluid flow velocity of less than 100 cm/s. Another assumption often invoked in the quantitative description of CVD reactors is the Boussinesq approximation. When the approximation is made, it means that only the gravity force is affected by the density change. In a broader sense, the approximation also means constant physical properties except for density in the buoyancy (gravity) term.

6.3 FREE CONVECTION AND FLOW STABILITY

The flow in CVD reactors is stable and laminar in the absence of a large temperature difference between the susceptor and reactor wall. The large temperature difference typical in CVD reactors, however, can give rise to flow driven by buoyancy. Flow due to a body force, such as buoyancy, is called free convection as opposed to the usual forced convection associated with bulk fluid flow, e.g., from an imposed pressure gradient. This means that flow due to free convection can occur in an otherwise quiescent fluid, as in an enclosed container, if there is a significant temperature difference between two regions.

Examination of a horizontal reactor with fully developed velocity and temperature profiles can bring out the effects of free convection due to the large temperature difference present in CVD reactors, in particular epitaxial reactors. With the Boussinesq approximation, one can write the following conservation equations for a binary system (dilute gas mixture) at steady state (Bird *et al.*, 1960):

Continuity:
$$\frac{\partial u}{\partial x} + \frac{\partial v}{\partial y} = 0, \qquad \frac{\partial w}{\partial z} = 0 \tag{6.15}$$

Momentum:
$$\rho\left(u\frac{\partial u}{\partial x} + v\frac{\partial u}{\partial y}\right) = -\frac{\partial P}{\partial x} + \mu\left(\frac{\partial^2 u}{\partial x^2} + \frac{\partial^2 u}{\partial y^2}\right) \tag{6.16}$$

$$\rho\left(u\frac{\partial v}{\partial x} + v\frac{\partial v}{\partial y}\right) = -\frac{\partial P}{\partial y} + \mu\left(\frac{\partial^2 v}{\partial x^2} + \frac{\partial^2 v}{\partial y^2}\right) + \rho(T)g \tag{6.17}$$

$$\rho\left(u\frac{\partial w}{\partial x} + v\frac{\partial w}{\partial y}\right) = -\frac{\partial P}{\partial z} + \mu\left(\frac{\partial^2 w}{\partial x^2} + \frac{\partial^2 w}{\partial y^2}\right) \tag{6.18}$$

Mass:
$$u\frac{\partial C}{\partial x} + v\frac{\partial C}{\partial y} + w\frac{\partial C}{\partial z} = D_{AB}\left(\frac{\partial^2 C}{\partial x^2} + \frac{\partial^2 C}{\partial y^2} + \frac{\partial^2 C}{\partial z^2}\right) \tag{6.19}$$

Energy:
$$\rho C_p\left(u\frac{\partial T}{\partial x} + v\frac{\partial T}{\partial y} + w\frac{\partial T}{\partial z}\right) = k\left(\frac{\partial^2 T}{\partial x^2} + \frac{\partial^2 T}{\partial y^2} + \frac{\partial^2 T}{\partial z^2}\right) \tag{6.20}$$

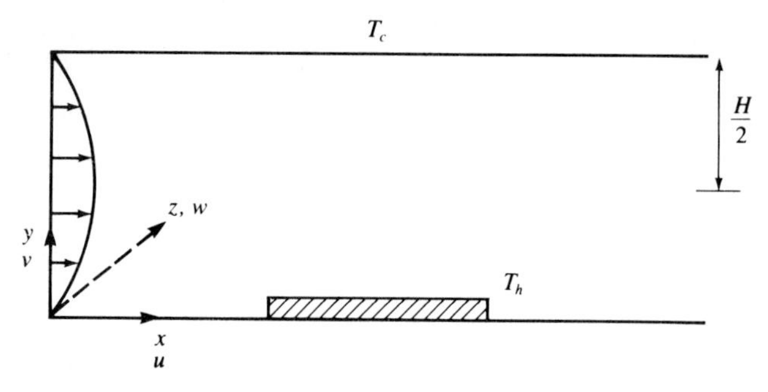

FIGURE 6-6
Coordinate system for balance equations.

The coordinate system and the velocities are shown in Fig. 6-6 and the notation can be found at the end of this chapter. The origin is on the centerline of the bottom midplane. Note that $\partial u/\partial z = \partial v/\partial z = \partial w/\partial z = 0$ for fully developed flow (Takahashi *et al.*, 1972). Additional assumptions are negligible viscous dissipation, thermal diffusion, radiation, and homogeneous phase reactions. The thermal diffusion (Soret effect) can be accounted for by adding the term, $(k_T C D_{AB} \nabla^2 \ln T)$, to the right-hand side of Eq. (6.19), where k_T is the thermal diffusion ratio (Bird *et al.*, 1960). Thermal diffusion has been reported (Jenkinson and Pollard, 1984) to play a role in a barrel reactor. However, no definitive experimental evidence has been put forward for the relative significance of thermal diffusion compared to other effects. The boundary conditions are zero velocities on all solid surfaces, $T = T_h$ at $y = 0$, $T = T_c$ at $y = H$, $u = u(y)$ on the side walls, which is parabolic, and a linear temperature profile with respect to y at $x = 0$, zero concentration gradients on all surfaces except on the susceptor, $C = C_{\text{in}}$ at $x = 0$, $\partial C/\partial y = \partial T/\partial y = 0$ at the reactor outlet, and symmetry conditions (zero gradients) about the plane $z = 0$. On the susceptor surface, thus one has

$$-D_{AB}\frac{\partial C}{\partial y} + \left(-k_T C D_{AB}\frac{\partial \ln T}{\partial y}\right) = r_c \tag{6.21}$$

where r_c is the intrinsic rate of film growth in moles per area. The thermal diffusion part is given in parentheses. The equations can be made dimensionless by normalizing the coordinates with respect to the height H, the velocities with respect to $\mu/\rho H$, the concentration with respect to the inlet concentration, and the pressure with respect to $\mu^2/\rho H^2$. The dimensionless temperature T' is given by $(T - T_0)/(T_h - T_c)$ where T_0 is $(T_h + T_c)/2$.

Continuity:
$$\frac{\partial U}{\partial X} + \frac{\partial V}{\partial Y} = 0, \qquad \frac{\partial W}{\partial Z} = 0 \tag{6.22}$$

Momentum:
$$U\frac{\partial U}{\partial X} + V\frac{\partial U}{\partial Y} = -\frac{\partial P}{\partial X} + \nabla_1^2 U \tag{6.23}$$

$$U\frac{\partial V}{\partial X}+V\frac{\partial V}{\partial Y}=-\frac{\partial P}{\partial Y}+\nabla_1^2 V+\mathrm{Gr}\,T' \tag{6.24}$$

$$U\frac{\partial W}{\partial X}+V\frac{\partial W}{\partial Y}=-\frac{\partial P}{\partial Z}+\nabla_1^2 W \tag{6.25}$$

Mass:

$$U\frac{\partial c}{\partial X}+V\frac{\partial c}{\partial Y}+W\frac{\partial c}{\partial Z}=\frac{1}{\mathrm{Sc}}\nabla^2 c \tag{6.26}$$

Energy:

$$U\frac{\partial T'}{\partial X}+V\frac{\partial T'}{\partial Y}+W\frac{\partial T'}{\partial Z}=\frac{1}{\mathrm{Pr}}\nabla^2 T' \tag{6.27}$$

where the Laplacians $\nabla_1^2=\partial/\partial X^2+\partial/\partial Y^2$, $\nabla^2=\partial/\partial X^2+\partial/\partial Y^2+\partial/\partial Z^2$, and the Prandtl, Schmidt, and Grashof numbers are defined in Table 6.1. Equation (6.21) becomes

$$-\frac{\partial c}{\partial Y}\left(-k_T\frac{c}{T'}\frac{\partial T'}{\partial Y}\right)=R_c \qquad \text{where } R_c=\frac{Hr_c}{D_{AB}C_{in}} \tag{6.28}$$

If the problem is for developing flow, the continuity equation is $\nabla\cdot v=0$ ($v=[U, V, W]$) and the left-hand sides of Eqs. (6.23) through (6.25) are $v\cdot\nabla U$, $v\cdot\nabla V$, and $v\cdot\nabla W$, with ∇_1^2 replaced by ∇^2 in the right-hand sides with uniform velocity and temperature at the inlet for the boundary conditions. Note in Eq. (6.24) that only the first-order term in the expansion of $\rho(T)$ in a Taylor series is used for the expression Gr T' (see Prob. 6.15). The equations can be condensed by introducing stream functions (see Prob. 6.16).

The purpose of writing the rather complete conservation equations is not to delve into the solution, which is a formidable numerical problem by itself, but rather to bring out the Schmidt and Grashof numbers and the aspect ratios, which enter into the picture through boundary conditions. The aspect ratios are the reactor width and length normalized with respect to the height. The question of interest for CVD reactors is under what conditions the flow is stable and laminar. Since the Prandtl and Schmidt numbers for almost all gases of interest are nearly constant and less than unity (the Prandtl number is around 0.7 to 0.77 for H_2 and N_2 in the temperature range of 500 to 1400 K and increases with temperature), the Grashof number and the aspect ratios should determine the flow stability, i.e., absence of vortices.

One might wonder at this point why the Rayleigh number is used for the conditions under which free convection is absent instead of the Grashof number, when in fact the conservation equations contain the Grashof number. Pellew and Southwell (1940) considered the condition of no free convection for two infinite parallel plates at different temperatures for a very large Prandtl number. For this one-dimensional problem at a very large Prandtl number, the transient version of Eqs. (6.16) and (6.17) can be cast in another dimensionless form with the thermal velocity α/L as the normalizing factor for the fluid velocity, resulting in the whole left-hand side of Eqs. (6.23) and (6.24) being divided by Pr. Those terms can be set to zero for very large Pr. The Grashof number in Eq. (6.24) is replaced with the

Rayleigh number in this description. They arrived at the following condition:

$$\text{No free convection if Ra} \leq 1708 \tag{6.29}$$

An interesting test of the criterion was made by Giling (1982) based on laser interference holography. For both cylindrical ($W/H \sim 1$) and rectangular ($W/H \sim 2$) horizontal reactors, a stable laminar flow seemed to result, provided the velocity and temperature profiles are fully developed (at 1400 K and 200 cm/s, the thermal entrance length is 15 cm at atmospheric pressure) for all velocities up to a Reynolds number of 274, for which the Rayleigh number is less than 660. Examples of the holograph pictures for air-cooled and top water-cooled cells are shown in Fig. 6-7. The lines in the pictures are the isotherms in the cross section. Note that the top water-cooled case is much closer to the ideal case of

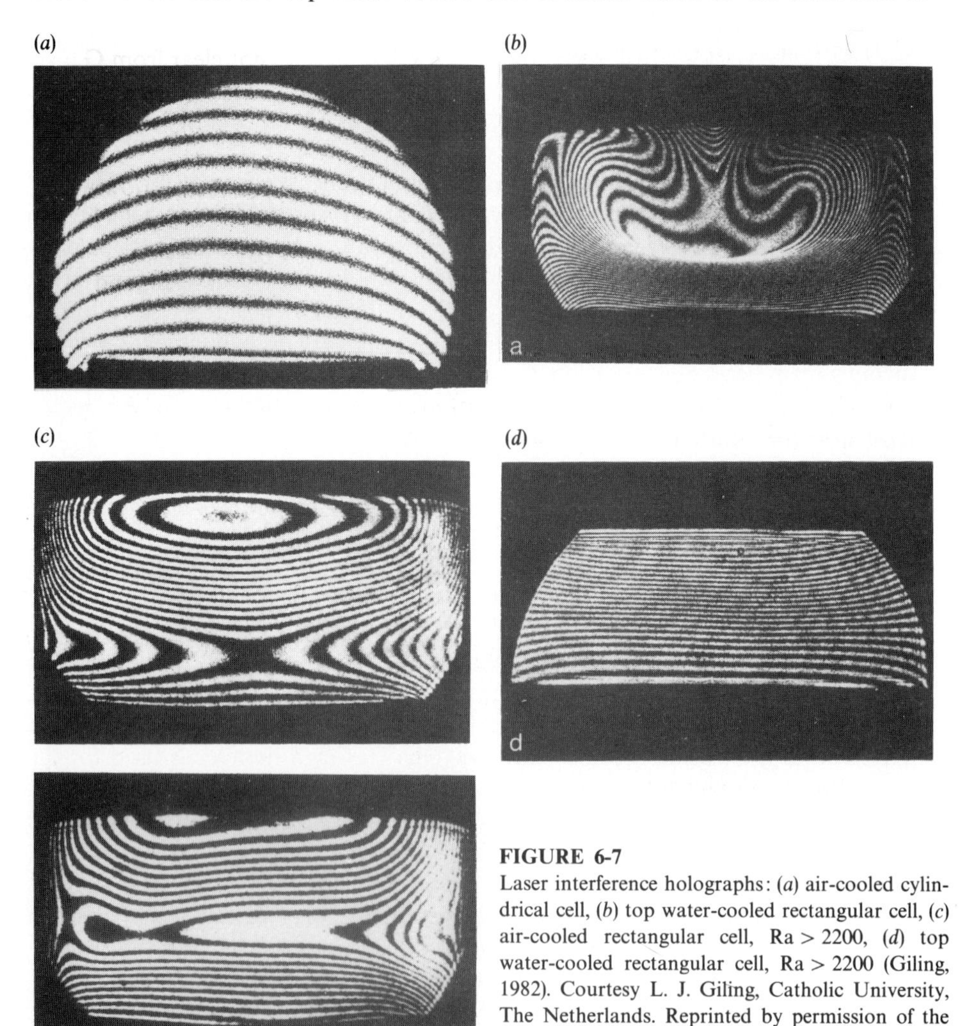

FIGURE 6-7
Laser interference holographs: (*a*) air-cooled cylindrical cell, (*b*) top water-cooled rectangular cell, (*c*) air-cooled rectangular cell, Ra > 2200, (*d*) top water-cooled rectangular cell, Ra > 2200 (Giling, 1982). Courtesy L. J. Giling, Catholic University, The Netherlands. Reprinted by permission of the publisher, The Electrochemical Society, Inc.

the two parallel plates (Fig. 6-7*b*) than the air-cooled case when the heat loss is through the whole circumference (Fig. 6-7*a*). When the two plane temperatures are well defined, as in Fig. 6-7*b*, a very desirable flow pattern results. Note in this regard that the holograph represents all cross sections along the reactor length compressed into a single cross section. The holographs for which the Rayleigh number is larger than 2200 are also shown in Fig. 6-7. The flow was not stable, as Fig. 6-7*c* and *d* shows. The experimental results also showed that at low Rayleigh numbers, there appeared to be no visible boundary layer in the vicinity of the hot susceptor but that at high Rayleigh numbers, there appeared to be a boundary layer, which is not necessarily stable because of the turbulent layer above the boundary layer. It is noteworthy that numerous results available in the heat transfer literature for laminar flow over heated horizontal plates do not appear to give any useful insights into CVD reactor behavior.

The effects of fluid velocity on the flow stability were not clear from Giling's experimental results. It is often thought that Gr/Re^2 (or $Gr/Re^{2.5}$; e.g., Sparrow *et al.*, 1959) determine the extent to which the free convection dominates a flow. The domination increases with an increasing value of Gr/Re^2. For the conditions relevant to CVD reactors, however, the criterion has not yet been shown to be useful. It is quite apparent that for a Reynolds number less than 100, which is the range of interest for CVD reactors, the velocity has very little, if any, effect on the flow stability for fully developed flow, constant top-wall temperature, and below the critical Rayleigh number. In addition to Giling's experimental results, the work by Ostrach and Kamotani (1977), based on a rectangular channel with water-cooled top, shows that the flow is stable and laminar for a Reynolds number less than 100 and that the Nusselt number is unity up to the critical Rayleigh number of approximately 1700.

Example 6.3. The experimental temperature profiles above a susceptor in a rectangular horizontal reactor (7.5 cm wide × 5 cm high) obtained by Ban (1978) are shown in Fig. 6-8 for three velocities in a helium environment. The temperature profiles correspond to the back end of the susceptor. The dimensions of the susceptor at 1000 °C were 5(width) × 2(height) × 15(length) cm. The reactor was air cooled and the top-wall temperature ranged from below 400 to over 600 °C, increasing in the direction of flow. Assuming that the top-wall temperature was 600 °C at the end of the susceptor, calculate the Rayleigh number and Reynolds number. One may evaluate the physical properties at the arithmetic mean temperature. The following physical properties at 1000 K may be used for the properties at the mean temperature (T_m) of 1073 K. All properties are in the units of kg, m, s, K, and J.

Gas	C_p	$10^3\rho$	$10^6\mu$	10^3k
He	5197	48.7	44.6	360
H_2	14,990	24.5	20.1	432
N_2	1168	341.4	40.0	62.6

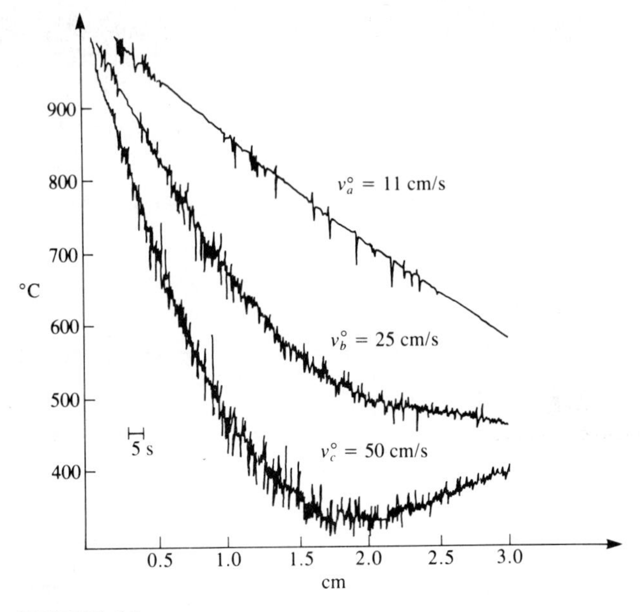

FIGURE 6-8
Temperature profiles reported by Ban (1978).

Calculate the thermal entrance length. Comment on the measured temperature profiles in the light of the work of Giling and Eq. (6.29). Note that the temperature profile above the susceptor is linear if the flow is stable and laminar.

Solution. The Rayleigh number (Table 6.1) is

$$\mathrm{Ra} = \frac{\rho^2 C_p g H^3 \,\Delta T}{\mu k T_m} = \frac{(48.7 \times 10^{-3})^2(5197)(9.8)(0.03)^3(400)}{(44.6 \times 10^{-6})(360 \times 10^{-3})(1073)}$$

$$= 76$$

Using the hydraulic diameter for the whole cross section for which d in Table 6.1 is 6 cm,

$$\mathrm{Re} = \frac{(48.7 \times 10^{-3})u(0.06)}{44.6 \times 10^{-6}} = 65.5u \text{ (m/s)} = \begin{cases} 7.2 \text{ for } u = 0.11 \text{ m/s} \\ 16.4 \text{ for } u = 0.25 \text{ m/s} \\ 32.8 \text{ for } u = 0.50 \text{ m/s} \end{cases}$$

The thermal entrance lengths are

$$l_h = 0.28H \text{ Re} = 0.84 \text{ Re} = \begin{cases} 6.0 \text{ cm for } u = 0.11 \text{ m/s} \\ 13.8 \text{ cm for } u = 0.25 \text{ m/s} \\ 27.6 \text{ cm for } u = 0.50 \text{ m/s} \end{cases}$$

The results and the temperature profile in Fig. 6-8 indicate that the flow is almost stable and laminar at the Rayleigh number of 76 for the third velocity of 11 cm/s, since the temperature profile is almost linear for that case. It appears that higher velocities than 11 cm/s tend to destabilize the flow, particularly in view of the fact that the thermal entrance length for the velocity of 25 cm/s is shorter than the susceptor length. More importantly, the critical Rayleigh number for an air-cooled horizontal reactor appears to be much less than the 1708 value given by Eq. (6.29), which was derived for a constant upper-wall temperature. When the top side is water cooled, as in the experimental work of Giling and that of Ostrach and Kamotani, the criterion has been shown to hold. This in turn reinforces the importance of maintaining the top-wall temperature constant for a much larger critical Rayleigh number than is possible without cooling.

The barrel reactor, which is a vertical version of the horizontal reactor, should give better flow stability than the horizontal reactor since the flow is in the same direction as gravity. For an inclined barrel, only the component of gravity in the direction of flow will be present. The rotation rate, used to obtain uniform exposure of the wafers to the fluid, should be such that it does not cause rotation-induced vortices. The flow instability due to rotation (Schlichting, 1960) is

$$\mathrm{Ta} = \mathrm{Re}_i \left(\frac{d}{R_i}\right)^{1/2} \geq 41.3 \qquad \text{where } \mathrm{Re}_i = \frac{\rho U_i d}{\mu} \tag{6.30}$$

where Ta is the Taylor number, Re_i is the Reynolds number based on the peripheral velocity U_i ($=\omega R_i$, where ω is the angular velocity) of the inner cylinder, d is the gap between the two vertical cylinders, and R_i is the radius of the inner cylinder. The onset of vortices occurs when the Taylor number exceeds 41.3 for two concentric vertical cylinders, with the inner cylinder rotating and the outer one at rest.

Example 6.4. A barrel reactor operating at an average temperature of 1000 K with hydrogen as a carrier gas has a barrel of radius of 10 cm at the top and 12 cm at the bottom with a corresponding gap of 3 and 2 cm, respectively. Calculate the critical peripheral velocity at both ends. Also calculate the corresponding angular velocities.

Solution. From Eq. (6.30),

$$\mathrm{Ta} = \frac{\rho U_i d}{\mu}\left(\frac{d}{R_i}\right)^{1/2} \geq 41.3 \tag{A}$$

Thus, the critical peripheral velocity $(U_i)_c$ follows from Eq. (A):

$$(U_i)_c = 41.3 \frac{\mu}{\rho d}\left(\frac{R_i}{d}\right)^{1/2}$$

$$= 41.3 \frac{\mu}{\rho}\left(\frac{R_i}{d}\right)^{1/2} \frac{1}{d} \tag{B}$$

At the top, $R_i/d = 10/3 = 3.33$; at the bottom, $R_i/d = 12/2 = 6$.

Using the physical properties of H_2 at 1000 K given in Example 6.3,

$$\frac{\mu}{\rho} = \frac{20.1 \times 10^{-6}}{24.5 \times 10^{-3}} = 0.82 \times 10^{-3}\ \mathrm{m^2/s}$$

From Eq. (B),

$$(U_i)_c = 338.7\ (\text{cm}^2/\text{s})\left(\frac{R_i}{d}\right)^{1/2}\frac{1}{d} = \begin{cases} 206\ \text{cm/s (top)} \\ 415\ \text{cm/s (bottom)} \end{cases}$$

The angular velocity ω is given by $\omega = U_i/R_i$. Thus, the critical angular velocities are

$$\omega_c = \begin{cases} 20.6\ \text{s}^{-1}\ \text{(top)} \\ 34.6\ \text{s}^{-1}\ \text{(bottom)} \end{cases}$$

For the LPCVD reactors in Fig. 6-1, stability is much less of a problem. The typical Reynolds number is of the order of unity and the Rayleigh number is very small due to the low pressure. Furthermore, the hot-wall arrangement means that the maximum temperature difference is much smaller than in cold-wall reactors. Therefore, the only concern here has to do with the intrusion of the streamlines from the bulk flow in the annulus formed by the wall and the edges of the mounted wafers into the space between two adjacent wafers. At a Reynolds number of 25, this intrusion is of the order of one interwafer spacing (Middleman and Yeckel, 1986). It is expected, however, that the intrusion is negligible at the typical LPCVD reactor Reynolds number of unity.

It is of interest, in view of the importance of the Rayleigh number, to know how the dimensionless number changes with the design and operating variables. As evident from the definition of the number in Table 6.1, it is proportional to ρ^2 and thus P^2. It is also approximately proportional to $T^{-3.4}$, assuming C_p increases linearly with T at high temperatures, which is typically the case. Thus, the Rayleigh number decreases with decreasing total pressure and increasing temperature. It is important to recognize that the number is proportional to H^3, where H is the distance between the susceptor (or wafer) and the reactor wall. Therefore, it is desirable to have the reactor height as small as is practically possible to achieve flow stability. The Rayleigh number is also proportional to the temperature difference, ΔT.

Various forces at work in a CVD reactor can have a stabilizing or destabilizing effect on the flow depending on the relative directions of these forces with respect to the gravitational force. Various situations are shown (Ostrach, 1983) in Fig. 6-9. The arrows indicate the directions of heat and fluid flow. As shown in the figure, when the heat and fluid flow are in the same direction as the direction of gravity, a stabilizing effect results. The cases B_1 and B_2 correspond to a barrel reactor. The cases H_1 and H_2 correspond to the horizontal reactor. Cases V_1 and V_2 are for the vertical reactor (e.g., Houtman *et al.*, 1986), similar to the pancake reactor. The case V_2 is more stable than case V_1, indicating that the temperature is a more important variable than the flow.

6.4 INTRINSIC KINETICS AND TRANSPORT EFFECTS

In order to carry out kinetic studies and to clearly delineate the effects of deposition conditions on the growth behavior, it is necessary to know the rate of depo-

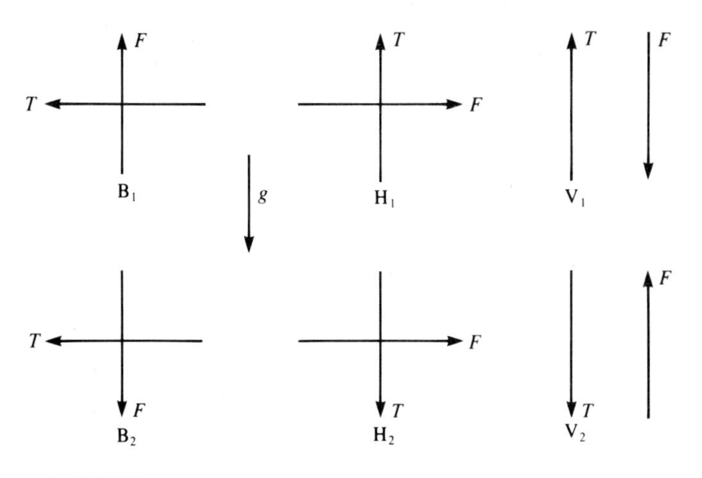

FIGURE 6-9
Stabilizing (B_2, H_2, V_2) and destabilizing (B_1, H_1, V_1) effects of various forces (Ostrach, 1983).

sition unaffected by transport effects, i.e., the intrinsic kinetics of deposition. An observed rate of deposition that is affected by heat, mass, and momentum transport can only lead to erroneous conclusions if used for the intrinsic kinetics. Unlike catalytic reactions involving supported catalyst particles or pellets (e.g., Lee, 1985), the intrinsic kinetics cannot be obtained simply by increasing fluid velocity, as evident from the previous section.

The intrinsic kinetics of deposition, however, can be obtained in a reactor that is well defined in terms of the transport effects. It is evident from the previous section that the Nusselt number for heat transfer is unity, if the top wall of a rectangular, horizontal reactor is kept constant by water-cooling and if the Rayleigh and Reynolds numbers, respectively, are less than 1700 and 100, and if the flow is fully developed thermally. When these conditions are met, one has (e.g., Ostrach and Kamotani, 1977)

$$\mathrm{Nu}_h = \frac{hH_{eq}}{k} = 1 \tag{6.31}$$

where H_{eq} is the rectangular channel height and h is the heat transfer coefficient. The heat transfer coefficient is based on the temperature difference between hot and cold walls. If the heat generated by the deposition is negligible, which is realistic in view of the very small amount of deposition, the heat balance for the rectangular channel yields a linear temperature profile, which has been found to be true experimentally for $\mathrm{Ra} < 1708$ and $\mathrm{Re} < 100$ with fully developed flow (Ostrach and Kamatoni, 1977). One then has

$$-k \left. \frac{dT}{dy} \right|_{\text{cold wall}} = h(T_h - T_c) = h\,\Delta T \tag{6.32}$$

Because of the linear temperature profile, the temperature gradient is $(-\Delta T/H)$ and the Nusselt number relationship of Eq. (6.31) results. For concentric hot and cold walls, as in a barrel reactor, the effective distance H_{eq} is given (see Prob. 6.17) by

$$H_{eq} = R_o \ln\left(\frac{R_o}{R_i}\right) \tag{6.33}$$

where R_o is the cold (outer) wall radius and R_i is the barrel radius (for a straight barrel).

Although the thermal diffusion ratio was used in Eq. (6.21) for the mass flux due to thermal diffusion, it is the usual practice (Hirschfelder *et al.*, 1954) for gaseous systems to write the thermal flux as $(\alpha_t + 1)C(\partial \ln T/\partial y)$ where α_t is the thermal diffusion factor. The Sherwood number can be derived (see Prob. 6.18) to give

$$\mathrm{Sh} = \frac{k_m H_{eq}}{D_{\mathrm{AB}}(T_h)} = \frac{20}{7} - (\alpha_t + 1)\left(\frac{C_s}{C_b - C_s}\right)\left(\frac{T_h - T_c}{T_h}\right) \tag{6.34}$$

The last term represents the contribution from the thermal diffusion. It is seen that the contribution is negligible in the mass-transfer-limited regime or when $C_s \ll C_b$, since α_t is of the order of unity. In such cases, the mass Nusselt number is 20/7. The mass transfer coefficient follows from Eq. (6.34):

$$k_m = \frac{20D_h}{7H_{eq}} - \frac{D_h(\alpha_t + 1)}{H_{eq}}\left(\frac{C_s}{C_b - C_s}\right)\left(\frac{T_h - T_c}{T_h}\right) \quad \text{where } D_h = D_{\mathrm{AB}}(T_h) \tag{6.35}$$

and where H_{eq} is H for the horizontal reactor and is given by Eq. (6.33) for the straight barrel reactor. For the barrel-type reactor with a hot wall, Wahl (1984) reported that the Sherwood number follows a typical relationship, which is

$$\mathrm{Sh} = 1.02\ \mathrm{Re}^{1/2}\ \mathrm{Sc}^{1/3}$$

The observed (measured) rate of deposition is at the wafer surface concentration and temperature. Since the surface temperature can be measured, the observed rate can be described in terms of the intrinsic kinetics provided the surface concentration is known. Since the observed rate is equal to the rate of mass transfer across the gas-solid film, one has

$$k_m(C_b - C_s) = r_G \tag{6.36}$$

where C_b is the bulk fluid concentration of the source gas, C_s is the surface concentration, and r_G is the rate of deposition in moles per time per area. It follows from Eq. (6.36) that the surface concentration is given by

$$C_s = C_b - \frac{r_G}{k_m} \tag{6.37}$$

Note that r_G is the measured rate, which is a known quantity. Thus, the surface concentration can be calculated from Eqs. (6.35) and (6.37). Since the measured rates are now in terms of the known surface concentration and temperature, the intrinsic kinetics can be determined on the basis of C_s versus $r_G(C_s)$.

Example 6.5. Suppose that the rate of deposition obtained in a reactor satisfying the conditions for the Sherwood number given by Eq. (6.34) is 2.0 μm/min at 1000 K and atmospheric pressure. Hydrogen is the carrier gas and the height of the rectangular channel is 3 cm. Assuming that the source gas in the feed is 1 percent by volume and the diffusivity of source species is 1 cm^2/s, calculate the surface concentration of the source gas. Assume the molecular weight is 102 g/mol and the density of the deposited material is 2.8 g/cm^3. Neglect thermal diffusion.

Solution. Since the molal growth rate r_G is related to the linear growth rate G by

$$r_G = \frac{G\rho_s}{M_w}$$

one has

$$r_G = \frac{(2.0 \times 10^{-4}/60) \times 2.8}{102} = 9.148 \times 10^{-8} \text{ mol/(cm}^2\text{·s)}$$

From Eq. (6.35), $k_m = 20D_h/7H = 0.952$ cm/s. From the ideal gas law,

$$C_b = \frac{p_i}{RT} = \frac{1 \times 0.01}{82 \times 1000} = 12.2 \times 10^{-8} \text{ mol/cm}^3$$

From Eq. (6.37),

$$C_s = 12.2 \times 10^{-8} - \frac{9.148 \times 10^{-8}}{0.952} = 2.59 \times 10^{-8} \text{ mol/cm}^3$$

An experimental reactor suitable for the determination of intrinsic kinetics is shown in Figure 6-10. It is the reactor configuration proposed by Giling (1982) combined with the feeding arrangement of Ostrach and Kamotani (1977). The feed section assures a uniform feed and promotes fully developed flow. The inlet section of the reactor is also heated so that, by the time the fluid enters the susceptor section, the flow is fully developed thermally. The top wall of the rectangular channel is water cooled to maintain a constant top-wall temperature. The after-heater section assures no sudden temperature change in the reactor.

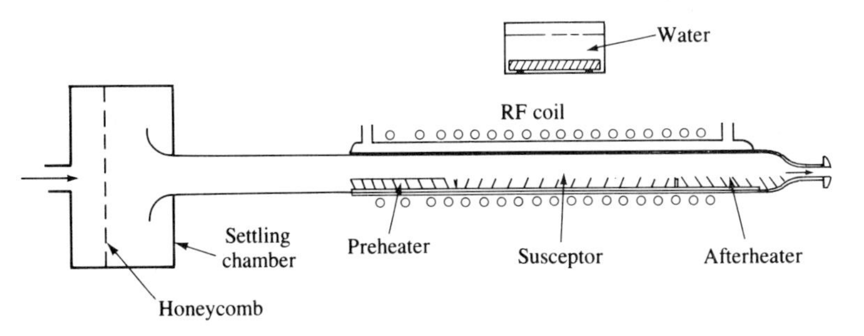

FIGURE 6-10
An experimental reactor for intrinsic kinetics.

The reactor can be operated in the stable laminar regime by making the Rayleigh number less than 1700.

The mass transfer rate in Eq. (6.36) deserves some comments. In a cold-wall reactor, no reaction takes place near the reactor wall because of low temperature, which is at about 300 K for a top water-cooled horizontal reactor and somewhat higher for an air-cooled barrel reactor. C_b is the feed concentration of the source species such as dichlorosilane. The mass transfer rate is the rate at which the source species is supplied to the wafer surface. As such, the surface concentration is an equivalent source-species concentration at the substrate surface, since gas phase reactions would lead to various adspecies precursors in the vicinity of the surface. Note that the kinetics are also expressed in terms of the source-species concentration, although they are based on adspecies.

Example 6.6. Suppose that an experimental reactor is to be fabricated for intrinsic kinetics studies, such as the one in Fig. 6-10. The top wall is maintained at 300 K and the susceptor temperature is to be varied between 1000 and 1400 K at atmospheric pressure. At the maximum velocity of 120 cm/s, the Reynolds number with hydrogen carrier gas ranges from 30 to 40. Specify the preheater length and the channel height for the Nusselt number given by Eq. (6.34).

Solution. The thermal entrance length l_h is

$$l_h = 0.28H \text{ Re} \tag{A}$$

The Nusselt number is unity for the given conditions if

$$\text{Ra} = \frac{\rho^2 C_p g H^3 \, \Delta T}{\mu k T_m} < 1708$$

Since

$$\text{Ra} \propto T^{-3.4} \, \Delta T$$

a conservative number will be that at 1000 K. Therefore,

$$\frac{\rho^2 C_p q H^3 \Delta T}{\mu k T_m} = \frac{(24.5 \times 10^{-3})^2 (14{,}990)(9.8)(700) H^3}{(20.1 \times 10^{-6})(432 \times 10^{-3})(650)} < 1708$$

or $$1094 \times 10^7 H^3(\text{m}) < 1708$$

or $$H^3(\text{cm}) < 156$$

Thus,

$$H_{\text{max}} = 5.38 \text{ cm} \tag{B}$$

The smaller the height, the smaller the thermal entrance length becomes and the better it is for the flow stability. Choosing an H of 3 cm, one has, from Eq. (A),

$$l_h = 0.28 \times 3 \times 40 = 33.6 \text{ cm}$$

Example 6.7. Suppose that the experimental reactor in Example 6.6 yields the following data with hydrogen carrier gas at a total pressure of one atmosphere:

Feed source gas, vol %	Substrate temperature, K	Rate of deposition, μm/min
0.2227	1300	0.85
0.0961	1300	0.52
0.0258	1300	0.20
0.2729	1200	0.325
0.1386	1200	0.213
0.0447	1200	0.092
1.417	1100	0.088
0.7187	1100	0.078
0.3502	1100	0.063

Determine the intrinsic kinetics. Assume that $\alpha_t = 1$, $T_c = 300$ K, $D_{AB} = 1$ cm^2/s at 1100 K, and the conversion factor from G (μm/min) to r_G [mol/(cm^2·s)] is 3.51×10^{-8}. Note that the bulk of the feed is carrier gas (more than 99 percent by volume) and that the concentration in the reactor is maintained constant at the desired level.

Solution. The relationships needed to solve the problem are

$$r_G\ [\text{mol}/(\text{cm}^2\cdot\text{s})] = 3.51 \times 10^{-8}\ G\ (\mu\text{m/min}) \tag{A}$$

$$C_b = \frac{p_s}{RT} = \frac{0.01x\ (\text{vol }\%)}{82T(\text{K})} \tag{B}$$

$$C_s = C_b - \frac{r_G}{k_m} \tag{C}$$

$$D_{AB}(T) = 1 \times \left(\frac{T}{1100}\right)^{1.75} \tag{D}$$

where k_m is given by Eq. (6.35). Use of Eqs. (A) through (D) for the calculations of C_s and r_G leads to the following:

C_b, mol/cm^3 × 10^8	C_s, mol/cm^3 × 10^8	T_s, K	r_G, mol/(cm^2·s) × 10^8
2.089	0.536	1300	2.984
0.900	0.303	1300	1.825
0.242	0.116	1300	0.702
2.56	1.02	1200	1.140
1.30	0.606	1200	0.749
0.419	0.236	1200	0.322
13.92	7.93	1100	0.309
7.06	4.11	1100	0.274
3.44	2.07	1100	0.221

Trial-and-error calculations are required to solve Eq. (C) for C_s. Now that the concentration and temperature, at which the deposition rate is obtained, are known, C_s and T_s can be related to r_G. The kinetics in the nth-order form, $r_G = kC^n$, would not fit the data well. This suggests that the following could be tried:

$$r_G = \frac{kC}{1 + KC}$$

Thus,

$$r_G(T_s, C_s) = \frac{k_s C_s}{1 + K_s C_s} \tag{E}$$

where the subscript s is for the surface condition. Equation (E) can be rewritten in the following form for a linear plot:

$$\frac{1}{r_G} = \frac{1}{k_s}\left(\frac{1}{C_s}\right) + \frac{K_s}{k_s} \tag{F}$$

For a linear plot, one has, from the above table,

$1/C_s$, cm³/mol × 10^{-8}	T_s, K	$1/r_G$, cm²·s/mol × 10^{-8}
1.87	1300	0.335
3.30	1300	0.548
8.62	1300	1.425
0.98	1200	0.877
1.65	1200	1.335
4.24	1200	3.106
0.126	1100	3.236
0.243	1100	3.650
0.483	1100	4.525

The linear plots based on the above table yield the following:

T_s, K	k_s, cm/s	K_s, cm³/mol × 10^{-8}
1300	6.18	0.182
1200	1.48	0.350
1100	0.27	0.758

Since $k = k_0 \exp(-E/RT)$ and $K = K_0 \exp(Q/RT)$, ln k_s versus $1/T_s$ and ln K_s versus $1/T_s$ plots can be made to determine the preexponential factors, the activation energy, and the heat of adsorption Q. These plots yield the following:

$$k = 1.74 \times 10^8 \exp\left(-\frac{44{,}600}{RT}\right)$$

$$E_a = 44.6 \text{ kcal/mol}$$

$$K = 7120 \exp\left(\frac{20{,}400}{RT}\right)$$

$$Q = 20.4 \text{ kcal/mol}$$

Any correlation for the kinetics should at least satisfy the Arrhenius relationship. As illustrated in the example, the experimental data, as a minimum, should have three concentrations at each of three different temperatures.

The example just considered reveals an interesting point. The second table in the example shows that the bulk fluid concentration of the source gas is 7.22×10^{-8} mol/cm^3 (first entry in the table) whereas the concentration at which the deposition takes place, i.e., the surface concentration, is 0.536×10^{-8} mol/cm^3 at 1300 K, which is much less than the bulk concentration. In contrast, the bulk concentration is not that much higher than the surface concentration at 1200 K. It is relatively easy to understand the difference when the deposition follows first-order kinetics as in polysilicon growth at a relatively high temperature. At steady state, the rate of mass transfer from the bulk fluid to the substrate surface should be equal to the rate of deposition:

$$k_m(C_b - C_s) = r_G = k_s C_s \tag{6.38}$$

for the first-order kinetics. It follows that the surface concentration is given by

$$\frac{C_s}{C_b} = \frac{k_m}{k_s + k_m} \tag{6.39}$$

and the rate of deposition is given by

$$r_G = k_s C_s = \left(\frac{1}{1/k_s + 1/k_m}\right) C_b \tag{6.40}$$

If $k_s \gg k_m$ or $1/k_s \ll 1/k_m$, the rate of deposition is simply given by $k_m C_b$. When the condition is satisfied, the resistance to mass transfer $(1/k_m)$ is much larger than the "resistance" to reaction $(1/k_s)$, which means that the deposition is controlled by mass transfer (resistance). Therefore, it should appear in this case that the deposition rate is almost independent of temperature since the mass transfer coefficient is a weak function of temperature. In the other extreme where $k_s \ll k_m$ or $1/k_s \gg 1/k_m$, the chemical kinetics controls the deposition, and the deposition rate becomes heavily dependent on temperature, because of the Arrhenius (exponential) temperature dependence of the rate constant. In this extreme case, the rate r_G becomes simply $k_s C_b$. It is often stated that mass transfer controls if the rate of mass transfer is much smaller than the rate of deposition, and vice versa. What is meant here is not the actual rates but rather the "potential" rates since, at steady state, the two rates are the same. If the potential rate of deposition is much higher than the potential rate of mass transfer, this potential rate of deposition cannot be sustained because of insufficient mass transfer. The actual deposition rate is at the mass transfer rate, and is said to be mass-transfer-limited (controlled). As the temperature increases, deposition becomes more mass-transfer-controlled, as evident from Example 6.7.

The question of whether mass transfer or surface reaction is rate limiting can always be resolved, even without knowledge of the kinetics, provided the actual rate of deposition is known. The maximum rate of mass transfer is $k_m C_b$ since it is given by $k_m(C_b - C_s)$. Therefore, one can readily determine which one controls the deposition by simply examining the ratio of $(r_G)_m/k_m C_b$:

$$\text{Surface kinetics controls} \quad \text{if} \quad (r_G)_m/k_m C_b \ll 1 \tag{6.41}$$

$$\text{Mass transfer controls} \quad \text{if} \quad (r_G)_m/k_m C_b \doteq 1 \tag{6.42}$$

where $(r_G)_m$ is the measured rate of deposition. It follows from Eq. (6.41) that the rate of deposition is proportional to concentration (apparent first order) regardless of the kinetics involved when the deposition is controlled by mass transfer.

It is important to recognize that the relative rates, and not the absolute magnitude of the rates, determine which rate process controls the deposition. A case in point is the use of low pressure for LPCVD reactors. The reason why the LPCVD reactors yield a better thickness uniformity than reactors operated at atmospheric pressure is often cited to be due to the enhancement of diffusion (mass transfer) afforded by the low pressure and thus enhanced mass transfer, since the diffusivity is inversely proportional to pressure [Eq. (6.12)]. If the rate of mass transfer is high, the deposition is more likely to be surface-kinetics-controlled, and it is more likely to yield better uniformity. Uniform thickness results if the deposition is controlled by surface kinetics and if bulk fluid concentration and temperature are uniform. The reasoning is valid in almost all cases as the current shift to LPCVD reactors amply demonstrates, since the rate of deposition (surface kinetics) also decreases with decreasing total pressure, resulting in an increased rate of mass transfer and a decreased rate of surface reaction. On the other hand, there would be no advantage in using a surface-kinetics-controlled deposition if it is operated in the region where the rate of surface reaction is inversely proportional to the pressure, which would be rare. The point in this case is that both the rate of mass transfer and the rate of surface reaction increase with decreasing pressure. The net result is that there is no change in the relative rates, although both rates are reduced in magnitude, and no advantage can be gained with the use of an LPCVD reactor. Another major reason why LPCVD reactors yield better uniformity, which often goes unrecognized, is the flow stability afforded by the low pressure due to the corresponding decrease in the Rayleigh number. Since the Rayleigh number is proportional to the square of total pressure, a reduction from atmospheric pressure to 1 torr, typical of LPCVD reactor operation, results in a decrease of the Rayleigh number by a factor of 760^2 $(= 5.8 \times 10^5)$.

The determination of the intrinsic kinetics of deposition based on the reactor in Fig. 6-10 can proceed as follows:

1. Obtain the rates of deposition over the ranges of concentration and temperature of interest.
2. Determine the surface concentration with the aid of Eq. (6.35):

$$C_s = C_b - \frac{r_G}{k_m} \tag{6.43}$$

3. Postulate the macroscopic deposition steps and derive the corresponding kinetics as detailed in the previous chapter. Express the kinetics in terms of the feed concentration.
4. Determine the rate and equilibrium constants based on the data and the derived kinetics.
5. Repeat steps 3 and 4 until the Arrhenius relationships for the temperature dependence are satisfied.

6.5 REACTOR DESIGN

The major objective of a CVD reactor design for given source gases is to achieve thickness uniformity within a wafer and from wafer to wafer. The requirement for this uniformity is likely to be a maximum 2 μm thickness deviation within a wafer for epitaxial film growth (Benzing, 1984). As the device size decreases, the maximum allowable thickness variation will also decrease because of the more stringent requirements of lithographic processes. There is a lower bound on the temperature because of the better material and electrical properties that result when a higher temperature is used. On the other hand, there is also an upper bound on the temperature because of the accompanying dopant redistribution that may occur at higher temperatures. This upper bound is approximately 700 °C for silicon and 450 °C for gallium arsenide.

The second objective is to maximize the number of wafers processed per time, although single wafer processing appears to be preferred in some fabrication steps. The rate of deposition and the reactor size determine the throughput. The costs associated with the source gases are a minor consideration relative to the two major objectives.

Within these broad objectives, there are also design objectives specific to particular applications and needs. One of them, which is applicable to almost all cases, is to have a susceptor design that will minimize the wafer bending. This has to do with finding the susceptor design that will yield nearly uniform temperature across the wafer surface. Minimization of the defects generated due to radial temperature gradients in the substrate is the main incentive, but the thickness uniformity also depends on the radial temperature profile. Formation of insulating and passivating layers and metallization often involve deposition over stepped surfaces rather than planar surfaces. The deposited layer in such cases should conform to the shape of the step, known as step coverage. Lastly, whenever a buried layer that is doped heavily is present, as in the bipolar junction transistor structure, dopants evaporate in the course of further processing and deposit on the downstream wafers. This (lateral) autodoping is another uniformity problem that has to be dealt with.

The design of CVD reactors in general involves satisfying various constraints and requirements of the design objectives. This is in contrast with the usual chemical and catalytic (heterogeneous) reactor design, where conservation equations are written and the solutions lead directly to the specification of the

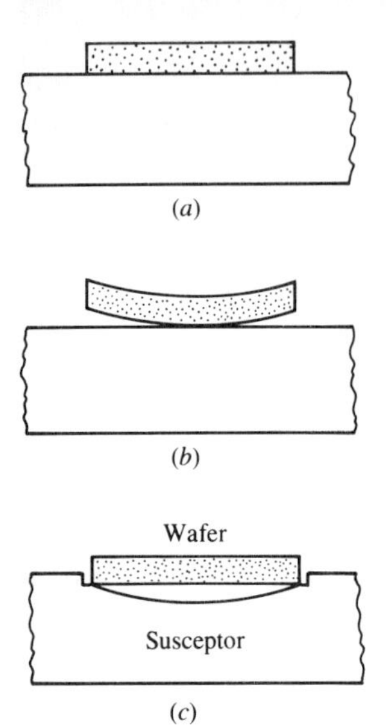

FIGURE 6-11
Shapes of wafer on susceptor.

reactor size. For each of the constraints, however, conservation equations can be written for the CVD reactors to obtain the conditions that satisfy a particular requirement(s), as was done for the free convection and flow stability in Sec. 6.3.

Consider the susceptor design first since this can be done separately from the overall reactor design. When a wafer on a susceptor is heated by an rf heater (induction heating), the bottom of the wafer is at a higher temperature than the surface of the wafer exposed to gas, since the surface loses heat by convection and radiation. The center of the wafer is at the highest temperature both at the bottom and top surfaces. This temperature difference causes the wafer to warp, and a slight separation of the wafer edges from the susceptor occurs. This in turn causes a higher temperature difference and eventually leads to the disc shape shown in Fig. 6-11*b*. The radial temperature gradients result in stresses that cause significant dislocation motion or slip. Robinson *et al.* (1982) have resolved this problem by providing a gas cavity between the wafer and the susceptor, as shown in Fig. 6-11*c*. The fact that a gas is a much poorer heat conductor than a solid and the fact that induction heating is dependent on thickness have apparently been used to specify the segment of a circle defining the gas cavity. Exact specification of the segment requires solving a two-dimensional heat conduction equation. Although numerical calculations would be required, the cavity design is dependent on the wafer thickness–radius ratio and the desired uniform wafer surface temperature. While induction heating requires a proper susceptor design, the radiant heating based on tungsten-halide lamps that are directed to the wafer surface does not cause any significant wafer bending (Hammond, 1978).

Because of the ever-increasing demand on the uniformity from device miniaturization, it becomes increasingly more important to have a stable laminar flow in the reactor to obtain uniform mass and heat transfer rates in the deposition region. For cold-wall reactors, the design requirements are to have a cold wall at a uniform temperature and a Rayleigh number less than 1708 with a Reynolds number less than 100. The outer wall of the barrel reactor is at a uniform temperature. In the case of a horizontal reactor, however, the top wall of a rectangular channel has to be water cooled for a constant wall temperature. The requirement on the Rayleigh number is almost always satisfied for typical LPCVD reactors operating much below atmospheric pressure. For atmospheric CVD reactors, however, the Rayleigh number is a major design consideration. Another overall requirement for the CVD reactor is that the wafer-to-wafer thickness variation be minimized, the ideal case being the uniform thickness for all wafers processed. The heating arrangement for this axial uniformity determines whether a reactor is operated as an isothermal (constant temperature throughout the deposition region of the susceptor) or nonisothermal reactor. If the reactor is operated as an isothermal reactor, the requirement for axial uniformity is that the source gas concentration be practically uniform throughout the deposition region. Since the source gas is consumed by the deposition, this means that the amount of the feed has to be in great excess of what is actually needed for the deposition. The primary factor that determines the thermal mode of operation should be the magnitude of the flow rate required for concentration uniformity. If the magnitude is high enough to push the flow into the turbulent regime, a nonisothermal reactor should be used instead. The temperature is gradually increased toward the outlet so that the reactant depletion can be compensated for by the higher temperature to maintain the uniform rate of deposition. When the surface area for actual deposition is relatively a small fraction of the wafer area, as in CVD metallization and epitaxial deposition for the buried layer, the isothermal mode of operation may be preferred. Another factor in choosing the mode of operation is the cost associated with the wasted (unused) source gas. Even when the total wafer area is involved, as in epitaxial GaAs over Si or superlattice film (repeating films of two semiconductor layers each 10 nm or so thick), the value attached to the film to be used as a substrate is high enough that the isothermal mode can be preferred. In such cases, the number of wafers has to be limited to meet the Reynolds number requirement for stable flow.

6.5.1 Isothermal Reactors

Consider the axial uniformity in an isothermal (constant susceptor temperature) reactor. The isothermal reactor is synonymous with constant concentration operation, since only when the reactants are depleted would the temperature need to be manipulated to maintain axial uniformity. With uniform heat and mass transfer at the susceptor established, the design requirement is then that the source gases are depleted only slightly by the deposition such that the concentration is practically uniform throughout the entire deposition region. If the total rate of deposition per area per time is r_t and the molar flow rate of the source gas at the reactor inlet is F_o, then the ratio of the outlet (C_e) to inlet (C_o) concentration of the species is

$$\frac{F_o - r_t S}{F_o} = \frac{C_e}{C_o} \tag{6.44}$$

where S is the surface area on which the deposition takes place. Note that F_o can be expressed as

$$F_o = \frac{P Q_o y_o}{R T_o} \tag{6.45}$$

where P is the total pressure, R is the gas constant, Q_o, T_o, and y_o, respectively, are the volumetric flow rate, temperature, and the mole fraction of the species at the reactor inlet. Equation (6.45) along with the ideal gas law leads to Eq. (6.44). For a uniform concentration in the reactor, C_e should approach C_o. The degree to which C_e approaches C_o is a design degree of freedom that can be set by the designer. If this degree of approach is denoted by B, say 0.99 for a 99 percent approach, one has

$$\frac{C_e}{C_o} > B \tag{6.46}$$

Use of Eqs. (6.45) and (6.46) in Eq. (6.44) yields

$$\frac{(1 - B) Q_o P y_o}{R T_o r_t S} > 1 \tag{6.47}$$

The amount of deposit on surfaces other than the wafers is usually small, even for hot-wall reactors. Further, this effect can be compensated for by making B closer to unity. Since $S = nA$ and $r_t = r_G = G\rho_s/M_w$, Eq. (6.47) can be rewritten as

$$\frac{(1 - B) Q_o P y_o M_w}{R T_o n A \rho_s G} > 1 \tag{6.48}$$

Here n is the number of wafers, A is the exposed surface area of a wafer, G is the linear growth rate (length/time), ρ_s is the density of the solid film, and M_w is its molecular weight.

Example 6.8. A constant-temperature (900 K) atmospheric barrel reactor is to be used for epitaxial film growth of gallium arsenide by MOCVD. Assuming a 1 : 5 feed ratio of $Ga(CH_3)_3$ and AsH_3 with 1 percent source gases in hydrogen, calculate the volumetric feed rate for B of 0.99. There are eighteen 2-inch diameter wafers and the expected growth rate is 0.2 μm/min. Calculate the corresponding Reynolds number for a reactor diameter of 30 cm and an almost straight barrel diameter of 24 cm. The gas properties may be taken as those for hydrogen. The hydrogen viscosity at the temperature is 2×10^{-4} poise. The density and molecular weight of GaAs are 5.32 g/cm^3 and 144.6, respectively.

Solution. Solving Eq. (6.48) for Q_o, one has

$$Q_o > \frac{RT_o nA\rho_s G}{(1 - B)Py_o M_w}$$

$$= \frac{82 \times 900 \times 18 \times 3.14 \times (2.54 \times 1)^2 \times 5.32 \times (0.2 \times 10^{-4})}{(1 - B) \times 1 \times 0.01 \times 144.6}$$

$$= 1.98 \times 10^5 \text{ cm}^3/\text{min}$$

The density of the feed is

$$\rho = \frac{PM}{RT} = \frac{1 \times 2}{82 \times 900} = 2.71 \times 10^{-5} \text{ g/cm}^3$$

The fluid velocity u is

$$u = \frac{Q_o}{A_c} = \frac{1.98 \times 10^5/60}{3.14(15^2 - 12^2)} = 13 \text{ cm/s}$$

Therefore, the Reynolds number is

$$d = \frac{\pi(15^2 - 12^2) \times 4}{2\pi(15 + 12)} = 6$$

$$\text{Re} = \frac{\rho u d}{\mu} = \frac{2.71 \times 10^{-5} \times 13 \times 6}{2 \times 10^{-4}} = 15.6$$

For cold-wall reactors such as horizontal and barrel reactors, Eq. (6.48) is sufficient for both axial (wafer-to-wafer) and radial (within wafer) thickness uniformity. For hot-wall reactors with stacked wafers, as in the LPCVD reactor in Fig. 6-1, the wafer-to-wafer spacing has to be considered as well to guarantee radial uniformity. If the radial uniformity is desired with a maximum deviation of B_r, say 0.01 for 1 percent maximum deviation, the following condition (Hong and Lee, 1985) has to be satisfied:

$$\frac{RG\rho_s/(M_w C_o D_{AB})}{1.50 \tanh (1.203\, s/R)} < B_r \tag{6.49}$$

where R is the wafer radius, D_{AB} is the diffusivity of the source species, and s is the distance between two adjacent wafers. For small arguments of the hyperbolic tangent $\tanh(x)$, say $x < 0.4$, $\tanh(x)$ is almost equal to x. Thus, the radial uniformity is the same as long as R^2/s is the same when $1.203s/R$ is less than 0.4. It is seen that the wafer spacing has to increase in proportion to the square of wafer diameter if the same radial uniformity is to be maintained for different wafer diameters.

Equation (6.49) can be rewritten as

$$\tanh (1.203\, s/R) > \frac{RG\rho_s}{1.5 B_r C_o D_{AB} M_w} \tag{6.49a}$$

Note that the maximum value of the hyperbolic tangent is unity. If the right-hand side of Eq. (6.49a) is larger than unity, the condition cannot be satisfied, which means that the radial uniformity cannot be met within B_r whatever the interwafer spacing s may be. Physically, this means that either the deposition rate or the wafer diameter is too large to give uniformity, regardless of spacing. When the supply of the source species is by diffusion only, which occurs at a fixed rate at steady state, it becomes depleted before it reaches the wafer center if either the wafer diameter or the deposition rate is too high. It is important to recognize that the deposition rate in the typical LPCVD reactor is limited by Eq. (6.49*a*); i.e., the deposition rate cannot exceed a certain limit if radial uniformity is to be realized within B_r.

The condition of Eq. (6.49) may have to be relaxed for the cooling requirement at the end of deposition, which is considered later in this section. Note that a thermal equilibrium is usually established in the reactor before the source gas is introduced.

The design of an isothermal reactor involves the following relationships:

$$l_h = 0.28H \text{ Re} \tag{6.9}$$

$$\text{Ra} = \frac{C_p \rho^2 g H^3 \, \Delta T}{k \mu T_m} < 1708 \tag{6.29}$$

$$\frac{k_m H_{eq}}{D_h} = \frac{20}{7} - (\alpha_t + 1)\left(\frac{C_s}{C_b - C_s}\right)\left(\frac{T_h - T_c}{T_h}\right) \tag{6.35}$$

$$\frac{(1 - B)Q_o P y_o}{R_g T_o r_G A n} > 1 \tag{6.47}$$

$$\text{Re} = \frac{\rho u d}{\mu} < 100 \quad \text{for } d = \text{hydraulic diameter for barrel and } H \text{ for horizontal reactor} \tag{6.50}$$

$$r_G = r_G(C_s, T_s) \qquad k_m(C_b - C_s) = r_G(C_s, T_s) \tag{6.51}$$

where H_{eq} is the effective distance between hot and cold regions. In addition, Eq. (6.49) is needed for the stacked wafer configuration and Eq. (6.30) for the rotating barrel reactor. It is implied that the top wall is cooled to a constant temperature for the rectangular horizontal reactor and that the source gas content in the feed is low for LPCVD reactors. The reason for this will be considered later in this section. Equations (6.9), (6.29), and (6.50) [and Eq. (6.30) for the rotating barrel reactor] are for stable laminar flow. Equation (6.47) [and Eq. (6.49) in the case of stacked wafer configuration] assures the axial and radial uniformity. Equation (6.51) is for the deposition kinetics considered in the previous chapter.

The design problem then becomes one of maximizing the deposition rate and the number of wafers processed, which is essentially equivalent to maximizing the production of wafers, while minimizing the reactor size. It should be clear from Eqs. (6.9), (6.20), and (6.35) that H_{eq}, which is the effective distance between hot and cold regions, should be as small as practically possible since a reduction in H_{eq} results in a small thermal entrance length (smaller reactor), smaller Rayleigh number, and larger mass transfer coefficient (smaller reactor). The only conflicting factor is the condition of Eq. (6.50), since a smaller H_{eq} leads to a larger fluid velocity and this larger Reynolds number for a given volumetric flow rate. This, however, can be partially compensated for by making the width of a rectangular channel or the diameter for a barrel reactor as large as practical. For the maximum throughput, the deposition rate should be as high as practically possible. This can be accomplished by setting the concentration and temperature as high as the particulars of the deposition allow. In general, higher temperatures and lower source gas concentrations lead to better material and electrical properties when the side effects of high-temperature processing can be made negligible. Therefore, the increased rates for higher concentrations should be tempered against the adverse effects on the desired properties. Although a higher temperature leads to better properties, the constraints due to dopant redistribution, possible evaporation of the substrate material (e.g., As in GaAs), and susceptor and susceptor coating limit the extent to which the temperature can be raised.

For cold-wall reactors, the design can proceed as follows:

1. Find the maximum allowable temperature and concentration by the constrants on the side effects.
2. From Eqs. (6.35) and (6.51), find the corresponding deposition rate with the aid of the intrinsic kinetics. The thermal diffusion effect may be neglected initially so that the Sherwood number is 20/7 in Eq. (6.35). An initial guess of H_{eq} must be made.
3. Find the required volumetric flow rate from Eq. (6.47) for the given deposition area and the number of wafers.
4. Find the cross-sectional area for the gas flow that satisfies Eq. (6.50). If it cannot be satisfied, reduce the number of wafers to be processed in the third step until the condition is satisfied.
5. Choose as small an H (distance between susceptor and cold wall) as possible while satisfying Eq. (6.29). This in turn yields the entrance length [Eq. (6.9)] and the width (for horizontal reactors) or the reactor diameter (for barrel reactors). Steps 2 through 5 have to be repeated until a satisfactory value of H is chosen.
6. The number of wafers to be processed and the manner in which they are arranged on the susceptor determine the reactor length.

The design involves satisfying a number of constraints and requirements. As such, it leads to acceptable ranges of the reactor dimensions. Further optimization would depend on the particulars of the susceptor and the operating costs involved in heating the unused portion of the susceptor that corresponds to the thermal entrance length.

Example 6.9. Design a horizontal and a barrel reactor for epitaxial GaAs growth on a 3-inch diameter silicon wafer at atmospheric pressure capable of processing 40 wafers. Assume that the maximum temperature allowed is 1000 K, and the feed consists of 0.1% TMG and 4% AsH_3 with H_2 as the carrier gas. These concentrations are typical of GaAs heteroepitaxy. The kinetic expression is

$$r_{GaAs} = \frac{k_s C}{1 + K_s C} \qquad \text{where } C = \text{concentration of TMG} \tag{A}$$

$$k_s = 2.16 \times 10^9 \exp\left(-\frac{44{,}600}{RT_s}\right)$$

$$K_s = 7120 \exp\left(\frac{20{,}400}{RT_s}\right)$$

Note that AsH_3 is present in excess (40 times the concentration of TMG) such that it is practically constant and therefore one can assume that the rate constants contain the concentration. Assume that the barrel walls are slanted just enough to support the wafers so that the barrel can be treated as an annulus for design purposes. Assume that the susceptor thickness is 0.5 cm for the horizontal reactor and that D_h is 0.35 cm²/s. Use a cold-wall temperature of 300 K and the physical properties of H_2 for the mixed fluid (see Example 6.3). Also assume constant reactor concentration.

Solution

(a) Horizontal reactor. At 1000 K (T_s), the rate constants are

$$k_s = 2.61 \times 10^9 \exp(-22.3) = 0.54 \text{ s}^{-1}$$

$$K_s = 7120 \exp(10.2) = 1.916 \times 10^8 \text{ cm}^3/\text{mol}$$

Choose an H of 1.0 cm. With the thermal diffusion effect neglected, one has, from Eq. (6.35),

$$k_m = \frac{20D_h}{7H} = 1 \text{ cm/s} \tag{B}$$

The TMG concentration at 1000 K and atmospheric pressure is

$$C_b = \frac{P_{TMG}}{RT} = \frac{0.001}{82 \times 1000} = 1.22 \times 10^{-8} \text{ mol/cm}^3 \tag{C}$$

The surface concentration can be obtained from Eq. (6.51):

$$k_m(C_b - C_s) = r_G = \frac{k_s C_s}{1 + K_s C_s} \tag{D}$$

Solving this for C_s,

$$C_s = \frac{-(k_s - K_s k_m C_b + k_m) + [(k_s - K_s k_m C_b + k_m)^2 + 4K_s k_m^2 C_b]^{1/2}}{2K_s k_m}$$

$$= \frac{-(0.54 - 1.916 \times 10^8 \times 1.22 \times 10^{-8} + 1) + [(-0.798)^2 + 4 \times 1.916 \times 10^8 \times 1.22 \times 10^{-8}]^{1/2}}{2 \times 1.916 \times 10^8}$$

$$= \frac{0.798 + 3.16}{3.832 \times 10^8} = 1.033 \times 10^{-8} \text{ mol/cm}^3$$

Therefore, the rate of deposition [Eq. (D)] is

$$r_G = \frac{0.54 \times 1.033 \times 10^{-8}}{1 + 1.916 \times 10^8 \times 1.033 \times 10^{-8}} = 0.187 \times 10^{-8} \text{ mol/(cm}^2\text{·s)} \quad \text{(E)}$$

For the volumetric flow rate, one has, from Eq. (6.47),

$$Q_o > \frac{R_g T_o r_G An}{(1-B)Py_o} = \frac{r_G An}{(1-B)C_b}$$

$$= \frac{0.187 \times 10^{-8} \times 3.14(2.54 \times 1.5)^2 n}{0.01 \times 1.22 \times 10^{-8}}$$

$$= 698.6n \text{ cm}^3\text{/s} \quad \text{(F)}$$

The velocity is given by Q_o/HW, where W is the width of the channel. Thus, Eq. (6.50) becomes

$$\text{Re} = \frac{\rho Q_o}{\mu W} < 100$$

or

$$\text{Re} = \frac{2.45 \times 10^{-5} \times 698.6n}{2 \times 10^{-4}\, W} = \frac{85.5n}{W} < 100 \quad \text{(G)}$$

From Eq. (6.29),

$$\text{Ra} = \frac{C_p \rho^2 g H^3 \,\Delta T}{k\mu T_m} < 1708$$

or

$$\text{Ra} = \frac{(14{,}990)(24.5 \times 10^{-3})^2(9.8)(H)^3(300)}{(432 \times 10^{-3})(20.1 \times 10^{-6})(650)} = 4.687 \times 10^6\, H^3 \text{ (m)}$$

$$= 4.687 \times H^3 \text{ (cm)} < 1708 \quad \text{(H)}$$

Equation (H) shows that any reasonable choice of H will satisfy Eq. (6.29). For the choice of 1 cm for H, one is then left with Eq. (G). For 3-inch wafers, the width should be at least in multiples of 3 inch diameter and the number of multiples should be an integer divisor of 40 so as not to waste the susceptor area (see Fig. 6-12). The value of n closest to but not exceeding 40 wafers [Eq. (G)] is that corresponding to 5 times the wafer diameter, or $W = 2.54 \times 3 \times 5 = 38.1$ cm. Then Eq. (G) yields

$$n < 44.6$$

For $n = 40$ and $W = 38.1$ cm, the Reynolds number is 89.8. Therefore, the thermal entrance length [Eq. (6.9)] is

$$l_h = 0.28 \times 1 \times 89.8 = 25.1 \text{ cm}$$

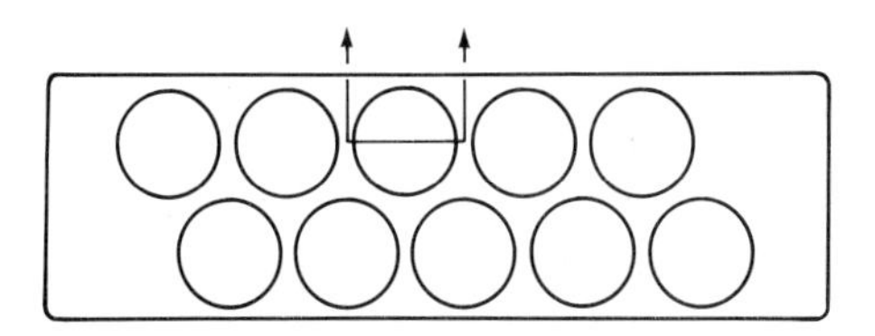

FIGURE 6-12
An example of wafer placement.

The portion of the susceptor used solely for preheating (no wafers) should be minimized. Therefore,

$$A_h = l_h W = 38.1 \times 25.1 = 958 \text{ cm}^2$$

The following table summarizes the results obtained for various H values:

H, cm	W, cm	Preheater area, cm^2	$r_G \times 10^8$, mol/(cm^2·s)	$C_s \times 10^8$	Re
1	38.1	958	0.187	1.03	89.8
2	38.1	901	0.176	0.868	84.5
3	38.1	901	0.177	0.875	84.5
4	30.5	779	0.152	0.611	91.2

From the table, one may choose a smaller H for a higher rate of growth. For the choice of 1 cm for H, the reactor height is 1.5 cm since the susceptor is 0.5 cm high. For this choice, the reactor dimensions are: 1.5 cm $(H) \times 38.1$ cm $(W) \times (25.1 + Y)$ cm, where Y is the length corresponding to 40 wafers placed in five rows. The actual width and length will be slightly larger than the above dimensions since some spaces are needed at the susceptor edges and ends (see Fig. 6-12). The volumetric flow rate for the reactor [Eq. (F)] is 466 cm^3/min. The linear growth rate ($G = r_G M_w/\rho_s$) is 30.4 nm/min. This is a typical rate of GaAs growth on Si.

(b) Barrel reactor. In this case, H_{eq} is given by Eq. (6.33) and it is the distance between the outer wall and the almost straight barrel. For small values of H, H_{eq} is slightly larger than H. The barrel reactor is inherently more stable than the horizontal reactor. Therefore, the criterion for the horizontal reactor can be used to obtain a conservative result, which means that the value of H can take on a range of values that satisfy Eq. (H). The only differences from the horizontal reactor are the calculations of the Reynolds number and the geometry of the barrel. If it is assumed that the number of the side walls, each at least 3-in wide for the 3-in wafer, is even and the barrel can be treated as a cylinder (with a slight loss of accuracy for the Reynolds number), the barrel diameter d_b depends on the number of side walls, m, as follows:

m	6	8	10	12	14
d_b, inches	6	7.84	9.71	11.59	13.48

where d_b can be obtained from the geometry of the top of the barrel for m-sided walls:

$$\cos\left(\frac{180 - 360/m}{2}\right) = \frac{1.5}{d_b/2} \tag{I}$$

Equation (G) can be written for this case as given below, since the velocity is given by $Q_o/\pi[(d_b/2 + H)^2 - (d_b/2)^2]$ and the equivalent diameter is given by $2H$:

$$\mathrm{Re} = \frac{2\rho Q_o H}{\mu\pi(d_b H + H^2)} \tag{J}$$

Using Eq. (F) in Eq. (J),

$$\mathrm{Re} = \frac{2 \times 2.45 \times 10^{-5} \times 698.6nH}{2 \times 10^{-4} \times 3.14(d_b H + H^2)}\left(\frac{r_G}{0.187 \times 10^{-8}}\right)$$

$$= \frac{54.5n}{d_b + H}\left(\frac{r_G}{0.187 \times 10^{-8}}\right) < 100 \tag{K}$$

A table similar to the one in part (a) above for the horizontal reactor can be prepared for chosen values of H. For an H of 1 cm, r_G is 0.187×10^{-8} and one has

$$\mathrm{Re} = \frac{54.5n}{d_b + 1} < 100 \tag{L}$$

Note that only the allowable values of d_b given in the table above can be used in Eq. (L). For the maximum n of 40, d_b has to be 9.71 inches to satisfy Eq. (L) since m is even. For this, Re = 98.7 and the entrance length is 23.8 cm. Following similar procedures, one can arrive at the following table:

H, cm	d_b, cm	Preheater area, cm²	$r_G \times 10^8$	Re
1	24.7	1843	0.187	84.9
2	19.9	3279	0.176	93.7
3	19.9	4703	0.177	89.6
4	15.2	4933	0.152	92.3

The obvious choice of H for the values in the above table is 1 cm. The corresponding barrel diameter is 24.7 cm (minimum) and it has ten side walls. For 40 wafers, this means that each side wall will contain four wafers. The value of H may need to be adjusted depending on the barrel angle required for containing the wafers vertically.

Hot-wall reactors are typified by the stacked wafer configuration and the use of low total pressure, as was shown in Fig. 6-1. In the typical low-pressure range of less than 10 torr, the Rayleigh number is very low since it is proportional to the square of the total pressure. Further, the temperature difference between the hot wall and the wafers is quite small compared to the cold-wall reactor. Therefore, one does not need to be concerned with free convection. A reduction in the deposition rate of nearly one order of magnitude is the trade-off made to obtain better uniformity. The low-pressure reactor also reduces contamination by undesired impurities. Although the linear fluid velocity in an LPCVD reactor is an order of magnitude higher than the velocity in the atmospheric reactor, the

Reynolds number is typically of the order of unity or less. In this Reynolds number range, the intrusion of the main streamlines into the interwafer space should be negligible. Therefore, only Eqs. (6.48) and (6.49) are required for the design along with Eqs. (6.35) and (6.51). In the absence of definitive results, one may insist for the absence of the streamlines in the interwafer space that

$$\text{Re} < 1 \tag{6.52}$$

When the concentration is constant in the annulus formed by the LPCVD reactor wall and the stacked wafer, and Eq. (6.49) is satisfied, the equivalent separation H_{eq} for the mass transfer coefficient is half the interwafer spacing. Further, the fluid in the interwafer space is stagnant so that the Sherwood number is unity. Therefore, one has

$$k_m = \frac{D_{\text{AB}}}{s/2} \tag{6.53}$$

Note that H_{eq} is the distance between the plane of highest concentration and that of lowest concentration. Equations (6.51) and (6.53) can be used for the desired growth rate. Equation (6.49) then specifies the interwafer spacing s for radial uniformity. Equation (6.48) in turn yields either the number of wafers that can be processed with axial uniformity for a given feed rate or the desired feed rate for a given number of wafers.

Specifications of reactor size and operating conditions can proceed as follows for a given number of wafers to be processed:

1. Use Eq. (6.49*a*) to find the feed content of the source species that gives a value less than unity for the given desired deposition rate. If the content thus found is unacceptable because of contamination, etc., the desired deposition rate has to be reduced.
2. For the chosen feed content of the source species, calculate the interwafer spacing, s, as a function of deposition rate. Note that the interwafer spacing is a measure of the reactor size. Particulars on fixed and operating costs may be utilized to obtain the optimal set of s versus r_G.
3. Once the set is chosen, calculate the rate constant that yields the chosen rate of deposition from the intrinsic kinetics and Eqs. (6.51) and (6.53). If the temperature is too high for the particular step of the fabrication, find the total pressure at which the temperature is acceptable.
4. Find the volumetric flow rate that yields axial uniformity to within the allowed tolerance from Eq. (6.48).
5. One may check the Reynolds number for Eq. (6.52).

Example 6.10. A salesman for a hot-wall isothermal LPCVD reactor makes the following claims for aluminum deposition based on an organic aluminum compound. Two hundred 5-inch wafers are to be stacked on a boat in a circular tube, each slot containing two wafers, so that only the circuit sides of the two wafers and not the back sides are exposed to the source gas. This arrangement can be treated as 100 equivalent stacked wafers for the deposition design. The deposition is only on 10 percent of the wafer area. The kinetic expression is

$$r_G = kC$$

where C is the metal organic concentration. The salesman claims that a deposition rate of 50 nm/min is possible with 1 percent (by volume) source species in hydrogen in the total pressure range of 1 to 10 torr and that both radial and axial uniformity variations are within 1 percent. You know that the diffusivity at atmospheric pressure in the temperature range of interest around 600 K is 1 cm²/s and that the conversion factor from G (nm/min) to r_G [mol/(cm²·s)] is 5×10^{-11}. Check his claim. If he is not right, specify the operating and design conditions. Assume that the maximum content of the source species that is acceptable is 10 percent by volume.

Solution. The concentrations of the source species at 1 and 10 torr total pressures are

$$C = \frac{p}{RT} = \frac{p}{82 \times 600} = \frac{0.01/760}{82 \times 600} = 2.67 \times 10^{-10} \text{ mol/cm}^3 \text{ at 1 torr}$$

At 10 torr, $C = 2.67 \times 10^{-9}$ mol/cm³. The desired deposition rate is

$$(r_G)_d = 5 \times 10^{-11} \times 50 = 2.5 \times 10^{-9} \text{ mol/(cm}^2\text{·s)}$$

The diffusivity is

$$D_{AB} = \begin{cases} 1 \times 760/10 = 76 \text{ cm}^2\text{/s at 10 torr} \\ 1 \times 760/1 = 760 \text{ cm}^2\text{/s at 1 torr} \end{cases}$$

If the variations in thickness are within 1 percent both radially and axially, $B = B_r = 0.01$. Now Eq. (6.49a) can be used to check the claim. In terms of r_G, Eq. (6.49a) is

$$\tanh(1.203s/R) > \frac{0.522 \times r_G}{1.5B_r C_b D_{AB}} \tag{A}$$

$$= \frac{(2.5 \times 2.54)(2.5 \times 10^{-9})}{1.5 \times 0.01 C_b D_{AB}} = 5.22$$

Since C_b is proportional to total pressure but D_{AB} is inversely proportional to it, $C_b D_{AB}$ is a constant for 1 percent source species whether the pressure is 1 or 10 torr. It is seen that the claim is not right since tanh (x) cannot be larger than unity.

Since R is fixed and B_r (or the tolerance in the radial thickness variation) should be small, the variables to choose for achieving uniformity are r_G and C_b. For the maximum value of C_b corresponding to 10 percent, or $C_b = 2.67 \times 10^{-9}$ at 1 torr and 2.67×10^{-8} at 10 torr, the desired deposition rate has to be decreased. Note that there is a trade-off between the desired deposition rate and the reactor size, since a higher deposition rate leads to a larger interwafer spacing s, which is equivalent to the reactor size for a fixed number of wafers. For both pressures, one has

$$\tanh(1.203s/R) > \frac{0.522 \times r_G}{2.5 \times 10^{-9}} \tag{B}$$

Equation (B) yields the following for s versus r_G:

$r_G \times 10^9$, mol/(cm²·s)	G, nm/min	s, cm
0.25	5	0.275
0.50	10	0.55
1.00	20	1.11
1.50	30	1.60
2.00	40	2.34
2.50	50	3.08

Particulars of fixed and operating costs, if desired, can be used to arrive at the desired growth rate. To proceed further, choose 3.08 cm for s, which corresponds to 2.5×10^{-9} mol/(cm²·s) or 50 nm/min.

For the s value of 3.08 cm, the mass transfer coefficient [Eq. (6.53)] is

$$k_m = \frac{D_{AB}}{s/2} = \frac{76}{3.08/2} = \begin{cases} 49.4\ \text{cm}^2/\text{s at 10 torr} \\ 494\ \text{cm}^2/\text{s at 1 torr} \end{cases}$$

For first-order kinetics, the effective rate constant including the mass transfer effect follows directly from Eq. (6.39):

$$r_G = \frac{1}{1/k_m + 1/k_s} C_b$$

or

$$2.5 \times 10^{-9} = \frac{2.67 \times 10^{-9}}{1/49.4 + 1/k_s} \ (1\ \text{torr})$$

which yields the desired k_s value of 0.954 cm/s. Likewise, one has the desired k_s value of 0.094 cm/s at 10 torr. The operating temperature ranges from the temperature corresponding to k_s of 0.094 to that for k_s of 0.954. The highest temperature allowed for the metallization will in turn determine the operating total pressure. The only item still to be resolved is the volumetric feed rate required for axial uniformity. In terms of r_G and C_b, Eq. (6.48) can be rewritten as

$$Q_o > \frac{r_G S}{(1 - B)C_b} \tag{C}$$

The total deposit area S is

$$S = 0.1(\pi R^2) \times 2 \times 100 = 2532.2\ \text{cm}^2$$

since only 10 percent of the area is involved in the deposition. To proceed further, it is assumed that 1 torr is the operating total pressure, for which $C_b = 2.67 \times 10^{-9}$. Therefore, from Eq. (C),

$$Q_o > \frac{2.5 \times 10^{-9} \times 2532.2}{0.01 \times 2.67 \times 10^{-9}}$$

$$= 2.37 \times 10^5\ \text{cm}^3/\text{s}$$

This corresponds to 9355 cm³/min at standard temperature and pressure (300 K and 1 atm) or 9355 sccm. At the reactor temperature, the fluid velocity is 955 cm/s for a 7-inch reactor diameter. Since ρ is of the order of 10^{-8} and μ is of the order of 10^{-4} for H_2, which is the typical carrier gas, the approximate Reynolds number is

$$\mathrm{Re} = \frac{du\rho}{\mu}$$

$$= \frac{3.5 \times 2.54 \times 955 \times 10^{-8}}{10^{-4}} = 0.85$$

which is of the order of unity. Note that streamline intrusion into the interwafer space of the order of one interwafer spacing that was reported at a Reynolds number of 25 is based on the Reynolds number defined as $\rho Q_o/\pi(R + R_t)\mu$, where R_t is the tube radius. The Reynolds number calculated in this way is 0.5.

It is clear from the example in conjunction with Eq. (6.49*a*) that the maximum rate of deposition that can be attained is severely limited, unless the feed content of the source species is high. In fact, in some LPCVD processes the feed content is quite high, anywhere from 1 to 100 percent. This in turn introduces another problem inherent in the pumping required to maintain a low reactor pressure. When the source-species content is high, the volume change (usually increase) due to the overall reaction becomes significant and the pressure toward the reactor outlet becomes higher than that in the inlet area, as reported by Brown and Kamins (1979) for a twofold increase in the total pressure, which is particularly severe if the temperature is increased along the reactor to compensate for the reactant depletion. Most of the reactions for CVD result in a volume increase due to the reactions, e.g.,

$$SiH_4 \longrightarrow Si(c) + 2H_2$$

which results in a two to one molar volume increase due to the overall reaction stoichiometry. In general, one can write the volume change as

$$\frac{P_o Q_o}{F_o T_o} = \frac{PQ}{FT} \tag{6.54}$$

which follows from the ideal gas law. Here F is the total molar flow rate and the subscript o is for the inlet condition. For the following general form of reaction:

$$aA(g) + bB(g) \longrightarrow cC(g) + dD(g) + gG(c) \tag{6.55}$$

the change in the number of moles due to the reaction is given by

$$F = F_o(1 + \alpha y_o x) = F_o(1 + qx) \tag{6.56}$$

where α is defined as

$$\alpha = \frac{c}{a} + \frac{d}{a} - \left(\frac{b}{a} + 1\right) \tag{6.57}$$

and y_o is the mole fraction of the source species in the feed and x is its fractional conversion. Note that only gaseous species are involved in the calculation of α

and that the lower-case letters are the stoichiometric coefficients. Equations (6.54) and (6.56) yield

$$\frac{P}{P_o} = \frac{TQ_o}{T_o Q}(1 + qx) \tag{6.58}$$

If the feed rate and the pumping rate are held constant, which is the typical practice, it can be seen from Eq. (6.58) that the pressure can easily increase with the conversion of the source species, when the mole fraction of the source species y_o is high. For the isothermal reactor being considered, $T = T_o$. Therefore, the reactor pressure can be made constant if the pumping rate Q_p (Q) is set by

$$Q_p = Q_o(1 + qx_t) \tag{6.59}$$

where x_t is the conversion at the end of the reactor and Eq. (6.58) has been used. An isothermal reactor implies almost constant concentration operation. Therefore, the pumping rate is simply equal to the feed rate since x_t is close to zero. For nonisothermal reactors, however, it should be a major consideration.

At the end of deposition, the reactor is cooled while feeding only inert gas. For LPCVD reactors with the stacked wafer configuration, there can exist steep temperature gradients within a wafer due to the cooling, the wafer edge being at the lowest temperature and the center at the highest. The thermal stresses generated by the gradients can lead to wafer deformation. The dominant mechanism of heat transport at high temperatures is radiation. The approximate temperature distribution within a wafer in the course of radiative cooling (Hu, 1969) when there are a large number of wafers is given by

$$\frac{T}{T_i} = [1 + 6f(\beta)t_r]^{-1/3} \tag{6.60}$$

$$f(\beta) = 1 - \frac{a}{2}\left\{1 + \frac{\beta'^2 - (1 + \beta^2)}{[(\beta'^2 + 1 - \beta^2)^2 + 4\beta^2]^{1/2}}\right\} \tag{6.61}$$

where $t_r = \dfrac{tE_m S_s T_i^3}{\rho_s C_p l}$

$$\beta' = \frac{R}{s}$$

$$\beta = \frac{r}{s}$$

and where E_m is the emissivity, a is the absorbance (usually equal to E_m), S_s is the Stefan-Boltzmann constant [$=1.36 \times 10^{-12}$ cal/(cm^2·s·K^4)], l is the wafer thickness, T_i is the temperature at the beginning of the cooling, and t is time. The angular and radial tensile stresses can be calculated assuming the wafer to be isotropic. The maximum stress is of interest and is the angular stress at the edge

(Hu, 1969), which is given by

$$\frac{S_{max}}{\alpha' E T_i} = -\frac{T}{T_i} + \frac{2}{\beta'^2} \int_0^{\beta'} \beta \frac{T(\beta, t)}{T_i} \, d\beta \tag{6.62}$$

where S_{max} is the maximum stress, α' is the linear thermal expansion coefficient, and E is Young's modulus. Actual data on deformation and the calculated maximum stress may be used to obtain the wafer spacing necessary for no deformation. Note that β' is the ratio of wafer radius to interwafer spacing. A typical value of β' is 5 (Hu, 1969). Note also that the approximate solution of Eq. (6.60) is valid up to a value of unity for $kl/E_m S_s T_i^3 s^2$.

The above results are valid for radiative cooling, which takes place in an LPCVD reactor arrangement such as the one in Fig. 6-1. If the ports along the reactor are used to introduce an inert gas in parallel with the wafer along with the axial flow, as in the arrangement in Fig. 6-2*b*, the dominant heat transfer mechanism can be changed to convection rather than radiation. A much better temperature profile should result that relieves the constraint on the interwafer spacing due to radiative cooling effects.

The use of side ports for the purpose of better radial thickness uniformity can lead to a flow stability problem. Thus, its use should be primarily for enriching the depleted reactant by bleeding additional source gas into the main flow in the annular region between the wafers and the tube wall.

The use of low pressure is not restricted to hot-wall reactors. The quality of epitaxial films that are typically grown in a cold-wall reactor is generally improved at a lower rate of adsorption and higher temperature. A low-pressure reactor is conducive to obtaining these desired conditions. The advantages discussed for the hot-wall LPCVD reactor also apply.

6.5.2 Nonisothermal Reactors

One drawback inherent in the isothermal reactor is the waste of chemicals. While this may be justified for epitaxial films and epitaxial substrate growth as in heteroepitaxy, it may not be for other film depositions. In fact, hot-wall LPCVD reactors are typically operated nonisothermally. If the source species are allowed to deplete along the reactor length for economical reasons, then the corresponding decrease in the rate of deposition should be counteracted by raising the reactor temperature. Given the kinetics and the mass transfer effects, the desired reactor temperature trajectory can be calculated. Even in the ideal case, however, realizing the precise temperature control as a function of the reactor position is difficult in the light of some uncertainty in the relationships for the overall rate of deposition. In practice, therefore, the reactor is divided into zones of constant temperature. The number of constant temperature zones is typically three (refer to Fig. 6-2*b*). Since the source species is more depleted toward the reactor outlet, the temperature is successively higher for successive zones. Even when the waste of the source species is not a major consideration, a nonisothermal reactor can be used to obtain a uniform rate of deposition.

The sensitivity of the rate of deposition to temperature is much higher than the sensitivity to concentration. In polysilicon deposition, for instance, a one-degree change in temperature is equivalent to a two to three percent change in concentration. Because of the temperature sensitivity, it is of interest to know the temperature profile within a zone of constant wall temperature.

In terms of the temperature profile, there can be two modes of operation. The first one involves maintaining an isothermal fluid temperature in each heating zone. The second involves making the temperature rise linearly in the direction of flow. It can be shown (Koopman and Lee, 1988) that the second mode of operation is not practical unless the fluid is preheated. For the first mode, the isothermality in later heating zones is guaranteed, if isothermality is achieved in the first heating zone, since the temperature rise experienced by the fluid entering the heated section of the reactor is maximum in the first heating zone. The fluid at near room temperature enters the first heating zone cold and must make a large temperature change. When the fluid enters the second heating zone from the first it is very near the new temperature and makes only a small change. The isothermality in the first heating zone depends on the following dimensionless number Ko:

$$\mathrm{Ko} = \frac{16 S_B^2 R^2 F_{wr}^2 T_w^6}{ks(\rho C_p)_f u_a} \tag{6.63}$$

where k is the fluid thermal conductivity, S_B is the Stefan-Boltzmann constant, T_w is the wall temperature, and the radiation shape factor from wafer face to reactor, F_{wr} (Siegal and Howell, 1981), is given by

$$F_{wr} + F_{ww} = 1 \tag{6.64}$$

$$F_{ww} = \frac{X - (X^2 - 4)^{1/2}}{2} \qquad \text{where } X = 2 + \left(\frac{s}{R}\right)^2 \tag{6.65}$$

The isothermality is assured to within the μ_c fraction of $(T_w - T_i)$ after z_c (=distance from the first wafer/s) number of wafers, if the following condition is satisfied:

$$\mathrm{Ko} > B(R_u)^{1.66} \qquad \text{where } R_u = \frac{4R^2 S_B F_{wr} T_w^3}{ks} \tag{6.66}$$

and where B is given by

$$\log B = \log\left[\ln\left(\frac{1.41879 - 0.14225\beta_c + 0.00474\beta_c^2}{\mu_c}\right)\right]$$

$$-\log z_c - 2\log(15.56282 - 11.08739\beta_c + 2.18814\beta_c^2)$$

where $\beta_c = R_t/R$. The above relationships are valid in the following ranges: $1.25 < R_t/R < 3$, $R/s < 32$.

Example 6.11. Polysilicon growth based on 23 percent by volume of silane in N_2 at 0.7 torr was reported by Rosler (1977). The total flow rate in the LPCVD reactor was 650 sccm and the first two heating zones were kept at 643 °C and the last one at 647 °C for 111 three-inch wafers. His data showed that in the first two zones, the deposition rate is uniform except at the entrance, which indicates that there was little reactant depletion. In the last heating zone, the film thickness after 25 minutes of deposition increased from 0.49 nm to 0.52 μm toward the end wafer. The inter-wafer spacing was 0.48 cm and the tube diameter (ID) was 13 cm. Determine whether the first heating zone is isothermal within 0.10 K of the wall temperature after the third wafer. Assume the fluid inlet temperature to be 25 °C. Use C_p of 0.25 cal/(g·K), k of 6×10^{-4} cal/(s·cm·K) and μ of 10^{-4} poise for the mixture. Calculate the fluid velocity at which the isothermality cannot be maintained.

Solution. The volumetric flow rate at the reactor conditions [Eq. (6.54)] is

$$Q = Q_o \frac{P_o}{P} \frac{T}{T_o} = \frac{650}{60} \frac{760}{0.7} \frac{920}{399} = 3.61 \times 10^4 \text{ cm}^3/\text{s}$$

The velocity in the annulus between the wafers and the tube wall is

$$u_a = \frac{3.61 \times 10^4}{\pi(R_t^2 - R^2)} = 415 \text{ cm/s}$$

The density is 3.5×10^{-7} g/cm^3 [$= PM/RT = (0.7/760) \times 28.9/(82 \times 920)$]. From Eq. (6.64), one has

$$F_{ww} = \frac{2.016 - (2.016^2 - 4)^{1/2}}{2} = 0.881$$

since $X = 2.016$. Thus, $F_{wr} = 1 - 0.881 = 0.119$. The dimensionless number (Ko) follows from Eq. (6.63) and the values of R_u and B follow from Eq. (6.66):

$$\text{Ko} = \frac{16 \times (1.355 \times 10^{-12})^2 (3 \times 2.54/2)^2 (0.119)^2 (916)^6}{(6 \times 10^{-4})(0.48)(3.5 \times 10^{-7} \times 0.25)(415)}$$

$$= 341$$

$$R_u = \frac{4(3 \times 2.54/2)^2 (1.355 \times 10^{-12})(0.119)(916)^3}{(6 \times 10^{-4})(0.48)}$$

$$= 25$$

For a 0.1 K approach to the wall temperature, $\mu_c = 0.1/(T_w - T_i) = 0.1/(643 - 25) = 0.000162$. For the third wafer, $z_c = 2$. When these values are used for B with $\beta_c = 13/(3 \times 2.54) = 1.706$, the value of B is 0.489. Thus, the criterion of Eq. (6.64) can be written as follows:

$$341 > 0.489(25)^{1.66} = 102.3$$

Since the criterion is satisfied, the first heating zone is isothermal, so all three zones are isothermal. From the results obtained so far, the criterion of Eq. (6.66) can be written as follows:

$$\frac{341 \times 415}{u_a} > 102.3$$

Solving this for u_a yields

$$u_a < 1383$$

Thus, the isothermality cannot be maintained for a fluid velocity larger than 1383 cm/s.

Although the variable heating, nonisothermal reactor can be modeled and the corresponding optimal temperature control for axial uniformity can also be devised, the concentration in the reactor cannot be allowed to change too much. If it is allowed to change significantly, the interwafer spacing also has to be a function of the axial length for the radial uniformity [refer to Eq. (6.49*a*)]. Therefore, the best mode of nonisothermal operation is simply to have several heating zones of constant temperature, with almost uniform concentrations within each heating zone.

6.5.3 Molecular Flow Reactors

It is anticipated that in the near future the thickness of deposited film will become of the order of 10 nm, as device sizes decrease and material properties improve, rather than a few micrometers which is common today. For such thin films, the rate or the speed of deposition is likely to be a minor consideration compared to the number of wafers that can be processed. In view of the fact that there is no thickness uniformity problem when the flow is molecular (refer to Sec. 6.2), it is very likely that the CVD reactors are going to be operated in a very low pressure regime (VLPCVD) where molecular flow dominates. According to Eq. (6.4), the mean free path of a molecule at 600 K is approximately given by

$$\lambda\,(\text{cm}) = \frac{10^{-2}}{P(\text{torr})} \tag{6.67}$$

When r/λ is much smaller than unity [Eq. (6.7)], molecular flow dominates over laminar flow. Here r is one half of the maximum distance between two walls, which is the radius of a tube for example. For more than 99 percent of the total flow to be due to molecular collisions, r/λ should be less than 0.069 [Eq. (6.7)]. This means that a mean free path larger than 7.5 cm is required for molecular flow in a square channel of 1 cm per side, which in turn means total pressure must be less than 1.3×10^{-3} torr. The flow would then be dominated by molecular collisions with walls.

In the molecular flow regime, the flow is by collisions of molecules with walls, which is a random process. Therefore, the thickness of deposited film is uniform everywhere as long as all wafers are exposed to the source species in the same manner. A reactor with flat wafer placement is shown in Fig. 6-13*a*. With equal access for all wafers to the source species, the deposition should be uniform. No mass transfer effects should exist. For the stacked wafer placement (Fig. 6.13*b*), however, the pressure should be lower than 10^{-3} torr, because of the wafer size. Further, radial uniformity is not necessarily guaranteed. Two conditions have to be satisfied for radial uniformity. The first is that the mean free path

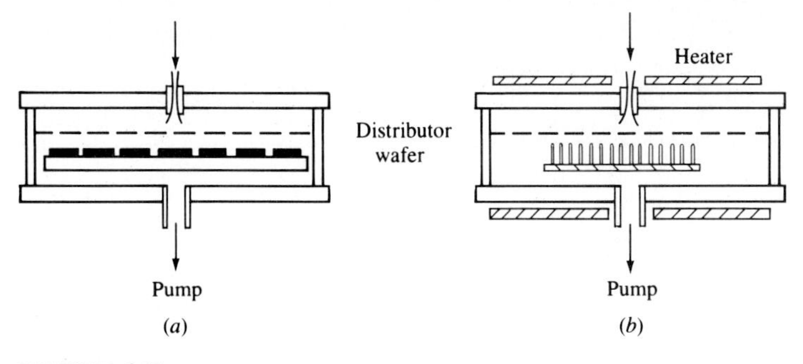

FIGURE 6-13
Reactor arrangements for molecular flow.

should be longer than the wafer radius. The second is that the rate of surface migration of adspecies has to be much higher than the rate of adsorption of adspecies. The first requirement stems from the fact that the longest path a molecule has to travel is the path shown in Fig. 6-14. This distance α_e is given by

$$\alpha_e = \frac{S}{\sin \Omega} \tag{6.68}$$

where the angle Ω in Fig. 6-14 is given by

$$\tan \Omega = \frac{S}{R} \tag{6.69}$$

Only when the mean free path is longer than α_e would the molecule reach the center of the wafer. Even with the mean free path longer than α_e, the rate at which the molecules strike the wafer surface is the highest at the wafer edge and the lowest at the center. Therefore, the radial thickness cannot be uniform unless the rate of surface migration of the adspecies is high enough that the thickness becomes uniform from surface migration. The first requirement is almost automatically satisfied at very low pressure, but the second is not. The only variable

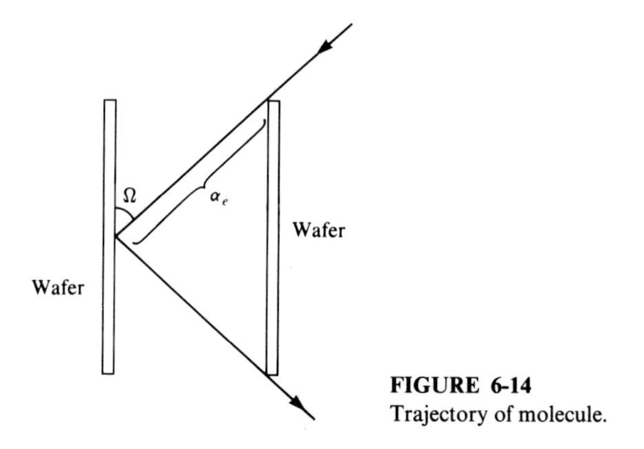

FIGURE 6-14
Trajectory of molecule.

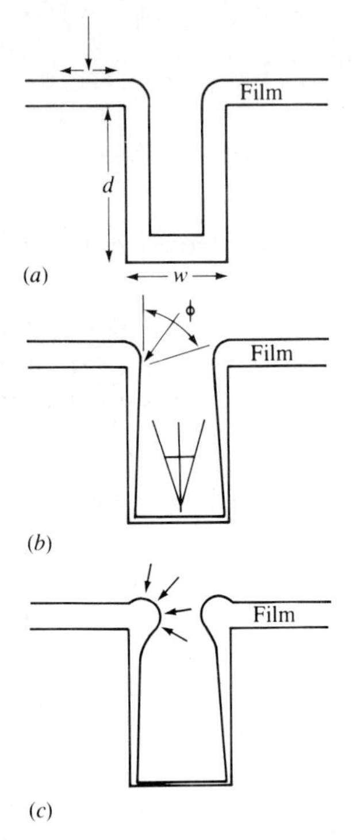

FIGURE 6-15
Three different types of step coverage (Adams, 1983).

that can have an influence on the surface migration is the reactor temperature. The rate of migration increases with temperature, while the rate of adsorption decreases. Perhaps the most important decision is the choice of adspecies and precursor, which implicitly determines the two relative rates.

The conformal step coverage discussed in the introduction is determined by the two rate requirements discussed above, although it is not directly related to molecular flow. Consider the three different types of step coverage (Adams, 1983) shown in Fig. 6-15. Figure 6-15*a* shows the uniform (conformal) step coverage. This coverage results when both requirements are met. The second case (Fig. 6-15*b*) is the step coverage resulting when the first requirement is satisfied but the second is not. The third case (Fig. 6-15*c*) is the coverage resulting when the first requirement is not satisfied. These figures are schematics for silicon dioxide deposition based on tetraethylorthosilane (TEOS). Referring to Fig. 6-15*b*, the minimum mean free path to satisfy the first requirement is that given by Eqs. (6.68) and (6.69) with s replaced by the step width w and R by the step height d. For film deposition in which step coverage is a problem, the main variable of interest is the operating pressure. The pressure required to satisfy the first requirement follows from Eqs. (6.4), (6.68), and (6.69):

$$\lambda = \frac{kT}{2^{1/2} P \pi \alpha^2} > \alpha_e \tag{6.70}$$

Equation (6.70) gives the maximum pressure at which a reactor can be operated for conformal step coverage.

Example 6.12. Adams (1983) reported experimental data for various step coverages for vertical grooves that are 5 μm wide and 50 μm deep. The silicon dioxide deposition was carried out at 700 °C and 0.3 torr. Determine the maximum pressure that satisfies the requirement of Eq. (6.70). Use the following for the mean free path:

$$\lambda\,(\text{cm}) = \frac{5 \times 10^{-3}}{P(\text{torr})} \tag{A}$$

Solution. According to Eq. (6.69),

$$\tan \Omega = \frac{5}{50} = 0.1$$

Thus, $\Omega = 5.7°$. From Eq. (6.68),

$$\alpha_e = \frac{5}{\sin(5.7°)} = 50.3\ \mu\text{m}$$

From Eq. (A),

$$\lambda = \frac{5 \times 10^{-3}}{P} \frac{973}{300}$$

Thus, according to Eq. (6.70),

$$\frac{1.62 \times 10^{-2}}{P} > 50.3 \times 10^{-4}$$

or

$$P_{\text{max}} = 3.2 \text{ torr}$$

6.5.4 Further Considerations and Continuous Reactors

The susceptor in a horizontal reactor is often tilted to enhance mass transfer near the outlet of the reactor. The barrel reactor in Fig. 6-1 has built-in tilting from the barrel shape. For reactors with well-defined hot and cold regions, negligible free convection, and fully developed thermal flow, the mass transfer coefficient is inversely proportional to the separation between the reactor wall and the substrate [Eq. (6.35)]. Thus, the mass transfer enhancement by tilting can be interpreted in terms of Eq. (6.35). It is therefore important that the barrel reactor have an entrance region similar to the one shown in Fig. 6-10 for the horizontal reactor.

When a film deposition is performed onto a heavily doped layer, dopant escapes from the heavily doped layer (evaporates) into the main fluid stream during prebake and subsequent film growth. The dopant carried downstream by the fluid in turn deposits onto other lightly doped regions. This undesired autodoping is termed lateral autodoping (refer to Chap. 7) and is a typical problem in bipolar junction transistor ICs. In a stable flow, the dopant is mostly confined to the immediate vicinity of the susceptor. Therefore, a susceptor with perforations in the areas between wafers and with light suction applied there can significantly reduce lateral autodoping. This light suction can also have a significant stabilizing effect on the flow (Schlichting, 1960).

With the need to grow quantum-dimension layers, e.g., for optoelectronics applications based on III-V compound semiconductors, a process called atomic layer epitaxy (Suntola and Anton, 1977) has been devised. A reactor system for the atomic layer epitaxy (ALE) is shown in Fig. 6-16 along with a detailed setup for the rotating substrate. As shown in the figure, the substrate is exposed alternately to different source gases (e.g., exposure to AsH_3 followed by the exposure to trimethylgallium for GaAs in Fig. 6-16) so that, if desired, the growth takes place layer by layer depending on the speed of the rotation and synchronization of the rotation with the substrate exposure to source gases. One unique feature of ALE is that the growth rate is determined by the number of exposure cycles rather than gas composition (Bedair, 1987). Impurity problems associated with

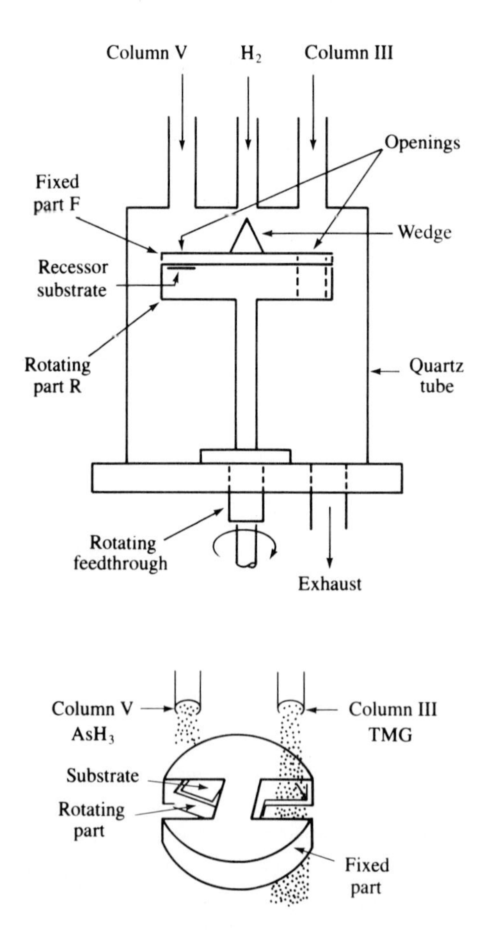

FIGURE 6-16
A reactor for atomic layer epitaxy and setup for the exposure of substrate to source gases (Bedair, 1987).

the switching and evacuation of the source gas per cycle have yet to be solved for this process.

Although only batch reactors are currently used in practice for IC fabrication steps, it is quite possible that continuous reactors will eventually be used for parts of the fabrication steps or for the whole fabrication. A wafer would be introduced into the reactor/fabricator at one end, and the finished ICs would be withdrawn at the other end. A continuous reactor for epitaxial film growth has already been proposed (Bellavance and Anderson, 1983). The continuous reactor consists of a series of chambers through which a single slice susceptor is passed, with each chamber having an independent gas flow associated with it.

If anything should be clear from this chapter, it is that we should not confine our interest only to the reactors currently in practice but rather open our minds to alternative reactors and reactor designs with the ultimate goal of a continuous reactor for wafer-to-fabricated IC, or whatever the best final design may be.

NOTATION

a	Absorbance
a, b, c, d	Stoichiometric coefficients
A	Wafer area (L^2)
c	C/C_{in}
C	Concentration (mol/L^3); mean velocity of a molecule (L/t)
C_b	Bulk fluid concentration (mol/L^3)
C_e	Exit concentration
C_o	Inlet concentration
C_p	Specific heat (E/MT)
C_s	Concentration at solid surface (mol/L^3)
d	Equivalent diameter in Reynolds number (L); gap between inner and outer cylinders (L)
d_b	Barrel diameter (L)
D_{AB}	Diffusivity of species A in B (L^2/t)
D_h	D_{AB} evaluated at T_h
E_a	Activation energy (E/mol)
E_m	Emissivity
F	Fraction of total flow due to molecular flow; shape factor in Eq. (6.63)
F_o	Inlet molar velocity (mol/t)
Fr	Froude number defined in Table 6.1
g	Gravity (L^2/t)
G	Linear film growth rate (L/t)
h	Film heat transfer coefficient (E/tT)
H	Channel height
H_{eq}	Equivalent channel height
k	Boltzmann constant (E/T); thermal conductivity (E/LtT); rate constant

k_m	Mass transfer coefficient (L/t)
k_T	Thermal diffusion ratio
K	Equilibrium constant
Ko	Dimensionless Koopman number
l	Tube length; wafer thickness (L)
l_h	Entrance length required for fully developed temperature profile (L)
l_v	Entrance length required for fully developed velocity profile (L)
L	Characteristic length in Table 6.1 (L)
m	Mass of a molecule (M)
M	Molecular weight (M/mol); Mach number
M_w	Molecular weight of a film material (M/mol)
n	Number of wafers
Nu	Nusselt number defined in Table 6.1
p	Partial pressure (P)
P	Total pressure
P'	$(P_1 + P_2)/2$
P_1	Inlet total pressure
P_2	Outlet total pressure
Pr	Prandtl number defined in Table 6.1
q	αy
Q	Volumetric flow rate (L^3/t); heat of adsorption (E/mol)
Q_m	Molecular flow rate (L^3/tP)
Q_o	Inlet flow rate (L^3/t)
Q_{tr}	Transition flow rate (L^3/tP)
Q_v	Pressure flow rate (L^3/tP)
r	Radial coordinate (L)
r_c	Intrinsic rate of film growth (mol/L^2t)
r_G	Observed rate of film growth (mol/L^2t)
r_t	Total rate of deposition (mol/L^2t)
R	Gas constant (E/mol T); wafer radius (L)
R_c	Dimensionless rate defined in Eq. (6.28)
R_i	Inner cylinder radius (L)
R_o	Outer cylinder radius (L)
R_t	Tube radius (L)
Ra	Rayleigh number defined in Table 6.1
Re	Reynolds number defined in Table 6.1
s	Spacing between two wafers (L)
S	Total surface area on which total deposition takes place
S_s	Stefan-Boltzmann constant
Sc	Schmidt number defined in Table 6.1
Sh	Sherwood number defined in Eq. (6.34)
t	Time
t_r	Normalized time defined in Eq. (6.61)
T	Temperature
T_c	Cold-wall temperature (T)

T_h	Hot-wall temperature (T)
T_i	Inlet temperature (T)
T_m	$(T_c + T_h)/2$
T_o	$(T_c + T_h)/2$
T_w	Wall temperature (T)
T'	$(T - T_o)/(T_h - T_c)$
Ta	Taylor number defined by Eq. (6.30)
u	Fluid velocity (L/t); x-component velocity (L/t)
u_a	Velocity in annulus (L/t)
U_i	Peripheral velocity (L/t)
U, V, W	u, v, w normalized with respect to $\mu^2/\rho H^2$
v, w	y- and z-component velocity, respectively
W	Width
x	Axial coordinate; conversion
x_t	Outlet conversion
X, Y, Z	x, y, z normalized with respect to H
y	Coordinate perpendicular to xy plane; reactor distance
Z	Factor defined by Eq. (6.7)
Z_e	Desired entrance length in Eq. (6.65)

Greek letters

α	Molecular diameter (L); thermal diffusivity (L^2/t) in Table 6.1; quantity defined by Eq. (6.57)
α_e	Quantity defined by Eq. (6.68)
α_t	Thermal diffusion factor in Eq. (6.34)
α'	Linear thermal expansion coefficient
α_1	Quantity defined by Eq. (6.64)
β	Parameter in Lennard-Jones relationship (L); quantity defined in Eq. (6.61)
β'	Quantity defined in Eq. (6.61)
λ	Mean free path of a molecule (L); fluid thermal conductivity
μ	Viscosity (M/Lt)
ρ	Density (M/L^3)
ρ_s	Solid density
ω	Angular velocity (L/t)
Ω	Angle defined in Fig. 6-12
Ω_1, Ω_2	Parameters in Lennard-Jones relationship

Subscripts

a	Atmospheric pressure
b	Bulk fluid
c	Cold
e	Low pressure
f	Fluid
h	Hot; heat

m	Mass
s	At solid surface
in	Reactor inlet

Units

E	Energy (ML^2/t^2)
L	Length
M	Mass
P	Pressure (M/Lt^2)
t	Time
T	Temperature

PROBLEMS

6.1. The susceptor in a horizontal reactor is usually tilted, the typical angle the susceptor makes with the bottom plane being 3°. The tilting angle in a barrel reactor is typically 15°. The difference in the angle is mainly due to the susceptor length. The barrel reactor length is that corresponding to two to three wafers whereas the horizontal reactor may have 20 wafers in the direction of flow. For the horizontal reactor, the Rayleigh number is reduced by the factor of cos Ω, where Ω is the tilting angle. For a square channel of 3 cm, express the distance between the susceptor and the upper wall as a function of the susceptor length. Assume the total susceptor length to be 50 cm and the angle to be 3°. For a thermal entrance length of 16.8 cm, express the Reynolds number as a function of the susceptor length. Express the Rayleigh number affected by the tilting as a function of the susceptor length.

6.2. Suppose for the whole susceptor length that the flow is stable and laminar such that Eq. (6.35) applies. Express the mass transfer coefficient as a function of the susceptor length z for the conditions in Prob. 6.1. Assume $D_{AB} = 1$ cm^2/s. Calculate the mass transfer coefficient at both ends of the susceptor.

6.3. The epitaxial deposition of silicon based on silane follows first-order kinetics with respect to silane concentration, if the rate of etching is negligible at temperatures higher than 1000 °C. Suppose that the intrinsic rate constant at 1000 °C is 14.5 cm/s. For the horizontal reactor in Prob. 6.1, calculate the extent of depletion of the reactant allowable, while maintaining the same growth rate at both ends of the susceptor. Do the same at the midpoint. Discuss whether it is sufficient for axial uniformity that the growth rate be the same at both ends.

6.4. The barrel reactor in Fig. 6-1 is similar to a horizontal reactor with a tilted susceptor in that each wall of the barrel is like a horizontal reactor put in a vertical orientation. Assuming that the flow is stable and laminar, calculate the mass transfer coefficient at both ends of the barrel. The diameter of the barrel at the top is 20 cm and is 25 cm at the bottom. The outer wall diameter is 28 cm. Assume the diffusivity to be 1 cm^2/s.

6.5. For the barrel reactor in Prob. 6.4, calculate the surface concentration of the source species at both ends of the barrel for the following kinetics for a film deposition at 800 K:

$$r_G[\text{mol}/(\text{cm}^2\cdot\text{s})] = \frac{0.27C}{1 + 0.758 \times 10^{11}C}$$

Assume the bulk concentration of the source species to be uniform at 10^{-7} mol/cm^3. As the concentration is lowered, the rate should approach first-order kinetics. Determine the apparent rate constant at the bottom of the barrel.

6.6. Consider the use of the experimental reactor in Fig. 6-10 for determining intrinsic kinetics. If the mass transfer coefficient is small such that the rate of deposition is controlled by the mass transfer, the measured rate should be very close to $k_m C_b$ regardless of kinetics. This situation should be avoided since no information on the kinetics can be obtained from the experimental data. What can you do in designing the reactor to avoid this situation?

6.7. The measured rates at atmospheric pressure for a certain film deposition are given below. The data are from an experimental reactor, such as the one in Fig. 6-10. Obtain the intrinsic rate of deposition. The mass transfer coefficient for the reactor is 1 cm/s at 800 K.

C_b, mol/cm^3	T_s, K	r_G, mol/(cm^2·s) × 10^8
3.41	900	1.410
2.00	900	1.000
1.21	900	0.707
8.35	800	0.352
4.25	800	0.249
2.18	800	0.176
20.16	750	0.159
10.11	750	0.113
5.08	750	0.080

6.8. Consider an LPCVD reactor operating at 0.7 torr and 643 °C with 23% SiH_4 in N_2 for polysilicon deposition (Rosler, 1977). For the case of 110 3-inch wafers stacked on a boat, calculate the minimum flow rates in sccm (cm^3/min at 1 atm and 300 K) that are required for the exit concentration to be 98 and 2 percent of the inlet concentration. The rate of deposition is 0.019 μm/min [1.58×10^{-7} mol/(cm^2·min)].

6.9. Suppose that a maximum of 2 percent variation in radial thickness can be allowed for 5-inch wafers in a typical LPCVD reactor with stacked wafers. Determine the maximum rate of deposition one can have. Assume that the total pressure is 1 torr and the temperature is 950 K with 20 percent source species in an inert carrier. Assume $D_{AB} = 1$ cm^2/s at atmospheric pressure. Discuss whether the answer depends on the total pressure.

6.10. Determine the maximum rate one can have for an s/R ratio of 0.1 in Prob. 6.9.

6.11. The tungsten deposition based on WF_6 and H_2 follows one half-order kinetics with respect to hydrogen. The overall reaction is

$$WF_6 + 3H_2 \longrightarrow W(s) + 6HF$$

Suppose for an LPCVD reactor that the total pressure is 10 torr and 0.3 torr for WF_6 and that the conversion of WF_6 is 50 percent. Calculate the ratio of the outlet to inlet volumetric flow rate at the standard state that can assure almost no secondary backflow, i.e., constant total pressure. Assume the reactor to be isothermal.

6.12. Suppose for the tungsten deposition of Prob. 6.11 that the observed rate is 200 nm/min at 678 K. Determine the mass transfer coefficient at which the deposition is completely dominated by mass transfer.

6.13. According to Hu (1969), the plastic deformation of a wafer occurs when the reduced tensile stress given by Eq. (6.62) exceeds 0.05 for a certain dislocation density of the wafer. Further, his conclusion was that the maximum stress occurs at $t_r = -1$. Choose the value of β' (R/s) that does not lead to plastic deformation. Assume $a = 1.0$ and $T_i = 1000$ K.

6.14. Suppose that for a 256-megabit memory chip, the gate width has to be 0.1 μm. In the metallization for the gate formation, this means that the width of the step is 0.1 μm. Calculate the largest step height that can be etched out for conformal step coverage at 0.1 torr. Assume that the mean free path for the species of interest is given by

$$\lambda(\text{cm}) = \frac{0.015}{P(\text{torr})}$$

6.15. In a fluid that is dilute with respect to source species, the density change is mainly due to temperature. Expanding the density in a Taylor series:

$$\begin{aligned}\rho &= \rho' - \left.\frac{\partial \rho}{\partial T}\right|_{T'} (T - T') + \text{higher-order terms} \\ &= \rho' - \rho'\beta'(T - T')\end{aligned}$$

where the thermal expansion coefficient β' is given by

$$\beta' = \left.\frac{\partial \rho}{\partial T}\right|_{T'} \rho' = -\frac{\rho'}{T'} \qquad \text{(ideal gas law)}$$

Use the results to derive Eq. (6.24) from Eq. (6.17).

6.16. The pressure terms in Eqs. (6.23) and (6.24) can be eliminated by subtracting the partial derivative of Eq. (6.24) with respect to X from the partial derivative of Eq. (6.23) with respect to Y. By defining the vorticity α by

$$\alpha = \frac{\partial V}{\partial X} - \frac{\partial U}{\partial Y} = -\nabla_1^2 \Omega$$

and the stream function Ω by

$$U = \frac{\partial \Omega}{\partial Y} \qquad \text{and} \qquad V = -\frac{\partial \Omega}{\partial X}$$

show that the elimination of the pressure terms leads to

$$U\frac{\partial \alpha}{\partial X} + V\frac{\partial \alpha}{\partial Y} = -\text{Gr}\frac{\partial T'}{\partial X} + \nabla_1^2 \alpha$$

Explain what one can do about the pressure term in Eq. (6.25).

6.17. Consider two infinite concentric cylinders with an inner radius of R_i and an outer radius of R_o, each maintained at two constant temperatures. The steady-state heat balance in the annulus is

$$\frac{1}{r}\frac{d}{dr}\left(r\frac{dT}{dr}\right) = 0$$

Show that the solution is given by

$$T = a \ln r + b$$

where a and b are constants. Show also that the heat flux, $-k\, dT/dr$, is given by $k\, \Delta T / R_o \ln (R_o/R_i)$, where ΔT is the temperature difference. If the heat transfer coefficient h is defined by $h\, \Delta T$ for the flux, show that H_{eq} in Eq. (6.33) is equivalent to H for a rectangular channel.

6.18. Consider the experimental reactor in Fig. 6-10. Since the velocity and temperature are fully developed before the fluid reaches the heated susceptor, the temperature profile is linear in y (y is normal to the bottom plane) and the velocity profile is parabolic above the susceptor when the conditions for stable flow are satisfied. The velocity profile is given by

$$u = \frac{6u_f}{H}\left(y - \frac{y^2}{H}\right) \tag{A}$$

where u_f is the average feed velocity and H is the channel height. The concentration profile may be expressed as follows:

$$C = a_0(x) + a_1(x)y + a_2(x)y^2 \tag{B}$$

Noting that at $y = 0$, $C = C_s$, $\partial C/\partial y = 0$ at $y = H$, and the mixing cup concentration over y, C_b, equals $\int uC\, dy/(\int u\, dy)$, one can rewrite Eq. (B) as follows:

$$C = C_s + \frac{20}{7H}(C_b - C_s)y - \frac{10}{7H^2}(C_b - C_s)y^2 \tag{C}$$

where C_s is the concentration at the susceptor surface. The observed rate of growth R_G is given by

$$R_G = D\left[\frac{\partial C}{\partial y} + (\alpha_t + 1)\frac{C}{T}\frac{\partial T}{\partial y}\right]_{\text{susceptor}} \tag{D}$$

The mass transfer coefficient is defined by

$$R_G = k_m(C_b - C_s) \tag{E}$$

Show that the mass transfer coefficient is that given by Eq. (6.35).

6.19. Consider the steady-state mass balance for the source species in the annulus between stacked wafers and the tube wall in an LPCVD reactor (Jensen and Graves, 1983):

$$-\frac{\partial}{\partial z}\left(D_{AB}\frac{\partial C}{\partial z}\right) - \frac{1}{r}\frac{\partial}{\partial r}\left(D_{AB}\, r\frac{\partial C}{\partial r}\right) + u\frac{\partial C}{\partial z} = 0 \tag{A}$$

with the radial boundary conditions:

$$-D_{AB}\left.\frac{\partial C}{\partial r}\right|_{R_t} = \left.r_G\right|_{R_t} \tag{B}$$

$$D_{AB}\left.\frac{\partial C}{\partial r}\right|_{R} = \left.\left(\frac{R}{s}\mu + \alpha\frac{R_t}{R}\right)r_G\right|_{R} \tag{C}$$

where the effectiveness factor μ is defined by

$$\mu = \frac{2\int_0^R r r_G \, dr}{R^2 r_G|_R} \tag{D}$$

Here α is the area of the wafer boat relative to the reactor tube area. If x is the conversion of the source species, show that

$$C = \frac{(1-x)C_o}{1+E_v x}$$
$$u = u_o(1+E_v x)\frac{T}{T_o} \tag{E}$$

where C_o is the inlet concentration and u_o is the inlet velocity. If one defines the average concentration by

$$\langle\cdot\rangle = \frac{\int_0^{2\pi}\int_R^{R_t} r\cdot dr\, d\Omega}{\int_0^{2\pi}\int_R^{R_t} r\, dr\, d\Omega} \tag{F}$$

show that Eq. (A), upon integration in accordance with Eq. (F), yields

$$D_o\frac{d}{dz}\left(B\frac{dx}{dz}\right) + u_o\frac{dx}{dz} = \frac{2r_G}{(R_t^2 - R^2)C_o}\left[R_t(1+\alpha) + \frac{R^2}{s}\mu\right] = 0 \tag{G}$$

where

$$B = \frac{1+E_v}{(1+E_v x)^2}\left(\frac{T}{T_o}\right)^{0.75} \tag{H}$$

with the aid of $D_{AB} = D_o(T/T_o)^{1.75}$, D_o being D_{AB} at the entrance. The amount of deposition on the tube wall and the boat is usually negligible. Further, under conditions satisfying radial uniformity, $\mu = 1$. If the axial dispersion is also neglected, Eq. (G) reduces to

$$u_o\frac{dx}{dz} = \frac{2r_G}{(R_t^2 - R^2)C_o}\frac{R^2}{s} \tag{I}$$

6.20. For an LPCVD reactor with constant fluid temperature in each heating zone, the annular region between the wafers and the tube wall can be modeled as a series of continuously stirred tank reactors (CSTR), each CSTR being that formed by two adjacent wafers. Write a mass balance for the source species. Use the result to express the concentration in a matrix form for any point (separate CSTR) in the reactor.

REFERENCES

Adams, A. C.: in S. M. Sze (ed.), *VLSI Technology*, chap. 3, McGraw-Hill, New York, 1983.

Ban, V. S.: *J. Electrochem. Soc.*, vol. 125, p. 317, 1973.

Bedair, S. M.: in *Conference Proceedings on Compound Semiconductor Growth, Processing and Devices*, Gainesville, Fla., p. 130, 1987. For more details see M. A. Tischler and S. M. Bedair, *Appl. Phys. Lett.*, vol. 48, p. 1681, 1986.

Bellavance, D., and R. N. Anderson: *J. Electrochem. Soc.*, vol. 130, p. 101C, 1983. For more details, see D. H. Westphal, *J. Crystal Growth*, vol. 65, p. 105, 1983.

Benzing, W. C.: in McD. Robinson *et al.* (eds.), *Chemical Vapor Deposition*, p. 373, The Electrochemical Society, Pennington, N.J., 1984.

Bird, R. B., W. E. Stewart, and E. N. Lightfoot: *Transport Phenomena*, Wiley, New York, 1960.

Blech, I. A., D. B. Fraser, and S. E. Haszko: *J. Vac. Sci. Technol.*, vol. 15, p. 13, 1973.

Brown, W. A., and T. I. Kamins: *Solid State Technol.*, p. 51, July 1979.

Chang, H. R., and B. J. Baliga: in McD. Robinson *et al.* (eds.), *Chemical Vapor Deposition*, p. 315, The Electrochemical Society, Pennington, N.J., 1984.

Douglas, E. C.: *RCA Engr.*, vol. 26(2), p. 8, 1980.

Giling, L. J.: *J. Electrochem. Soc.*, vol. 129, p. 634, 1982.

Hammond, M. L.: *Solid State Technol.*, vol. 21, p. 68, 1978.

Hirschfelder, J. O., C. F. Curtiss, and R. B. Bird, *Molecular Theory of Gases and Liquids*, Wiley, New York, 1954.

Hong, J. C., and H. H. Lee: *J. Crystal Growth*, vol. 71, p. 711, 1985.

Houtman, C., D. B. Graves, and K. F. Jensen: *J. Electrochem. Soc.*, vol. 133, p. 961, 1986.

Hu, S. M.: *J. Appl. Phys.*, vol. 40, p. 4413, 1969.

Hwang, G. J., and K. C. Chang: *Trans. ASME, J. Heat Transfer*, vol. 95, p. 72, 1973.

Hyrayama, M., H. Miyoshi, N. Tsubouch, and H. Abe: *IEEE J. Solid State Circuits*, vol. SC-17, p. 133, 1982.

Jenkinson, J. P., and R. Pollard: *J. Electrochem. Soc.*, vol. 131, p. 2911, 1984.

Jensen, K. F.: in McD. Robinson *et al.* (eds.), *Chemical Vapor Deposition*, The Electrochemical Society, Pennington, N.J., 1984.

Kern, W., and G. L. Schnable: *IEEE Trans. Elect. Dev.*, vol. ED-26, p. 647, 1979.

Koopman, D. C., and H. H. Lee: unpublished results, 1988.

Lee, H. H.: *Heterogeneous Reactor Design*, Butterworth Publishers, Stoneham, Mass., 1985.

Meyerson, B. S., E. Ganin, D. A. Smith, and T. N. Nyguyen: *J. Electrochem. Soc.*, vol. 133, p. 1232, 1986.

Middleman, S., and A. Yeckel: *J. Electrochem. Soc.*, vol. 133, p. 1951, 1986.

Ostrach, S.: *J. Fluids Engng*, vol. 105, p. 5, 1983.

——— and Y. Kamotani: in D. B. Spalding and N. Afgan (eds.), *Heat Transfer and Turbulent Buoyant Convection*, p. 729, Hemisphere Publishing Corporation, Washington, D.C., 1977.

Pearce, C. W.: in S. M. Sze (ed.), *VLSI Technology*, chap. 2, McGraw-Hill, New York, 1983.

Pellew, A., and R. V. Southwell: *Proc. R. Soc. Lond.*, vol. A176, p. 312, 1940.

Robinson, McD., C. C. Chang, R. B. Marcus, G. A. Rozgonyi, L. E. Katz, and C. L. Paalneck: *J. Electrochem. Soc.*, vol. 129, p. 2858, 1982.

Rosler, R. S.: *Solid State Technol.*, p. 63, April 1977.

Schlichting, H.: *Boundary Layer Theory*, McGraw-Hill, New York, 1960.

Siegel, R., and J. R. Howell: *Thermal Radiation Heat Transfer*, 2d ed., McGraw-Hill, New York, 1981.

Sparrow, E. M., R. Eichhorn, and J. L. Gregg: *Phys. Fluids*, vol. 2, p. 319, 1959.

Suntola, T., and M. J. Anton: US Patent 4-058-430, 1977.

Takahashi, R., Y. Koga, and K. Sugawara: *J. Electrochem. Soc.*, vol. 119, p. 1406, 1972.

Wahl, G.: in McD. Robinson *et al.* (eds.), *Chemical Vapor Deposition*, p. 60, The Electrochemical Society, Pennington, N.J., 1984.

CHAPTER

7

INCORPORATION AND TRANSPORT OF DOPANTS

7.1 INTRODUCTION

Incorporation of dopants into a semiconductor material, commonly referred to as doping, is the means by which junctions are formed and is essentially what makes a semiconductor function as an integrated circuit. Thermal diffusion and ion implantation are the methods used to introduce a controlled amount of chosen impurities (dopants) into selected regions of a semiconductor. The more recent method of ion implantation is often preferred because of its capability to control the number of implanted dopant atoms. For silicon, for example, precise dopant concentrations can be obtained from 10^{14} to 10^{21} atoms/cm^3.

Dopant incorporation by thermal diffusion usually involves a two-step processing of predeposition followed by "drive-in." Predeposition is a process in which a dopant in the gas phase is introduced into the semiconductor by being in contact with the solid surface at high temperature. Since the dopant concentration in the gas phase is kept constant, the solid surface becomes saturated with the dopant atom to the solid solubility, while surface dopant atoms diffuse into the solid interior. This predeposition is followed by drive-in in which the semiconductor is heated in an inert environment to cause a redistribution of the predeposited dopant to a desired dopant profile.

There are three different ways of carrying out the predeposition. The first, shown in Fig. 7-1, involves introducing a dopant-carrying gas stream to a furnace

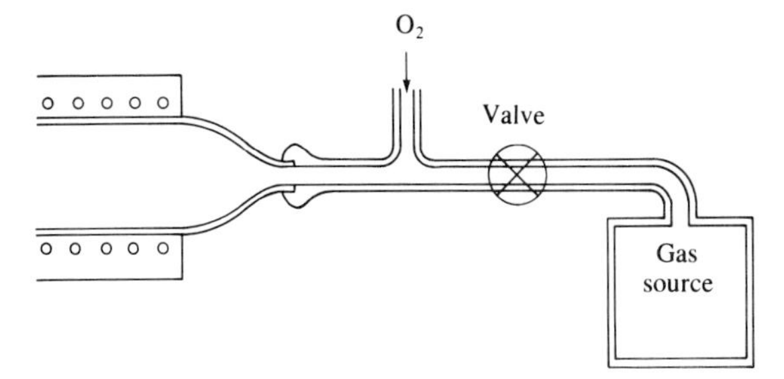

(*a*) Gaseous source

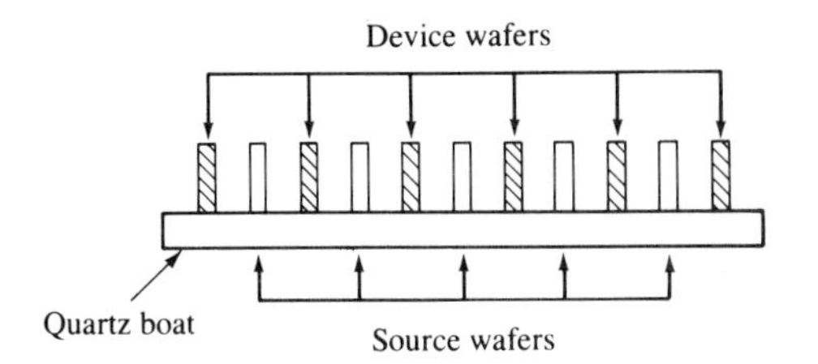

(*b*) Use of source wafer

FIGURE 7-1
Predeposition methods (Gise and Blanchard, 1979).

containing a quartz boat on which wafers are stacked upright as in the LPCVD reactor. Although Fig. 7-1*a* only shows a gas dopant source, the dopant source can also be solid or liquid. Chemical reaction between the carrier gas (usually nitrogen) and the solid source produces the dopant when the source is solid. For liquid sources, a carrier gas bubbled through a liquid compound containing the dopant produces a wet gas stream saturated with the dopant. In most cases, oxygen is added to the gas flow to ensure that the dopant reaches the wafer surface as an oxide. Another method uses wafers made of a compound of the desired dopant. These source wafers are usually prepared by oxidizing them. They are placed as shown in Fig. 7-1*b* in a quartz boat and then introduced to a furnace. An inert gas is used for the predeposition. The third method involves the use of a doped layer of an oxide such as silicon dioxide (for silicon wafers) that is bonded to the wafer surface. This can be accomplished by depositing a doped layer of oxide by a low-temperature CVD process or by bonding a doped layer to the wafer surface by a spinning-on technique, in much the same way as the spinning-on coating of a photoresist. The wafers are loaded into a furnace with the dopant layer already on the front side of the wafer. During the predeposition, the required amount of dopant diffuses into the semiconductor.

Masking against dopant diffusion is usually accomplished with silicon dioxide. Only the area not masked is involved in the predeposition. The thickness required to mask against the diffusion is dependent on the type of dopant, time, and temperature, but is usually less than 1 μm. The drive-in is carried out in an oxidizing atmosphere to regrow a protective oxide layer over the freshly diffused region.

Ion implantation offers a number of technological advantages such as speed and reproducibility of the doping process, and exact controllability of the number of doping atoms introduced. It is particularly useful for forming shallow junctions with high doping requirements. In ion implantation, ionized-projectile atoms are introduced into solid targets with enough kinetic energy (3 to 500 keV) to penetrate beyond the surface regions, anywhere from 100 angstroms up to micrometers. The high energy used is perhaps the main distinction between ion implantation and particle (particularly ion) beam lithography (treated in Chap. 8). A schematic of a typical commercial ion implant system is shown in Fig. 7-2. As shown, a gas source of dopant such as BF_3 or AsH_3 is energized at a high potential to produce an ion plasma, containing dopant ions. The analyzer magnet then selects only the ion species of interest and rejects other species. The desired ion species is then injected to the accelerator through a revolving slit, where its velocity is increased. The *x*- and *y*-scan plates direct the beam to the target wafers. The wafer feeding arrangement (denoted simply as wafer in Fig. 7-2) is external to the whole machine in one package such that wafers can be loaded and unloaded through a feeder. Because of the nature of ion implantation, the implantation is basically independent of solid solubility limits, temperature during implantation, and concentration of dopant at the semiconductor surface. The doping atoms introduced have a concentration profile that is generally described by a Gaussian distribution. Although ion implantation in this chapter is treated only in relation to doping, it is not difficult to imagine other various

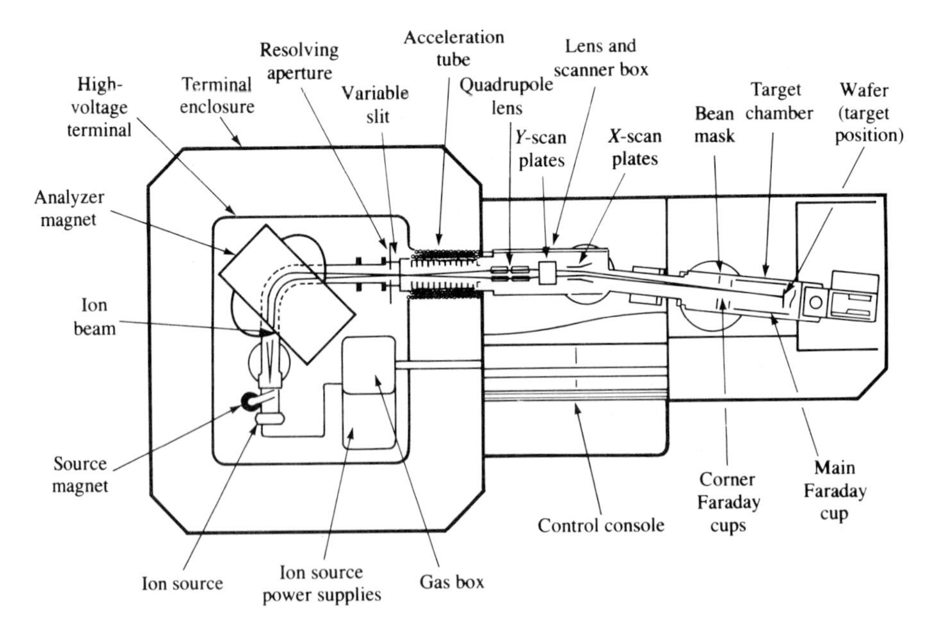

FIGURE 7-2
Schematic illustration of a medium current implanter (Ryssel and Ruge, 1986).

applications. One example is the current practice of forming an insulator below the semiconductor surface by deeply implanting oxygen ions and then annealing.

Ion implantation, however, does have its problems. Because of the bombardment involved with heavy particles, radiation damage occurs, which in turn causes changes in the electrical properties of semiconductor. The radiation damage is the major problem with ion implantation. Most of the doping atoms do not come to rest after implantation on regular lattice sites, and are therefore not electrically active. By a suitable annealing method, the crystal lattice is restored and the introduced dopant atoms brought to electrically active lattice sites by diffusion. Side effects of implantation such as ion channeling are another source of problems.

Both thermal diffusion and ion implantation methods involve diffusion of dopant atoms in the semiconductor during and after dopant incorporation. Therefore, the nature of diffusion will be examined first before considering the subject of dopant incorporation.

7.2 NATURE OF DIFFUSION IN SOLIDS

Diffusion in solids can be visualized as movement of atoms through vacancies and interstitial regions of the crystal lattice. Lattice atoms vibrate around their equilibrium sites in much the same way as in the case of surface diffusion considered in Chap. 5. As the temperature is raised, some lattice atoms occasionally acquire sufficient energy to overcome the binding energy and leave the lattice sites, thereby creating lattice vacancies and interstitials, shown in Fig. 7-3. When a neighboring host or impurity atom migrates to the vacancy site, the diffusion is referred to as diffusion by the vacancy mechanism. When an interstitial atom

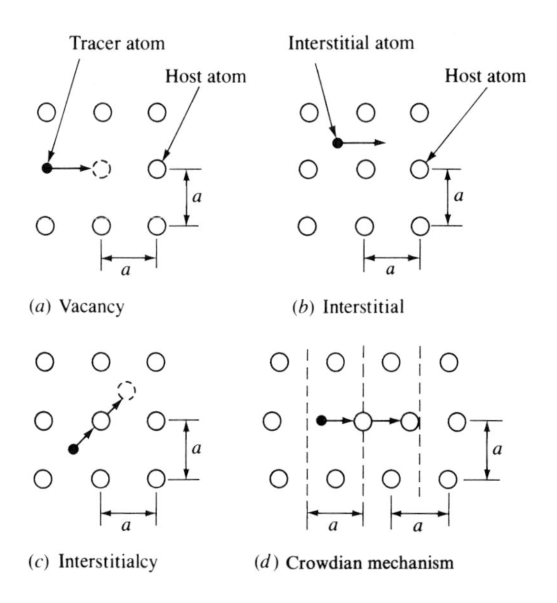

FIGURE 7-3
Models of diffusion mechanism with a the lattice constant (Tuck, 1974).

moves from one interstitial site to another without occupying a lattice site, interstitial diffusion is said to have taken place. Extensions of interstitial diffusion are also shown in Fig. 7-3, which involve simultaneous movement of two atoms. Diffusion is called self-diffusion or impurity diffusion depending on whether a migrating atom is a host or impurity atom.

Because of smaller binding energy associated with interstitial atoms, the activation energy required for diffusion of interstitial atoms is lower than for diffusion of lattice atoms via the vacancy mechanism. An atom smaller than the host atom often moves interstitially. Group I and VIII elements have small ionic radii compared with silicon. They diffuse fast in silicon and they are considered to diffuse by an interstitial mechanism. On the other hand, group III and V elements are considered to diffuse predominantly by the vacancy mechanism. They are slow diffusers. Diffusion in gallium arsenide is believed to take place by movement on sublattices, i.e., the gallium sublattice and arsenic sublattice. Group II elements, which are p type, are believed to move along the gallium sublattice by interstitial mechanisms. Group VI elements, which are n type, are believed to move along the arsenic sublattice by the vacancy mechanism. They are slow diffusers. Group IV elements, which can be either n or p type depending on the sublattice on which they are preferentially located, are usually assumed to move on both sublattices. They are extremely slow diffusers.

As was the case in surface diffusion, the flux of atoms J in solid can also be described by Fick's law:

$$J = -D \frac{\partial C}{\partial x} \tag{7.1}$$

where D is the diffusivity and C is the concentration. Because of the impurities (dopant atoms) present in a semiconductor, diffusion and thus the diffusivity depends strongly on the impurity content in the solid. For intrinsic semiconductors (those with an impurity concentration less than the intrinsic carrier concentration n_i), the impurity effect is negligible and the diffusivity can be considered as that of the intrinsic material. This intrinsic diffusivity D_i can be expressed as follows:

$$D_i = D_0 \exp\left(-\frac{E}{k_B T}\right) \tag{7.2}$$

where D_0 is the preexponential factor dependent on the vibrational frequency of atoms in lattice or interstitial sites and E is the binding energy at the site.

Point defects such as vacancies and interstitial atoms play a dominant role in the diffusion of impurities in semiconductors (Boltaks, 1963; Frank *et al.*, 1984) along with the charge state of the point defects. A vacancy or an interstitial atom can be neutral or charged by exchanging electrons. Thus, a point defect can be neutral, which is called a neutral point defect, or negatively (acceptor point defect) or positively (donor point defect) charged. Even for intrinsic semiconductors, a distinction is made as to which type of point defect dominates the diffusion. Because of the extensive work available for silicon, the intrinsic diffusi-

TABLE 7.1
Intrinsic diffusivity of B, P, As, and Sb (Fair, 1981)

Unit	B(D_i^+)	P($D_i^\times$)	As(D_i^-)	Sb($D_i^\times$)
D_0, cm²/s	0.76	3.85	22.9	0.214
E, eV	3.46	3.66	4.10	3.65

vities are denoted by $D_i^\times$ (neutral), D_i^- (acceptor point effect), and D_i^+ (donor point defect) to signify the dominance of a particular type of point defect in the diffusion. The diffusivities are given in Table 7.1. For gallium arsenide, however, less information is available for the various types of diffusion. The diffusivities are summarized in Table 7.2.

Diffusivity is usually defined as the root mean square distance an atom travels per unit time (Sec. 5.2). If one assigns parts of the distance travelled to the distance due to neutral point defects and that due to charged point defects, one has

$$D_i = D_i^\times + D_i^+ + D_i^- + D_i^{2-} + \cdots \tag{7.3}$$

Under extrinsic conditions, however, the concentration of various point defects changes due to the excess amount of impurities. The neutral point defect concentration should remain the same, since it is not affected by the electric field induced by the impurity concentration gradient. Since diffusion is the movement of impurity atoms along point defects, the diffusivity, or the distance travelled, should be weighted according to the change in the point defects' concentrations when excessive impurity atoms are present (Shaw, 1973). Thus, one has

$$D = D_i^\times + D_i^+ \frac{[P^+]}{[P^+]_i} + D_i^- \frac{[P^-]}{[P^-]_i} + D_i^{2-} \frac{[P^{2-}]}{[P^{2-}]_i} + \cdots \tag{7.4}$$

TABLE 7.2
Diffusivities of impurities in GaAs (Shaw, 1973)

Impurity	D_0, cm²/s	E, eV
Au	2.9×10^1	2.64
Be	7.3×10^{-6}	1.2
Cr	4.3×10^3	3.4
Cu	3×10^{-2}	0.53
Li	5.3×10^{-1}	1.0
Mg	2.6×10^{-2}	2.7
Mn	6.5×10^{-1}	2.49
O	2×10^{-3}	1.1
S	1.85×10^{-2}	2.6
Se	3.0×10^3	4.16
Sn	3.8×10^{-2}	2.7

where it has been recognized that $[P^\times]_i = [P^\times]$. The point defects' concentrations are denoted by brackets and the subscript i is used to designate the intrinsic semiconductor. For charged point defects, one has

$$P^\times + re^- \rightleftharpoons P^{-r} \tag{7.5}$$

At equilibrium, one can write

$$K_r = \frac{[P^{-r}]_i}{n_i^r[P^\times]}$$

where n_i is the intrinsic carrier concentration. Under intrinsic conditions, $p = p_i = n_i$ such that

$$K_r = \frac{[P^{-r}]_i}{n_i^r[P^\times]_i}$$

Combining the two equations yields

$$\frac{[P^{-r}]}{[P^{-r}]_i} = \left(\frac{n}{n_i}\right)^r \tag{7.6}$$

since $[P^\times]_i = [P^\times]$. Similar expressions can be obtained for different charged point defects. Using Eq. (7.6) and the similar expressions in Eq. (7.5) gives

$$D = D_i^\times + D_i^+\left(\frac{p}{n_i}\right) + D_i^-\left(\frac{n}{n_i}\right) + D_i^{2-}\left(\frac{n}{n_i}\right)^2 + \cdots \tag{7.7}$$

Example 7.1. For silicon, the bandgap energy E_g varies from about 1.11 eV at 300 K to about 1.16 eV at 0 K. Neglecting this change and also the $T^{3/2}$ dependence, the temperature dependence of intrinsic carrier concentration for any semiconductor is approximately given by

$$\frac{n_i(T)}{n_i(300)} = \exp\left[-\frac{E_g}{2k}\left(\frac{1}{T} - \frac{1}{300}\right)\right]$$

For silicon at 300 K, $n_i = 1.5 \times 10^{10}$ cm^{-3}. As shown in Table 7.1, the diffusion of boron in silicon is dominated by acceptor point defects. Calculate the boron diffusivity at 1000 K in silicon doped with boron at 10^{17} cm^{-3}.

Solution. If the acceptor point defect dominates, Eq. (7.7) reduces to

$$\begin{aligned} D &= D_i^+\left(\frac{p}{n_i}\right) \\ &= \left[0.76 \exp\left(-\frac{3.46}{k_B T}\right)\right]\left(\frac{10^{17}}{1.5 \times 10^{10}}\right) \\ &= 0.76 \exp\left[-\frac{3.46}{(8.63 \times 10^{-5})(1000)}\right]\left(\frac{10^7}{1.5}\right) \\ &= 1.96 \times 10^{-11} \text{ cm}^2/\text{s} \end{aligned}$$

An effect much less important than impurity effects is the enhancement of diffusion by the electric field between the impurity atoms ionized at diffusion temperature and the electrons and holes (Hu and Schmidt, 1968). For an n-type impurity as an example, the flux in one dimension is given by

$$J = -D\frac{dN}{dx} + \mu n E \tag{7.8}$$

where N is the impurity concentration, μ is the electron mobility, and E is the electric field. Using the Einstein relationship and

$$E = -\frac{k_B T}{q}\frac{1}{n}\frac{dn}{dx} \tag{7.9}$$

one has

$$J = -D\left(1 + \frac{dn}{dN}\right)\frac{dN}{dx}$$

$$= -Dh\frac{dN}{dx} \tag{7.10}$$

where $h = 1 + dn/dN$ is the field enhancement factor. For an n-type impurity, one has

$$\frac{n}{n_i} = \frac{N}{2n_i} + \left[\left(\frac{N}{2n_i}\right)^2 + 1\right]^2 \tag{7.11}$$

such that

$$\frac{dn}{dN} = \frac{1}{2}\left\{1 + \left[1 + \left(\frac{2n_i}{N}\right)^2\right]^{-1/2}\right\} \tag{7.12}$$

It can be seen that dn/dN approaches unity as N or the impurity concentration becomes much larger than n_i. Therefore, the maximum value of h is 2. Note that the same relationship holds for a p-type semiconductor if n is replaced by p. In general, then, the diffusivity can be expressed as follows:

$$D = h\left[D_i^{\times} + D_i^{+}\left(\frac{p}{n_i}\right) + D_i^{-}\left(\frac{n}{n_i}\right) + D_i^{2-}\left(\frac{n}{n_i}\right)^2 + \cdots\right] \tag{7.13}$$

An oxidizing environment significantly affects diffusion. As discussed in the previous section, doping by thermal diffusion is often carried out in the presence of oxygen, particularly during the drive-in step. This serves the dual purpose of providing an oxide layer for the next photoengraving process and creating steps in the surface of the silicon wafer, since the oxide growth rate in the window area is much greater than that under the already oxidized regions, permitting alignment of patterns. The diffusivity under oxidizing environment is often correlated (Taniguchi *et al.*, 1980; Lin *et al.*, 1981) as follows:

$$D_m = D_i + \Delta D \tag{7.14}$$

$$\Delta D = \alpha\left(\frac{dX}{dt}\right)^n \tag{7.15}$$

where D_m is the diffusivity as modified by the presence of oxygen, ΔD is the corresponding change under oxidizing conditions, α is a proportionality constant, the exponent n is between 0.4 and 0.6, and dX/dt is the oxidation rate. Another enhancement (sometimes retardation) effect is lateral enhanced diffusion (Kennedy and O'Brien, 1965; Gibbon *et al.*, 1972) at an oxide or silicon-nitride edge. Elastic strain of the lattice near the window "edges" has been cited as the reason for the enhancement.

Diffusion of impurities in polycrystalline materials is also of interest in microelectronics processing. A case in point is the doping of polycrystalline silicon used as a gate material. Because the least resistance to diffusion is along grain boundaries, the overall diffusivity is higher than the diffusivity in a single crystal and depends heavily on the grain structure. Nevertheless, a single diffusivity can be assigned to a given material to estimate the dopant distribution. Typical diffusivities are in the range of 10^{-13} to 10^{-14} cm^2/s (Tsai, 1983).

Diffusion in silicon oxide is also of interest, since it is used as a mask against impurity diffusion. As might be expected from the fact that it is used as a mask, the diffusivities are much lower in silicon oxide than in silicon. Diffusivities of various elements in silicon oxide are given in Table 7.3 along with those in silicon for some of the elements.

7.3 DOPANT INCORPORATION

Two different methods can be used to introduce a dopant into a semiconductor. These are thermal diffusion and ion implantation. The diffusion process alone is sufficient to analyze doping by thermal diffusion. On the other hand, some basics of ion beams and its scattering in solids have to be understood before doping by ion implantation can be understood.

Consider the doping by thermal diffusion discussed in Sec. 7.1 which involved a two-step process to obtain a desired dopant profile. During predeposi-

TABLE 7.3
Diffusivities of various elements in SiO_2 and Si (Tsai, 1983)

	SiO_2		Si	
Element	D_0, cm^2/s	E, eV	D_0, cm^2/s	E, eV
B	7.23×10^{-6}	2.38	—	
Ga	1.04×10^{5}	4.17	—	
P	5.73×10^{-5}	2.30	—	
As	67.25	4.7	—	
Sb	1.31×10^{16}	8.75	—	
Au	8.2×10^{-10}	0.8	1.1×10^{-3}	1.12
Pt	1.2×10^{-13}	0.75	1.6×10^{2}	2.18
Na	6.9	1.3	1.6×10^{-3}	0.76

tion, the semiconductor is either exposed to a gas stream containing excess dopant at low temperature in order to obtain a surface region saturated with the dopant or a dopant is diffused into a thin surface layer from a solid dopant source coated onto the semiconductor surface. In the drive-in step, the dopant in the thin surface layer of the semiconductor is diffused into the interior at high temperature to obtain the desired dopant profile. Regardless of the steps, therefore, the problem of determining the profile reduces to two different cases, either diffusion with constant surface concentration or diffusion with constant total dopant present. Thermal diffusion for gallium arsenide is often carried out with doped cap layers of oxides (e.g., Ghandhi and Field, 1981) to inhibit arsenic evaporation at the doping temperature.

The thickness of the doped layer is small such that a one-dimensional conservation equation in the direction perpendicular to the surface, x, can be used. For this one-dimensional case, one has

$$\frac{\partial C}{\partial t} = -\frac{\partial J}{\partial x} = \frac{\partial}{\partial x}\left(D\frac{\partial C}{\partial x}\right) \tag{7.16}$$

where Eq. (7.1) has been used. For the constant surface concentration case, the semiconductor surface is saturated with the dopant from the gas phase. Thus, one of the boundary conditions is

$$C(0, t) = C_s \tag{7.17}$$

where C_s is the saturation concentration. Far away from the surface, the dopant concentration is zero, i.e., $C(\infty, t) = 0$. Also, there is no dopant initially in the solid, i.e., $C(x, 0) = 0$. The solution of Eq. (7.17) with the initial and boundary conditions is

$$C(x, t) = C_s \operatorname{erfc}\left[\frac{x}{(4Dt)^{1/2}}\right] \tag{7.18}$$

where the diffusivity has been assumed to be constant. The complementary error function, $\operatorname{erfc}(y) = 1 - \operatorname{erf}(y)$, is a tabulated function and can be found in the Appendix.

For the constant total dopant case, the only difference is in one of the boundary conditions, which is

$$\int_0^\infty C(x, t)\,dx = Q_o \tag{7.19}$$

where Q_o is the total amount of dopant initially present per unit exposed area. In the case of predeposition followed by drive-in, Q_o is the amount predeposited per unit area. The solution of Eq. (7.16) for this case is

$$C(x, t) = \frac{Q_o}{(\pi D t)^{1/2}} \exp\left(-\frac{x^2}{4Dt}\right) \tag{7.20}$$

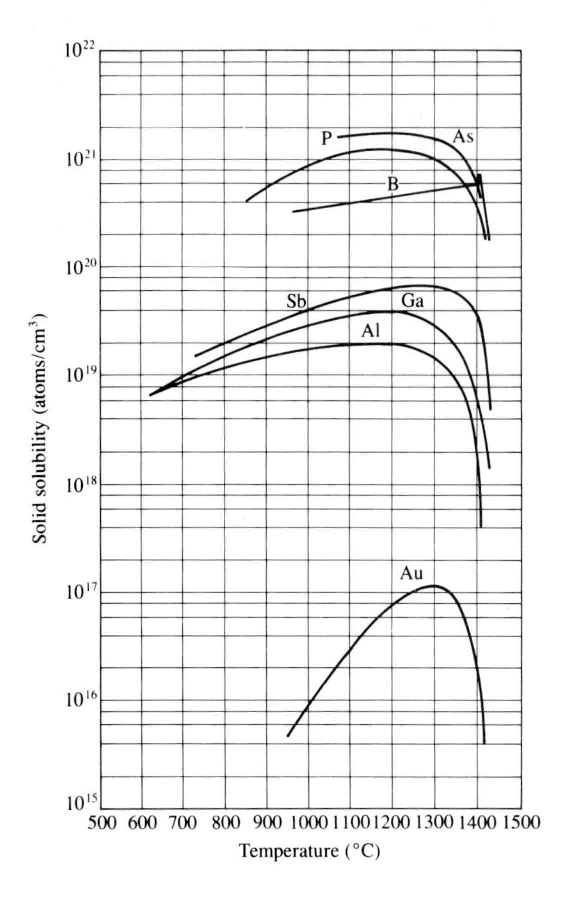

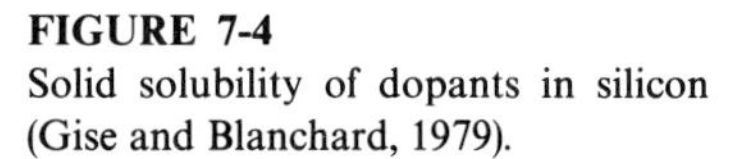
FIGURE 7-4
Solid solubility of dopants in silicon (Gise and Blanchard, 1979).

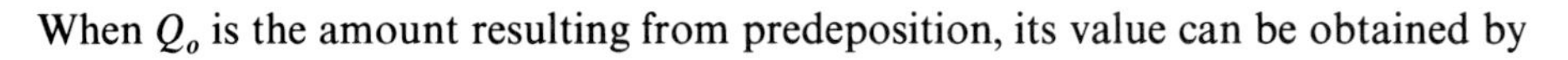
When Q_o is the amount resulting from predeposition, its value can be obtained by

$$Q_o = 2C_s\left(\frac{D_1 t_1}{\pi}\right)^{1/2} \tag{7.21}$$

where the subscript 1 is for predeposition values.

The saturation solid solubility C_s is relatively well known for impurities in silicon. The solubility curves are shown in Fig. 7-4 as a function of temperature. As seen from the figure, the solubilities in general increase with temperature, reach a maximum, and fall off rapidly as the melting point of the host material, which is silicon in this case, is approached. This behavior is commonly referred to as retrograde solid solubility characteristics.

Example 7.2. A predeposition is carried out for 15 minutes for an *n*-type silicon wafer (10^{17} atoms/cm^3) at 950 °C using a gas containing excess boron. Determine the junction depth and the distance from the surface at which the dopant concentration is one half of the surface saturation concentration. At 950 °C, the saturation concentration is 3.8×10^{20} atoms/cm^3.

Solution. The junction depth is determined by the point of transition from p-type to n-type silicon. Therefore, the junction depth x_j is determined from

$$C_s \operatorname{erfc}\left[\frac{x_j}{(4Dt)^{1/2}}\right] = 10^{17} \tag{A}$$

where Eq. (7.18) has been used. For boron, the diffusivity (Table 7.1) is given by

$$\begin{aligned} D &= 0.76 \exp\left(-\frac{3.46}{k_B T}\right) \\ &= 0.76 \exp\left[-\frac{3.46}{(8.36 \times 10^{-5})(1223)}\right] \\ &= 1527 \times 10^{-15} \text{ cm}^2/\text{s} \end{aligned}$$

Equation (A) can be rewritten as

$$\operatorname{erfc}(y) = \frac{10^{17}}{3.8 \times 10^{20}} = 2.632 \times 10^{-4}$$

From the table for the erfc function in the Appendix, $y = 2.58$. Therefore, one has

$$2.58 = y = \frac{x_j}{(4Dt)^{1/2}}$$

Solving this equation for x_j yields

$$\begin{aligned} x_j &= 2.58 \times (4 \times 1.527 \times 10^{-15} \times 900)^{1/2} \\ &= 6.05 \times 10^{-6} \text{ cm} = 0.0605\ \mu\text{m} \end{aligned}$$

The point at which $C = C_s/2$ can also be obtained from Eq. (7.18):

$$\frac{1}{2} = \operatorname{erfc}\left[\frac{x}{(4Dt)^{1/2}}\right] = \operatorname{erfc}(y)$$

Again, from the table for the erfc function, $y = 1.386$. Thus,

$$\begin{aligned} x &= 1.386(4Dt)^{1/2} = 1.386(4 \times 1.527 \times 10^{-15} \times 900)^{1/2} \\ &= 3.25 \times 10^{-6} \text{ cm} = 0.0325\ \mu\text{m} \end{aligned}$$

Example 7.3. To form a *pn* junction 1.28 μm below the wafer surface in Example 7.2, a drive-in diffusion is to be carried out. Determine the doping schedule, i.e., time and temperature for the drive-in diffusion. Determine the time required for the diffusion at 1250 °C.

Solution. The amount of dopant introduced during the predeposition is given by Eq. (7.21):

$$\begin{aligned} Q_o &= 2C_s\left(\frac{D_1 t_i}{\pi}\right)^{1/2} = 2 \times 3.8 \times 10^{20}\left(1.527 \times 10^{-15} \times \frac{900}{\pi}\right)^{1/2} \\ &= 5.03 \times 10^{14} \text{ atoms/cm}^2 \end{aligned}$$

Using this value of Q_o in Eq. (7.20) for the junction 1.28 μm away, x_j yields

$$10^{17} = \frac{5.03 \times 10^{14}}{(\pi y)^{1/2}} \exp\left(-\frac{x_j^2}{4y}\right) \qquad \text{where } y = Dt$$

Solving this equation for y by trial and error gives Dt of 9×10^{-9} cm^2. Since the profile is determined entirely by Dt, a choice of temperature (and thus D) automatically yields the time. For boron diffusion at 1250 °C,

$$D = 0.76 \exp\left(-\frac{3.46}{8.36 \times 10^{-5} \times 1523}\right)$$

$$= 1.193 \times 10^{-12} \text{ cm}^2/\text{s}$$

Since $Dt = 9 \times 10^{-9}$, the required time is

$$t = \frac{9 \times 10^{-9}}{D} = 7.54 \times 10^3 \text{ s} = 126 \text{ min}$$

Several cases of diffusion are of interest in introducing a dopant into a semiconductor. One such case is the predeposition from an oxide of thickness L_x having a dopant concentration of C_x that is spinned onto a semiconductor. Another is diffusion from a narrow slot of width w that would be of practical importance in shallow-diffused devices. Still another is the case in which the diffusivity is a function of time, which would result when the temperature is varied as a function of time. A similar case is one in which the diffusivity is a function of impurity concentration, which is applicable when the impurity concentration already in the semiconductor is much higher than the intrinsic carrier concentration. The expressions for the dopant profiles for these cases are summarized in Table 7.4. In the table, m is the impurity segregation coefficient at the Si-SiO_2 interface, which is often expressed in an Arrhenius form (Colby and Katz, 1976). For the last case, only transformed versions of the diffusion equation [Eq. (7.16)] are given, which can be solved as for Eq. (7.18) or Eq. (7.20).

Although theoretical results based on models as discussed in the previous section are available for diffusivity, particularly for the dopants for silicon, it is often necessary to determine diffusivity experimentally for specific applications. The problem is that of determining the diffusivity, given experimental measurements of the dopant profile. When experiments are carried out at constant surface concentration, the case of $D = D(C)$ in Table 7.4 applies. The transformed equation can be used to arrive at the following (see Prob. 7.11):

$$D(\beta) = -2\left[\frac{\beta C + \int_\beta^\infty C(\beta)\, d\beta}{dC/d\beta}\right] \qquad \text{where } \beta^2 = \frac{x^2}{4t} \tag{7.22}$$

Example 7.4. A doping carried out at constant surface concentration C_s for 50 minutes at 1100 K gives the following dopant profile:

TABLE 7.4
Some solutions for $C(x, t)$

Diffusion from a doped oxide source†	$C(x,t) \doteq \dfrac{C_x (D_m/D)^{1/2}}{1+\alpha} \operatorname{erfc}\left[\dfrac{x}{(4Dt)^{1/2}}\right]$ where $\alpha = \dfrac{(D_m/D)^{1/2}}{m}$
Diffusion from a slot of width w‡	$C(r,t) \doteq \dfrac{Q_s w}{2\pi Dt} \exp\left(-\dfrac{r^2}{4Dt}\right)$ where $r^2 = x^2 + y^2 + z^2$; Q_s = surface density, atoms/cm^2
$D = D(t)$†	$\dfrac{\partial C}{\partial T} = \dfrac{\partial^2 C}{\partial x^2}$ where $T = \int_0^t D(t)\,dt$
$D = D(C)$‡	$-2\beta \dfrac{dC}{d\beta} = \dfrac{d}{d\beta}\left(D \dfrac{dC}{d\beta}\right)$ where $\beta^2 = \dfrac{x^2}{4t}$ for $C = \begin{cases} C_s & \text{at } x = 0 \quad (\beta = 0) \\ 0 & \text{at } t = 0 \quad (\beta = \infty) \end{cases}$ (Boundary conditions must be expressible in terms of β)
Diffusion for the initial condition; $C = C_o$ for $0 < x < h$ and $C = 0$ for $x > h$‡	$C(x,t) = \dfrac{C_o}{2}\left[\operatorname{erf}\dfrac{h-x}{(4Dt)^{1/2}} + \operatorname{erf}\dfrac{h+x}{(4Dt)^{1/2}}\right]$

† Ghandhi (1983).
‡ Carslaw and Jaeger (1959).

x, μm	0	0.5	1.0	1.5	2.0	...
C, cm^{-3}	9×10^{19}	5×10^{19}	7.3×10^{18}	3×10^{17}	4×10^{15}	

For the purpose of illustrating the use of Eq. (7.22), explain how D at $C = 5 \times 10^{19}$ can be determined from the above data.

Solution. A table of β versus C has to be prepared first. From the definition of β, one has

$$\beta = \frac{x}{(4t)^{1/2}} = \frac{x}{(4 \times 50 \times 60)^{1/2}} = 9.13 \times 10^{-3} x \tag{A}$$

Using Eq. (A), the data given in the table become:

β, cm/s	0	4.57×10^{-7}	9.13×10^{-7}	6.86×10^{-6}	1.83×10^{-5}
C, cm^{-3}	9×10^{19}	5×10^{19}	7.3×10^{18}	3×10^{17}	4×10^{15}

From the above table, $dC/d\beta$ at $\beta = 4.57 \times 10^{-7}$, which is for $C = 5 \times 10^{19}$, can be determined by numerical differentiation of the curve drawn from C versus β. Also, at $\beta = 4.57 \times 10^{-7}$, $\beta C = (4.57 \times 10^{-7})(5 \times 10^{19}) = 2.29 \times 10^{13}$. To evaluate the integral in Eq. (7.22), the profile should be known to the point where C is negligible. Since the complete profile is not given, let α be the integral integrated from $\beta = 4.57 \times 10^{-7}$ to the value of β at which C is negligibly small. Then, Eq. (7.22) can be written as

$$D(\beta = 4.57 \times 10^{-7}) = -2\left(\frac{2.29 \times 10^{13} + \alpha}{-9.05 \times 10^{25}}\right)$$

The diffusivity thus calculated is the value of D at $C = 5 \times 10^{19}$ cm^{-3} since the value is that at $\beta = 4.57 \times 10^{-7}$. Repeating the same procedures at various values of β yields the diffusivity as a function of C.

The diffusivity is often determined with the assumption that it is independent of concentration. In practice this is usually the case, particularly for gallium arsenide. When the diffusivity is assumed to be independent of concentration, sheet resistance measurements (e.g., Runyan, 1975) along with the surface concentration for erfc distribution [Eq. (7.18)] or Gaussian distribution [Eq. (7.20)] is sufficient to determine not only the diffusivity but also the junction depth. The sheet resistivity R_s is related to the dopant concentration as follows:

$$R_s = \frac{1}{q \int_0^{x_j} \mu C(x)\, dx} \tag{7.23}$$

where x_j is the junction depth and μ is the mobility. Average resistivity ρ is defined as

$$\rho = R_s x_j \tag{7.24}$$

Various curves such as that in Fig. 1-25 for the erfc and Gaussian distributions are available (Irvin, 1962). Given the surface concentration along with the background dopant concentration (N_B in Fig. 1-25), the average resistivity can be determined from the measured sheet resistivity as was done in Sec. 1-6. Then Eq. (7.24) yields the junction depth x_j. Once the junction depth is obtained, an estimate of the constant diffusivity can be made.

Example 7.5. Consider n-type doping by the predeposition method into a p-type semiconductor (10^{18} cm^{-3}) for 20 minutes at 1100 K. Suppose that the surface concentration is 3.1×10^{20} cm^{-3}. The sheet resistance measurement yields a value of 50 Ω. Find the junction depth x_j from Fig. 1-25. Determine the (constant) diffusivity at 1100 K.

Solution. From Fig. 1-25, the point defined by the background concentration of 10^{18} cm^{-3} and the surface concentration of 3.1×10^{20} cm^{-3} gives $R_s x_j$ of 10 μm. Therefore, the junction depth is given by

$$x_j = \frac{10}{50} = 0.2\ \mu\text{m}$$

For predeposition, the dopant profile resulting after 20 minutes at 1100 K is given by

$$C(x_j, t = 20 \text{ min}) = C_s \operatorname{erfc}\left[\frac{x_j}{(4Dt)^{1/2}}\right]$$

Since the background (p-type substrate) doping level is 10^{18} cm^{-3} at $x_j = 0.2$ μm, one has

$$10^{18} = 3.1 \times 10^{20} \operatorname{erfc}(y) \qquad y = \frac{0.2 \times 10^{-4}}{(4 \times 20 \times 60 \times D)^{1/2}} = \frac{2.887 \times 10^{-7}}{D^{1/2}}$$

or $\quad \operatorname{erfc}(y) = 0.323 \times 10^{-2}$

From the table of the erfc function in the Appendix, $y = 2.08$. Therefore, the diffusivity is given by the following:

$$D^{1/2} = \frac{2.887 \times 10^{-7}}{2.08} = 1.388 \times 10^{-7}$$

$$D = 1.926 \times 10^{-14} \text{ cm}^2/\text{s}$$

Many different methods can be used to determine the diffusivity if the dopant profile is known. For instance, two points of the profile can be used to obtain the ratio of concentrations from Eq. (7.18):

$$\frac{C_1}{C_2} = \frac{\operatorname{erfc}[x_1/(4Dt)^{1/2}]}{\operatorname{erfc}[x_2/(4Dt)^{1/2}]}$$

in which D is the only unknown, and C_1 and C_2 are the dopant concentrations at positions x_1 and x_2.

The dopant profile measurements can be made by a number of methods including the capacitance-voltage technique, the differential conductivity technique, SIMS (secondary ion mass spectroscopy), and the sheet resistance technique. For detailed diffusion studies, the SIMS method appears to be the method of choice (Tsai, 1983).

Although it is often desirable to determine the diffusivity experimentally for specific applications, many useful semi-empirical diffusivity relationships are available for silicon substrates. These are given in Fair and Tsai (1975) and Tasi *et al.* (1980) for As, Fair (1981) for P, and Fair (1981) and Ryssel *et al.* (1980) for B.

Introduction of a dopant into a semiconductor by ion implantation involves penetration of accelerated ions into the solid. As the ions penetrate the solid, the ions eventually lose their energy by collisions with the electrons and nuclei of the host material and come to rest. The manner in which the penetrating ions stop or scatter determines the dopant distribution resulting from the ion implantation. Although both elastic (simple interchange of kinetic energy with no change in the internal energies) and inelastic collisions occur with both electrons and nuclei, penetrating ions are stopped mainly by inelastic collisions with electrons (electronic stopping) and elastic collisions with nuclei (nuclear stopping) for the energy range of interest in ion implantation (< 1 MeV). For the calculation of stopping (Townsend *et al.*, 1976), one defines cross sections for electronic and nuclear stopping S_j as follows:

$$S_j = -\frac{1}{N}\left(\frac{dE}{dx}\right)_j \qquad \text{for } j = e \text{ for electron and } n \text{ for nuclei} \tag{7.25}$$

where N is the atomic density of the target (host) atoms and dE/dx is the rate of energy loss in the direction of x. If it is assumed that the electronic stopping is independent of the nuclear stopping, it follows from Eq. (7.25) that

$$-\frac{dE}{dx} = N[S_n(E) + S_e(E)] \tag{7.26}$$

The average range R of a particle of energy E can be obtained by integrating Eq. (7.26):

$$R = \frac{1}{N}\int_0^E \frac{dE}{S_n(E) + S_e(E)} \tag{7.27}$$

The cross section for electronic stopping S_e is proportional to the square root of the energy. In some instances, the cross sections are lumped into one total cross section, which is in turn correlated to a power of E.

The total path length R in Eq. (7.27) has to be projected on the incident direction of the ion beam (Fig. 7-5) for the projected average range R_p. The depth profile of implanted ions is obtained from R_p and from the standard deviation

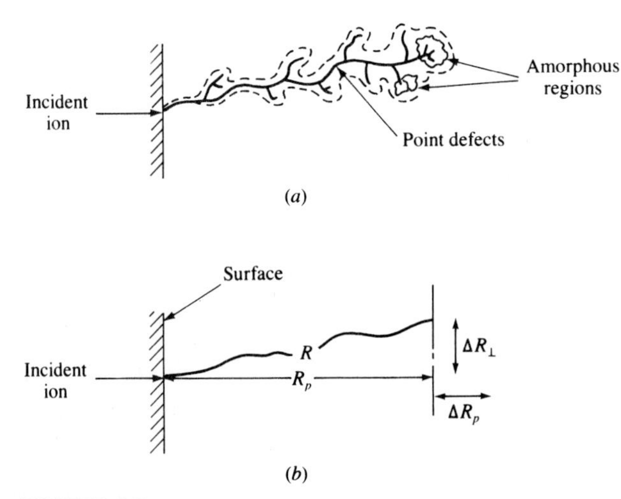

FIGURE 7-5
Disorder for typical implanted ions with corresponding ion range and straggles (ΔR_p and $\Delta R_\perp$) (Steidel, 1983).

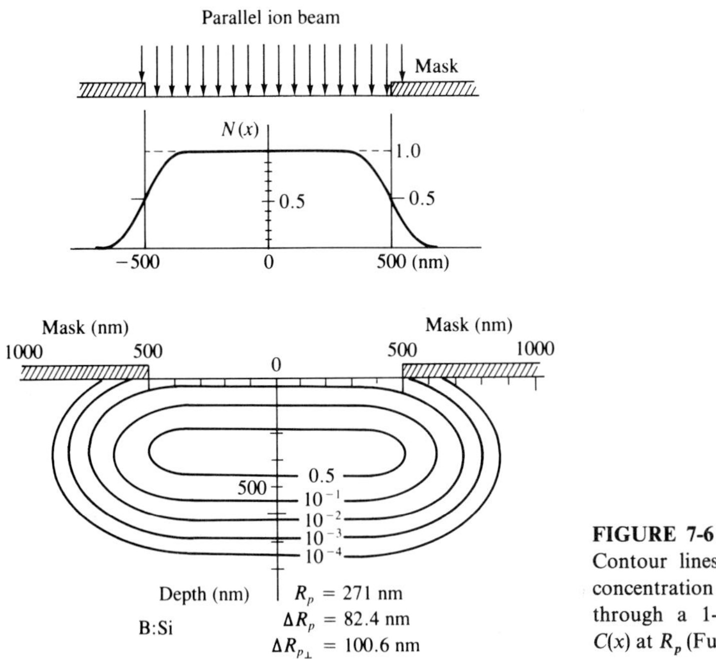

FIGURE 7-6
Contour lines of constant boron concentration after implantation through a 1-μm-wide mask and $C(x)$ at R_p (Furukawa *et al.*, 1972).

ΔR_p of a symmetric Gaussian distribution that is used to approximate the profile:

$$N(x) = \frac{N_i}{(2\pi)^{1/2}\Delta R_p} \exp\left[-\frac{(x - R_p)^2}{2\Delta R_p^2}\right] \tag{7.28}$$

where N_i is the total number of ions implanted. This profile from LSS theory (Lindhard and Scharff, 1961; Lindhard *et al.*, 1963) is derived under the assumption that Eq. (7.28) integrated from $-\infty$ to ∞ is the implanted dose (see Example 7.6). The dose N_i is obtained from the measurement of the current registered by a charge integrator connected to a metal conductor that is in good electrical contact with the wafer, and is given by

$$N_i\ (\text{atoms/cm}^2) = \frac{I\ dt}{mqA} \tag{7.29}$$

where I is the beam current (A) applied for time t (s), A is the aperture area, and m is the charge number (e.g., 1 for a singly ionized ion). Electrons pass through the integrator and neutralize the implanted charges as they come to rest in the solid. As shown in Fig. 7-5, there is also a lateral straggle $\Delta R_\perp$. If one implants through a small opening in a mask onto a semiconductor, lateral spreading cannot be ignored. In such cases, the distribution is given (Furukawa *et al.*, 1972) by

$$N(x, y) = \frac{N(x)}{(2\pi)^{1/2}\Delta R_\perp} \int_{-\infty}^{\infty}\int_{-\infty}^{\infty} \exp\left[-\frac{(y-\alpha)^2 - (d-\mu)^2}{2\Delta R_\perp^2}\right] d\alpha\, d\mu \tag{7.30}$$

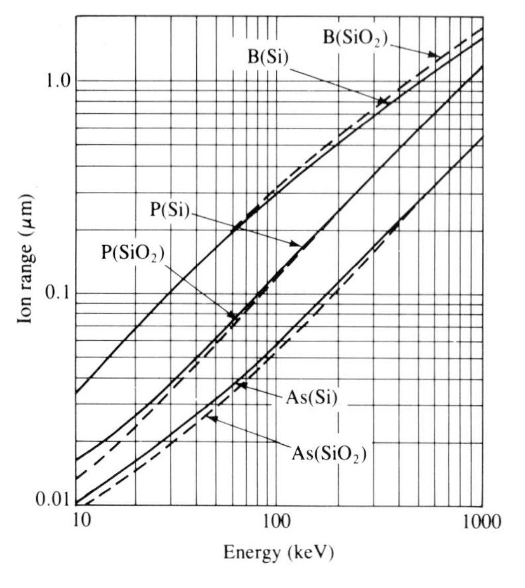

FIGURE 7-7
Projected range, R_p (Smith, 1977).

where $N(x)$ is that given by Eq. (7.28). When the mask thickness is much larger than R_p, the equation reduces to the following:

$$N(x, y) = \frac{N(x)}{2}\left[\operatorname{erfc}\left(\frac{y-a}{2^{1/2}\Delta R_{\perp}}\right) - \operatorname{erfc}\left(\frac{y+a}{2^{1/2}\Delta R_{\perp}}\right)\right] \tag{7.31}$$

where a is the half-width of the opening in the mask. An example of the lateral spread through a small opening is shown in Fig. 7-6. The values of the ion range and the ion straggles are given in Figs. 7-7 and 7-8 for typical silicon dopants. Tables of the projected range and the corresponding range straggles are given in Biersack (1981) for various substrates and dopants.

All theories on the range in a solid are based on the assumption that the target is amorphous. However, almost all semiconductors are crystalline and thus have highly anisotropic properties. Because of the ordered arrangement of lattice atoms, ions can penetrate more deeply into the crystal along major axis and planes. This phenomenon is known as the channeling effect. Within the channels along the axis and planes, no nuclear collisions (nuclear stopping) occur. The stopping takes place electronically, and the range is proportional to the ion velocity. Because of the channeling, ion implantation is often carried out with an ion beam misoriented from the major axis by an angle of at least 7 to 10°. Both silicon and gallium arsenide behave nearly as if they were amorphous solids when a misoriented beam is used. The deviations from the theoretical profile for amorphous solids that might be caused by channeling or other effects are often accounted for and approximated by the use of the "four-moment" approach (Hofker, 1975).

In many cases, ions are implanted through a layer covering the underlying semiconductor. The top layer can be a mask, an insulator, or a passivating layer.

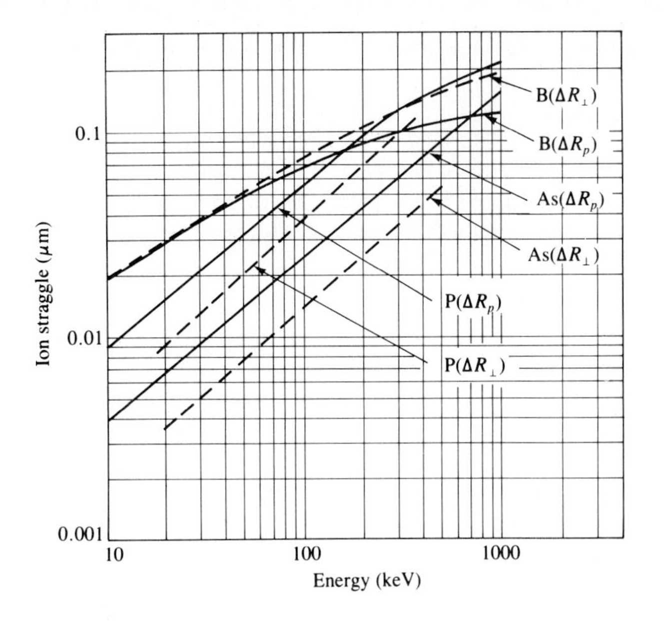

FIGURE 7-8
Calculated ion straggles, ΔR_p (vertical) and $\Delta R_\perp$ (transverse) (Smith, 1977).

If the two layers have similar ranges, they can be treated as a single layer and Eq. (7.28) applies. When they are different, the following approximation (Ishiwara *et al.*, 1975) can be used:

$$N_1 = \frac{N_i}{(2\pi)^{1/2}\Delta R_{p1}} \exp\left[-\frac{(R_{p1}-x)^2}{2\Delta R_{p1}^2}\right] \qquad 0 < x < t \qquad (7.32)$$

$$N_2 = \frac{N_i}{(2\pi)^{1/2}\Delta R_{p2}} \exp\left\{-\frac{[t+(R_{p1}-t)\Delta R_{p2}/\Delta R_{p1}-x]^2}{2\Delta R_{p2}^2}\right\} \qquad x > t \qquad (7.33)$$

where t is the thickness of the top layer and the subscripts 1 and 2, respectively, are for the top and bottom layer.

Example 7.6. Some characteristics of the ion distribution given by Eq. (7.28) are of interest. The position at which the implanted dopant concentration is the maximum is R_p, as can be shown by differentiating Eq. (7.28) with respect to x. The corresponding maximum concentration is given by

$$N_{\max} = \frac{N_i}{(2\pi)^{1/2}\Delta R_p} \qquad (A)$$

Since both R_p and ΔR_p increase with incident ion energy, the maximum concentration decreases with increasing energy. Therefore Eq. (7.28) can also be written as

$$N(x) = N_{\max} \exp\left[-\frac{(x-R_p)^2}{2\Delta R_p^2}\right] \qquad (B)$$

(*a*) Show that the amount of ion implanted that can be obtained by integrating Eq. (7.28) from zero to infinity is $N_i/2$.

(*b*) Measurements of the implanted ion profile can be used to find the parameters in Eq. (7.28). Hofker and Politiek (1980) reported the profile obtained after boron implantation into silicon at 50 keV. The dose was 10^{15} atoms/cm^2. The peak concentration was 8×10^{19} atoms/cm^3 at 0.19 μm. Determine N_i and ΔR_p from the data. Calculate $N(0)$. The experimental value of $N(0)$ was 5×10^{18} atoms/cm^3.

Solution

(*a*) The amount of ion implanted, N_T, that is obtained by integrating the equation from zero to infinity is

$$N_T = \int_0^\infty N(x)\, dx$$

$$= \frac{N_i}{(2\pi)^{1/2}\Delta R_p} \int_0^\infty \exp\left[-\frac{(x - R_p)^2}{2\Delta R_p^2}\right] dx \qquad \text{(C)}$$

Let $y = (x - R_p)/(2^{1/2}\Delta R_p)$. Then, $dx = 2^{1/2}\Delta R_p\, dy$. Therefore, Eq. (C) can be rewritten as

$$N_T = \frac{N_i}{\pi^{1/2}} \int_0^\infty \exp(-y^2)\, dy$$

$$= \frac{N_i}{2} \operatorname{erf}(\infty) = \frac{N_i}{2}$$

The following identities are of interest:

$$\operatorname{erf}(x) = \frac{2}{\pi^{1/2}} \int_0^x \exp(-t^2)\, dt$$

$$\operatorname{erfc}(x) = 1 - \operatorname{erf}(x)$$

$$\operatorname{erf}(\infty) = 1$$

(*b*) If the dose is 10^{15} atoms/cm^2, $N_i = 10^{15}$ atoms/cm^2. From Eq. (A),

$$N_{\max} = 8 \times 10^{19}\ \text{atoms/cm}^3 = \frac{10^{15}\ \text{atoms/cm}^2}{(2\pi)^{1/2}\Delta R_p}$$

Solving this for ΔR_p yields a value of 0.05 μm. The value of $N(0)$ can be obtained from Eq. (B):

$$N(0) = N_{\max} \exp\left(-\frac{R_p^2}{2\Delta R_p^2}\right)$$

$$= 8 \times 10^{19} \exp\left[-\frac{0.19^2}{2 \times (0.05)^2}\right]$$

$$= 5.9 \times 10^{16}\ \text{atoms/cm}^3$$

This calculated value of 5.9×10^{16} atoms/cm^3 compares with the experimental value of 5×10^{18} atoms/cm^3.

7.4 RADIATION DAMAGE AND ANNEALING

The crystal lattice undergoes significant structural changes as ions are bombarded into the crystal body. Atoms can be displaced from their lattice sites and the displaced atoms themselves can also displace other atoms such that a collisional cascade results. This displacement leads to the creation of vacancies and interstitial atoms, known as Frenkel defects, as well as complex lattice defects along the ion path (clusters). This damage is known as radiation damage. The damaged areas begin to overlap with increasing dose, and finally form an amorphous layer extending to a certain depth. Since the radiation damage is caused by the kinetic energy of the bombarding (penetrating) ions, it is natural that the damage is in general more severe for larger ions, higher doses, and higher energies.

Simple Frenkel defects usually result at low doses, when the ion mass is smaller than that of the target atom. These defects can have different charge states. In silicon, for instance, the defects are negatively charged in *n*-type material and neutrally charged in *p*-type material. At higher doses, locally amorphous zones, known as clusters, can form. The extent of radiation damage is also strongly dependent on implantation temperature. Significantly higher doses are required to obtain the same level of radiation damage at higher implantation temperatures. This is due to increased mobility of defects at higher temperature, which leads to a more orderly structure. In fact, the radiation damage can be cured either by subjecting the implanted substrate to a high temperature or by annealing. Thus, the implantation at high temperature corresponds to partial *in situ* annealing.

Simple defects can also lead to dislocation lines and loops. The dislocations can also result from the stress on the lattice caused by unannealed radiation damage. They can grow during annealing into undamaged regions. Dislocations anneal only at temperatures higher than 1000 °C, which is much higher than usual annealing temperatures.

Radiation damage has a significant effect on the electrical properties of semiconductors. This is not surprising in view of the fact that the electrical properties are largely determined by dopant atoms at lattice sites. Even a very small dose reduces the lifetime of charge carriers considerably. Thus, the desired electrical properties are restored only after annealing.

The total number of displaced atoms per incident ion N_d is approximately given (Kinchin and Pease, 1955) by

$$N_d = \frac{E_n}{2E_d} \tag{7.34}$$

where E_n is the incident energy available for nuclear stopping processes (area under the $dE/dx|_n$ curve) and E_d is the displacement energy of a lattice atom, which is approximately 14 eV for silicon (Novak, 1965) and ranges from 8 to 30 eV for other semiconductors. Values of N_d are given in Table 7.5 for silicon. It is

TABLE 7.5
N_d in silicon (Kinchin and Pease, 1955)

Ion	10 keV	50 keV	700 keV
Sb	357	1785	7,143
As	357	1785	28,500
B	606	609	620
Si	357	4280	4,290

interesting to observe that N_d is the same except for the light dopant (boron) at low energies (for example, 10 keV). The range distribution of radiation damage closely follows that of implanted ion. According to Sigmund and Sanders (1967), it follows the same Gaussian distribution of Eq. (7.28) with a different range (X_D in place of R_p) and straggle (ΔX_D in place of ΔR_p). These parameter ratios are given in Fig. 7-9. One quantity of practical interest in ion implantation is the critical dose, which is the dose needed to form an amorphous layer. This critical dose for silicon is shown in Fig. 7-10 for various dopants.

Annealing accomplishes electrical activation of the implanted ions. In addition, it brings about recrystallization of the lattice. When the ionic radius of the dopant atom is very similar to that of the host atom, simultaneous activation and recrystallization can take place. However, the temperature required for the activation can be different from that for recrystallization. For gallium arsenide, for instance, radiation damage is eliminated after annealing at 500 °C. However, temperatures higher than 700 °C are necessary for the electrical activation (Ryssel and Ruge, 1986). The electrical activation is determined by Hall and sheet resistivity measurements or simply by four-point probe measurements. Details on the measurement techniques can be found in Runyan (1975).

The behavior of annealing is usually studied by two different methods. Isochronal annealing is one in which the time of annealing is held constant but the

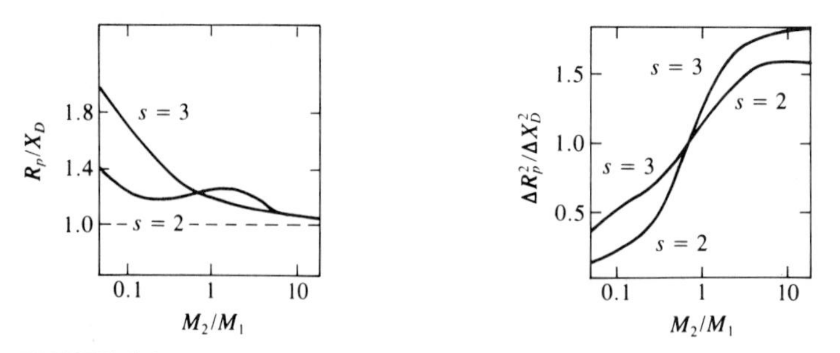

FIGURE 7-9
Ratio of R_p to radiation damage range X_D, and ratio of ΔR_p to radiation damage straggle (vertical) ΔX_D as a function of mass ratio M_2(target)/M_1(ion). A larger s corresponds to a lower energy transfer upon collision (Sigmund and Sanders, 1967).

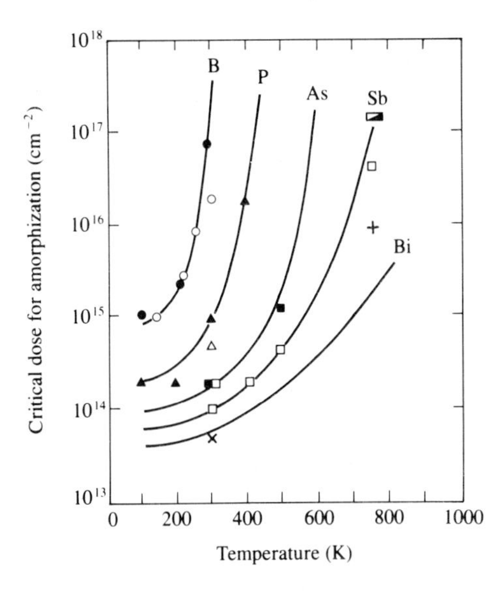

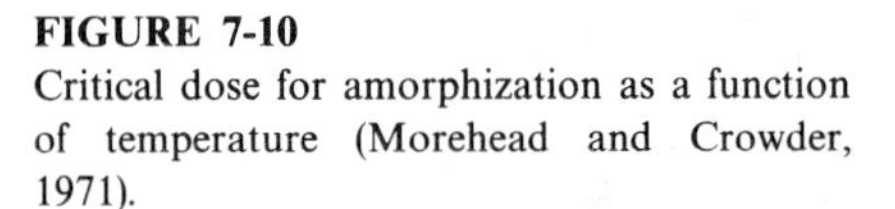

FIGURE 7-10
Critical dose for amorphization as a function of temperature (Morehead and Crowder, 1971).

temperature is varied. Isothermal annealing is one in which the time is varied while holding the temperature constant. Annealing is mainly studied by the isochronal method. For some dopants, such as arsenic in silicon, the isochronal curve is too steep to be useful and therefore the isothermal method is used.

Annealing for electrical activation is often described by phenomenological nth-order kinetics with respect to the electrically inactive ion concentration (e.g., Capelani *et al.*, 1974):

$$r_a = k'I^n \tag{7.35}$$

where r_a is the rate per unit volume, k' is the rate constant, and I is the concentration of electrically inactive ions. The electrically active ion concentration C is related to I as follows:

$$C(t, T) = C(0, 0) + I(0, 0) - I(t, T) \tag{7.36}$$

where $I(t, T)$ is the concentration of electrically inactive ions at time t and temperature T during annealing and $I(0, 0)$ is the same at the start of annealing. Since $C(0, 0)$ and $I(0, 0)$ constitute the sum of active and inactive ions present before annealing, the maximum possible concentration of active ions C_m is given by

$$C_m = C(0, 0) + I(0, 0) \tag{7.37}$$

Since the volume of the solid being annealed is constant, Eq. (7.35) can be rewritten as

$$\frac{dI}{dt} = kI^n \qquad \text{where } k = \frac{k'}{V} \tag{7.38}$$

and V is the volume of solid being annealed. Use of Eqs. (7.36) and (7.37) in Eq. (7.38) yields

$$\frac{dC}{dt} = -k(C_m - C)^n \tag{7.39}$$

Whether the annealing is carried out isochronally or isothermally, the temperature is constant at least for a period of time. Therefore, Eq. (7.39) can be solved as follows:

$$\ln \frac{C_m - C_o}{C_m - C} = kt \qquad \text{for } n = 1 \tag{7.40a}$$

$$(C_m - C)^{1-n} - (C_m - C_o)^{1-n} = (n-1)kt \qquad \text{for } n \neq 1 \tag{7.40b}$$

where C_o is the concentration of electrically active ions at time zero, which is much smaller than C_m and thus can be neglected.

The concentration of electrically active ions is inferred from resistivity measurements. Noting that C in Eq. (7.39) is an average concentration over a thickness to which ions are implanted, one can rewrite Eq. (7.23) as follows:

$$R_s = \frac{1}{\mu_e q x_j \left[\int_0^{x_j} C(x)\, dx/x_j \right]} = \frac{1}{\mu_e q x_j C} \tag{7.41}$$

since the quantity within the bracket is the average concentration C. In terms of average resistivity ρ [Eq. (7.24)], Eq. (7.41) becomes

$$\rho = \frac{1}{q\mu_e C} \qquad \text{or} \qquad S = q\mu_e C \tag{7.42}$$

where μ_e is an effective mobility over the thickness and S is the conductivity. Since the conductivity (inverse of ρ) is proportional to the average concentration, experimental data are often given in terms of the conductivity. The maximum concentration C_m is ideally that corresponding to the dose. However, not all implanted ions are necessarily activated by annealing. Therefore, C_m is often taken as that obtained when a plot of C versus T reaches a plateau at high temperatures.

Example 7.7. In isochronal annealing, only one sample is subjected to various temperatures but annealed for the same length of time at each temperature, and measurements are made. Let T_i and C_i be the temperature and concentration at the ith time the sample is annealed. Explain how one can determine the activation energy and preexponential factor from the isochronal annealing data.

Solution. Applying Eq. (7.40) to the intermittent annealing and measurement, one has, for instance, for first-order kinetics:

$$\ln \frac{C_m - C_{i-1}}{C_m - C_i} = k(T_i)t$$

Let the left-hand side of the above equation be Y and take the logarithm of both sides of the equation to obtain

$$\ln Y_i = -\frac{E_a}{RT_i} + \ln k_0 t \qquad \text{where } k = k_0 \exp\left(-\frac{E_a}{RT}\right)$$

Since the time is the same for all i, a plot of $\ln Y_i$ versus $1/T_i$ should yield the activation energy E_a from its slope and k_0 from the intercept. Similar procedures can be followed for $n \neq 1$. This is a form of integral kinetic analysis (e.g., Levenspiel, 1972). Therefore, various n values should be tried until a straight line (or the best fit) is obtained. For isothermal annealing, the usual procedures of Chap. 5 or those detailed in the above reference can be followed.

The annealing process is complex, as is ion implantation. An example of a complex annealing process for boron in silicon is shown in Fig. 7-11. At low doses, a slow increase in the electrical activation with temperature is seen. As the dose increases, however, a retrograde annealing (reverse annealing) results. According to Blamires (1970), the reverse annealing is due to ion-pair formation, which dissociates again at higher temperatures, and the atoms are incorporated individually into the crystal lattice. The steep increase in the electrical activation around 600 °C at high doses is usually attributed to recrystallization. Similar behavior is also observed for phosphorus in silicon, shown in Fig. 7-12. As indicated in the figure, the annealing behavior is represented by a combination of three separate reaction rates, each operative in the temperature range shown.

As device dimension and junction depth decrease, it has become necessary to activate the implanted ions in such a way that the implanted ions do not diffuse during annealing. This need has brought about rapid thermal annealing

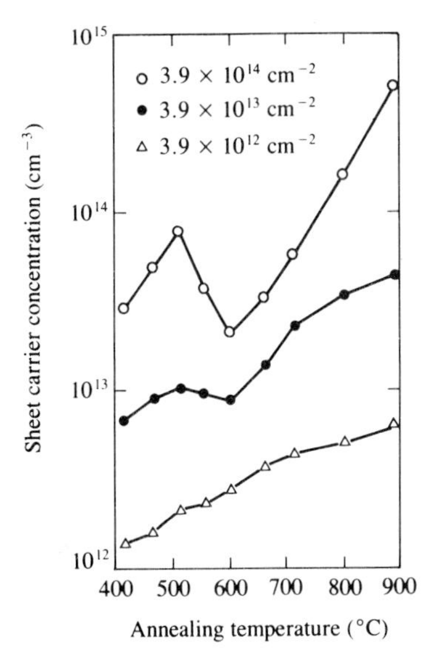

FIGURE 7-11
Isochronal annealing of boron-implanted silicon (Webber *et al.*, 1969).

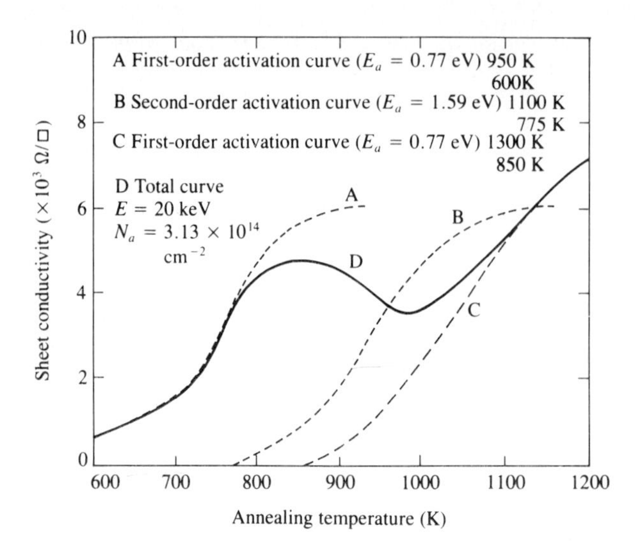

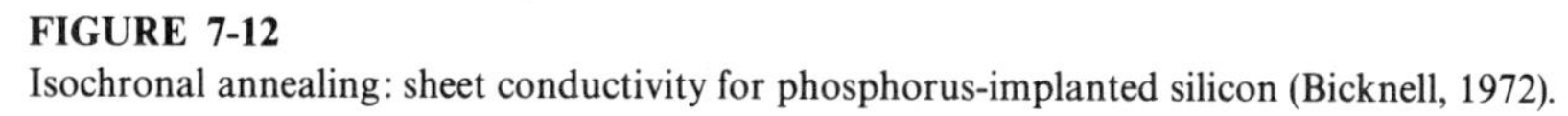

FIGURE 7-12
Isochronal annealing: sheet conductivity for phosphorus-implanted silicon (Bicknell, 1972).

(RTA) techniques. The distance S atoms travel by diffusion follows from the definition of diffusivity (Example 5.1) and is

$$S = (Dt)^{1/2} = \left[D_0 t \exp\left(-\frac{E}{k_B T} \right) \right]^{1/2} \tag{7.43}$$

where Eq. (7.2) has been used. The distance S has to be small to prevent redistribution of implanted ions by diffusion. The temperature must be set high enough to activate the electrically inactive ions, so only linked design freedom exists here. On the other hand, the time for minimum redistribution can be minimized, which is why it is called rapid annealing. It should be recognized at the same time that a certain amount of diffusion is necessary for the activation, since diffusion itself is the mechanism by which the atoms reorder their positions for activation. The distance required for the reordering might be several times the lattice constant.

Example 7.8. According to Fig. 7-12, the group of ions most difficult to activate belongs to the curve B with an activation energy of 1.59 eV, and the fraction of this group is approximately 30 percent of the total implanted ions if one uses the plateau values for each group. Determine the temperature required to activate 90 percent of the group B ions. Determine also the fraction of total ions still not activated at that temperature. Assuming that four times the silicon lattice constant is the distance for displaced atoms to diffuse for reordering of the lattice structure, calculate the (minimum) time for the annealing. The Boltzmann constant is 8.62×10^{-5} eV/K. Use the diffusivity given by $D(cm^2/s) = 0.76 \exp(-3.46/k_B T)$. The silicon lattice constant is 0.543 nm.

Solution. If the activation energy is 1.59 eV, the probable fraction of group B ions overcoming the energy barrier, f, is given by

$$f = \exp\left(-\frac{E}{k_B T}\right)$$

$$= \exp\left(-\frac{1.59}{8.62 \times 10^{-5} T}\right)$$

$$= 0.90$$

Solving this for T yields a temperature of 1.75×10^5 K. At this temperature, the fraction of group A and C ions activated is given by $\exp(-0.77/8.62 \times 1.75)$, which is 0.95. Since 90 percent of group B ions, which constitute 30 percent of the total ions, and 95 percent of groups A and C (70 percent of the total) are activated, the fraction of total ions activated, w, is given by

$$w = 0.3 \times 0.9 + 0.7 \times 0.95 = 0.935$$

Thus, 93.5 percent of the total ions are activated. The distance S the atoms have to diffuse is 4×0.543 nm $= 2.172$ nm. Using this value in Eq. (7.43), one has

$$2.172 \times 10^{-7}\ (\text{cm}) = \left[0.76t \exp\left(-\frac{3.46}{8.62 \times 1.75}\right)\right]^{1/2}$$

or

$$t = 7.81 \times 10^{-14}\ \text{s}$$

This example illustrates the point that the time required for the activation is very short. Although some vaporization should occur at the calculated temperature, the only limitation for rapid annealing is the ability to heat the substrate quickly. A heating method suitable for this very rapid thermal annealing is laser heating. The controllable time scale for a pulsed laser is down to 10^{-11} s, and it is 10^{-4} s for a continuous (cw) laser. The average temperature rise ΔT due to exposure to a laser beam with short wavelengths can be estimated (Bloemberger, 1979) as follows:

$$T = \frac{(1 - R)I_E U}{\rho C_v (2D_T U)^{1/2}} \tag{7.44}$$

where R is the reflectivity, C_v is the specific heat, U is the laser pulse duration, D_T is the thermal diffusivity, I_E is the pulse energy, and ρ is the density of the semiconductor. With pulsed laser annealing, melting of the surface of the irradiated semiconductor almost always occurs, followed by an epitaxial regrowth. With a cw laser, recrystallization takes place via solid phase epitaxy. Auston *et al.* (1979) expressed the regrowth rate by solid phase epitaxy as follows:

$$v = v_0 \exp\left(-\frac{E}{k_B T}\right) \tag{7.45}$$

where v_0 is 2.9×10^9 cm/s and $E = 2.7$ eV for (100) silicon. Laser annealing, while attractive, is not conducive to high throughput. Although limited in time scale (only down to 1 s), radiation heating with tungsten-halogen lamps allows arrays of wafers to be heated for annealing (Sedgewick, 1986).

During annealing, implanted atoms redistribute. If dopants do not escape out of the exposed surface during annealing, the dopant profile after annealing (Ryssel and Ruge, 1986) is given by

$$C(x, t) = \frac{N_i}{[2\pi(\Delta R_p^2 + 2Dt)]^{1/2}} \exp\left[\frac{-(x - R_p)^2}{2(\Delta R_p^2 + 2Dt)}\right]$$

$$\times \left(\frac{1}{2}\right)\left\{1 + \operatorname{erf} y_+ + \left[\exp\left(-\frac{4xR_p}{2\Delta R_p^2 + 4Dt}\right)\right](1 + \operatorname{erf} y_-)\right\} \quad (7.46)$$

$$y_\pm = \frac{R_p(4Dt)^{1/2}/(2^{1/2}\Delta R_p) \pm x(2^{1/2}\Delta R_p)/(4Dt)^{1/2}}{(2\Delta R_p^2 + 4Dt)^{1/2}}$$

Except in the immediate region near the surface, the first term before the ($\frac{1}{2}$) factor in Eq. (7.46) is sufficiently accurate for the profile. As was the case for doping by thermal diffusion, annealing is often carried out in an oxidizing environment, especially when a masking layer for the next fabrication step is to be produced simultaneously with annealing. In such cases, the dopant profile is determined by the following relationship (Prince and Schwettmann, 1974):

$$\frac{\partial C}{\partial t} = \frac{\partial}{\partial x}\left(D\frac{\partial C}{\partial x}\right) + \alpha\frac{dX}{dt}\frac{\partial C}{\partial x} \quad (7.47)$$

where the origin is at the substrate-oxide interface. Here α is the ratio of the thickness of the substrate consumed during the oxide formation to that of the oxide, which is approximately 0.44 for Si-SiO_2 (Atalla and Tannenbaum, 1960), and dX/dt is the oxide growth rate.

When a substrate is bombarded with high-energy, nondoping particles such as neutrons and protons during or after implantation, a significant increase in the rate of diffusion results. This effect is known as radiation-enhanced diffusion (Seeger and Chick, 1968; Ryssel and Ruge, 1986). Defects (vacancies in ideal cases) produced by the bombardment increase the diffusivity, since most doping elements diffuse via a vacancy mechanism. Relatively high temperatures are necessary during irradiation to avoid the formation of stable defects. In the case of silicon, the temperature must be over 750 °C.

Although not related to dopant incorporation, ion implantation damage is an effective method of gettering, as is other damage gettering such as sandblasting and sound abrasion. Gettering is a process of removing undesired impurities and defects from junction regions. Other effective gettering techniques are heavy phosphorus doping, which causes formation of ion pairs such as P^+Cu^-, and intrinsic gettering, in which oxide precipitates, such as SiO_x, play the role of a sink for capturing undesired impurities.

7.5 DOPANT REDISTRIBUTION AND AUTODOPING

After dopants are incorporated into substrates to form junctions, they undergo many processing steps for device fabrication. Although efforts are made to use low-temperature processing techniques to minimize redistribution of incorporated dopants, the dopants still redistribute during the course of further processing. This redistribution problem can become serious enough to receive careful attention when an epitaxial film is grown on top of the doped area, because of the high temperature required for epitaxy. A consideration of this severe case of dopant redistribution leads to other dopant redistributions that can arise in various situations.

When a film is grown on a doped substrate, the dopant diffuses into the growing film in the course of the film growth. This phenomenon is referred to as autodoping (Basseches *et al.*, 1961; Srinivasan, 1980). The autodoping behavior is shown in Fig. 7-13. As shown in the figure, the dopant in the substrate diffuses out into the growing epitaxial layer, designated as Epi. At the same time, during the growth, the dopant in the substrate evaporates into the growing film. Such doping is referred to as lateral autodoping as opposed to the redistribution caused by the solid phase diffusion, called outdiffusion. It is noted that the autodoping also leads to the unintentional doping of the film in between the doped regions in Fig. 7-13, denoted as the "off" position (nondiffused area) in the figure.

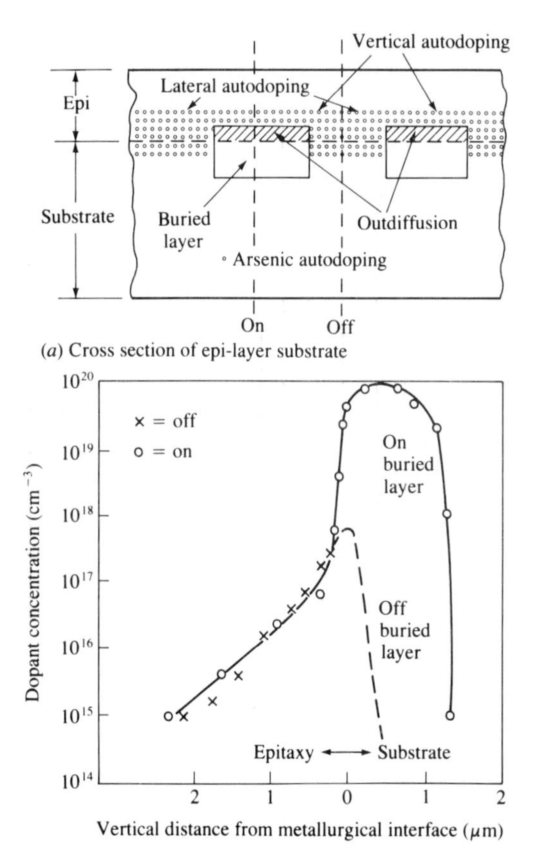

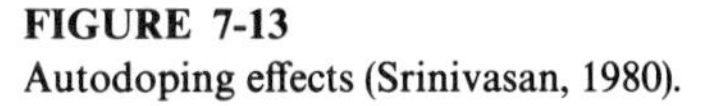
FIGURE 7-13
Autodoping effects (Srinivasan, 1980).

In the process of growing an epitaxial layer onto a substrate with a heavily doped buried layer as in BJT, the first step is to prebake the substrate so as to establish stabilized thermal and flow conditions for the epitaxial deposition to follow. This prebake is not unique to the epitaxial deposition and can represent other similar situations in fabrication steps. In this prebake, the dopant profile in the buried layer changes. Because of dopant evaporation taking place at the buried layer surface exposed to flowing gas, the dopant is carried away by the flowing gas and deposits onto the nondiffused substrate. The lateral autodoping onto the nondiffused substrate is complicated by the fact that the dopant concentration in equilibrium with the surface concentration changes with position and flow conditions. The dopant distributions resulting from the prebake are summarized in Table 7.6. The distributions are based on the assumption that the profile in the buried layer of length L is uniform at C_o initially. In the table, C_1 is the dopant profile in the buried layer, C_2 is the same in the layer below the buried layer, and C_n is the profile in the nondiffused regions. In the nondiffused regions, the profiles depend on the surface concentration of dopant in the gas phase $(C_s)_g$. Further, the surface concentration depends on the mass transport in the reactor, i.e., whether the nondiffused regions are in the upstream or downstream with respect to the position of the buried layer in the reactor. The region in the upstream is exposed to the dopant transported by gas phase diffusion only; the region in the downstream is exposed to the dopant transported by diffusion as well as convective flow. The proportionality constant α in the table is a factor relating the feed velocity U to the velocity near the substrate, which has to be chosen depending on the radial velocity profile pertinent to the flow regime.

TABLE 7.6
Dopant distribution resulting from prebake (Lee, 1986)

$$\frac{C_1}{C_o} = 0.5\,(\text{erfc}\; y_1 - \text{erfc}\; y_2) + \text{erf}\; y_3 + Y$$

$$\frac{C_2}{C_o} = 0.5\,(\text{erfc}\; y_1 - \text{erfc}\; y_4) - \text{erf}\; y_3 + Y$$

$$y_1 = (L + x)/(4Dt)^{1/2} \qquad y_2 = (L - x)/(4Dt)^{1/2}$$

$$y_3 = x/(4Dt)^{1/2} \qquad y_4 = -y_2$$

$$Y = \exp\,(y_5)[\text{erfc}\; y_6 - \exp\,(hL/D)\;\text{erfc}\; y_7]$$

$$y_5 = (hx + h^2 t)/D \qquad y_6 = y_3 + h(t/D)^{1/2}$$

$$y_7 = y_1 + h(t/D)^{1/2}$$

$$h = k_m/K_e$$

$$C_n = K_e (C_s)_g\; \text{erfc}\; [X/(4Dt)^{1/2}]$$

$$\frac{(C_s)_g}{(C_s)_b} = \begin{cases} \exp\,[-(a_1 - b_0)z] & \text{downstream} \\ \exp\,[-(a_1 + b_0)z] & \text{upstream} \end{cases}$$

$$a_1 = [(\alpha U)^2 + 4D_g k_m]/2D_g$$

$$b_0 = \alpha U/2D_g$$

After the prebake during which thermal and gas flow conditions are stabilized, reactants and dopant are introduced into the reactor for epitaxial deposition. The dopant redistribution during epitaxial deposition is distinctly different from that during the prebake in that the growing solid surface moves with time because of deposition. The one-dimensional conservation equation is

$$\frac{\partial C}{\partial t} = D \frac{\partial^2 C}{\partial x^2} \tag{7.48}$$

One of the boundary conditions is given by

$$-D \frac{\partial C}{\partial x} = k_m[(C_s)_g - C_\infty] + VC \qquad \text{at } x = x_f;\ x_f = Vt \tag{7.49}$$

where V is the constant deposition rate, x_f is the moving boundary given by Vt and C is the concentration in the bulk gas flow. It states that the flux at the moving boundary is the same as the amount transferred by mass transfer at the gas-solid interface plus the amount transferred by the deposition (Grove *et al.*, 1965). Noting that the surface concentration of the dopant in the solid, $C(0, t)$, is assumed to be in local equilibrium with the gas phase surface concentration, $(C_s)_g$, such that

$$C(0, t) = K_e(C_s)_g \tag{7.50}$$

where K_e is the equilibrium constant, Eq. (7.49) can be rewritten as

$$-D \frac{\partial C}{\partial x} = (h + V)C - k_m C \qquad \text{where } h = \frac{k_m}{K_e} \tag{7.51}$$

As shown in Fig. 7-14, the origin for the coordinate is at the epi-substrate interface. The initial condition is that resulting from the prebake. Although exact solutions are available, approximate solutions under the condition of $Vt \gg (Dt)^{1/2}$, which is usually satisfied, are useful. The approximate solutions (Lee, 1986) are

$$C(t, x) = C_i(0, x)\left(W + \frac{1 - W}{2} \operatorname{erfc} y_3\right) \qquad \text{where } y_3 = \frac{x}{(4Dt)^{1/2}} \tag{7.52}$$

and

$$W = \frac{k_m C_\infty}{(h + V)C_o}$$

which is applicable to both diffused and nondiffused regions. It is noted, however, that $C_i(0, x)$ is C_1 in Table 7.6 for the diffused region whereas it is C_n in Table 7.6 for the nondiffused region, and that t in the equations for C_1 and C_n is the duration of the prebake, but x as used in Eq. (7.52) is applicable to the surface of the epitaxial layer. The maximum lateral autodoping, which is the maximum

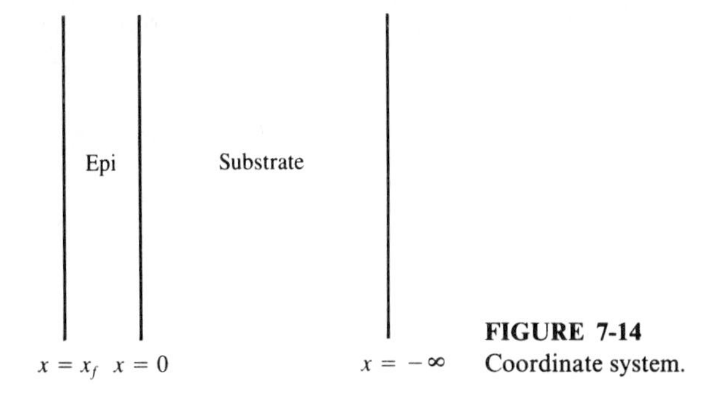

FIGURE 7-14
Coordinate system.

dopant concentration in the nondiffused region, is approximately given by

$$(C_n)_{\max} = K_e(C_s)_g\left(W + \frac{1 - W}{2}\right) \tag{7.53}$$

The dopant distribution that results when no dopant is present in the feed can be obtained by simply setting C_∞ or W equal to zero. If the epitaxial layer is to be doped with an opposite type of dopant relative to the substrate, then the dopant distribution in the epitaxial layer (Grove *et al.*, 1965), C_e, is given by

$$C_e = \frac{C_f}{2}\,[1 + \operatorname{erf} y_3 + \exp(z_1)\operatorname{erfc}(z_2)] \tag{7.54}$$

where $z_1 = (V/D)(Vt - x)$
$z_2 = (2Vt - x)/(4Dt)^{1/2}$

The corresponding dopant distribution of thc opposite type, emanating from the substrate into the epitaxial layer being deposited, is given by Eq. (7.52) with $W = 0$. The junction forms at the point where $C(t, x_j) = C_e(t, x_j)$. A significant redistribution can also occur during thermal oxidation of native substrates. The distribution occurs not only due to thermal diffusion but more importantly due to segregation of dopant at the substrate-oxide interface. This segregation is represented by a segregation coefficient m (at the interface) defined as follows:

$$m = \frac{\text{equilibrium dopant concentration in substrate}}{\text{equilibrium dopant concentration in oxide}} \tag{7.55}$$

Four different cases of dopant redistribution are shown in Fig. 7-15 (Grove *et al.*, 1964). They can be considered in two groups: in one, the oxide has a tendency to take up ($m < 1$) dopant, and in the other, it has a tendency to reject ($m > 1$) dopant. In each case the situation can be different for slow and for fast diffusion in the oxide. The dopant concentration on the substrate side of the interface, $(C_s)_i$, is given (Grove *et al.*, 1964) by

$$\frac{(C_s)_i}{C_o} = \frac{1 + (C_x/C_o)\lambda}{1 + (1/m - \alpha)\pi^{1/2}\exp(\alpha^2 y_b^2)\operatorname{erfc}(\alpha y_b)y_b + \lambda/m} \tag{7.56}$$

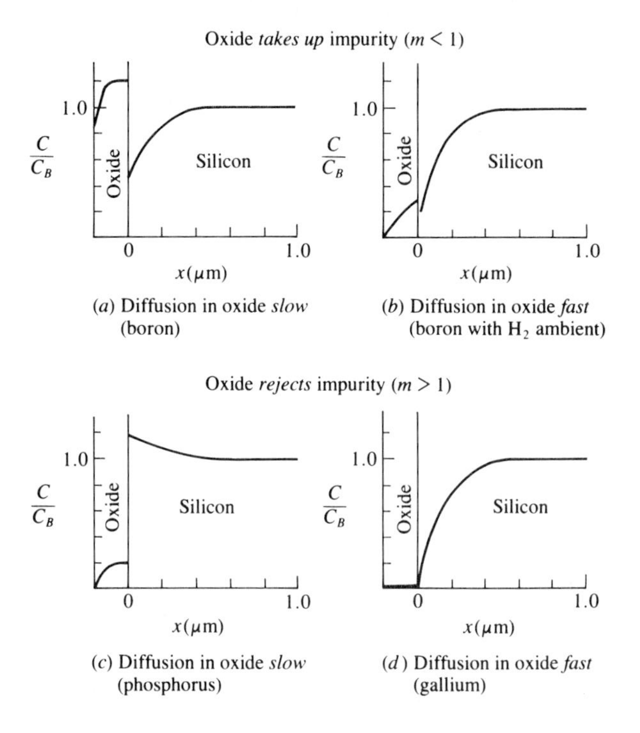

FIGURE 7-15
Four cases of impurity redistribution in silicon due to thermal oxidation (Grove *et al.*, 1964).

where $\lambda = r \exp[(\alpha^2 r^2 - 1)y_x^2] \operatorname{erfc}(\alpha y_b)/\operatorname{erf}(y_x)$

$$r = (D_x/D)^{1/2}$$
$$y_b = (B/4D)^{1/2}$$
$$y_x = (B/4D_x)^{1/2}$$

and C_o is the initial dopant concentration in the substrate, which is assumed constant, C_x is the dopant concentration in the oxide at the gas-oxide interface, D_x is the diffusivity in the oxide, B is the parabolic rate constant (see Chap. 5, gas-solid reactions) for the oxidation, and α is the ratio of the thickness of substrate consumed during oxidation to the oxide thickness, which is 0.45 for silicon. The segregation coefficient is 10 for P, Sb, and As and 0.3 for B in Si (Grove *et al.*, 1964).

NOTATION

a	Half-width of a window (L)
A	Area (L^2)
B	Parabolic rate constant for oxidation ($L/t^{3/2}$)
C	Solid phase dopant concentration (atoms/L^3, or L^{-3})
C_m	Maximum C defined by Eq. (7.37) (atoms/L^3)

C_e	Dopant concentration in epitaxial layer given by Eq. (7.54)
C_f	Dopant concentration in gas phase
C_n	C in epitaxial layer
C_o	Constant background C in substrate
C_s	Solid surface value of C
C_v	Specific heat
C_x	C in oxide layer
C_∞	Gas phase concentration of dopant in bulk fluid flow (molecules/L^3)
$(C_i)_g$	Gas phase concentration of dopant at interface
$(C_s)_g$	Gas phase dopant concentration at solid surface
D	Diffusivity (L^2/t)
D_i	Intrinsic diffusivity
$D_i^\times$, D_i^-, D_i^+	Intrinsic diffusivity due to neutral ($\times$), negative, and positive ions
D_g	Gas phase diffusivity
D_m	Diffusivity defined by Eq. (7.14)
D_x	Diffusivity in oxide
D_T	Thermal diffusivity
D_1	Diffusivity during predeposition
E	Activation energy (E); electric field (V/L)
E_n	Incident energy available for nuclear stopping
E_d	Displacement energy of a lattice atom
h	Enhancement factor in Eq. (7.10) (dimensionless); k_m/K_e (L/t)
I	Current (I); concentration of electrically inactive ion (atoms/L^3)
I_E	Pulse energy (E)
J	Flux (molecule/L^2)
k	Rate constant
k_B	Boltzmann constant
k_m	Mass transfer coefficient (L/t)
K_e	Equilibrium constant defined in Eq. (7.50) (dimensionless)
K_{eq}	Equilibrium constant defined in Prob. 7.12
K_r	Reaction equilibrium constant
m	Charge number in Eq. (7.29); segregation coefficient defined by Eq. (7.55)
n	Electron concentration (electron/L^3 or L^{-3}); reaction order
n_i	Intrinsic value of n
N	Impurity concentration (L^{-3}) in Eq. (7.8); atomic density of target atoms in Eq. (7.25); ion concentration after implantation (L^{-3}) in Eq. (7.28)
N_i	Dose (atoms/L^2 or L^{-2})
N_0	Number of atoms displaced per incident ion
p	Hole concentration
q	Electric charge
Q_o	Total amount of dopant after predeposition (atoms/L^2)
Q_s	Surface density in Table 7.4 (atoms/L^2)

r	Radius defined in Table 7.4 (L)
R	Ion range defined by Eq. (7.27) (L); reflectivity in Eq. (7.44)
R_p	Projected value of R (L)
R_s	Sheet resistivity ($\Omega \cdot$cm)
ΔR_p	Standard deviation of R_p in x direction (L)
$\Delta R_\perp$	Standard deviation of R_p in y direction (L)
S	Distance atoms travel by diffusion (L)
S_e	Electronic stopping cross section (L^2E)
S_j	Quantity defined by Eq. (7.25) (L^2E)
S_n	Nuclear stopping cross section (L^2E)
t	Time (t)
t_1	Time for predeposition
T	Temperature (T)
ΔT	Temperature rise due to exposure to laser beam [Eq. (7.44)]
u	Laser pulse duration (t)
U	Bulk fluid velocity (L/t)
v	Regrowth rate given by Eq. (7.45) (L/t)
v_0	Preexponential quantity in Eq. (7.45) (L/t)
V	Epitaxial growth rate (L/t)
w	Width of a window (L)
W	Quantity defined in Eq. (7.52) (dimensionless)
x	Coordinate in the direction perpendicular to solid surface
x_j	Junction depth (L)
X	Oxide thickness (L)
y	Coordinate perpendicular to x (width coordinate)
y_1, y_2, y_3, y_4	Quantities defined in Table 7.6
Y	Quantity defined in Table 7.6
z_1, z_2	Quantities defined in Eq. (7.54)

Greek letters

α	Ratio of substrate thickness consumed during oxidation to oxide thickness; quantity defined in Table 7.4; proportionality constant in Table 7.6
β	$x/(4t)^{1/2}$
λ	Quantity defined in Eq. (7.56)
μ	Mobility (L^2/Vt)
μ_e	Effective mobility
ρ	Average resistivity (ΩL)

Units

E	Energy
I	Current
L	Length
V	Voltage

t	Time
T	Temperature
Ω	Ohm

PROBLEMS

7.1. According to Fair (1981), the diffusivities of dopants in silicon are given as follows:

$$D_{\text{As}} = 2D_i^- \frac{n}{n_i}$$

$$D_{\text{B}} = D_i^+ \frac{p}{n_i}$$

$$D_{\text{P}} = D_i^\times + D_i^{2-}\left(\frac{n}{n_i}\right)^2 \qquad \text{where } D_i^{2-} = 44.2 \exp\left(-\frac{4.37}{k_B T}\right)$$

Assuming that $n/n_i = 10^7/1.5$ and $p/n_i = 10^7/1.5$, which are also assumed to be independent of temperature, calculate the temperatures for As and P at which the diffusivities are the same as that calculated in Example 7.1 for boron. Give the dopant that is most difficult to diffuse.

7.2. When predeposition is carried out for dopant incorporation by thermal diffusion, the gas phase dopant concentration is kept constant at a high level so that the semiconductor surface is saturated with dopant. Suppose that the predeposition is carried out with phosphorus at 1040 °C for 30 minutes, and the intrinsic diffusivity is 6×10^{-14} cm^2/s. Calculate the amount of phosphorus incorporated through an area of 5 cm^2. If the predeposition is to be less than 1 percent, what would be the minimum amount of boron that has to be present initially?

7.3. Suppose that the substrate in Prob. 7.2 is already lightly doped with phosphorus at 10^{14} atoms/cm^3 and that a *pn* junction is to be formed by carrying out the drive-in at the same temperature as in Prob. 7.2 at a distance 2 μm away from the surface. Determine the time required for the drive-in.

7.4. The sheet resistivity is often expressed as follows:

$$R_s = \frac{1}{q\mu_e \int_0^{x_i} C\, dx} \tag{A}$$

where the definition of the effective mobility μ_e follows from Eq. (A) and Eq. (7.23). When the dopant surface concentration, which is approximately 2.54×10^{20} atoms/cm^3 in Prob. 7.3, is much higher than the background doping (10^{14} in Prob. 7.3), the integral in Eq. (A) is approximately the same as Q_o if one excludes the background doping. Therefore, Eq. (A) can be approximated as follows:

$$R_s = \frac{1}{q\mu_e(Q_o - x_j C_B)} \tag{B}$$

where C_B is the background dopant concentration. Note that the concentration in Eq. (A) is that for electrically active dopant, i.e., the concentration minus C_B. Suppose for the sample in Prob. 7.3 that a four-point probe measurement yields an average resistivity of 0.02 Ω·cm. Calculate the effective hole mobility μ_e for the sample.

7.5. When the diffusivity is constant, measurements of junction depth and background doping concentration (from resistivity) are used to determine the constant diffusivity. Suppose a predeposition is carried out for 20 minutes into a background doping of 10^{16} atoms/cm^3 and the junction depth is 0.5 μm for the surface dopant concentration of 10^{20} atoms/cm^3. Determine the constant diffusivity.

7.6. For ion implantation of boron into silicon at 100 keV, determine the maximum ion concentration for a dose of 10^{15} cm^{-2}. Where does the maximum occur and what is the spread of ion distribution in the direction perpendicular to the incident direction compared to that in the incident direction?

7.7. Suppose that the sample in Prob. 7.6 is annealed for 30 minutes at 1000 °C ($D = 10^{-13}$ cm^2/s). Calculate the junction depth for a background doping of 10^{15} atoms/cm^3.

7.8. For silicon, the displacement energy of a lattice atom is approximately 14 eV. Calculate the incident energy for nuclear stopping E_n for Sb using Table 7.5. Obtain a relationship between E_n and the incident energy. Your result should show that the fraction of the incident energy consumed by nuclear stopping is small. Assuming that the incident energy is entirely spent by electronic stopping, obtain a relationship for the range.

7.9. The analysis of isothermal annealing data can be carried out in the same manner as for reaction kinetics by integral or differential analysis (Levenspiel, 1972). In the annealing literature, however, a method similar to the half-life method has been used (Ryssel and Kranz, 1975). Let t_g be the time at which the electrically active ion concentration reaches g times the maximum concentration C_m. One can get a table of t_g versus temperature from various isothermal runs. From Eq. (7.40), one has

$$G_1 = kt_g; \qquad G_1 = \ln \frac{1}{1-g} \qquad \text{for } n = 1 \tag{A}$$

$$G_n = kt_g; \qquad G_n = \frac{C_m^{1-n}}{n-1}\left[(1-g)^{1-n} - 1\right] \qquad \text{for } n \neq 1$$

since C_o is negligible. Equation (A) can be rewritten as

$$\ln G_i = \ln k_0 + \ln (t_g) - \frac{E}{k_B T} \qquad \text{for } i = 1 \text{ or } n$$

or

$$\ln (t_g) = \ln \left(\frac{k_0}{G_i}\right) - \frac{E}{k_B T} \tag{B}$$

According to Eq. (B), a plot of ln (t_g) versus $1/T$ should give the activation energy from the slope. Ryssel and Kranz (1975) gave the following data for the annealing of As in Si for $g = 0.9$:

t_g, min	23	205	1000
$1000/T$, K^{-1}	1.315	1.385	1.426

Determine the activation energy. Discuss ways of determining the order n.

7.10. According to Yeh and Armstrong (1961), the surface concentration of boron in silicon is 2.5×10^{20} atoms/cm^3 when the substrate is exposed to a B_2O_3 partial pressure of 0.01 torr at 1100 °C in the region where Henry's law applies. According to our definition of K_e, we have

$$C_s = K_e(C_s)_g$$

Determine the equilibrium constant K_e. Typical values of k_m and epitaxial growth rate are 5 cm/s and 1 μm/min, respectively. Determine the dopant concentration in the bulk gas flow that will make W in Eq. (7.52) larger than 0.1. Assume that the initial concentration in the substrate is 10^{16} atoms/cm^3.

7.11. For the dopant, the profile resulting from a predeposition can be written as follows:

$$-2\beta \frac{dC}{d\beta} = \frac{d}{d\beta}\left(D \frac{dC}{d\beta}\right) \qquad \text{(A)}$$

which is the fourth entry in Table 7.4. Derive Eq. (7.22) from Eq. (A).

7.12. Only a fraction of implanted ions is activated by annealing unless the ion dose is low (e.g., less than 10^{16} cm^{-2} for As in Si) or the annealing temperature is high (e.g., greater than 1000 °C). According to the model proposed by Tsai *et al.* (1980), arsenic atoms form clusters of three atoms that are partially active when the concentration is larger than 10^{20} cm^{-3}. Relationships for the clusters are proposed as follows:

$$3As^+ + e^- \underset{}{\overset{\text{high } T}{\rightleftharpoons}} {As_3}^{2+} \xrightarrow{\text{room } T} As_3 \qquad \text{(A)}$$

The equilibrium for the first path is

$$K_{eq} = \frac{[{As_3}^{2+}]}{n[As^+]^3} \qquad \text{(B)}$$

The carrier concentration at the annealing temperature is

$$n = [As^+] + 2[{As_3}^{2+}] \qquad \text{(C)}$$

The equilibrium constant K_{eq} is given by

$$K_{eq} = 1.26 \times 10^{-70} \exp\left[\frac{2.06(\text{eV})}{kT}\right] \qquad \text{(D)}$$

Show that the total amount of ions implanted per unit volume N_v is given by

$$N_v = x + \frac{3K_{eq}X^4}{1 - 2K_{eq}X^3} \qquad \text{(E)}$$

where X is the concentration of As^+. For N_v of 1.84×10^{20} atoms/cm^3, calculate $[As^+]$ at 800 and 1000 °C.

REFERENCES

Attala, M. M., and E. Tannenbaum: *Bell Syst. Tech. J.*, vol. 39, p. 933, 1960.

Auston, D. H., J. A. Golovhenko, A. L. Simons, R. E. Shlusher, P. R. Smith, C. M. Surco, and T. N. C. Venkatesan: in S. D. Ferris, H. J. Leamy, and J. M. Poate (eds.), *Laser Solid Interactions and Laser Processing 1978*, p. 11, American Institute of Physics, New York, 1979.

Basseches, H., R. C. Manz, C. O. Thomas, and S. K. Tung: in J. B. Schroeder (ed.), *AIME Semiconductor Metallurgy Conference, LA*, vol. 15, Interscience Publishers, New York, 1961.
Bicknell, R. W.: *Phil. Mag.*, vol. 26, p. 273, 1972.
Biersack, J. P.: *Nucl. Inst. Heth.*, vol. 182/183, p. 199, 1981.
Blamires, N. G.: *European Conference on Ion Implantation*, p. 52, Reading, Stevenage, England, 1970.
Bloemberger, N.: in S. D. Ferris, H. J. Leamy, and J. M. Poate (eds.), *Laser Solid Interactions and Laser Processing 1978*, p. 52, American Institute of Physics, New York, 1979.
Boltaks, B. J.: *Diffusion in Semiconductors*, Academic Press, New York, 1963.
Cappelani, F., G. Restelli, and I. Spinoni: *J. Phys. C.: Solid State Phys.*, vol. 1, p. 650, 1974.
Carslaw, H. S., and J. C. Jaeger: *Conduction of Heat in Solids*, Oxford University Press, 1959.
Colby, J. W., and L. E. Katz: *J. Electrochem. Soc.*, vol. 123, p. 409, 1976.
Fair, R. B.: in F. F. Y. Wang (ed.), *Impurity Doping Processes in Silicon*, North-Holland, New York, 1981.
——— and J. C. C. Tsai: *J. Electrochem. Soc.*, vol. 122, p. 1689, 1975.
Frank, W., U. Gosele, H. Mehrer, and A. Seager: in G. E. Murch and A. S. Nowick (eds.), *Diffusion in Crystalline Solids*, chap. 2, Academic Press, New York, 1984.
Furukawa, S., H. Matsumura, and H. Ishiwara: *Jap. J. Appl. Phys.*, vol. 11, p. 134, 1972.
Ghandhi, S. K.: *VLSI Fabrication Principles*, Wiley-Interscience, New York, 1983.
——— and R. J. Field: *Appl. Phys. Lett.*, vol. 38, p. 267, 1981.
Gibbon, C. F., E. J. Povilonis, and D. R. Ketchow: *J. Electrochem. Soc.*, vol. 119, p. 767, 1972.
Gise, P. E., and R. Blanchard: *Semiconductor and Integrated Circuit Fabrication Techniques*, Reston Publishing Company, Reston, Va., 1979.
Grove, A. S., O. Leistiko, and C. T. Sah: *J. Appl. Phys.*, vol. 35, p. 2695, 1964.
———, A. Roder, and C. T. Sah: *J. Appl. Phys.*, vol. 36, p. 802, 1965.
Hofker, W. K.: Phillips Research Reports Supplement 8, 1975.
——— and J. Politiek: *Philips Tech. Rev.*, vol. 39, p. 1, 1980.
Hu, S. M., and S. Schmidt: *J. Appl. Phys.*, vol. 39, p. 4272, 1968.
Irvin, J. C.: *Bell Syst. Tech. J.*, vol. 41, p. 387, 1962.
Ishiwara, H., S. Furukawa, J. Yamada, and M. Kawamura: in S. Namba (ed.), *Proceedings on Ion Implantation in Semiconductors*, Plenum Press, New York, 1975.
Kennedy, D. P., and R. R. O'Brien: *IBM J. Res. Dev.*, vol. 9, p. 3, 1965.
Kinchin, G. H., and R. S. Pease: *Rep. Prog. Phys.*, vol. 18, p. 1, 1955.
Lee, H. H.: *J. Electrochem. Soc.*, vol. 133, p. 2416, 1986.
Levenspiel, O.: *Chemical Reaction Engineering*, 2d ed., Wiley, New York, 1972.
Lin, A. M. R., D. A. Antoniadis, and R. W. Dutton: *J. Electrochem. Soc.*, vol. 128, p. 1131, 1981.
Lindhard, J., and H. Scharff: *Phys. Rev.*, vol. 124, p. 128, 1961.
———, M. Scharff, and H. Schiott: *Kgl. Dan. Vidensk. Selsk. Mat.-Fys. Medd.*, vol. 33, p. 14, 1963.
Morehead, F. F., and B. L. Crowder: in F. H. Eisen and L. T. Chadderton (eds.), *First International Conference on Ion Implantation, Thousand Oaks*, Gordon and Breach, New York, 1971.
Novak, R. L.: *Bull. Am. Phys. Soc.*, vol. 8, p. 235, 1965.
Prince, J. L., and F. N. Schwettmann: *J. Electrochem. Soc.*, vol. 121, p. 705, 1974.
Runyan, W. R.: *Semiconductor Measurements and Instrumentation*, McGraw-Hill, New York, 1975.
Ryssel, H., and H. Kranz: *Appl. Phys.*, vol. 7, p. 11, 1975.
——— and I. Ruge: *Ion Implantation*, Wiley-Interscience, New York, 1986.
———, K. Muller, K. Haberger, R. Henkelmann, and F. Jahael: *Appl. Phys.*, vol. 22, p. 35, 1980.
Seeger, A., and K. P. Chick: *Phys. Stat. Sol.*, vol. 29, p. 455, 1968.
Sedgewick, T.: *Reduced Temperature Processing for VLSI*, p. 49, The Electrochemical Society, Pennington, N.J., 1986.
Steidel, T. E.: in S. M. Sze (ed.), *VLSI Technology*, chap. 6, McGraw-Hill, New York, 1983.
Shaw, D. (ed.): *Atomic Diffusion in Semiconductor*, Plenum, New York, 1973.
Sigmund, P., and J. B. Sanders: in P. Glotin (ed.), *Proceedings of International Conference on Applied Ion Beams Semiconductor Technology*, p. 215, Grenoble, 1967.
Smith, B.: "Ion Implantation Range Data for Silicon and Germanium Development Technology," Research Studies, Forest Grove, Oregon, 1977.

Srinivasan, G. R.: *J. Electrochem. Soc.*, vol. 127, p. 1334, 1980.

Taniguchi, K., Kurosawa, and M. Kashiwagi: *J. Electrochem. Soc.*, vol. 127, p. 2243, 1980.

Townsend, P. D., J. C. Kelly, and N. E. W. Hartly: *Ion Implantation, Sputtering and Their Applications*, Academic Press, 1976.

Tsai, J. C. C.: in S. M. Sze (ed.), *VLSI Technology*, chap. 5, McGraw-Hill, New York, 1983.

———, F. F. Morehead, and J. E. E. Baglin: *J. Appl. Phys.*, vol. 51, p. 3230, 1980.

Tuck, B.: *Introduction to Diffusion in Semiconductors*, vol. 16, p. 119, IEE Monograph Services, London, 1974.

Webber, R. F., R. S. Thorm, and L. N. Large: *Int. J. Electronics*, vol. 26, p. 163, 1969.

Yeh, T. H., and W. Armstrong: Electrochemical Society Meeting, Abstract 69, Indianapolis, Spring 1961.

CHAPTER 8

PATTERN GENERATION, TRANSFER, AND DELINEATION

8.1 LITHOGRAPHY

Design of an integrated circuit eventually leads to specification of the circuit elements in terms of the length, width, and depth of each element such as doped regions, isolated (insulated) regions, conducting regions, and so on. The "blueprint" containing the specifications and the layout of the device elements is referred to as composite layout. Elements overlay in the device structures considered in Chap. 1 (e.g., Fig. 1-13), so it is necessary to specify the order that overlaying elements are to be processed for device fabrication and also specify the patterns of the layout corresponding to each step of the fabrication.

The minimum number of patterns required for fabrication is the number of mask levels. Each level represents the mask that can be used to transfer the mask pattern to the wafer surface. Figure 8-1 shows a composite layout of a minimum geometry bipolar transistor (Colclaser, 1980). The corresponding mask levels are shown in Fig. 8-2 in the order to be fabricated. Note that the first mask level is for the mask defining the pattern for the buried layer, although the dimensions are not given. The masks in Fig. 8-2 are to the scale of the composite layout in Fig. 8-1. The second mask is for isolating the buried layer from the rest of the device structure to be fabricated by subsequent processing. An examination of all

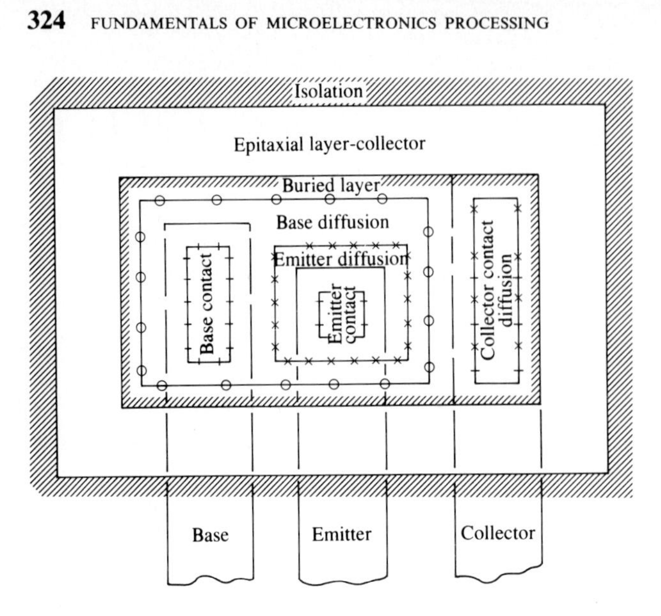

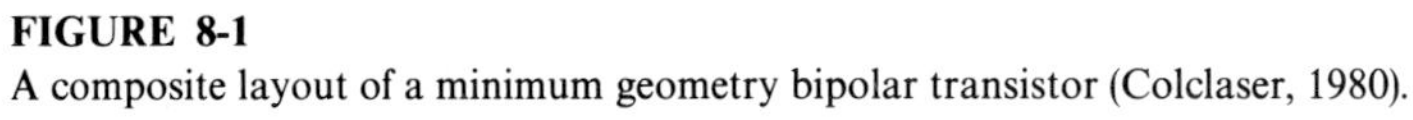

FIGURE 8-1
A composite layout of a minimum geometry bipolar transistor (Colclaser, 1980).

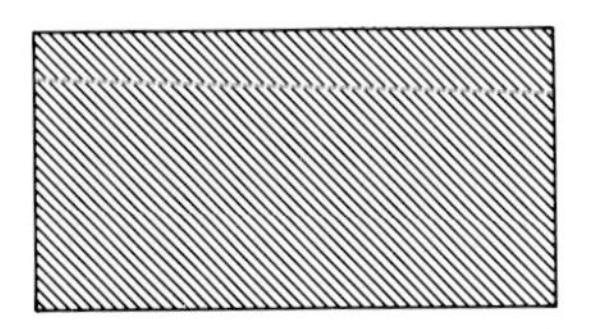

(*a*) Buried layer

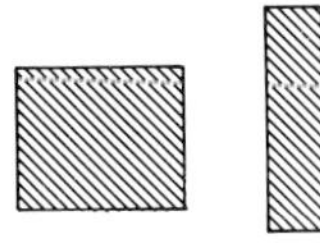

(*d*) Emitter

(*e*) Contacts

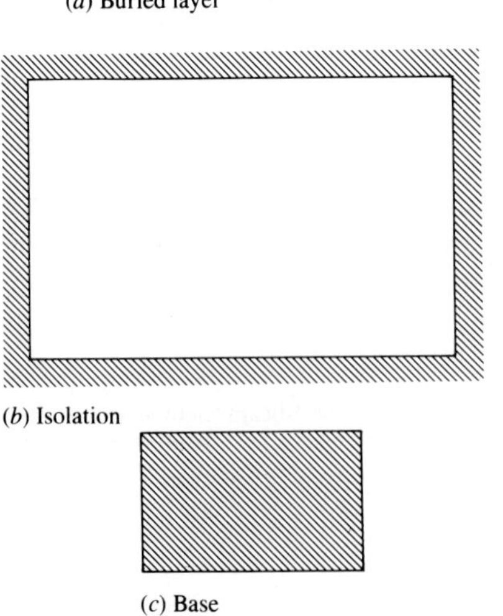

(*b*) Isolation

(*c*) Base

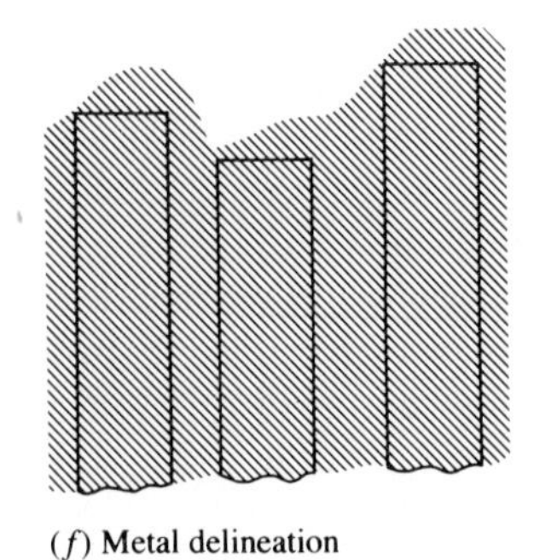

(*f*) Metal delineation

FIGURE 8-2
Individual masks for the transistor of Fig. 8-1 (Colclaser, 1980).

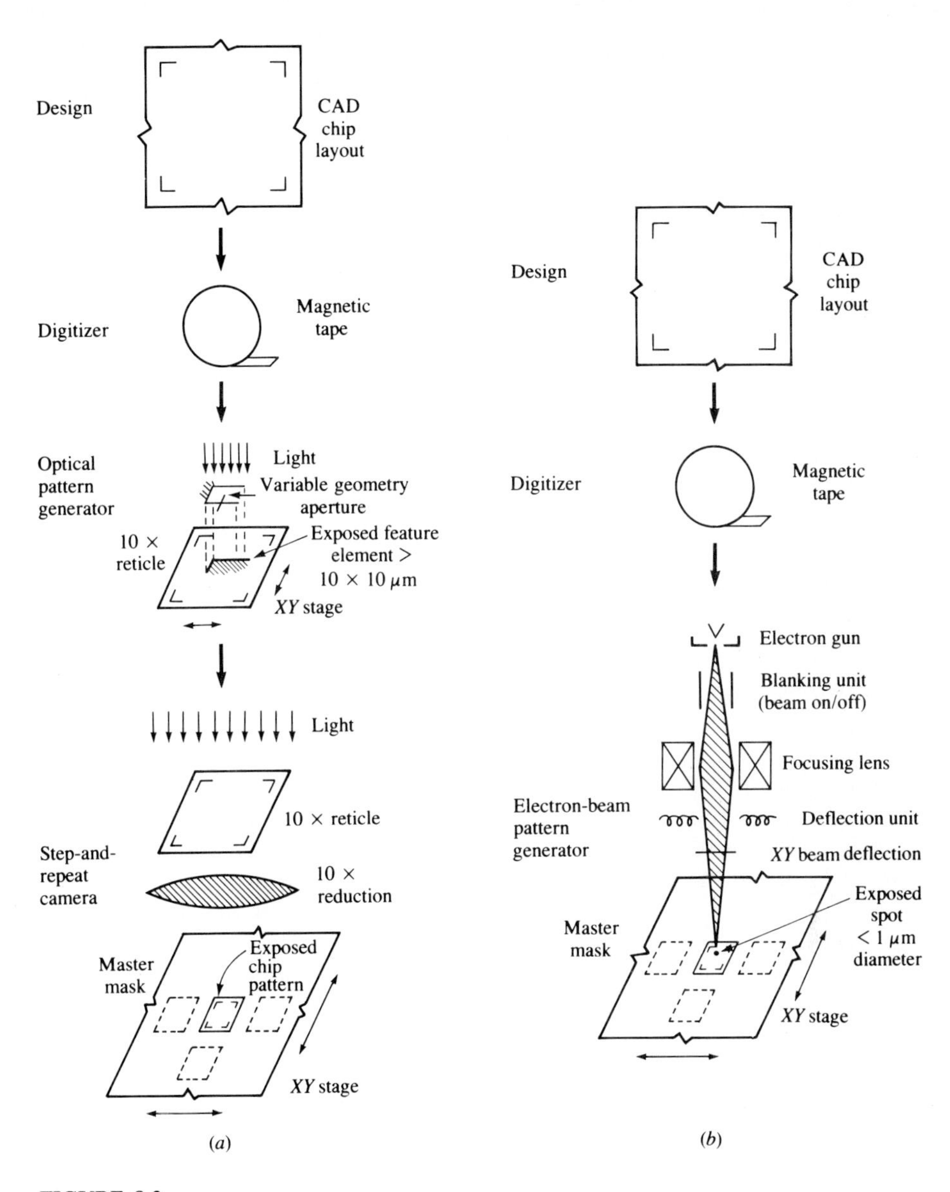

FIGURE 8-3
(*a*) Schematic representation of photolithographic mask fabrication; (*b*) e-beam method (Ballantyne, 1980).

the mask levels in Fig. 8-2 should reveal that the device fabrication proceeds, in general, from the lowest to the uppermost layer of the fabricated device, the uppermost being the metal delineation representing the metallization pattern.

The composite layout and the mask patterns were once made by drawing an oversized picture and then reducing the size successively by photographic reduction to the actual miniature scale. Today computer-aided design (CAD) is used primarily with the aid of a software library to design the circuit on a CRT. Then the final design is digitized and stored on a tape to generate the composite layout and mask patterns. These patterns are registered on masks made from glass plates coated with either an emulsion or a hard film of chromium, chromium oxide, or silicon. The process of transferring a pattern onto a substrate, which can be a wafer or a mask material, is called lithography. The mask material of the coated glass plate with the transferred pattern is called a master mask or simply mask.

The mask can be fabricated by photolithographic or electron-beam (e-beam) methods. Schematic examples of the methods are shown in Fig. 8-3. The pattern is usually delineated either photographically, in high-resolution emulsion on glass plates, or photolithographically (Fig. 8-3*a*) on chromium films on glass plates. Both these and the e-beam methods involve the use of a resist. A resist is a material that is sensitive to either light or e-beam such that a pattern can be engraved into the resist film, in much the same way as conventional photographic development. Because of the poor resolution of lines inherent in thick emulsions (4 to 6 μm), chromium films are preferred. The mask fabrication is similar to the usual lithography involved in device fabrication. However, better resolution is afforded by the much thinner resist film and the absence of steps (thus planar surface) in the mask fabrication.

Although the resist is also used in the mask fabrication, the name is derived from the fact that the portion of the surface covered by the resist in device fabrication is protected from (resistant to) penetration of undesired materials to the protected surface, whether the penetration is by diffusion, chemical reaction, or ion implantation. Almost all resists are of a polymeric material, which upon sensitization by light, x-ray, or e-beam, changes its chemical structure in such a way that the sensitized portion either dissolves (positive resist) or remains intact (negative resist) when placed in a suitable solvent. These two types of resist are shown in Fig. 8-4, which illustrates the e-beam mask fabrication.

Once the masks are made, the resist material is spin-coated onto the wafer. Then the resist on the wafer is exposed to a sensitizing source through the mask so that the mask pattern is transferred onto the resist. The resist film thickness l resulting from the spin coating (Thompson and Bowden, 1983) is often correlated as follows:

$$l = \frac{KC^{\beta}\mu^{\gamma}}{S^{\alpha}} \tag{8.1}$$

where C is the concentration of the resist solution (percent solids), S is the spinning speed (r/min), K is a constant, μ is the viscosity on the polymeric material, and α, β, and γ are constants. The sensitized resist is then developed in a solvent so that the desired pattern emerges after the development for the actual device fabrication steps such as film deposition, ion implantation, and so on, that are specified by the mask and masking level. In essence, repetition of lithographic processing constitutes the entire device fabrication procedure as shown in Fig.

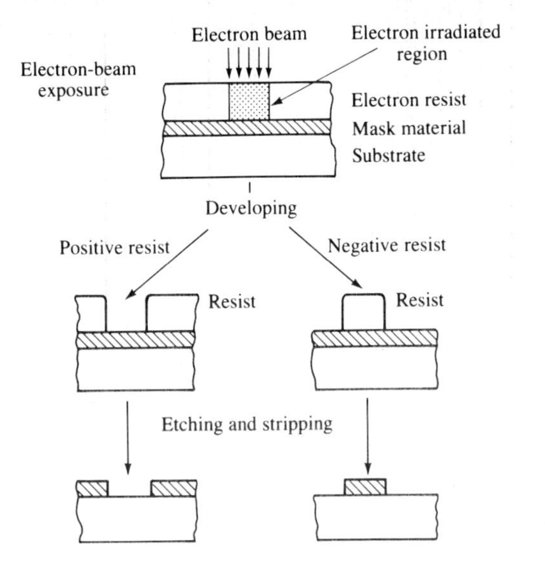

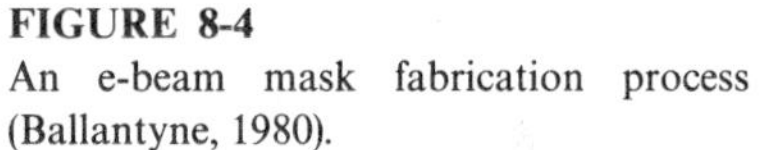

FIGURE 8-4
An e-beam mask fabrication process (Ballantyne, 1980).

8-5. The portion within the box represents the processing steps for the lithography, each cycle representing one mask level. Each time the cycle is repeated, the new mask pattern has to be aligned to the pattern already present on the wafer.

An analysis of the growth trend in the number of components per chip over a 14-year period prior to 1975 (Moore, 1975) indicates an increase by a factor of 64,000, of which a factor of 32 was attributed to improvements in lithography, a factor of 20 to the use of larger chips, and the remaining factor of 100 to improved circuit design and layout. Obviously lithography has been a key to the miniaturization of integrated circuits. More important, however, the ultimate limitation to miniaturization, even when inherent limitations in device design are resolved, is the ability to clearly delineate high-resolution patterns by lithography.

Photolithography based on ultraviolet light (300 to 400 nm wavelength) has been the main lithography method and can deliver nearly 0.5 μm resolution under practical conditions. Electron beam lithography, which is typically being used for mask fabrication, is likely to be the main means by which practical lithography is carried out in the future. Lithography based on soft x-rays (0.1 to 5 nm wavelength), which almost eliminates the diffraction effects that limit the resolution in photolithography, suffers from poor contrast and low energy but could fill the gap between photolithography and e-beam lithography. The major problem in e-beam lithography is electron scattering. This has led to lithography based on ions or ion-beam lithography. Although there are still major problems yet to be solved, it has already been demonstrated in laboratory work that 10 nm resolution is possible even with e-beam.

The basics governing the various lithographies are very complex. Nevertheless, an understanding of some basics of optics and particle beams, theoretical and approximate, is needed for subsequent chapters. These principles are treated next.

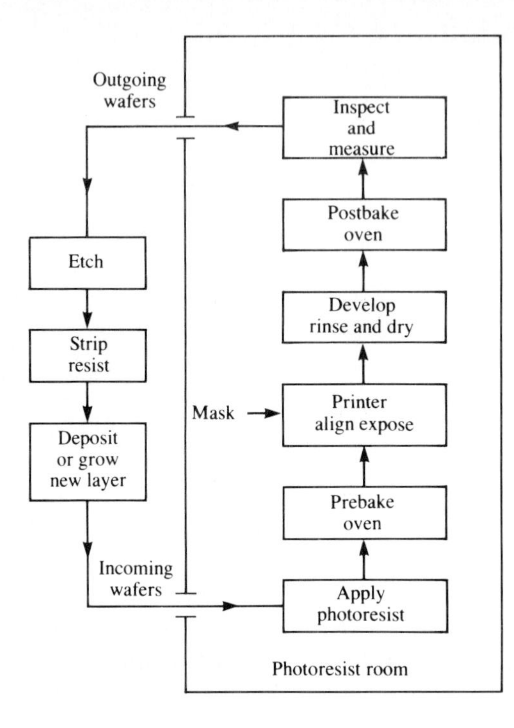

FIGURE 8-5
Lithographic processing sequence (McGillis, 1983).

8.2 BASICS OF OPTICS AND PARTICLE BEAMS

When a beam of light strikes a flat, shiny surface (a plane mirror), it is reflected. The law of reflection is that the incidence angle (the angle between the incident beam and a line normal to the mirror surface) is equal to the reflection angle (the angle between the reflected beam and the normal line). Light also travels through a transparent material. When a beam of light enters a transparent material at an oblique angle to its surface, the phenomenon of refraction occurs, i.e., bending of the beam. This is due to the fact that the light always travels in a material at a speed less than the speed of light in vacuum. A measure of this speed is called the refractive index n:

$$n = \frac{C}{v} \tag{8.2}$$

where C is the speed of light in vacuum and v is the speed of light in the transparent material. The law of refraction or Snell's law states that when a light beam passes from one medium (with refractive index n and incidence angle Ω) to another (with refractive index n' and refraction angle Ω', which is the angle between the refractive beam and a line normal to the interface), it satisfies the following relationship:

$$n \sin \Omega = n' \sin \Omega' \tag{8.3}$$

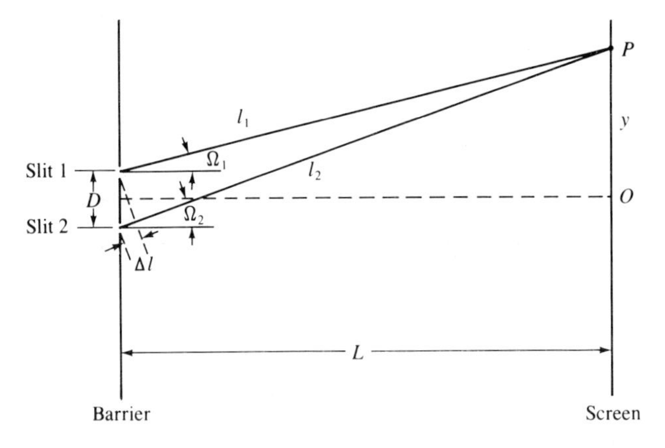

FIGURE 8-6
Two-slit diffraction behavior (Eisberg and Lerner, 1981).

The most important optical behavior in photolithography is diffraction, the phenomenon of bending around corners due to the wave property of light. Consider the diffraction pattern that results when a light beam passes through two slits in a barrier, as shown in Fig. 8-6. The light travels to the right with speed v and wavelength λ. The wavelength is the distance between two crests with a trough in between in a wave. As the light wave passes through the two slits, it diffracts. Consider two particular components of the diffracted beam out of the two slits, one reaching the point P and another reaching the point O on the screen in Fig. 8-6. The diffracted beam out of the slit 1 that reaches the point O travels the same distance as that out of the slit 2 that reaches the point O. Therefore, the two crests out of the two slits are in phase or superpose constructively at the point O. Suppose, on the other hand, the diffracted beam out of the slit 2 travels a length of l_1 and that of the slit 1 a length of l_2 such that the difference, $(l_2 - l_1)$, is one half the wavelength λ. Then the crest of the wave, for instance, out of the slit 1 arrives at the point P whereas the trough of the wave out of the slit 2 arrives there, the latter being out of phase by one half the wavelength. Therefore, their magnitudes cancel each other and they superpose destructively. In general, therefore, destructive superposition (node) occurs whenever the path difference Δl $(= l_2 - l_1)$ satisfies the following condition:

$$\Delta l = (j + \tfrac{1}{2})\lambda \qquad \text{for } j = 0, \pm 1, \pm 2, \pm 3, \ldots \tag{8.4a}$$

Constructive superposition (antinode) takes place when the following condition is satisfied:

$$\Delta l = j\lambda \tag{8.4b}$$

When the distance L between a barrier with slits and a screen is much larger than the slit separation d (refer to Fig. 8-6), the diffraction is called Fraunhofer diffraction; otherwise it is called Fresnel diffraction. In the case of Fraunhofer diffraction, the paths of lengths l_1 and l_2 are very nearly parallel so that the angles Ω_1

and Ω_2 are very nearly equal. Then one has with the angle Ω in Fig. 8-6:

$$\Delta l = d \sin \Omega \tag{8.5}$$

Further, $\cos \Omega$ is nearly unity for very small Ω such that $\sin \Omega = \tan \Omega$ $(= y/l)$, leading to the following approximate relationship:

$$y = \frac{L(j + \frac{1}{2})\lambda}{d} \quad \text{(minima or nodes)} \tag{8.6a}$$

and

$$y = \frac{Lj\lambda}{d} \quad \text{(maxima or antinodes)} \tag{8.6b}$$

The same relationship holds (Jenkins and White, 1976) for multislit (N slits) diffraction for the maxima. An important difference, however, is that amplitude of the maxima is much greater for the multislit case. This is due to the fact that N waves are constructively superposed instead of only two. Also, the distance from a maximum to the next minimum is smaller than it is when there are two slits by the factor of $2/N$. The Fraunhofer diffraction patterns for two slits and six slits are shown in Fig. 8-7 for the same slit separation, screen distance, and wavelength for light of intensity I. The Fraunhofer diffraction limits the resolution of projection photolithography.

The Fresnel diffraction limits the resolution of proximity photolithography in which the mask is close to the wafer surface, as illustrated in Fig. 8-8. The

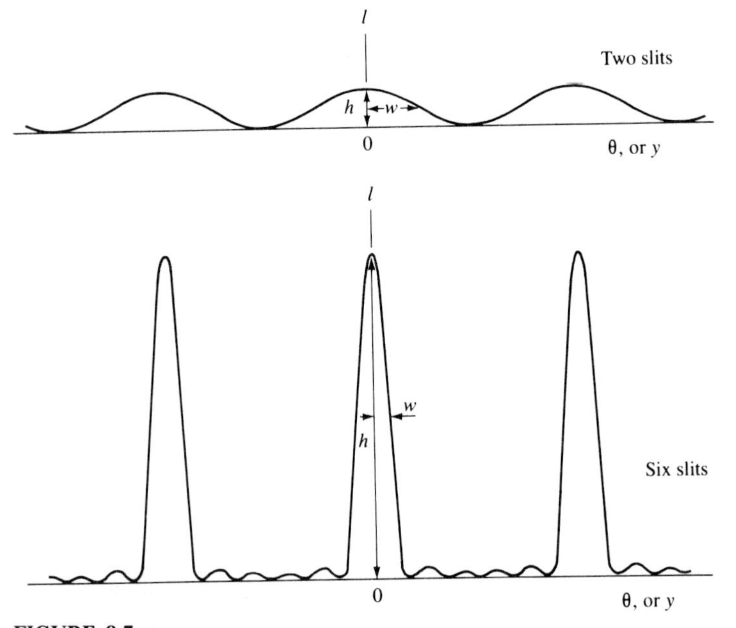

FIGURE 8-7
Diffraction patterns for two and six slits (Eisberg and Lerner, 1981).

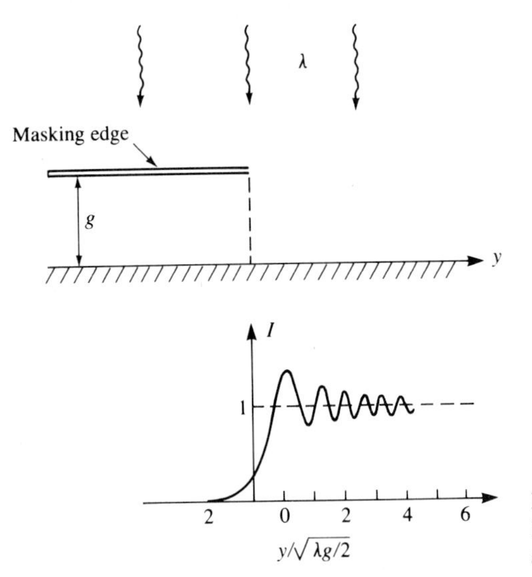

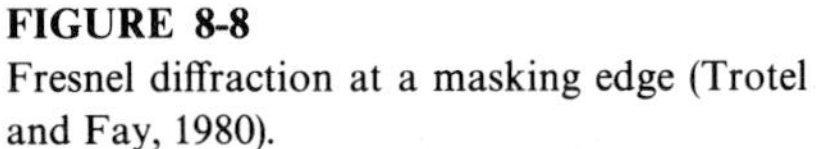

FIGURE 8-8
Fresnel diffraction at a masking edge (Trotel and Fay, 1980).

intensity of the beam on the wafer surface depends on the dimensionless variable, $y/(\lambda g/2)^{1/2}$ (e.g., Jenkins and White, 1976), if the focal length is taken as the gap g. Here λ is the coordinate shown in Fig. 8-8.

As with any electromagnetic wave, the intensity of a light wave is proportional to the square of its amplitude. The amplitude of the maxima of an N-slit diffraction pattern is proportional to N because of N combining waves that superpose in phase. Therefore, the intensity of the maxima is proportional to N^2. The fact that the distance from a maximum to the next minimum is proportional to $1/N$ together with the same maxima positions as indicated by Eq. (8.6b) for arbitrary N means that the width of the maxima is proportional to $1/N$. Although w in Fig. 8-7 is called simply the width, its full name is half-width at half-maximum intensity. A multislit device with a large value of N is called a diffraction grating and the spacing d between its slits is called the grating spacing. Combination of Eqs. (8.4) and (8.5) leads to the grating formula:

$$d \sin \Omega = j\lambda \tag{8.7a}$$

For the general case of a beam incident in arbitrary angle i, the grating formula (Jenkins and White, 1976) is

$$d(\sin i + \sin \Omega) = j\lambda \tag{8.7b}$$

where the integer j is known as the order.

Constructive and destructive superposition, or interference, also take place when a light beam passes through a medium and then is reflected at the interface adjoining the medium to another medium, e.g., at the interface between a resist and a substrate in photolithography. This interface also leads to standing waves as in diffraction, as shown in Fig. 8-9. It can be shown (Cuthbert, 1977) for perfect reflection at the interface (substrate surface in Fig. 8-9) that the antinodes

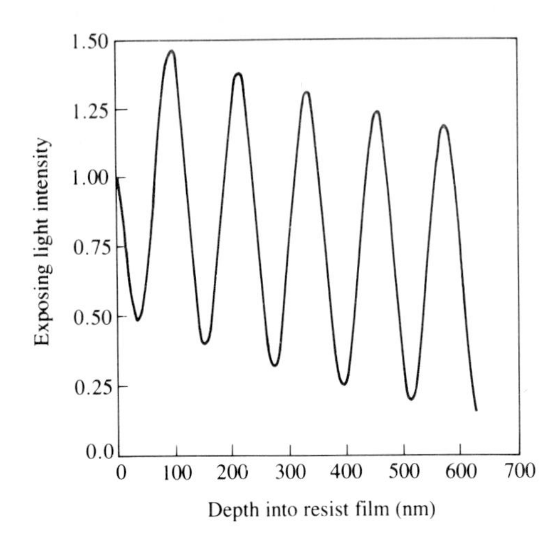

FIGURE 8-9
Light intensity distribution (standing wave) in a resist (Dill *et al.*, 1975a).

(maxima) and nodes (minima) are given by

$$n(l-z)=\begin{cases}\dfrac{(2j-1)\lambda}{4} & \text{(maxima)} \quad (8.8a)\\[2ex] \dfrac{j\lambda}{2} & \text{(minima)} \quad (8.8b)\end{cases} \qquad \text{for } j=1,2,3,4,\ldots$$

where l is the resist film thickness and z is the distance into the resist.

When the ratio of wavelength to slit width (λ/d) is very small, the diffraction effect is also very small and most of the light is not diffracted. In such cases, ray optics, as opposed to the wave optics considered so far, can be used. The basic law of ray optics is that light travels in straight rays, providing it travels through uniform material. This ray optics (also called geometric optics) is the basis for describing image formation by lenses. Central to the image formation is a quantity called focal length f shown in Fig. 8-10. All paraxial rays (rays that are near the centerline axis) emitted from a point on an "object" plane parallel to the plane of a converging lens will converge to a point on another parallel plane, the image plane, no matter where the point of emission is located. This occurs when the object distance, s, is greater than the focal length, f. The relationship between the object and image distances (s and s') and the focal length f in Fig. 8-10 is (e.g., Eisberg and Lerner, 1981)

$$\frac{1}{s}+\frac{1}{s'}=\frac{1}{f} \tag{8.9}$$

Note that the magnification (h'/h in the figure) can be determined from s and s' using triangle similarity.

There are a number of quantities obtainable from the focal length and the lens aperture D. The numerical aperture (NA) often used in describing image formation is given by $D/2f$ and the F number is given by f/D. For the role of the numerical aperture, consider the refractive lens imaging system shown in Fig.

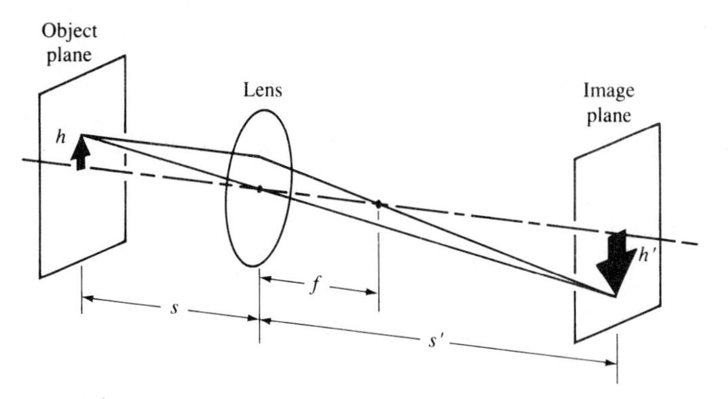

FIGURE 8-10
Image formation by a lens (Eisberg and Lerner, 1981).

8-11 (Thompson and Bowden, 1983). When light passes through a mask and has been diffracted, it is next imaged onto a wafer using an objective lens. The aperture in the figure is such that the objective lens can collect light from angles smaller than $2\alpha_o$, where α_o is the maximum cone angle of rays reaching the lens from the object point (mask). For a given magnification, the numerical aperture $(\mathrm{NA})_o$ is related to α_o as follows:

$$(\mathrm{NA})_o = n \sin \alpha_o$$

where n is the refractive index in image space and is usually equal to unity. Similarily, the numerical aperture of the condenser lens in Fig. 8-11 is given by

$$(\mathrm{NA})_c = n \sin \alpha_c$$

Numerical aperture is thus a measure of the "acceptance" angle of a lens. It is important to recognize that images are formed by the intersection of rays of light

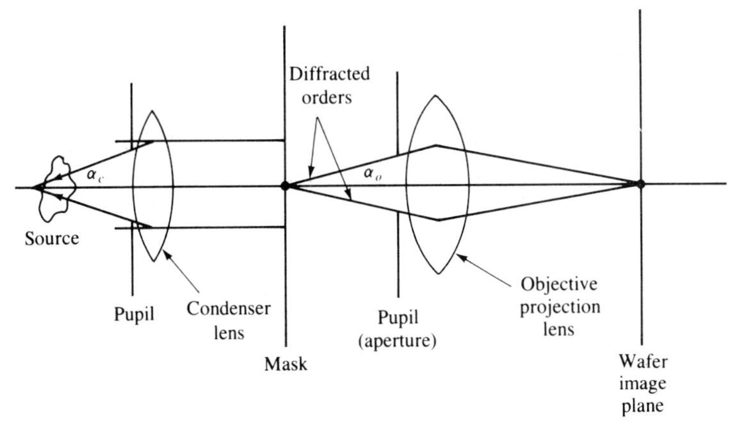

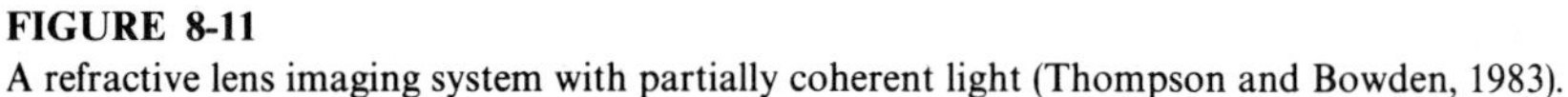

FIGURE 8-11
A refractive lens imaging system with partially coherent light (Thompson and Bowden, 1983).

which emanate from a point (Fig. 8-10) and that the pattern information is contained in the diffracted light (Fig. 8-11). The undiffracted or zero-order beam [$j = 0$ in Eq. (8.7*a*)] from an edge or grating constitutes only a single ray and at the very least a second ray emanating from the edge is needed to reconstruct an image of the edge. Therefore, the larger the numerical aperture of the projection lens is, the greater is the amount of diffracted information that can be collected and subsequently imaged.

A light beam is termed coherent when the angular range of light waves incident on the barrier with slits is small. Completely coherent illumination occurs when all beams pass through a slit in one angle. A beam is termed incoherent when the angular range of incident waves is large. Coherency is quantified in terms of the degree of coherency (Thompson and Bowden, 1983). It is defined in Fig. 8-12 for two common types of illumination where M is the magnification and the subscripts o, c, and s are for objective lens, condenser, and source, respectively. The degree of coherency is zero for a coherent source since it is a point

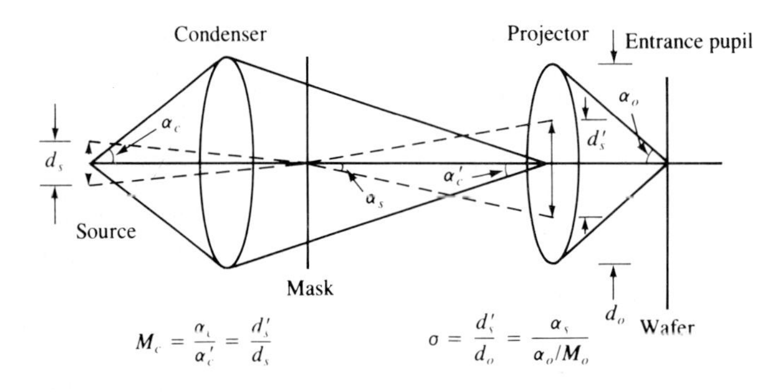

(*a*) Kohler

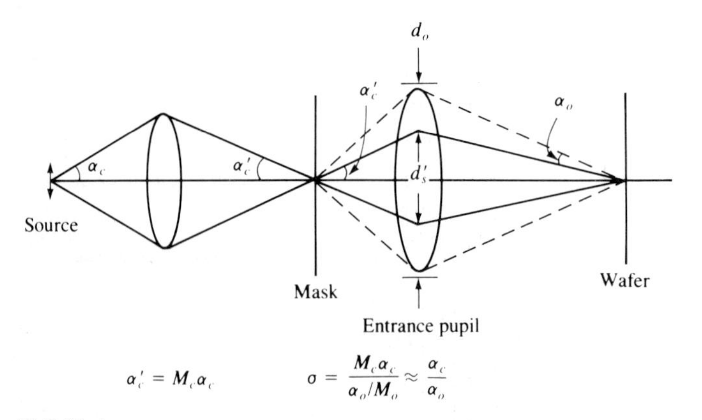

(*b*) Critical

FIGURE 8-12
Degree of coherency for two common types of illumination system (Thompson and Bowden, 1983).

source, i.e., zero diameter. The degree of coherency is infinite for an incoherent source. In reality, however, $\sigma \leq 1$ since all light collected from any real source is always imaged within the entrance pupil. In practice, partially coherent sources are used.

The basic laws governing light optics also apply to electron optics provided the square root of electrical potential is used in place of the index of refraction. The index of refraction is a measure of light velocity; the square root of the potential is a measure of electron velocity. In light optics, the brightness or luminous intensity is defined as the flux per unit solid angle emitted by a luminous source. By analogy the brightness B of a source of charged particles is defined as the current density J per unit solid angle Ω:

$$B = \frac{J}{\Omega} \tag{8.10}$$

If the current is emitted from (or converges toward) a small area through a cone of included half-angle α, the brightness can be approximated (Herriott and Brewer, 1980) by

$$B = \frac{J}{\pi\alpha^2} \tag{8.11}$$

if α (in radians) is small.

Electrons or a beam of electrons can be generated (extracted) from a crystalline material by giving sufficient energy to the conduction electrons such that they may overcome the potential well of the crystal lattice. Simple heating of a metal accomplishes this extraction and is called thermionic emission, the basic source of electrons for most practical vacuum electronic devices. Extraction can also be accomplished by applying an intense external electric field to the metal surface and this is termed field emission. When a beam of electrons is focused and directed to the surface of a resist material for imaging, the electrons entering the resist are scattered by interaction (collisions) with the atoms of the resist material. This scattering can be divided roughly into two classes: forward and backward scattering. Since most of the electrons (Greeneich, 1980) are forward scattered through small angles (less than 90°) from their original direction, this effect merely broadens the incident beam. Some electrons experience large-angle scattering (approaching 180°), causing these electrons to return to the surface. Electrons are scattered both elastically and inelastically. Elastic scattering results only in a change of direction of the electrons while inelastic collisions result in energy loss. Consequently, the incident electrons will spread out as they penetrate until all of their energy is lost or until they exit the solid as a result of backscattered deflections. The simulation of electron trajectories is very complex and a Monte Carlo method (Kyser and Viswanathan, 1975) is available for this purpose. The simulation results for 100 electrons are shown in Fig. 8-13. Note that most of the energy is absorbed in the silicon substrate.

For approximate results, it is sufficient to know the distance electrons penetrate and their radial distribution at any depth. An approximate maximum

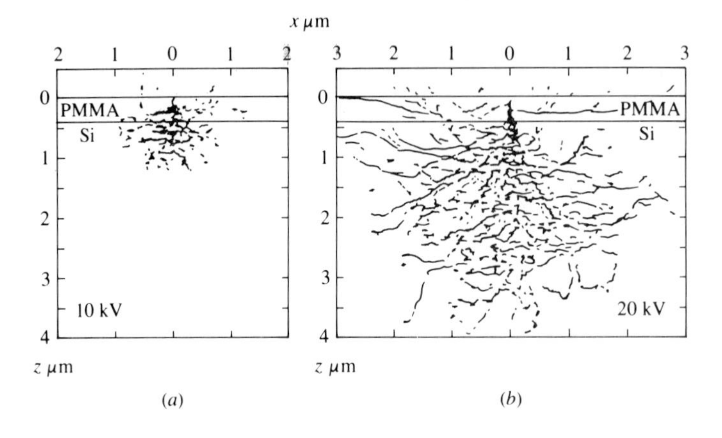

FIGURE 8-13
Monte Carlo simulated electron trajectories for 100 electrons scattered in a PMMA resist on a Si substrate (Kyser and Viswanathan, 1975).

penetration depth is known as the Grun range R_G, given (Everhart and Hoff, 1971) by

$$R_G \text{ (cm)} = \frac{4.6 \times 10^{-6}}{\rho \text{ (g/cm)}} E_o^{1.75} \text{ (keV)} \tag{8.12}$$

where E_o is the incident energy and ρ is the density. Based on the Grun range, Everhart and Hoff (1971) give the following for the average absorbed energy per volume E_v as a function of penetration depth z in the material:

$$E_v = \frac{DE_o}{qR_G} \lambda(f) \tag{8.13}$$

where

$$\lambda(f) = 0.74 + 4.7f - 8.9f^2 + 3.5f^3 \qquad \text{where } f = \frac{z}{R_G} \tag{8.14}$$

D is the incident charge per unit area, known as dose, and q is the electronic charge. The radial distribution at any depth for a point beam source can be approximated (Chang, 1975; Parikh and Kyser, 1978) by

$$E_v(r; z) = k(z)\left[\exp\left(-\frac{r^2}{\beta_f^2}\right) + \mu_e\left(\frac{\beta_f^2}{\beta_b^2}\right)\exp\left(-\frac{r^2}{\beta_b^2}\right)\right] \tag{8.15}$$

where β_f and β_b are the characteristic half-widths of the forward and backscattered distributions and

$$\mu_e = \frac{I_b}{I_f} \qquad \text{and} \qquad I_i = 2\pi \int_0^\infty r \exp\left(-\frac{r^2}{\beta_i^2}\right) dr \tag{8.16}$$

TABLE 8.1
Coefficients for the Gaussian approximation of Eq. (8.15) (Greeneich, 1980)

Substrate	Energy	Resist thickness	β_b or β_{bs}	β_{bd}	μ_e	μ_{ds}
Si	10	0.5	0.65	—	0.51	—
Si	15	0.5	1.14	—	0.51	—
Si	15	1.0	1.41	—	0.52	—
Si	25	0.5	2.6	—	0.51	—
Si	25	1.0	2.9	—	0.49	—
Si	25	1.5	2.9	—	0.52	—
Si	40	0.5	6.0	—	0.42	—
Si	40	1.0	6.0	—	0.45	—
Si	40	1.5	6.2	—	0.44	—
Cu	10	0.5	0.23	0.8	0.60	0.66
Cu	15	0.5	0.33	1.0	0.60	0.19
Cu	25	1.0	0.77	3.0	0.65	0.19
Cu	40	1.0	1.43	3.6	0.63	0.16
Au	10	0.5	0.16	0.5	0.65	0.9
Au	15	0.5	0.16	0.8	0.76	0.24
Au	25	1.0	0.37	1.4	0.79	0.28
Au	40	1.0	0.64	4.0	0.82	0.15

The backscattering effect given by the second term in Eq. (8.15) is mostly by the scattering from the substrate. In general, the parameters are determined from a Monte Carlo simulation (Hawryluk *et al.*, 1974) or experimental data. A relationship for β_f is shown in Fig. 8-12 where the number of elastic events P_e is given by

$$P_e = \frac{400z\ (\mu m)}{E_o\ (\text{keV})} \tag{8.17}$$

Values of the parameters β_b and μ_e are given in Table 8.1.

The dose-position function $\lambda(f)$ is relatively constant for small values of the normalized position f, goes through a maximum, and then decreases rapidly with f. The radial distribution of the energy dissipated is dominated by desired forward scattering, when the incident energy E_o is large, as should be evident from Eq. (8.17) and Fig. 8-14. Therefore increasing the incident energy (beam voltage) is desirable to a certain extent. There are, however, two limitations. The first is that the rate of energy dissipated per electron decreases as the beam energy is increased. Most of the energy is absorbed by the substrate when the energy is high (see Prob. 8.2). The second limiting aspect is the proximity effect discussed in Sec. 8.4. As the energy increases, there is more interaction between closely spaced features (proximity effect), limiting the resolution. In any event, a reduction in the resist thickness leads to a reduction in backscattering and thus to a smaller feature size. The resolution is usually defined in terms of the minimum feature that can be repeatedly exposed and developed in at least 1 μm of resist.

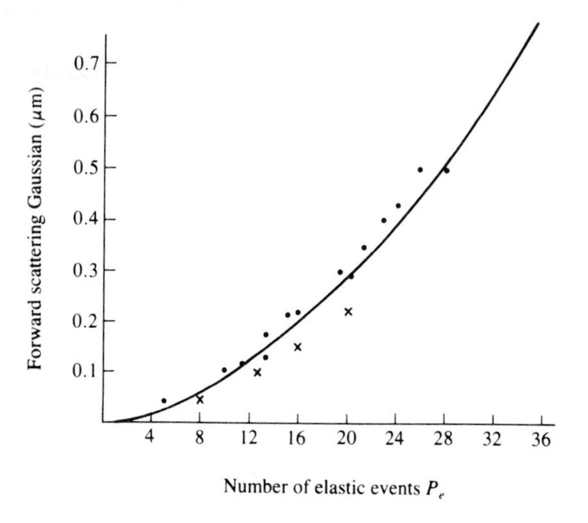

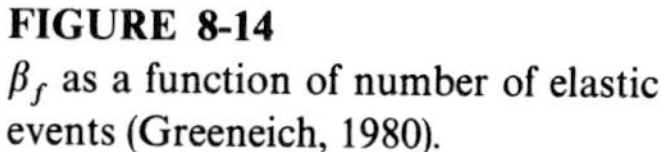

FIGURE 8-14
β_f as a function of number of elastic events (Greeneich, 1980).

Undesired scattering can be minimized if particles of higher mass than electrons are used. This has led to the use of ion beams. In fact, backscattering can be made almost absent with an ion beam for silicon substrates with a resist. Ionization by electron impact is the primary technique for generating ions. Electrons for ionization can be created by thermionic emission or can result from the discharge itself. These electrons are accelerated by the use of dc or rf fields and confined by the use of magnetic fields. An ion is generated when the energy transferred to a molecule by an accelerated electron exceeds the ionization energy for that molecule. Ions can also be generated by an electric field, termed field ionization.

As in electron-beam lithography, the quantities of interest in ion-beam lithography are the distance ions travel in a solid (resist and substrate) and the distribution of the energy imparted by the penetrating ions. The two main mechanisms of energy transfer are nuclear collisions and electronic interactions (Brodie and Muray, 1982). The energy loss by nuclear collisions results from collisions between nuclear charges and the target atoms. The second energy loss mechanism involves the interaction of the fast ion with the lattice electrons. An approximate ion range R_i resulting from the two energy loss mechanisms (Brodie and Muray, 1982) is

$$R_i = 2kE_o\left(1 - \frac{4kk'E_o}{3}\right) \tag{8.18}$$

where E_o is the incident energy and k and k' are given (Schwartz and Helms, 1979) by

$$k\ (\text{Å/eV}) = \left(\frac{0.018}{N}\right)\left[\frac{(Z_1^{2/3} + Z_2^{2/3})^{1/2}}{Z_1 Z_2}\right]\left(\frac{M_1 + M_2}{M_1}\right) \tag{8.19}$$

$$k'\ (\text{Å}^{-1}) = 0.328(Z_1 + Z_2)M_1^{-1/2} \tag{8.20}$$

Here Z and M are the atomic number and the mass, the subscripts 1 and 2, respectively, are for the ion and lattice atoms, and N is the atomic density (particles/Å^3). The first term in Eq. (8.18) represents the nuclear collisions and the second the electronic interactions. Since the backscattering can be neglected, the radial energy distribution can be represented only by the first term in Eq. (8.15). Here, again, Monte Carlo simulations (Karapiperis *et al.*, 1981) similar to the electron-beam calculations can be used.

The resolution-limiting diffraction effects in photolithography can be reduced by switching from the usual ultraviolet (300 to 400 μm) to deep-ultraviolet light whose wavelength is half as large, or almost eliminated by using soft x-rays whose wavelengths are several orders of magnitude smaller. This can be seen from Eqs. (8.3) and (8.4). If the wavelength λ is very small, Δl in the equations that define the maxima and minima is so small that there is almost no distinction between the nodes and antinodes and the interference effects become negligible. Soft x-rays, those with wavelengths between 0.1 and 5 μm, behave differently from lower energy photons in that they are neither significantly reflected nor refracted by any material. Therefore, they are not suitable for projection lithography. For this reason, the only currently feasible form of x-ray lithography is shadow lithography in which the pattern, formed of regions opaque and transparent to x-rays, is transferred to a nearby resist layer, as shown in Fig. 8-15.

The soft x-rays suffer negligible scattering as they pass through materials. They essentially move in straight lines until captured by an atom with ejection of a photoelectron (Wittels, 1980). The material is most transparent to x-rays when their wavelength is slightly longer than a "critical wavelength" and most opaque when it is shorter. The x-ray beam flux (energy per area per time) I decreases exponentially with penetration depth z:

$$I = I_0 \exp(-\mu z) \tag{8.21}$$

where μ is an absorption coefficient. The energy dissipated in a resist layer is proportional to the energy removed from the beam:

$$E_v(z) = \mu I(z) = \mu I_0 \exp(-\mu z) \tag{8.22}$$

The absorption coefficient, which is dependent on wavelength, is given in Fig. 8-16 for various materials (Spiller and Feder, 1977).

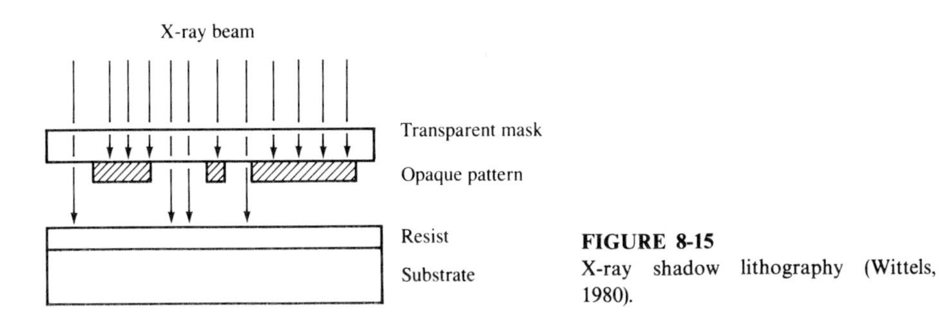

FIGURE 8-15
X-ray shadow lithography (Wittels, 1980).

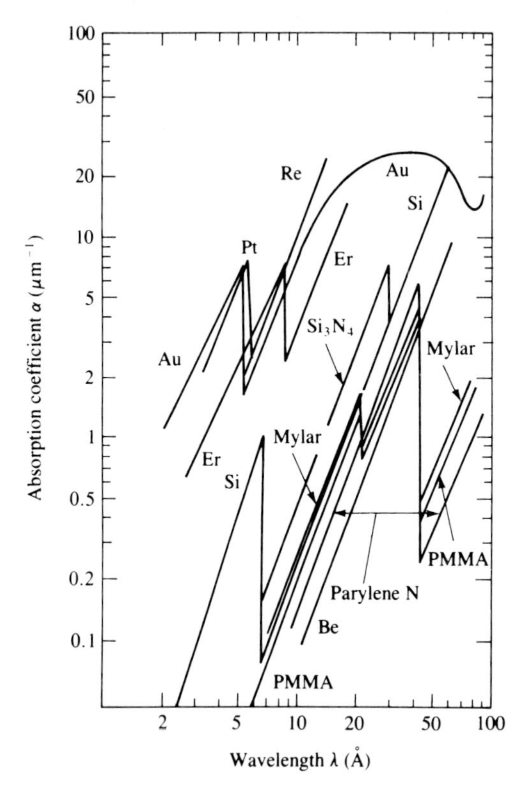

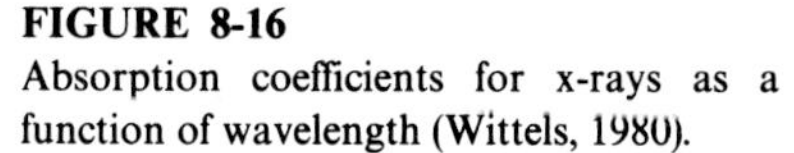

FIGURE 8-16
Absorption coefficients for x-rays as a function of wavelength (Wittels, 1980).

8.3 ILLUMINATION AND PATTERN TRANSFER

The first step in mask and device fabrication is to expose a resist (or an emulsion in the case of the photographical method) to a beam of monochromatic light, x-ray, electrons, or ions so that the pattern image can be transferred to the resist, which contains a component sensitive to the beam. The resolution and delineation of the pattern to be transferred depends on the beam intensity, its spatial distribution and the beam width.

The methods of pattern transfer (imaging) vary with the beam source. Contact, proximity, or projection printing can be used for light beams. Because of the poor contrast with soft x-rays, shadow printing is used, which is a form of proximity printing, as shown in Fig. 8-15. The three different methods used for the optical lithography are illustrated in Fig. 8-17 along with the corresponding typical optical intensity pattern (Skinner, 1973). The oldest method is contact printing (imaging) in which the mask and the resist are in direct contact. The major drawback on the contact method is poor yield. Defects in the mask can be caused by small particles on the wafer surface, which leave scratches, etc. These altered mask sites are copied along with the original pattern on subsequent wafers. However, very high resolution up to 0.5 μm is possible. Other photolitho-

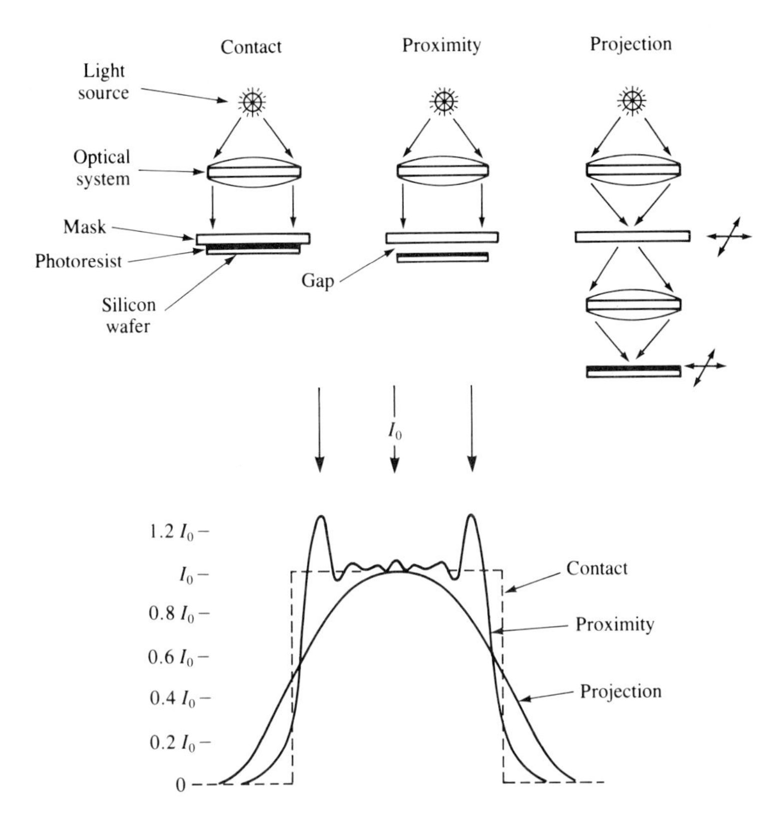

FIGURE 8-17
Schematics of three methods of optical lithography and the corresponding intensity patterns (McGillis, 1983).

graphic techniques can approach but not exceed its resolution capabilities. The defect problem can be avoided by providing a gap (10 to 25 μm) between the mask and resist, which is the essence of the proximity printing. The resolution is proportional to $(\lambda g)^{1/2}$, where λ is the monochromatic light wavelength and g is the gap (McGillis and Fehrs, 1975) and is on the order of a few micrometers. Projection printing involves projecting an image of the pattern onto the resist surface. Since the mask is far away from the resist-coated wafer, only a small portion of the mask is imaged at a time to achieve high resolution. This small image field necessitates scanning or stepping over the surface of the wafer and therefore the method is also called step-and-repeat printing. Resolution can be made close to that possible with contact printing.

Although masks are used for printing in general, maskless printing is also used. In maskless printing the image is generated on a resist directly without a mask. A typical example is mask fabrication with e-beams. Maskless printing should become a major method in time for wafer processing in spite of the slow speed with which a wafer can be processed. This is the major impediment to the

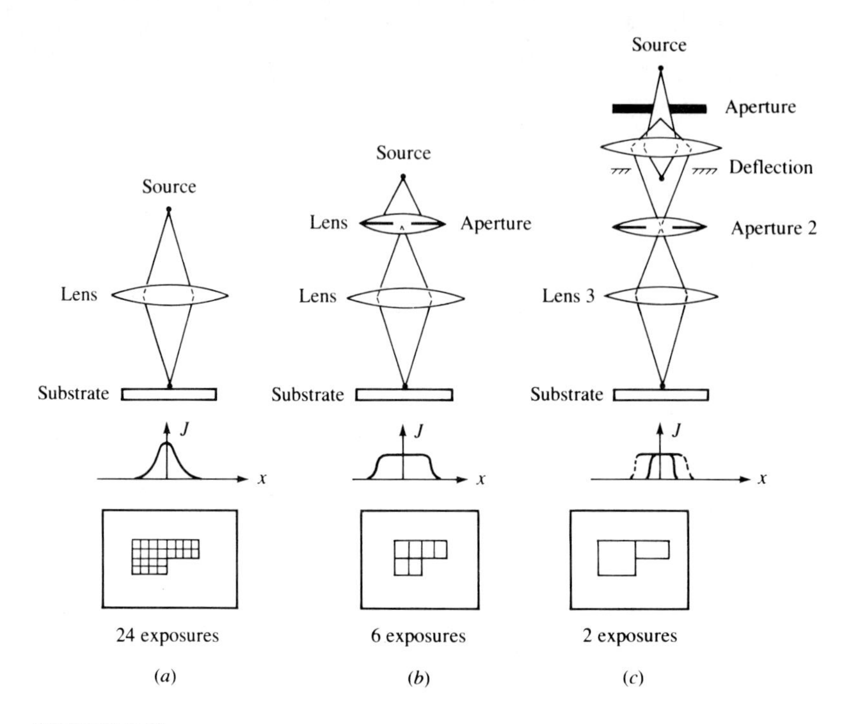

FIGURE 8-18
Spot-forming strategies in scanning e-beam lithography (Wittels, 1980).

maskless particle-beam lithography. In maskless printing a beam is focused to a spot the same size as or smaller than the minimum pattern size. With the combination of beam deflection and shuttering, individual spot exposures are built up into complete patterns that are stored on a tape in accordance with the digitized circuit layout. Different spot-forming strategies are used in scanning particle-beam lithography. These are shown in Fig. 8-18 (Wittels, 1980). In the simplest system (Fig. 8-18*a*), the source is imaged directly on the substrate, and the beam current density distribution is Gaussian. A spot size a small fraction of the minimum feature size is necessary for good pattern fidelity and, as seen in the figure, 24 exposures are necessary for the example. In the shaped beam case (Fig. 8-18*b*), the spot size is made equal to the minimum feature size. The Gaussian edge of the current density profile results only from lens aberrations and not the source characteristics. In a variable-shaped beam (Fig. 8-18*c*), the image of the first illumination-filled aperture is deflected by the intermediate deflection unit and then imaged onto the second aperture. Thus, variable shapes can be imaged on the wafer. The current density edge profile remains the same as that of the spot size.

Now let us consider photolithography in some detail. The limiting factor in the lateral dimension of a single pattern in photolithography is diffraction: Fresnel diffraction in the case of proximity printing and contact printing, and

Fraunhofer diffraction in the case of projection printing. The resolution depends on how much tolerance can be allowed in the intensity and the line width. According to Lin (1980), it is

$$\frac{W^2}{\lambda Z} \geq 0.5 \qquad \text{for proximity printing} \tag{8.23}$$

$$W \geq \frac{0.6\lambda}{\text{NA}} \tag{8.24a}$$

for projection printing

$$Z \leq \frac{0.8\lambda}{(\text{NA})^2} \tag{8.24b}$$

where W is the resolution and Z is a limit on the depth of focus over which image quality is not degraded. If the resist thickness is larger than Z, given by Eq. (8.24*b*), the latent image cannot be focused throughout the resist depth. The first relationship corresponds to the dimensionless distance in Fig. 8-8 that yields an intensity of approximately unity. For a coherent beam, $0.5\lambda/(\text{NA})$ corresponds to the half-width of the principal diffraction peak and $0.5\lambda/(\text{NA})^2$ is the defocusing aberration for the principal peak (also known as the depth of focus). The numbers in Eqs. (8.24) reflect the effect of partial coherence on the allowed tolerances.

When a beam of constant intensity is shined on a mask, the intensity is the same as the incident beam for contact printing, as shown in Fig. 8-17. Because of edge effects due to Fresnel diffraction in proximity printing (see Fig. 8-8), the intensity distribution on the wafer is that shown in Fig. 8-17 for a single slit.

An approach based on a transfer function called the modulation transfer function (MTF) is used in projection printing to relate the incident intensity (input) to the spatial distribution of the intensity of the image on a wafer (output) (see Prob. 8.17). This is similar to the transfer functions often used in the field of process control. Instead of the time frequency used in process control, the MTF uses a spatial frequency, defined as the number of slits or line pairs (each consisting of a slit opening and the spacing in between) per unit length. MTF is a measure of the accuracy of image transfer with respect to sharpness or contrast. Consider a model 1:1 projection printing, shown in Fig. 8-19 (Thompson and Bowden, 1983). The mask consists of a grid of periodic opaque lines and transparent spaces of equal widths. The diffraction of light at the edge of an opaque feature results in the projected pattern or image in the photoresist exhibiting gradual (rather than sharp) transitions from light to dark. Therefore, the edges of the projected feature appear blurred rather than sharp because of considerable light in the middle of the opaque feature. MTF of an exposure system is defined as the ratio of the modulation in the image plane to that in the object plane. The modulation M is defined in Fig. 8-19. For a given spatial frequency, the contrast of the projected image is greater for higher MTF values (see Prob. 8.17 for MTF calculations for a sinusoidal object, which is used to approximate a square wave object).

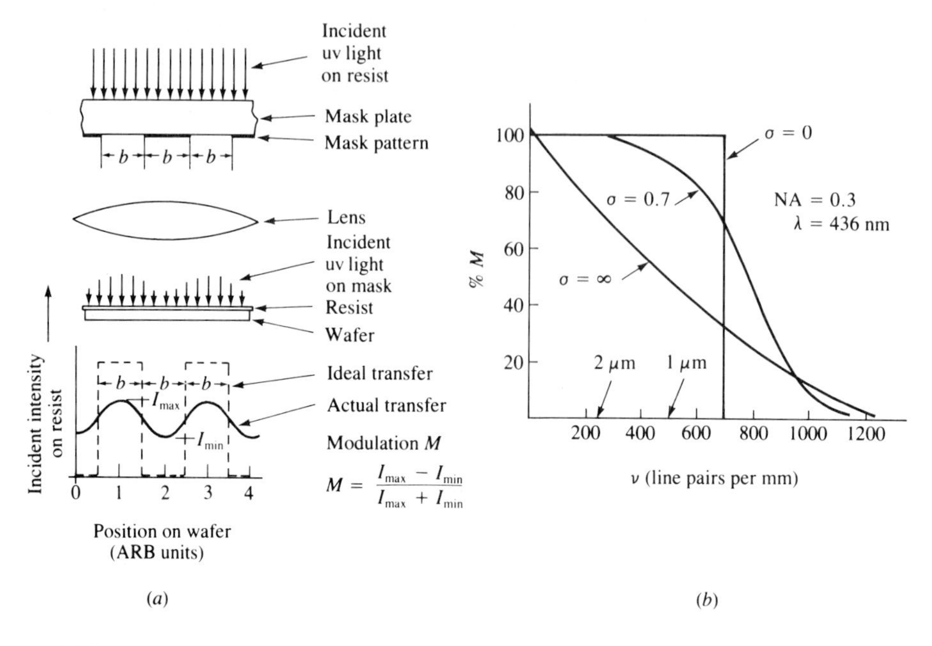

FIGURE 8-19
(*a*) Schematic of image transfer efficiency for a 1 : 1 projection (Bowden, 1981) and (*b*) MTF (%M) as a function of spacial frequency for three coherency factors (Thompson and Bowden, 1983).

The primary parameter in the MTF is the critical spatial frequency f_c (see Prob. 8.17) that defines the maximum number of slits allowable for imaging. The general grating formula of Eq. (8.7*b*) for primary diffraction ($j = 1$) can be rewritten as

$$\frac{\sin i + \sin \Omega}{\lambda} = \frac{1}{d} = f \tag{8.25}$$

where d (length of a pair of lines consisting of the slit opening and the spacing between slits) is $2b$ in Fig. 8-19. For image formation, it is required that $i, \Omega < \alpha$, where α is the angle that a focused ray makes with the screen (wafer). The maximum f or f_c is that corresponding to the case of $i = \Omega = \alpha$, and since $\sin \alpha$ in air is (NA), one has [from Eq. (8.25)]

$$f_c = \frac{2(\text{NA})}{\lambda} \tag{8.26}$$

This critical spatial frequency is the maximum number of repeating slits, each consisting of the slit opening plus the spacing between slits that can be contained per unit width without losing the image. At f_c, the MTF is zero. For most organic photoresists, the MTF has to be larger than 0.4 for good imaging. The MTF is very sensitive to spatial frequencies near the critical spatial frequency for a coherent beam but not for incoherent or partially incoherent beams as shown in Fig.

8-19*b*. Note that $\sigma = 0$ for a coherent beam and $\sigma = \infty$ for an incoherent beam. This is the major reason that a partially coherent beam is used in projection printing. A partially coherent source with a value of σ around 0.7 yields optimum pattern reproduction in conventional photoresists (Thompson and Bowden, 1983).

Example 8.1. Photolithography uses an ultraviolet (uv) beam with wavelength around 400 nm. Suppose that deep uv projection lithography uses a beam of 200 nm. For the beams of 200 and 330 nm, calculate the depth of focus, the minimum feature size that can be obtained, and the maximum possible number of lines that can be printed per unit width. Discuss the effect of the wavelength on the projection performance. Assume the numerical aperture (NA) to be 0.17.

Solution. From Eq. (8.24*a*), the minimum feature sizes are

$$W_{200} \geq \frac{0.6 \times 200\ \text{(nm)}}{0.17} = 707\ \text{nm} = 0.706\ \mu\text{m}$$

$$W_{330} \geq \frac{0.6 \times 330}{0.17} = 1165\ \text{nm} = 1.165\ \mu\text{m}$$

From Eq. (8.24*b*), the depths of focus are

$$Z_{200} \leq \frac{0.8 \times 200\ \text{(nm)}}{(0.17)^2} = 5.54\ \mu\text{m}$$

$$Z_{330} \leq \frac{0.8 \times 330}{(0.17)^2} = 9.14\ \mu\text{m}$$

The respective maximum possible spatial frequencies, or number of lines per unit width, follows from Eq. (8.26):

$$(f_c)_{200} = \frac{2(\text{NA})}{\lambda} = \frac{2 \times 0.17}{200\ \text{(nm)}} = 1.7\ \text{lines}/\mu\text{m} = 1700\ \text{lines/mm}$$

$$(f_c)_{330} = \frac{2 \times 0.17}{330} = 1030\ \text{lines/mm}$$

The following brief table summarizes these results:

λ, nm	**W, μm**	**Z, μm**	**f, lines/mm**
200	0.7	5.5	1700
330	1.2	9.1	1030

It is seen that as the wavelength decreases, the minimum feature size decreases with a corresponding increase in the number of lines that can be written per unit width. However, the resist thickness to which the feature can be imaged (Z) decreases with decreasing wavelength. Thus, for the 200-nm beam, the resist should not be thicker than 5.5 μm whereas it could be as thick as 9.1 μm for the 330-nm beam.

In scanning e-beam lithography, the current density of the Gaussian beam is given by

$$J = J_p \exp\left[-\left(\frac{r}{\beta}\right)^2\right] \tag{8.27}$$

where β is the standard deviation of the distribution and J_p is the source current density. The total current I in this beam is given by $I = \pi\beta^2 J_p$. For some purposes the actual Gaussian beam is treated as an equivalent beam of uniform current density, J_p, with a Gaussian beam diameter, d_G (Herriott and Brewer, 1980) such that the current is expressed as $I = \pi J_p d_G^2/4$, which means that $d_G = 2\beta$. The actual beam diameter d_G is given by

$$d_G = \frac{1}{\alpha}\left(\frac{I}{3.08B}\right)^{1/2} \tag{8.28}$$

where B is the brightness given by Eq. (8.11).

8.4 RESISTS

Resists are of a polymeric material that serves as a medium by which the device pattern can be imprinted onto the wafer or mask surface. As such, the resist is one of the main factors determining the resolution and delineation of the transferred pattern. Depending on the sensitizing (radiating) source, resists are termed photoresists, x-ray resists, e-beam resists, and so on. The same resist can sometimes be used for different sensitizing sources.

Although resists for wet processing are of interest here, there are also resists for dry processing in which the resists are etched away for desired patterns by colliding ions as in plasma etching (Chap. 10). Resists can also be classified into inorganic and organic resists. Inorganic resists have been used only recently (Yoshikawa *et al.*, 1976). An example is a thin layer of Ag_2Se on top of a film of Ge_xSe_{1-x} glass (Tai *et al.*, 1982). The species sensitive to light is Ag. Migration of Ag into the underlying glass upon exposure to light leads to the desired pattern.

Organic resists consist of either only one component or two components. Two-component resists are made from an inert resin that serves only as a binder and film-forming material. They contain in the resin a sensitizer molecule which in general is monomeric in nature and undergoes the chemical transformations that are responsible for imaging. Photoresists are invariably two-component resists. One-component resists are polymers that have radiation sensitivity. They are a single, homogeneous material. Most e-beam resists are one-component resists.

In the common negative resists, the exposure to radiation yields an insoluble, crosslinked polymer. In the process of dissolving the unexposed polymer to

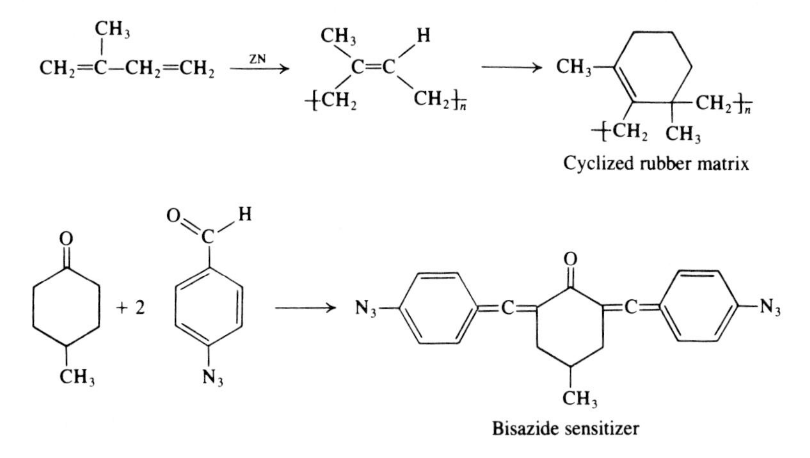

FIGURE 8-20
Formation of poly(*cis*-isoprene) and bisazide photosensitive compound.

obtain the final pattern, the crosslinked position swells and often distorts. For this reason, positive resists are usually preferred when feature sizes are in the submicrometer range.

The classical two-component negative photoresist is a cyclized polyisoprene synthetic rubber matrix with bisarylazide photoactive compound such as Kodak's KTFR. The poly(*cis*-isoprene) is produced by Zeigler-Natto polymerization of isoprene followed by proprietary treatments which cyclize the polymer. The photoactive compound is produced by a condensation reaction combining *para*-azido benzaldehyde with a substituted cyclohexanone as in Fig. 8-20.

The response of the sensitizer to incident light is generation of a crosslinked, three-dimensional network. The azide group emits a nitrogen molecule to form a highly reactive nitrene intermediate. The nitrene may combine with another nitrene to form an azo dye, enter a carbon-hydrogen bond to form an amine, remove a hydrogen atom from the polymer backbone to form an amine radical and a carbon radical, or enter the double bond of the polymer to form an aziridene linkage as in Fig. 8-21.

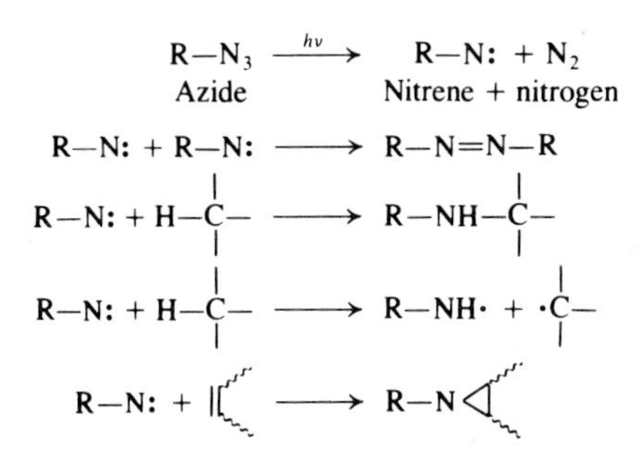

FIGURE 8-21
Crosslinking reactions of bisazide sensitizer and poly(*cis*-isoprene).

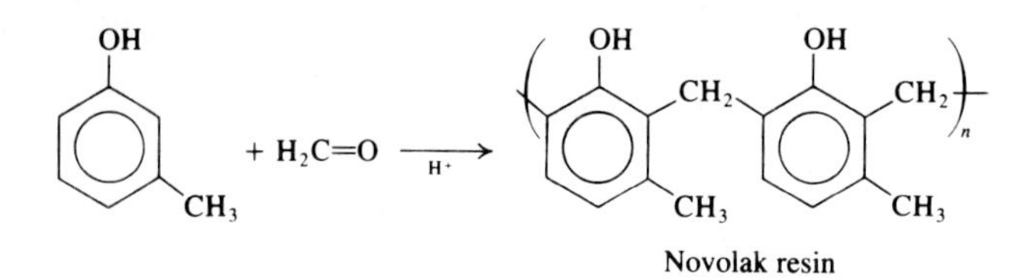

FIGURE 8-22
Novolac polymer formation.

The positive two-component photoresists are replacing the negative photoresists in many applications because of their higher resolution and better resistance to dry etching. The matrix portion of the positive resist is a novolac resin, while a diazoquinone is used for the photoactive compound. Novolac resin is a copolymer of a phenol and a formaldehyde, as shown in Fig. 8-22. The phenolic group of the novolac matrix imparts an acidic character which enables the resin to be dissolved in an aqueous base (Willson, 1983). The phenolic reactants used for production of novolac are a mixture of *meta*-cresol (60%), *para*-cresol (30%), xylenols, and *ortho*-cresol. The *meta*-cresol is much more reactive and generates a high molecular weight, very crosslinked polymer if it is the only phenolic monomer, while pure *para*-cresol produces a low molecular weight polymer that is unusable as a resist matrix. Variations in viscosity and developer solubility require extensive blending of production lots to produce a consistent product (Pamplone, 1984).

The diazonaphthoquinone sensitizer acts to inhibit solubility of the resist in aqueous base until it is exposed to light. Upon exposure, an excited state is produced which evolves nitrogen to form a carbene intermediate. The carbene then

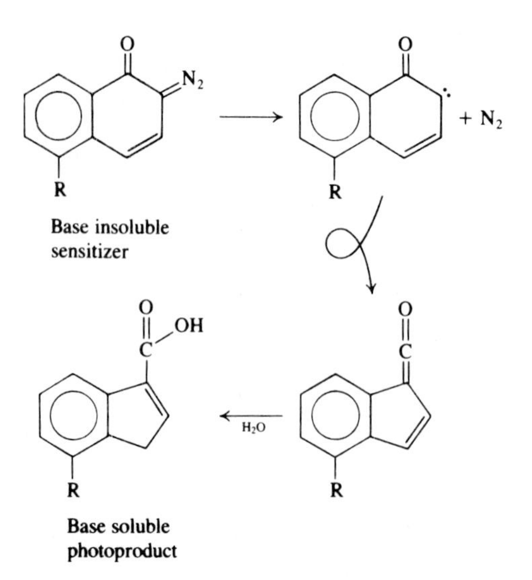

FIGURE 8-23
Transformation of diazonaphthoquinone photoactive compound to carboxylic acid. R is usually an aryl sulfonate.

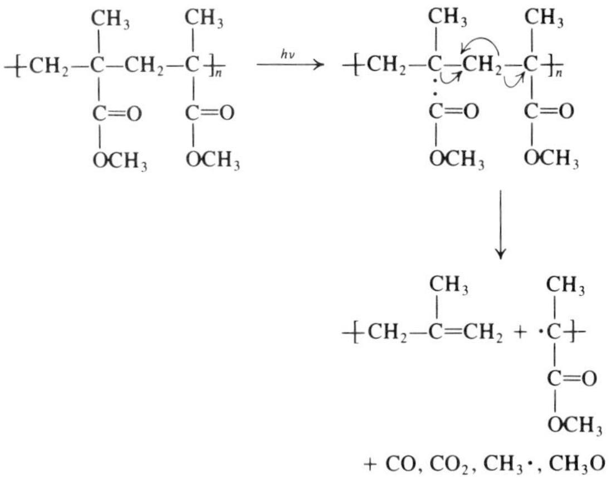

FIGURE 8-24
Mechanism of PMMA chain scission initiated by carbonyl–main chain carbon homolysis.

undergoes a Wolff rearrangement to produce a ketene. The ketene intermediate then adds water present in the resist film to produce an indenecarboxylic acid product that is soluble in the aqueous base developer (Willson, 1983) (Fig. 8-23).

The classical electron-beam resist is poly(methyl methacrylate), PMMA, a positive one-component resist. PMMA is produced from the methyl methacrylate monomer by a radical initiated polymerization catalyzed by azo-bis-isobutyronitrile. It has extremely high resolution but its relatively low sensitivity, 50 to 100 $\mu C/cm^2$ at 20 keV exposure, and poor resistance to plasma and reactive ion etching limit its usefulness (Willson, 1983).

The initial effect of radiation upon PMMA resist is homolysis of the bond between the main chain and carbonyl carbons to form a stable radical on the main chain (see Fig. 8-24). The chain is cleaved by beta scission rearrangement of the radical, resulting in an acyl-stabilized radical. The process may also be initiated by homolysis of the carbonyl carbon-oxygen sigma bond. Decarbonylation follows, resulting in the same tertiary radical. Wet development with an organic solvent that dissolves only the low molecular weight material resulting from chain scission is used to produce the image. The poly(olefin sulfones) are alternating copolymers of sulfur dioxide and an alkene. Despite the drawback of thermal instability, high sensitivity and low materials cost have attracted and maintained interest in these polymers.

Exposure of poly(olefin sulfones) to radiation causes scission of the carbon-sulfur bond followed by cationic and radical depolymerization to produce the olefin and sulfur dioxide monomers (see Fig. 8-25). Minor products include alkanes corresponding to loss of the side chain group and removal of the side chain radical by the olefin monomer (Bowmer and O'Donnell, 1982).

The most widely used negative electron-beam resist is COP, a copolymer of glycidylmethacrylate and ethyl acrylate. COP is produced by a radical copolymerization in benzene solvent with a benzoyl peroxide initiator.

The epoxy group in the glycidyl side chain of COP is the portion of the molecule that responds to radiation. Upon exposure, a cation, cationic radical,

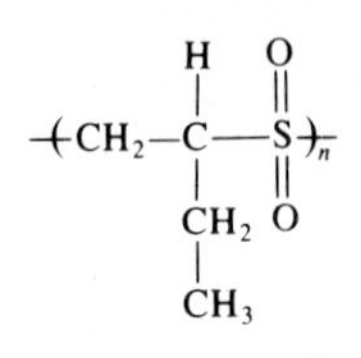

$$R{-}SO_2{-}R' \longrightarrow [RSO_2R']^{\dot{+}} + e^-$$

$$[RSO_2R']^{\dot{+}} \longrightarrow RSO_2^+ + \cdot R' \longrightarrow R^+ + SO_2$$

FIGURE 8-25
Radical and cationic depolymerization of poly(butene-1-sulfones), PBS.

anion, or anionic radical is formed which initiates a crosslinking chain reaction (see Fig. 8-26). The epoxy group is opened to generate the crosslinking bond and another reactive oxygen species. This chain reaction mechanism makes COP a relatively sensitive material but also causes a "dark reaction" which may continue after exposure (Willson, 1983). More details on resists can be found in Willson (1983).

The demands placed upon a potential resist material are great. Since the lithographic process is repeated in its entirety several times for each wafer, the sensitivity of the resist is a primary concern. Another major issue is contrast, as it has been shown that the contrast value and the resolution of a resist are directly related. Ideally, the limit of the feature size that can be produced will be dictated by the writing and etching procedures, not the exposure characteristics of the resist. Also, the resist must have adequate film-forming properties when spin-coated, such as adhesion to the substrate, and resistance to acid, base and reactive ion etching. The response of the resist to small, daily variations in process

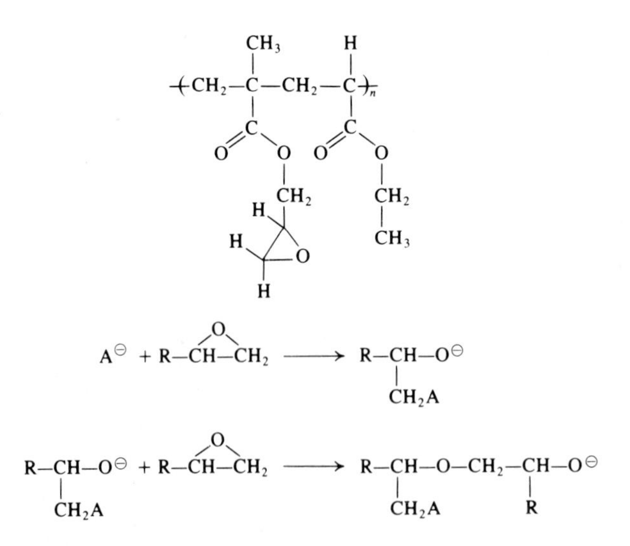

FIGURE 8-26
Glycidyl methacrylate-co-ethyl acrylate, COP. Anionic crosslinking mechanism.

conditions should be minimal, and a shelf-life of three or more months is considered adequate (Thompson and Bowden, 1983).

8.5 RESIST DEVELOPMENT

Depending on the types of resists, the exposed parts, when dissolved (developed) in a solvent, become either more readily soluble (positive resists) or less soluble (negative resists) than the unexposed parts. The reason for the difference in the dissolution rate is the change in the average molecular weight of polymer due to the exposure for e-beam resists. For photoresists, on the other hand, the reason is the transformation of the inhibitor that inhibits dissolution into a species soluble in a solvent. However, some photoresists dissolve by the same mechanism as for e-beam resists. The resists whose dissolution behavior owes their characteristics to changes in molecular weight (mostly e-beam resists) will be considered first.

There are two types of transformation that occur when the resists are exposed to radiation. Depending on the radiation-sensitive group in the resist, the polymer either goes through crosslinking resulting in a higher molecular weight polymer or decomposes through bond-breaking (scission) to a lower molecular weight polymer. The solubility of a polymer in a solvent, in general, decreases with increasing molecular weight. Therefore, the exposed part of the resist dissolves when the irradiation results in a decrease in the molecular weight. When the irradiation leads to an increase in the molecular weight, the exposed part remains intact while the unexposed part dissolves. The chemical transformation taking place upon irradiation requires a certain activation energy. Therefore, the amount of the polymer going through the transformation is determined by the amount of energy it receives and whether the energy source is photon (light), x-ray, electron, or ion. This in turn means that the energy (or radiation intensity) distribution, considered in Sec. 8.2, determines the shape of the resist when it is developed in a solvent.

The molecular weight distribution resulting from the irradiation varies with the position in the resist. The local, number-averaged molecular weight M for positive resists can be expressed (Herzog *et al.*, 1972; Greeneich, 1975) as

$$M = \frac{M_0}{1 + \mu(E_v)} \tag{8.29}$$

where M_0 is the initial number average molecular weight. The average number of scissions per molecule, μ, is a function of the absorbed energy per unit volume, E_v, and is given by

$$\mu = \frac{G E_v M_0}{m N_A} \tag{8.30}$$

where m is the mass density of the polymer, N_A is Avogadro's number, and G is the number of chain scissions per unit dissipated energy and is dependent only on the chemical nature of the polymer. Equation (8.29) is a simple mass balance, since the number of chain scissions per molecule, μ, generates $(1 + \mu)$ fragments (decomposed parts) of total weight M_0 and new average molecular weight M. For the case of a negative resist, G in Eq. (8.30) can be defined as the net number of average crosslinkings directly responsible for the molecular weight increase, since crosslinking and scission can take place simultaneously. Then one has, for the negative resist (see Prob. 8.14),

$$M = (1 + \mu)M_0 \tag{8.31}$$

For a given resist, the development (solubility) rate, usually expressed in length per time, is dependent on the type of solvent and its concentration. In much the same way as in the gas-solid reactions (Chap. 5), the development (penetration) front can be diffused or sharp (shell-progressive) depending on the relative rate of dissolution with respect to the rate of the penetration (diffusion) of the solvent. An ideal situation is one in which the development front moves in a shell-progressive manner with time. For a given resist, the solvent and its concentration can be chosen for the purpose (Ouano, 1984).

The development rate for positive polymeric resists is often correlated in a power-law form. The correlation of Greeneich (1975) is rewritten in the following form so that it can be applicable to both positive and negative resists:

$$R = R_0 + k_d\left[\left(\frac{M_0}{M}\right)^{\alpha} - 1\right] \tag{8.32}$$

where k_d is a rate constant in the form of an Arrhenius relationship and α is the order in a kinetic reaction sense. R_0 is the solubility rate of the resist in the absence of irradiation. In general, it is small for positive resists but large for negative resists. If a solvent and its concentration are chosen in such a way that the resist is almost impermeable to the solvent, the dissolution is limited to the surface being dissolved and the observed rate is that at the rate dictated by the intrinsic kinetics of dissolution. Under the conditions, the progression of the dissolution can be written (see Prob. 8.15) as follows:

$$\frac{dz}{dt} = R(z, x) \tag{8.33}$$

where x is the direction perpendicular to the resist depth coordinate z.

The concepts of contrast and sensitivity are used in evaluating a resist and the key parameter is energy dose, defined as photons or electrons per unit area. The dose is equal to the photon or electron intensity multiplied by the exposure time t_r and the typical units are coulombs per square centimeter for electrons and calories (ergs) per square centimeter for photons. The contrast Ω is defined as

$$\Omega = \left|\log\left(\frac{D_f}{D_i}\right)\right|^{-1} \tag{8.34}$$

where D_f is the extrapolated dose for full thickness, as shown in Fig. 8-27 for

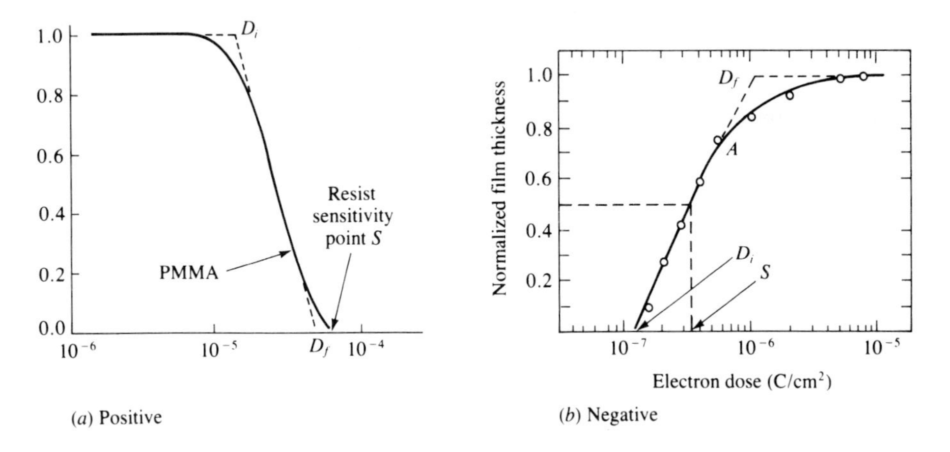

FIGURE 8-27
Determination of contrast for positive and negative resist (Greeneich, 1980).

both negative and positive resists, and D_i is the idealized minimum dose necessary for the dissolution. The contrast is a measure of the slope of the line representing the resist thickness remaining after development (normalized with respect to the final resist thickness) as a function of dose. The contrast is directly related to the resolution capability of a resist. The higher the contrast is, the better are the resolution and the edge definition of a resist.

The sensitivity S is the minimum dose that gives dimensional equality of clear and opaque features. For the negative resist in Fig. 8-27, this is taken as the dose at which the remaining resist thickness is 50 percent. For positive resists, all of the resist material must be removed to be useful and hence the sensitivity is that at the complete removal point, as shown in Fig. 8-27. Although the contrast, as defined by Eq. (8.34), has mostly been used in electron- and ion-beam lithography, the same definition can be used for photoresists. Noting that μ in Eqs. (8.29) and (8.31) is much larger than unity and that μ is proportional to the absorbed energy E_v, for a given resist, it follows from these equations that M_0/M is proportional to E_v for positive resists and is inversely proportional to E_v for negative resists. Since the dose D is proportional to the energy, it follows from Eq. (8.32) that the development rate due to the irradiation, which is $R - R_0 + k_d$, is proportional to D^α for positive resists and $D^{-\alpha}$ for negative resists. The contrast is given by the logarithm of the slope of the line for the fraction of resist dissolved plotted against the dose. Therefore, the order α is a measure of the contrast, the value being higher for higher contrast.

The definition of sensitivity, which can be determined experimentally, allows one to find a minimum feature size that can be obtained in an e-beam resist. For the satisfactory development of the resist, it has to receive a certain number of electrons per unit area for the necessary chemical transformation of polymers in the resist. If the minimum dimension for a line is denoted by l_p, the minimum number of electrons required (N_{min}) for the sensitivity S is

$$N_{min} = \frac{Sl_p^2}{q} \tag{8.35}$$

The area corresponding to l_p^2 is called a picture element or pixel. The minimum dimension, therefore, follows from Eq. (8.35):

$$l_p = \left(\frac{qN_{min}}{S}\right)^{1/2} \tag{8.36}$$

An important conclusion that can be drawn from Eq. (8.36) is that the minimum dimension is inversely proportional to the square root of the sensitivity. Since electron emission is a random process, a question arises as to the probability of the number of electrons that will strike a given surface. The limit imposed by a probabilistic argument is 200 for N_{min} (Greeneich, 1980). Then Eq. (8.36) yields

$$l_p = \left(\frac{200q}{S}\right)^{1/2} \tag{8.37}$$

where N_{min} is taken as 200.

Example 8.2. Poly(methyl methacrylate), or PMMA, is a positive resist. Greeneich (1975) correlated his extensive data on the development rate R as follows:

$$R = A + \frac{B}{M^{\alpha}} \tag{A}$$

For PMMA, he arrived at the following set of parameters for two types of solvents:

Developer† (solvent)	T, °C	A, nm/min	B	α
MIBK:IPA (1:3)	22.8	0	9.33×10^{13}	3.86
MIBK:IPA (1:3)	32.8	0	1.046×10^{15}	3.86
MIBK	22.8	8.4	3.154×10^{7}	1.50
MIBK	22.8	24.2	5.67×10^{7}	1.50

† MIBK = methyl isobutyl ketene, IPA = isopropyl alcohol.

Calculate the development rates at 22.8 and 32.8 °C for the mixed solvent (MIBK + IPA) at the uniform molecular weight, M, of 3.4×10^3. Calculate the development time for 310 nm PMMA resist.

Solution. According to Eq. (A),

$$R_{22.8} = \frac{9.33 \times 10^{13}}{(3.4 \times 10^3)^{3.86}} = 2.18 \text{ nm/min}$$

$$R_{32.8} = 24.44 \text{ nm/min}$$

Therefore the development rate depends strongly on temperature. Since the rate is uniform everywhere, Eq. (8.33) for the dissolution rate can be integrated with $z = 0$ at $t = 0$ to give

$$z = Rt \qquad \text{or} \qquad Z = Rt_f \tag{B}$$

For Z of 310 nm, one has, from Eq. (B),

$$t_f = \frac{310}{R}$$

Thus,

$$(t_f)_{22.8} = \frac{310}{2.18} = 142.2 \text{ min}$$

$$(t_f)_{32.8} = \frac{310}{24.44} = 12.68 \text{ min}$$

The strong dependence of the development time on temperature should be noted.

Example 8.3. Consider the PMMA in Example 8.2 (310 nm thick) coated onto a silicon substrate. An electron beam of 1 μm diameter of constant intensity is used such that the dose is 5×10^{-5} coulomb/cm^2. For the following data, calculate the development time at 22.8 °C with the mixed solvent. [Note that the energy density (and thus the molecular weight) varies with the depth. Assume the radial energy density to be uniform.]

$$E_o = 20 \text{ keV} \qquad \rho = 1.2 \text{ g/cm}^3$$

$$G = 19 \text{ keV}^{-1} \qquad M_0 = 2 \times 10^5$$

Solution. From Eq. (8.12),

$$R_G = \frac{4.6 \times 10^{-6}}{\rho \text{ (g/cm}^3)} E_o^{1.75} \text{ (keV)} = \frac{4.6 \times 10^{-6} \times 20^{1.75}}{1.2} = 7.25 \ \mu\text{m} \tag{A}$$

The axial energy density is given by Eq. (8.13):

$$E_v = \frac{DE_o}{qR_G} \lambda(f)$$

$$= \frac{5 \times 10^{-5}}{1.6 \times 10^{-19}} \times \frac{20}{7.25 \times 10^{-4}} \lambda(f)$$

$$= 8.62 \times 10^{18} \text{ (keV/cm}^3) \ \lambda(f) \tag{B}$$

The axial molecular weight distribution is given by Eq. (8.29):

$$\frac{M}{M_0} = \frac{1}{1 + \mu} \tag{C}$$

where
$$\mu = \frac{GE_v M_0}{mN_A} = \frac{19 \times 8.62 \times 10^{18} \lambda(f) \times 2 \times 10^5}{1.2 \times 6.02 \times 10^{23}}$$
$$= 44\lambda(f)$$

Since $\mu \gg 1$, one can rewrite Eq. (C) as

$$M = \frac{mN_A}{GE_v}$$
$$= \frac{1.2 \times 6.02 \times 10^{23}}{19 \times 8.62 \times 10^{18} \lambda(f)} = \frac{4411}{\lambda(f)} \tag{D}$$

From Eq. (8.33) and Eq. (A) in Example 8.2,

$$\frac{dz}{dt} = \frac{9.33 \times 10^{13}}{M^{3.86}} = 9.33 \times 10^{13} \left(\frac{\lambda}{4411}\right)^{3.86} \quad \text{(nm/min)} \tag{E}$$

where Eq. (D) has been used. In terms of f, $z = R_G f$. Thus, Eq. (E) can be rewritten as

$$\frac{df}{dt} = \frac{0.798}{R_G} \lambda^{3.86} = \frac{0.798}{7250} \lambda^{3.86}$$
$$= 1.10 \times 10^{-4}(0.74 + 4.7f - 8.9f^2 + 3.5f^3)^{3.86} \tag{F}$$

The time required follows from Eq. (F):

$$t_f = \int_0^{0.0428} \frac{df}{1.10 \times 10^{-4}(0.74 + 4.7f - 8.9f^2 + 3.5f^3)^{3.86}} \tag{G}$$

since the resist thickness normalized with respect to R_G is 0.0428. Numerical integration of Eq. (G) based on three points, for illustration, yields t_f of 815 minutes. If the uniform molecular weight assumed in Example 8.2 were 6139, which corresponds to the M value at the surface in this example, the required time would be 122 minutes at 22.8 °C for Example 8.2.

The edge shape of the dissolved resist in the above example is perfectly straight since the radial (lateral) energy density distribution was assumed to be uniform. In reality, however, the edge is not perfectly straight because of electron scattering. This is shown in Fig. 8-28 for three different doses (Hatzakis, 1975). The first is typical of the case of high doses, relative to the resist thickness. If the dose is high enough, scattering, which widens the opening toward the resist-substrate interface, is also sufficiently high to cause substantial scission of polymer even in the region dominated by backscattering. Thus, the region readily dissolves when developed, causing widening toward the interface. In the other extreme of low doses, not much scission occurs and the dissolution is mainly dependent on how long the resist is developed. Therefore, the shape in the bottom of Fig. 8-28 results. Note in this case that the resist thickness is considerably reduced due to prolonged development, compared to the other cases. A combination of the two extreme cases, in which both the energy dissipation distribution and development time contribute, leads to the straight shape shown in

FIGURE 8-28
Actual cross-sectional views of PMMA resist profiles at high (undercut), intermediate (vertical), and low (undercut) doses (Hatzakis, 1975).

the middle of Fig. 8-28. The shape of the developed resist can be simulated based on the Gaussian profile given by Eq. (8.15) (see Prob. 8.2).

Similar procedures, as used for the electron beam, can be followed for the development of a resist exposed to an ion beam. R_G in Eqs. (8.12) and (8.13) is replaced by R_i given by Eq. (8.18), and the penetration range function (depth-dose function), $\lambda(f)$, similar to Eq. (8.14) must be known. Because of negligible backscattering, only the first term in Eq. (8.15) for the forward scattering can be used for the lateral energy distribution.

Another major effect of the scattering of particle beams is called the proximity effect. Scattering leads to the overlapping of patterns in close proximity to one another. The proximity effect is strongly dependent on the electron energy since high-energy exposures have large exposure tails and these tails, when overlapped, lead to the effect. Procedures for correcting the proximity effect hinge on this recognition. Dose manipulations are used for the purpose (Parikh, 1980).

Unlike the electron-beam resist, a positive photoresist contains an inhibitor component that prevents dissolution in a solvent. The inhibitor readily absorbs photons and then goes through a chemical transformation upon exposure, leading to the disappearance of the inhibitor, and the resist becomes soluble in a suitable solvent. The typical solvents are dilute aqueous alkaline solutions. The negative photoresist, however, behaves much the same as the electron-beam resist in that the absorption of photons leads to crosslinking of polymers.

Because of the nature of the inhibitor, the concentration distribution of the inhibitor largely determines the development process. The rate of inhibitor concentration normalized with respect to the initial inhibitor concentration, M, is often assumed (Dill *et al.*, 1975a) to be dependent on the light flux I_m and to be of first order with respect to the concentration:

$$\frac{\partial M}{\partial t} = -k_i I_m M \tag{8.38}$$

where k_i is a pseudo rate constant. As the light passes through the resist, the intensity decreases and this change follows the Beer-Lambert law (Chap. 5):

$$\frac{\partial I_m}{\partial z} = -\alpha I_m \tag{8.39}$$

where α is the absorption constant and z is the resist depth coordinate. Since the inhibitor absorbs photons, the absorption constant is proportional to the inhibitor concentration:

$$\alpha = AM + B \tag{8.40}$$

where A and B are constants that are dependent on photoresist type. The rate of development is in turn dependent on the inhibitor concentration M. A correlation proposed by Dill *et al.* (1975b) is

$$R\ (\text{nm/s}) = \exp\,(a_1 + a_2 M + a_3 M^2) \tag{8.41}$$

where a_1, a_2, and a_3 are constants. More physically based models are given by Kim *et al.* (1984) and Hershel and Mack (1987).

Example 8.4. For a positive photoresist that is 0.545 μm thick on a silicon substrate, Dill *et al.* (1975b) gave the following parameters:

$$A = 0.54\ \mu\text{m}^{-1} \qquad B = 0.03\ \mu\text{m}^{-1} \qquad k_i = 0.014\ \text{cm}^2/\text{mJ}$$

The incident beam intensity I_0 was 57 mJ/(cm^2·s) for the wavelength of 436 nm and the constants for the development rate are

$$a_1 = 5.96 \qquad a_2 = -1.19 \qquad a_3 = -2.27$$

Assuming that the incident beam intensity is uniform laterally and 1 μm wide, calculate a minimum exposure time required for the inhibitor concentration at the resist-substrate interface to be less than 1 percent. The minimum time may be defined as the time corresponding to M of unity throughout the resist for the purpose of calculating $I_m(z)$. Describe a numerical method of calculating the required, actual exposure time for the 1 percent inhibitor concentration at the interface. Note that, at time zero,

$$M(z, 0) = 1$$

$$I_m(z, 0) = I_0 \exp[-(A + B)z]$$

and at $z = 0$,

$$I_m(0, t) = I_0$$

$$M(0, t) = \exp(-k_i I_0 t)$$

Solution. Combination of Eqs. (8.39) and (8.40) yields

$$\frac{\partial I_m(z, t)}{\partial z} = -[AM(z, t) + B]I_m$$

$$= -(0.54M + 0.03)I_m \tag{A}$$

Equation (8.38) is

$$\frac{\partial M(z, t)}{\partial t} = -0.014MI_m \tag{B}$$

For the minimum exposure time, M is unity throughout the resist for the purpose of calculating the intensity. Equation (A) integrated with $M = 1$ is

$$I_m(z) = I_0 \exp(-0.57z) \tag{C}$$

Use of Eq. (C) in (B) and integration of the resulting equation yields

$$M(z, t) = \exp\left[-\int_0^t 0.0141I_0 \exp(-0.57z)\, dt\right]$$

$$= \exp[-0.798t \exp(-0.57z)] \tag{D}$$

At the interface ($z = 0.584\ \mu$m), M is 0.01. Thus, from Eq. (C),

$$0.01 = \exp[-0.572t]$$

The minimum time, therefore, is 8.05 s.

The maximum time for the required condition of 1 percent M at the interface is that corresponding to $M = 0$ in Eq. (A), or $I(z) = I_0 \exp(-0.03)$. Use of this in Eq. (D) yields

$$M(z, t) = \exp[-0.798t \exp(-0.03z)]$$

$$= \exp(-0.014t)$$

Thus, the maximum time is 329 s. The actual time is between 8 s and 329 s.

For the actual time, Eqs. (A) and (B) are written in the following Euler form:

$$I_l(i+1, j) = -[0.54M(i, j) = 0.03]I_m(i, j)\,\Delta z + I_m(i, j) \qquad i = 0, 1, 2, \ldots \qquad \text{(E)}$$

$$M(i, j+1) = -0.014M(i, j)I_m(i, j)\,\Delta t + M(i, j) \qquad j = 0, 1, 2, \ldots \qquad \text{(F)}$$

Since the starting points $[M(i, 0), M(0, j); I_m(i, 0), I_m(0, j)]$ are known, a computer program based on Eqs. (E) and (F) should yield the actual time. The subscripts i and j, respectively, are the grid points for z and t. Equation (E) is first used to calculate the axial I_m profile at time zero ($j = 0$). Based on the profile, time is incremented by Δt and the axial M profile is calculated for the incremented time, which is then used in Eq. (E) for the corresponding I profile. These procedures can be repeated until the time at which M at the interface becomes 0.01. This time is the required exposure time.

Example 8.5. For the same problem in Example 8.4, calculate the time required for the development of the resist. Assume that the final inhibitor concentration profile at the end of the exposure is given by

$$M(z) = z$$

Solution. For the parameters given in Example 8.4, Eq. (8.41) is

$$R = \exp(5.96 - 1.19M - 2.27M^2) \qquad \text{(nm/s)}$$

$$= 10^{-3} \exp(5.96 - 1.19M - 2.27M^2) \qquad (\mu\text{m/s}) \qquad \text{(A)}$$

Use of the M profile in Eq. (A) yields

$$R\ (\mu\text{m/s}) = 10^{-3} \exp(5.96 - 1.19z - 2.27z^2) \qquad \text{(B)}$$

From Eq. (8.33),

$$\frac{dz}{dt} = R(z)$$

or

$$t_f = \int_0^{0.584} \frac{dz}{R(z)} \qquad \text{(C)}$$

Integration of the right-hand side of Eq. (C) based on an equidistant (seven-points) use of the trapezoidal rule yields

$$t_f = 3.07 \text{ s}$$

The development time reported by Dill *et al.* (1975b) is an order of magnitude larger, indicating that the exposure time was very short and that the actual $M(z)$ was much larger than that used in the example.

Although the incident beam imaged on the photoresist is uniform laterally in its intensity for contact printing and, to a certain extent, even for proximity printing, the intensity distribution is not uniform in the case of projection printing, as shown in Fig. 8-17. As illustrated in Prob. 8.17, this intensity distribution can be determined through the MTF. The intensity distribution caused by light diffraction is the reason why the width of the rectangular opening in Fig. 8-29 is

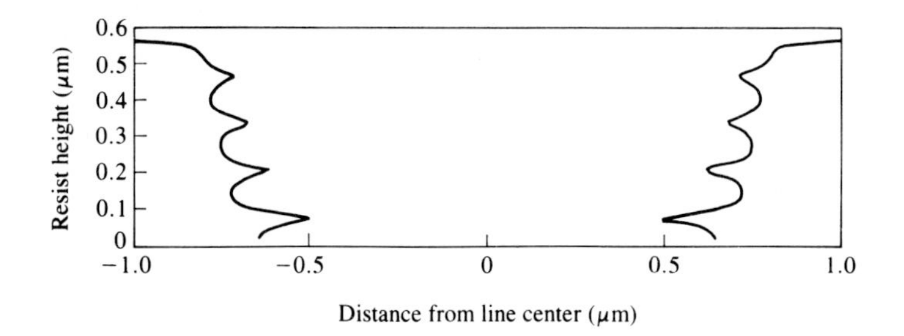

FIGURE 8-29
Positive photoresist edge profile after development (Dill *et al.*, 1975b).

larger toward the top. When light travels in a resist, it also reflects, and this reflection causes a standing wave within the resist (refer to Fig. 8-9) in much the same way as diffraction causes a standing wave (Cuthbert, 1977). The effect of this standing wave results in the wavy-edge shape shown in Fig. 8-29, which is a simulation result (Dill *et al.*, 1975b).

Example 8.6. The lateral intensity distribution of the incident light beam imaged on the resist surface by projection printing can be well approximated by a Gaussian distribution (see Prob. 8.12):

$$I_m = I_i \exp(-a_0 x^2) \tag{A}$$

where a_0 is a constant. Write a sufficient set of equations necessary to simulate the development profile for the incident beam.

Solution. For a given exposure time, one has

$$\frac{\partial M}{\partial t} = -k_i M(z, x, t) I_m(z, x, t) \tag{B}$$

$$\frac{\partial I_m}{\partial z} = -[AM(z, x, t) + B] I_m(z, x, t) \tag{C}$$

Equations (A), (B), and (C) need to be solved numerically to obtain the inhibitor concentration distribution $M(z, x)$ at the end of the exposure. With the distribution, Eq. (8.33) can be used to determine the development profile along with a development rate such as Eq. (8.41).

Example 8.7. The contrast given by Eq. (8.34) is a measure of the slope of the resist thickness still remaining after development (F) with respect to the dose (D), and as such it can be rewritten as

$$\Omega = \frac{\partial F}{\partial \ln D} \tag{A}$$

For the data given in Fig. 8-27 for a negative resist, determine the contrast line, i.e., a relationship between F and $\ln D$ for the straight line. Determine the exposure time corresponding to D_f and that for the actual full thickness. Assume for 10 keV that the current flux is 10^{-8} A/(cm^2·s). Note that $D_i = 1.2 \times 10^{-7}$ and $D_f = 10^{-6}$ coulomb(C)/cm^2.

Solution. Integrating Eq. (A),

$$F = 0.472 \ln D + b \tag{B}$$

The dose is given by

$$D = Jt \tag{C}$$

where J is the current flux. For $D = D_f = 10^{-6}$,

$$t = \frac{D}{J} = \frac{10^{-6}}{10^{-8}} = 100 \text{ s}$$

The actual dose for the full resist thickness in Fig. 8-27 is 5×10^{-6} C/cm^2. Thus, the corresponding time is

$$t = \frac{D}{J} = \frac{5 \times 10^{-6}}{10^{-8}} = 500 \text{ s}$$

Postexposure bake and low-temperature prebake (softbake) of resists can enhance the imaging and development characteristics (Batchelder and Piatt, 1983). In the case of positive resists, more residual water in the resist following softbake can eliminate potentially uncontrollable crosslinking. Elimination of standing waves is also possible with a proper bake.

8.6 EDGE SHAPE AND ALIGNMENT

The edge shapes of the resist after development are dominated by diffraction effects in photolithography and by scattering in electron beam lithography. A straight edge is the ideal shape in almost all cases. The edge shape, as determined by electron scattering and resist development, has been discussed in the previous section. As shown in Fig. 8-18, the pattern and the corresponding line width can be obtained by a combination of single beam spots, each of which is of Gaussian distribution in its intensity or energy. The spot diameter of the beam is given by Eq. (8.28) and can be as small as 0.05 μm, in the practical energy range, i.e., sufficient energy for polymer scission or crosslinking for the development. Although the edge shape can be controlled with proper dose and development it is of interest to know how the edge shape changes with the beam half-width. The characteristic parameters are the Gaussian beam half-width, β_g, and the half-width of forward scattering, β_f. A correlation based on an effective half-width, β_e, which is given by $(\beta_f^2 + \beta_g^2)^{1/2}$, is shown (Greeneich, 1980) in Fig. 8-30. The relative edge slope at the resist-substrate interface would be infinite for the ideal, straight edge. It is seen from the figure that a relatively higher slope results for a smaller β_e. The Gaussian half-width can be reduced by properly placing or scan-

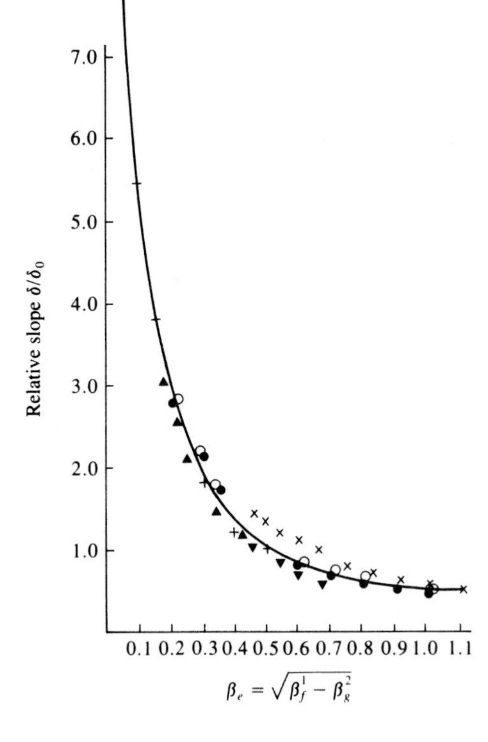

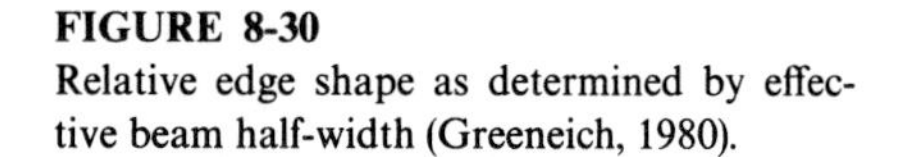

FIGURE 8-30
Relative edge shape as determined by effective beam half-width (Greeneich, 1980).

ning several single beams side by side with the edges overlapping so that the relatively flat center region becomes large. In fact, this is a typical practice in scanning e-beam lithography. According to Eq. (8.17) and Fig. 8-14, a smaller β_f results when the resist thickness is small and the incident energy is large. Because of the sensitivity of the linewidth to backscattering that dominates the scattering effects at high energy, the increase in the incident energy for a smaller β_f should be limited to the range in which forward scattering dominates. For negative electron resists, the resist thickness distribution after development by exposure to a single scanning Gaussian beam of radius r_0 is given (Heidenreich *et al.*, 1975) by

$$p = 0.434\Omega\left(1 + \ln\left\{\exp\left(-\frac{x^2}{r_0^2}\right) + \exp\left[-\left(\frac{2r_0 - x}{r_0}\right)^2\right]\right\}\right) \tag{8.42}$$

where p is the thickness normalized with respect to the initial resist thickness. The edge shape follows from this relationship.

In photolithography, the edge shape is dictated by Fresnel diffraction in proximity printing. The energy distribution and the corresponding shape of the developed resist have already been discussed.

Example 8.8. For negative resists, the value of the contrast usually lies between 0.6 and 1. For a Gaussian e-beam of 0.5 μm diameter and a resist of 0.9 contrast, calculate the edge length, i.e., the lateral length of the resist still remaining after development that is in excess of the beam diameter. The experimental value (Heidenreich *et al.*, 1975) is approximately 0.46 μm.

Solution. The length in excess of the beam diameter can be obtained by finding the point x at which the resist thickness is zero, that is, $p = 0$ in Eq. (8.42). Thus, one has, from Eq. (8.42),

$$0 = 0.43\Omega\left(1 + \ln\left\{\exp\left(-\frac{x^2}{0.25^2}\right) + \exp\left[-\frac{(1-x)^2}{0.25^2}\right]\right\}\right)$$

or

$$-1 = \ln\left\{\exp\left(-\frac{x^2}{0.0625}\right) + \exp\left[-\frac{(1-x)^2}{0.0625}\right]\right\}$$

A value of 0.74 for x satisfies the equation, which means that the excess length l_e is given by

$$l_e = 0.74 - 0.25 = 0.49\ \mu\text{m}$$

As discussed in the previous section, the straight-edge shape is accomplished by properly manipulating both the dose level and the development time. In projection printing photolithography, the light beam is reflected by mirrors for the edge control so that it arrives at an angle rather than perpendicularly to the resist surface. This leads to the energy profile in the resist that is wider at the resist-substrate interface than at the resist surface. Proper manipulation of development time then yields straight edges.

Each time a masking level is added for further processing, the new mask features have to be properly aligned to the pattern already in the resist from the previous masking level so that the device can be fabricated according to the composite layout. The same is true with maskless processing using particle beams. Another factor to consider is the level of tolerance that must be allowed for uncertainties in mask alignment and deviations of the resist image from the mask image. This tolerance is one of the design rules used to lay out the circuit. An estimate of the nesting tolerance can be made if the distributions of developed resist feature sizes and registration are known. Registration is a measure of how closely successive mask levels can be overlaid. If the standard deviation in the developed resist-feature size for mask level 1 is β_1 and that for the level 2 is β_2 with the registration standard deviation of β_r, the tolerance T is given by

$$T = 3\left[\left(\frac{\beta_1}{2}\right)^2 + \left(\frac{\beta_2}{2}\right)^2 + \beta_r^2\right] \tag{8.43}$$

where the distributions are assumed to be normal. For this tolerance, the probability that the edge of a developed (etched) feature from level 1 will touch a feature from masking level 2 is approximately only 0.1 percent (McGillis, 1983).

Although an optical microscope can be used to align a mask to the wafer pattern with the aid of alignment marks on the wafer, the accuracy is only of the order of 0.25 to 0.5 μm. For better accuracy and automation, however, laser interferometric alignment (e.g., Bouwhuis and Wittekock for photolithography, 1979; Alles *et al.*, 1975) and chip-by-chip e-beam alignment techniques are used.

8.7 OTHER CONSIDERATIONS, YIELD, AND ULTIMATE LIMITS

The fact that a thinner resist leads to a higher resolution has led to a technique of using multilayered resists. There are oxide or metal steps on a device that are 1 μm or so high. The first layer of the resist is applied to cover the previously patterned device topography on a silicon wafer. This leads to a planar surface. A very thin resist imaging layer is then spun on top of the planarizing layer. Only the top layer is used as a resist and the other layers are removed using the top resist as a mask. Another version of multilevel resists involve exposing and developing all resist layers. The resist layers are not necessarily of the same material (Saotome *et al.*, 1985). A form of multilayer structure was originally developed for the metal lift-off process (Havas, 1976). Multilayer structures are also used for patterning by reactive ion etching. In this case, the thin layer is used as a mask and the underlying thick layer then serves as an etch mask against reactive ion etching for subsequent pattern transfer into the underlying substrate (Moran and Maydan, 1979).

Defects in lithography are imperfections in a pattern. Some defects are fatal in that they lead to an inoperative device; others are cosmetic in nature. Types of defects commonly encountered on a process line are shown in Fig. 8-31. For the lithography based on masks, defects that are large enough to be resolved by the lithographic process will be transferred to the wafer and cause defects in devices.

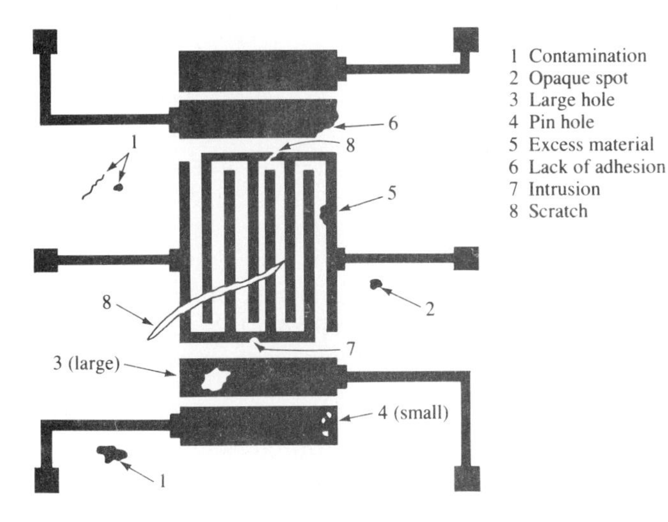

FIGURE 8-31
Types of defects (Ballantyne, 1980).

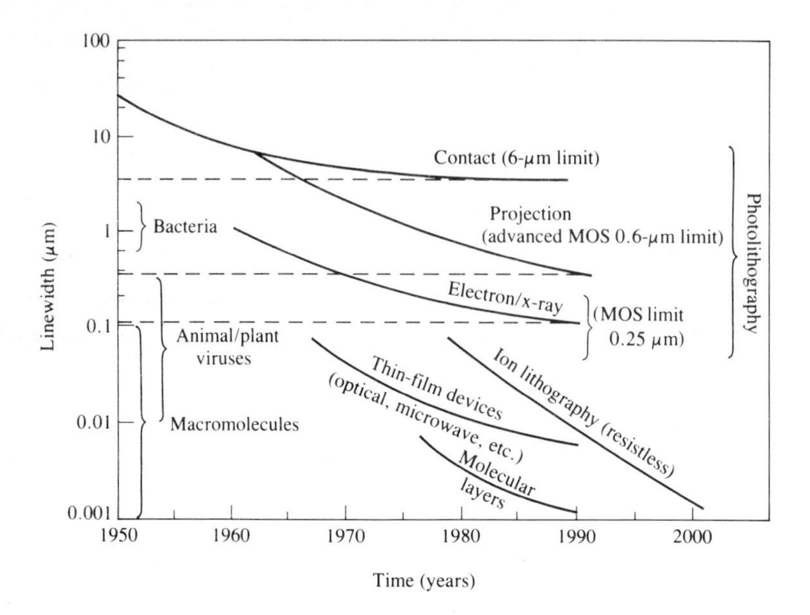

FIGURE 8-32
Linewidth of microelectronic devices and lithographies as a function of time (Brodie and Muray, 1982).

Further, each lithographic step generates additional defects which will combine to reduce the total process yield of good devices. A relationship (Price, 1970) between potential device yield Y and defect density (defects/area) is

$$Y = \prod_{i=1}^{n} (1 + X_i)^{-1} \tag{8.44}$$

where $X_i = D_i A_w$. The average defect density on the ith mask level is D_i, A_w is the wafer area, and n is the number of mask levels.

If it is assumed that X_i is the same for all levels, that is, $X = X_i$, Eq. (8.44) becomes

$$Y = (1 + X)^{-n} \tag{8.45}$$

One can see how rapidly chip yield decreases as the number of mask levels (n) increases. At $X = 0.3$, for instance, the yield decreases from 0.27 to 0.07 as the number of mask levels increases from 5 to 10 according to Eq. (8.45). The usual number of mask levels ranges from 7 to 10.

As the device size continues to decrease and the technology advances, the minimum feature size is going to be ultimately limited (Brodie and Muray, 1982) by the uncertainties dictated by the uncertainty principle of Heisenberg:

$$\Delta l \, \Delta p \geq \frac{h}{4\pi} \tag{8.46}$$

where Δl (μm) is the uncertainty (errors) in the coordinate of a particle, Δp is the uncertainty in its momentum, and h is Planck's constant. For photons, Eq. (8.47) can be rewritten as

$$\Delta l\ (\mu\text{m}) \geq \frac{hC}{4\pi E} = \frac{1.23}{E\ (\text{eV})} \tag{8.47}$$

Likewise, one can write similar equations for electrons and ions:

$$\Delta l\ (\mu\text{m}) = \begin{cases} \dfrac{h}{(2mE)^{1/2}4\pi} = \dfrac{1.22 \times 10^{-3}}{[E\ (\text{eV})]^{1/2}} & \text{(electrons)} \quad (8.48) \\[2ex] \dfrac{h}{(2mE)4\pi} = \dfrac{2.74 \times 10^{-5}}{[(M/M_p)E]^{1/2}} & \text{(ions)} \quad (8.49) \end{cases}$$

where C is the speed of light, E is the beam energy, m is the electron mass, M is the ion mass, and M_p is the mass of the proton. This position uncertainty limits the sharpness of the line edge. For photon beams in the visible range this line edge uncertainty is in the 0.5 μm range; for electron beams it is about 1 Å for $E = 10^3$ eV. However, it is on the order of 10 nm if one includes the low-energy electron contribution and the proximity effect (Wallmark, 1979). It appears that the ultimate limit on the linewidth is in the range of 20 nm. The limiting values on the linewidth and the corresponding technologies (Brodie and Muray, 1982) are shown in Fig. 8-32. In the physical limits of the 10 to 20 nm range indicated in the figure, one would encounter fundamental questions on the device physics and on the traditional planar approach that are now accepted and practiced.

NOTATION

a_1, a_2, a_3	Constants in Eq. (8.41)
A, B	Constants in Example 8.2; constants in Eq. (8.40)
A_w	Wafer area
B	Brightness given in Eq. (8.10)
C	Concentration of resist solution (percent solid); light speed in vacuum
d	Distance from the top of a slit to that of an adjacent slit
d_G	Gaussian beam equivalent diameter given by Eq. (8.28)
D	Dose (C/L^2), incident charge per unit area; lens aperture; defects per area
D_f	Extrapolated dose for full thickness in Fig. 8-18
D_i	Minimum dose necessary for polymer dissolution
E_o	Incident energy (E)
E_T	Threshold energy
E_v	Energy per volume
f	Z/R_G; focal length defined in Eq. (8.9) and Fig. 8-10; spatial frequency (L^{-1})
f_c	Critical spatial frequency given by Eq. (8.26)
g	Gap between mask and resist (L)

G	Number of chain scissions or crosslinking per dissipated energy (E^{-1})
h	Planck's constant
I	Current; intensity (E/t)
I_b, I_f	Quantities defined in Eq. (8.16)
I_l	Light energy flux
j	Integer
J	Current flux (A/L^2t), often referred to as current density
J_p	Maximum J in Eq. (8.10)
$k(z)$	Normalizing factor dependent on resist depth in Eq. (8.15)
k, k'	Constants given by Eqs. (8.19) and (8.20), respectively
k_d	Rate constant in Eq. (8.32)
k_0	Constant in Eq. (8.43)
K	Constant in Eq. (8.1)
l	Resist thickness; distance a light travels for imaging
l_e	Edge length
l_p	Minimum dimension for a line in Eq. (8.36)
Δl	Difference between two light beams
L	Distance between barrier with slits and screen
m	Mass density; edge slope in Eq. (8.43)
M	Molecular weight of polymer; normalized inhibitor concentration in Eq. (8.38)
M_0	Initial value of M
M_1, M_2	Mass of ion and lattice atoms, respectively
MTF	Modulation transfer function (see Prob. 8.17 for tabulated values)
n	Refractive index defined by Eq. (8.1)
N	Number of slits; atomic density (atoms/L^3)
N_A	Avogadro's number
N_{min}	Minimum number of electrons for the sensitivity S
NA	Numerical aperture ($D/2f$)
p	Normalized resist thickness
P	Total pressure
P_e	Number of elastic events given in Fig. 8-12
q	Electronic charge
r	Radius coordinate
R	Development rate (L/t)
R_G	Grun range given by Eq. (8.12) (L)
R_i	Ion penetration length
R_o	R at M_0
s, s'	Object and image distances in Fig. 8-10
S	Selectivity or the minimum dose at which the dimensional quality of clear and opaque features results; spinning speed (r/min)
t	Time
t_r	Exposure time
T	Tolerance given by Eq. (8.44); temperature
v	Light speed in a transparent material

w	Dose function given by Eq. (8.14)
W	Resolution (minimum feature size)
x	Coordinate perpendicular to resist depth coordinate z
y	Length coordinate in Fig. 8-6
Y	Fractional yield in Eq. (8.45)
z	Distance into resist
Z	Depth of focus in Eqs. (8.23) and (8.24)
Z_1, Z_2	Atomic number of ion and lattice atoms, respectively

Greek letters

α	Included half-angle for a cone in Eq. (8.11) (rad); dissolution order in Eq. (8.32)
β	Standard deviation
β_b	Half-width of backward scattering distribution
β_f	Half-width of forward scattering distribution
β_e	Effective half-width
λ	Wavelength; dose function given by Eq. (8.14)
μ	Absorption coefficient in Eq. (8.21); average number of scission or crosslinking per polymer molecule
ρ	Density
Ω	Solid angle in Eq. (8.10); contrast defined by Eq. (8.34)

Units

A	Ampere
C	Coulomb
E	Energy
L	Length
M	Mass
P	Pressure
t	Time
T	Temperature

PROBLEMS

8.1. The grating formula of Eq. (8.7) is the basis for the spatial frequency of mask lines that can be imaged on a resist with a given light beam. Note that d in the equation is the total length for a combination of clear and opaque lines on the mask and sin Ω can be taken as the numerical aperture. For the first-order (principle) diffraction of interest, $j = 1$. Calculate the minimum length of the repeating clear/opaque lines that can be imaged for a coherent beam of 200 nm wavelength and NA of 0.17. Explain what happens if the desired length is smaller than the minimum. Also explain what happens if it is larger than the minimum. Refer to Fig. 8-17.

8.2. Plot the average axial distribution of energy dissipated and the radial distribution at the resist-substrate interface for a silicon substrate and an e-beam resist for the following cases:

	A	B	C
Incident energy, keV	15	25	25
Resist thickness, μm	1.0	0.5	1.0

Draw conclusions on the effect of the incident energy and resist thickness on the distributions. The electron beam may be assumed to be a spot or a point. Use Table 8.1. For the averaging, use the following:

$$E_v \text{ from Eq. (8.13)} = \int_0^R \frac{rE_v(r, z) \text{ of Eq. (8.15) } dr}{R^2/2}$$

where R is 2 times β_b. For simplicity, use ρ of 1 g/cm^3 for the Grun range.

8.3. Calculate the maximum penetration of electron into silicon and compare to the ion range based on Ar^+ for E_o of 20 keV for both. Draw conclusions from the results. Use the atomic density of 10^{22} cm^{-3} and ρ of 2.33 g/cm^3 for silicon.

8.4. X-ray printing is usually done by shadow printing, which is a form of proximity printing. Shadow printing is the typical way of making replicas of masks. Calculate the minimum linewidth that can be projected for a gap of 5 μm and an x-ray of 10 nm wavelength. Determine the energy dissipated at the resist (PMMA)-substrate interface if the resist thickness is 0.5 μm. Assume the incident energy to be 20 mJ/cm^2. Use Fig. 8-14.

8.5. Feder *et al.* (1975) reported a feature of 10 nm using an x-ray of 4.48 nm wavelength. Calculate the gap between the mask and the resist.

8.6. Greeneich (1975) gives the following for the development rate of an electron-beam resist:

$$R \text{ (nm/min)} = \frac{1.046 \times 10^{15}}{M^{3.86}}$$

Calculate the time required to develop the resist for a uniform incident energy of 20 keV. The resist thickness is 0.31 μm. Use the following information:

$$\rho = 1.2 \text{ g/cm}^3 \qquad G = 19 \text{ keV}^{-1} \qquad M_0 = 2 \times 10^5 \qquad D = 5 \times 10^5 \text{ C/cm}^2$$

8.7. The selectivities for some e-beam resists are given below:

$$S = \begin{cases} 2 \times 10^{-5} \text{ (C/cm}^2\text{)} & \text{for PMMA} \\ 4 \times 10^{-7} \text{ (C/cm}^2\text{)} & \text{for COP} \end{cases}$$

Determine the ratio of the minimum features that can be obtained with the resists.

8.8. For the resist considered in Example 8.9, calculate the developed resist width in excess of the clear mask feature for a light beam of 0.3 μm and incident energy of 20 mJ/cm^2.

8.9. For the minimum and maximum exposure time considered in Example 8.4, calculate the corresponding development times for 1-μm thick resist. Use the parameters given in the example.

8.10. For negative e-beam resists, the number of crosslinking events per kiloelectronvolt is around 3; it is usually 19 keV^{-1} for positive resists. For a negative e-beam resist,

Thompson *et al.* (1978) reported the following:

Selectivity (minimum dose) $= 4 \times 10^{-7}$ (C/cm^2)

Dose for 50 percent development in the developed resist thickness versus dose curve $= 10 \times 10^{-7}$ (C/cm^2)

$M_0 = 2.29 \times 10^5$

$\Omega = 1.2$

For an incident energy of 20 keV, calculate the molecular weight at the resist-substrate interface for a resist of 0.6 μm thickness. Do the calculation at the two dose levels given above. Assume the resist density to be 1.2 g/cm^3. For the Grun range, use ρ of 2.33 g/cm^3.

8.11. Suppose only the first term in Eq. (8.46) is taken. Show that the usable throughout P for a given machine, i.e., the fractional number of good circuits in a wafer per unit time, can be written as follows:

$$P = \frac{1}{K'nt_e(1 + A_w D)^n}$$

where t_e is the exposure time. Note that the development time is not included in the formulation.

8.12. Babu and Barouch (1986) arrived at the following implicit solutions of Eqs. (8.38) through (8.40) for positive photoresists:

$$\frac{I_m}{I_0} = \frac{A(1 - M) - B \ln M}{A[1 - \exp(-Dk_i)] + BDk_i} \tag{A}$$

$$\int_{e^{-Dk_i}}^{M} \frac{dy}{y[A(1 - y) - B \ln y]} = z \tag{B}$$

where D is the dose given by $I_0 t$, t being the exposure time. For an ideal photoresist, the absorption coefficient α in Eq. (8.40) should be zero if no inhibitor is present, that is, $B = 0$. For this ideal case (B is usually quite small), the above equations can be manipulated along with Eq. (8.39) to give

$$M(z, t) = \frac{\exp(Az - Dk_i)}{1 + \exp(Az - Dk_i) - \exp(-Dk_i)} \tag{C}$$

which gives the axial inhibitor concentration profile for a given dose. Dill *et al.* (1975b) gave the following parameters for a positive photoresist:

$A = 0.86\ \mu\text{m}^{-1}$ $\quad k_1 = 0.018$ cm^2/mJ $\quad B = 0.07\ \mu\text{m}^{-1}$

$a_1 = 5.63$ $\quad a_2 = 7.43$ $\quad a_3 = -12.6$

Plot the inhibitor concentration profiles for the 0.6-μm thick resist for doses of 15, 60, and 240 mJ/cm^2. Set B equal to zero for the plot. Draw conclusions. Calculate the time required to develop the resist for the dose of 60 mJ/cm^2.

8.13. In projection printing photolithography, the lateral intensity distribution imaged on the resist surface is of Gaussian type when the linewidth is smaller than the depth of focus [Eq. (8.24*b*)] according to the rather rigorous simulation obtained by Lin

(1980). Suppose for a 1-μm linewidth that the lateral distribution is given by

$$I = I_i \exp(-bx^2) \tag{A}$$

(*a*) Show that for an incident light flux of 60 mJ/(cm^2·s), the lateral distribution for b of 2 (μm^{-2}) yields

$$I = 47.9 \exp(-2x^2) \qquad (\text{mJ/cm}^2) \tag{B}$$

Note that

$$\frac{I_0 w}{2} = I_i \int_0^{\infty} \exp(-2x^2)\, dx \tag{C}$$

where w is the linewidth.

(*b*) For the distribution given by Eq. (B), the equation equivalent to Eq. (C) in Prob. 8.12 is

$$M(z, x) = \frac{\exp(Az - D_i k_i X)}{1 + \exp(Az - D_i k_i X) - \exp(-D_i k_i X)} \tag{D}$$

where X is given by

$$X = \exp(-bx^2)$$

and where D_i is the dose corresponding to I_i. Using the parameters in Prob. 8.12, plot the developed resist profiles similar to the one in Fig. 8-21 as a function of development time for D_i of 60 mJ/cm^2 up to t of 4 s.

8.14. For a polymer to crosslink into an insoluble gel, a certain absorbed energy per unit volume of resist E_r is required so that, on the average, one crosslink per chain is formed. This energy can be related as follows:

$$E_r = \frac{\rho N_A}{G M_0}$$

where G is the average number of the crosslink events per unit energy. If a total of E energy per unit volume is absorbed, the total number of crosslink μ is given by

$$\mu = \frac{E_v}{E_r} = \frac{G M_0 E_v}{\rho N_A}$$

which is Eq. (8.30). Thus, the same μ applies to the negative resists. Since $(1 + \mu)$ number of polymer units with molecular weight of M_0 crosslink into one crosslinked polymer unit of molecular weight of M, one has

$$(\mu + 1)M_0 = M$$

which is Eq. (8.31). The value of μ is around 3 keV^{-1}. Assuming that the crosslinked polymer (gel) becomes insoluble at a dose corresponding to μ of 20, obtain a relationship for α from Eq. (8.32).

8.15. If R is the rate of development in length per time, a mole balance on the decomposed (positive resists) or crosslinked (negative resists) polymer is

$$\frac{dN}{dt} = \frac{RA\rho_M}{M} \tag{A}$$

where N is the number of moles of the polymer, A is the surface area irradiated, ρ_M is the polymer density, and M is the polymer molecular weight. On the other hand, N can also be expressed as

$$N = \rho_M \int_0^z \frac{A(\alpha)\, d\alpha}{M} \tag{B}$$

Derive Eq. (8.33) from the above two relationships.

8.16. When an electron beam is of Gaussian type rather than a spot (delta function), the approximate lateral distribution of the energy dissipated can be obtained by convolution. If β_g is the incident Gaussian half-width, the linear lateral distribution can be expressed (Greeneich, 1980) as

$$E_l(r, z) = \int_0^{2\pi} \int_0^{\infty} r' E_v(r', z) \exp\left(-r - \frac{r'^2}{\beta_g^2}\right) dr'\, d\theta \tag{A}$$

where $E_v(r', z)$ is that given by Eq. (8.15) with the half-widths corrected for the incident beam width as follows:

$$E_v(r, z) = k(z)\left[\exp\left(-\frac{r^2}{\alpha_f^2}\right) + \mu_e\left(\frac{\alpha_f^2}{\alpha_b^2}\right) \exp\left(-\frac{r^2}{\alpha_b^2}\right)\right] \tag{B}$$

where
$$\alpha_f = (\beta_f^2 + \beta_g^2)^{1/2}$$
$$\alpha_b = (\beta_b^2 + \beta_g^2)^{1/2}$$

For the dimensionally correct dissipation energy distribution, i.e., energy/volume as in Eqs. (8.13) and (8.15), one can average as in Prob. 8.2:

$$E_v(z) \text{ of Eq. (8.13)} = \int_0^{2\alpha_b} \frac{E_l\, dr}{4\alpha_b^2} \tag{C}$$

Write procedures for obtaining the values of $k(z)$ and the lateral distribution.

8.17. If I_i is the Fourier transform of the lateral distribution of the incident beam $I_i(x)$, defined by

$$I_i(f) = \int_{-\infty}^{\infty} I_i(x) e^{-ifx}\, dx = F[I_i(x)] \tag{A}$$

the image spatial frequency distribution $I_m(f)$ is given by

$$I_m(f) = I_i(f)\ \text{MTF}\ (f) \tag{B}$$

which follows from the definition of MTF. The MTF for a partially coherent beam (Goodman, 1968) is given by

$$\text{MTF}\ (f) = \frac{2}{\pi}\left\{\cos^{-1}\left(\frac{f}{f_c}\right) - \frac{f}{f_c}\left[1 - \left(\frac{f}{f_c}\right)^2\right]\right\}^{1/2} \tag{C}$$

where the critical spatial frequency f_c is that given by Eq. (8.26). The values of MTF are given in the table below as a function of f/f_c (Lin, 1980):

MTF	f/f_c	MTF	f/f_c
1.0000	0.0000	0.4000	0.4010
0.9000	0.0787	0.3000	0.5852
0.8000	0.1578	0.2000	0.6871
0.7000	0.2378	0.1000	0.8054
0.6000	0.3197	0.0000	1.0000
0.5000	0.4040		

The intensity distribution of the image projected onto the photoresist surface, $I_m(x)$, is obtained by the inverse Fourier transformation of Eq. (B):

$$I_m(x) = \frac{1}{2\pi}\int_{-\infty}^{\infty} I_m(f)e^{ifx}\,df = F^{-1}[I_m(f)] \qquad \text{(D)}$$

For the given incident beam, Eqs. (A) through (D) can be used to numerically calculate $I_m(x)$. Assuming that the incident beam is uniform in its intensity over the 1-μm slit, calculate the projected image intensity at $x = 0.1$ μm, that is, $I_m(0.1)$. For the purpose of calculation, use Δf of $0.1f_c$. The values of NA and λ are 0.45 and 435.8 nm.

REFERENCES

Alles, D. S., F. R. Ashley, A. M. Johnson, and R. L. Townsend: *J. Vac. Sci. Technol.*, vol. 12, p. 1252, 1975.

Babu, S. V., and E. Barouch: *IEEE Elect. Dev. Lett.*, vol. EDL-7, p. 252, 1986.

Ballantyne, J. P.: in G. R. Brewer (ed.), *Electron-Beam Technology in Microelectric Fabrication*, chap. 5, Academic Press, New York, 1980.

Batchelder, T., and J. Piatt: *Solid State Technol.*, p. 211, August 1983.

Bouwhuis, G., and S. Wittekock: *IEEE Elect. Dev.*, vol. ED-26, p. 723, 1979.

Bowden, M. J.: *J. Electrochem. Soc.*, vol. 128, p. 195c, 1981.

Bowmer, T. N., and J. H. O'Donnell: *J. Macromol. Sci.*, vol. A17(A), p. 243, 1982.

Brodie, I., and J. J. Muray: *The Physics of Microfabrication*, Plenum Press, New York, 1982.

Chang, T. H. P.: *J. Vac. Sci. Technol.*, vol. 12, p. 1271, 1975.

Colclaser, R. A.: *Microelectronics Processing and Device Design*, Wiley, New York, 1980.

Cuthbert, J. D.: *Solid State Technol.*, vol. 20, p. 59, August 1977.

Dill, F. H., W. P., Hornberger, P. S. Hauge, and J. M. Shaw: *IEEE Trans. Elect. Dev.*, vol. ED-22, p. 445, 1975a.

———, A. R. Neureuther, J. A. Tuttle, and E. J. Walker: *IEEE Trans. Elect. Dev.*, vol. ED-22, p. 456, 1975b.

Eisberg, R. M., and L. S. Lerner: *Physics: Foundations and Applications*, McGraw-Hill, New York, 1981.

Everhart, T. E., and P. H. Hoff: *J. Appl. Phys.*, vol. 42, p. 5837, 1971.

Feder, R., E. Spiller, and J. Topalian: *J. Vac. Sci. Technol.*, vol. 12, p. 1332, 1975.

Goodman, J. W.: *Introduction to Fourier Optics*, chap. 6, McGraw-Hill, New York, 1968.

Greeneich, J. S.: *J. Electrochem. Soc.*, vol. 122, p. 970, 1975.

———: in G. R. Brewer (ed.), *Electron-Beam Technology in Microelectronic Fabrication*, chap. 2, Academic Press, New York, 1980.

Hatzakis, M.: *J. Vac. Sci. Technol.*, vol. 12, p. 1275, 1975.

Havas, J. R.: *Electrochem. Soc. Extended Abstr.*, vol. 76-2, p. 743, 1976.
Hawryluk, R. J., A. M. Hawryluk, and H. I. Smith: *J. Appl. Phys.*, vol. 45, p. 2551, 1974.
Heidenreich, R. D., J. P. Ballantyne, and L. R. Thompson: *J. Vac. Sci. Technol.*, vol. 12, p. 1284, 1975.
Herriott, D. R., and G. R. Brewer: in G. R. Brewer (ed.), *Electron-Beam Technology in Microelectronic Fabrication*, chap. 3, Academic Press, New York, 1980.
Hershel, R., and C. A. Mack: in N. G. Einspruch (ed.), *VLSI Electronics*, vol. 16, *Lithography for VLSI*, chap. 2, Academic Press, Orlando, 1987.
Herzog, R. F., J. S. Greeneich, T. E. Everhart, and T. Van Duzer: *IEEE Trans. Elect. Dev.*, vol. ED-19, p. 629, 1972.
Jenkins, F., and H. White: *Fundamentals of Optics*, 4th ed., McGraw-Hill, New York, 1976.
Karapiperis, L., I. Adesida, C. A. Lee, and E. D. Wolf: *J. Vac. Sci. Technol.*, vol. 19, p. 1259, 1981.
Kim, D. J., W. G. Oldham, and A. R. Neureuther: *IEEE Trans. Elect. Dev.*, vol. ED-31, p. 1730, 1984.
Kyser, D. F., and N. S. Viswanathan: *IEEE Trans. Elect. Dev.*, vol. 12, p. 1305, 1975.
Lin, B. J.: in R. Newman (ed.), *Fine-Line Lithography*, chap. 2, North-Holland, Amsterdam, 1980.
McGillis, D. A.: in S. M. Sze (ed.), *VLSI Technology*, chap. 7, McGraw-Hill, New York, 1983.
——— and D. L. Fehrs: *IEEE Trans. Elect. Dev.*, vol. ED-22, p. 471, 1975.
Moore, G. E.: Technical Digest International Electrical Development Meeting, pp. 11–13, December 1975.
Moran, J. M., and D. J. Maydan: *J. Vac. Sci. Technol.*, vol. 16, p. 1620, 1979.
Ouano, A. C.: in T. Davidson (ed.), *Polymers in Electronics*, ACS Symposium Series 242, ACS, Washington, 1984.
Pamplone, T. R.: *Solid State Technol.*, p. 115, June 1984.
Parikh, M.: *IBM J. Res. Dev.*, vol. 24, p. 438, 1980.
——— and D. F. Kyser: IBM Research Report RJ2261, 1978.
Price, J. E.: *Proc. IEEE*, vol. 58, p. 1290, 1970.
Saotome, Y., H. Gokam, K. Saiga, M. Suzuki, and Y. Ohniske: *J. Electrochem. Soc.*, vol. 132, p. 909, 1985.
Skinner, J. G.: *Proc. Kodak Interface, '73*, p. 53, 1973.
Spiller, E., and R. Feder: in H. J. Queisser (ed.), *X-Ray Optics*, Springer-Verlag, Berlin, 1977.
Tai, K. L., E. Ong, and R. G. Vadimsky: *Proc. Electrochem. Soc.*, vol. 82 (9), p. 9, 1982.
Thompson, L. F., L. E. Stillwagon, and E. M. Doerries: *J. Vac. Sci. Technol.*, vol. 15, p. 938, 1978.
——— and M. J. Bowden: in L. F. Thompson *et al.* (eds.), *Introduction to Microlithography*, chap. 4, ACS Symposium Series 219, ACS, Washington, 1983.
Trotel, J., and B. Fay: in G. R. Brewer (ed.), *Electron-Beam Technology in Microelectronic Fabrication*, chap. 6, Academic Press, New York, 1980.
Wallmark, J. T.: *IEEE Trans. Elect. Dev.*, vol. ED-26, p. 135, 1979.
Willson, C. G.: in L. F. Thompson *et al.* (eds.), *Introduction to Microlithography*, chap. 3, ACS Symposium Series 219, ACS, Washington, 1983.
Wittels, N. D.: in R. Newman (ed.), *Fine-Line Lithography*, chap. 1, North-Holland, Amsterdam, 1980.
Yoshikawa, A., O. Ochi, H. Nagai, and Y. Mizyshima: *Appl. Phys. Lett.*, vol. 29, p. 677, 1976.

CHAPTER 9

PHYSICAL AND PHYSICOCHEMICAL RATE PROCESSES

9.1 INTRODUCTION

Physical processes are an important part of microelectronics processing. In general, much of metallization and dielectric material deposition is still carried out either by direct evaporation of a source solid (for deposition on a substrate) or by physical sputtering. Deposition based on the evaporation of a source solid(s) is sometimes referred to as physical vapor deposition (PVD), as opposed to chemical vapor deposition. Physical sputtering is a process in which ions accelerated by an applied electric field sputter (eject) atoms from a substrate surface by momentum transfer. As such, physical sputtering can be used for both etching and deposition: etching if the substrate is the target and deposition if it receives sputtered atoms from a target. An important application of evaporation is the technique known as molecular beam epitaxy (MBE). MBE is primarily used for compound semiconductors because of its ability to deliver well-defined thin-film layers at relatively low temperatures, which in turn allows for well-defined doping profiles.

Plasma processes are becoming more versatile and useful for device fabrication. Although physical sputtering processes (in which the ions are generated) may also involve plasmas, the term plasma process is usually reserved for one in which chemical processes eventually get involved in deposition or etching. The origin of the chemical activity, however, is the physical process in which reactive neutral species (or ions) are generated in a plasma by electrons colliding with molecules. Therefore, plasma processing is governed by physicochemical rate processes. The major impetus for plasma processing is the low-temperature environment that the plasma offers. Further, anisotropy can be realized much more easily by plasma etching than by any other etching technique due to the normal incidence of ions onto the substrate. Substrate surface modification through momentum transfer by colliding ions is a major reason for the enhanced etching by reactive species in the plasma. Because a plasma is formed when an electric field is applied between two electrodes or by an alternating voltage, electrical characteristics of the plasma process are one of the major factors, if not the sole factor, that govern the plasma process.

In this chapter, PVD (physical vapor deposition) is treated first. Since plasma is central to the rest of the chapter, the nature and characteristics of plasma are then treated for subsequent sections of physical sputtering, plasma deposition, and plasma etching.

9.2 EVAPORATION AND PHYSICAL VAPOR DEPOSITION

Evaporation in vacuum from free solid surfaces is often used to deposit metal for device interconnections. The evaporation is also used to deposit epitaxial films, better known as molecular beam epitaxy (MBE). The source solid is heated by resistance or induction heating (with a crucible that couples with the rf) and the evaporated molecules then deposit by condensation onto the cold surface of the substrate. The source solid can also be bombarded by a particle beam (electron or ion) to generate the vapor. Conventional sputtering, which involves acceleration of ions (usually Ar^+) through a potential gradient and the bombardment by these ions of a target or cathode, is another form of physical vapor deposition. Through momentum transfer, atoms near the surface of the target material become volatile and are transported as a vapor to the substrate.

The quantities of interest are the rates of evaporation and deposition. The latter depends strongly on the directionality of motion of evaporating molecules. The driving force for the evaporation is the saturation vapor pressure, p^*, on the solid surface that must be maintained at a given temperature. Counteracting the driving force is the rate of impingement (adsorption) from the vapor phase to the solid. Thus, the net number of molecules dN evaporating from a solid surface area A during the time dt is given by

$$\frac{dN}{A\,dt} = (2mk_B T)^{-1/2}(p^* - p) \tag{9.1}$$

where k_B is the Boltzmann constant, T is temperature, m is the mass of the molecule, p is the partial pressure of the evaporant in the gas phase, and p^* is the corresponding equilibrium pressure. The relationship is known as the Hertz-Knudsen equation (Glang, 1970) when the right-hand side of the equation is multiplied by a constant called the evaporation coefficient, which is less than unity. The maximum rate of evaporation, which is the rate for unit evaporation coefficient and zero pressure, is often used to represent the mass rate of evaporation V as follows:

$$V[g/(\text{cm}^2\cdot\text{s})] = 5.834 \times 10^{-2}\left(\frac{M}{T}\right)^{1/2} p^* \tag{9.2}$$

where M is the molecular weight and p^* is in torr. The total mass evaporation rate is that obtained by integrating over the solid surface area A:

$$V_t\ (\text{g/s}) = \int_A V\, dA \tag{9.3}$$

The direction in which molecules will be emitted from a solid surface can be modeled by considering the interaction between the solid surface and an individual molecule about to evaporate. Consider an isothermal enclosure with an infinitesimally small opening dA_e bounded by vanishingly thin walls as shown in Fig. 9-1. It can be shown (Glang, 1970) that the differential total mass rate of evaporation is given by

$$dV_t(\Omega) = V_t \cos\Omega\, \frac{d\omega}{\pi} \tag{9.4}$$

Since $dA_r = r^2\, d\omega/\cos\alpha$ according to the geometry in Fig. 9-1, one has, from Eq. (9.4),

$$\frac{dV_t}{dA_r} = \frac{V_t}{\pi r^2} \cos\Omega \cos\alpha = r_D \tag{9.5}$$

where r_D (mass per area per time) is the maximum rate of deposition and r is the distance between the source and substrate. This relationship is known as the

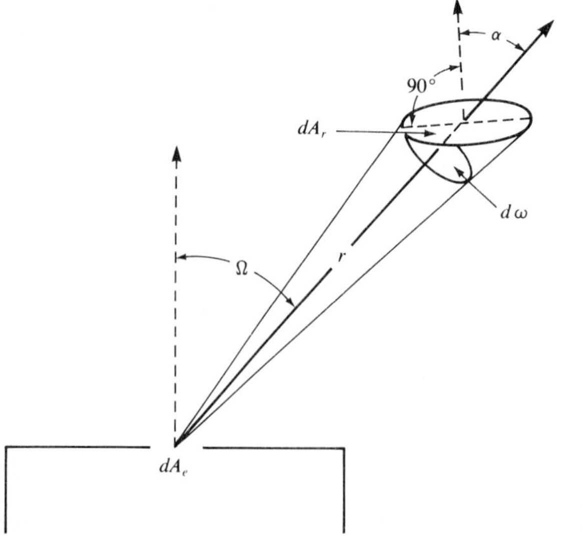

FIGURE 9-1
Surface element dA_r receiving deposit from a small area source (Glang, 1970).

cosine law of emission. The extension of this cosine law to emission from solid surfaces is generally taken to be permissible. For a point source, although of limited practical utility, Eq. (9.5) reduces to

$$\frac{dV_t}{dA_r} = \frac{V_t \cos \alpha}{\pi r^2} \tag{9.6}$$

The deposition rate at various points on a substrate plane above a small area source [Eq. (9.5)] can be expressed (Chang, 1970) as

$$\frac{r_D}{(r_D)_0} = \left[1 + \left(\frac{L}{r}\right)^2\right]^{-2} \tag{9.7}$$

where $(r_D)_0$ is the rate directly above the source at a distance r and r_D is the rate at the point L away from the center of the substrate plane. Likewise, one has, for a point source [Eq. (9.6)],

$$\frac{r_D}{(r_D)_0} = \left[1 + \left(\frac{L}{r}\right)^2\right]^{-3/2} \tag{9.8}$$

When the receiving surface is spherical with a radius of r_0, as shown in Fig. 9-2, $\cos \alpha = \cos \Omega = r/2r_0$ and Eq. (9.5) can be rewritten as

$$r_D = \frac{V_t}{4\pi r_0^2} \tag{9.9}$$

It is seen that the maximum deposition rate is the same everywhere on the spherical surface. This is the reason why planetary substrate-supporting systems (rotating spherical sections) are used in deposition chambers.

Example 9.1. At 3230 °C, the saturation pressure of tungsten (mp = 3380 °C) is 0.01 torr. Calculate the mass rate of evaporation. Assume the source area to be 1 cm^2 and the evaporation chamber pressure to be 10^{-4} torr. Also calculate the maximum

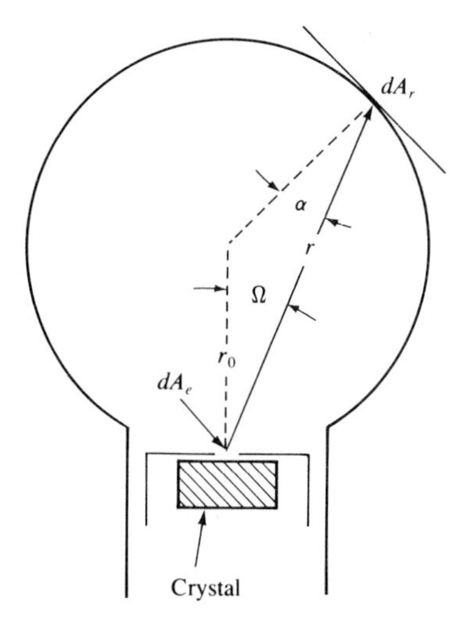

FIGURE 9-2
Evaporation from a small area source dA_e onto a spherical receiving surface (Glang, 1970).

rate of deposition on a substrate placed directly above and 2 cm away from the metal source. Determine the maximum deposition rate for a substrate 5 cm away from the center of the substrate plane.

Solution. The mass rate is given by Eq. (9.2):

$$V = 5.834 \times 10^{-2}\left(\frac{183.85}{3503}\right)^{1/2} \times 0.01$$

$$= 1.337 \times 10^{-4} \text{ g/(cm}^2\text{·s)}$$

For the source area of 1 cm^2, $V_t = 1.337 \times 10^{-4}$ g/s. For the substrate placed directly above the source, $\cos \alpha = \cos \Omega = 1$. Thus, one has, from Eq. (9.5),

$$r_D = \frac{V_t}{\pi r^2} = \frac{1.337 \times 10^{-4}}{3.14(2)^2} = 1.06 \times 10^{-5} \text{ g/(cm}^2\text{·s)}$$

Noting for the last part of the problem that $(r_D)_0 = 1.06 \times 10^{-5}$ g/(cm^2·s), one has, from Eq. (9.7),

$$r_D = (r_D)_0\left[1 + \left(\frac{L}{r}\right)^2\right]^{-2}$$

$$= 1.06 \times 10^{-5}[1 + (\tfrac{2}{5})^2]^{-2}$$

$$= 0.788 \times 10^{-5} \text{ g/(cm}^2\text{·s)}$$

Physical vapor deposition (PVD) occurs when evaporated source material is condensed onto a surface (substrate surface) that is cold relative to the evaporating surface. For silicon epitaxy by MBE, for example, the cold surface temperature ranges from below 800 to 850 °C; it is around 500 °C for GaAs. As in CVD film growth, the growth by PVD involves condensation of atoms (adatoms), subsequent migration of the adatoms on the surface, and incorporation of the adatoms into a crystal structure. The major difference is that PVD involves condensation (physical adsorption in a broad sense) whereas CVD involves mostly chemical adsorption. The consequence of this difference is that both adsorption and desorption have to be considered for the kinetics of CVD whereas only condensation onto a cold surface is needed in PVD. The type of crystal structure that results from PVD is entirely determined by the relative rate of condensation with respect to the rate of adatom migration. If the condensation rate is much higher than the surface migration rate, an amorphous structure would result, which is typical of the metallization by PVD. On the other hand, a single crystalline structure would result if the condensation rate is less than or equal to the (potential) rate of surface migration at a given temperature, as in the case in MBE. A polycrystalline structure forms in between these two extremes. Therefore, it is natural that the rate of deposition is the same as the net rate of condensation (often called the rate of impingement) when an amorphous structure is involved. Noting that the condensation rate is the limiting process in epitaxial deposition, one can conclude the same is true for epitaxial deposition. Therefore, it is sufficient to consider only the condensation rate in determining the rate of deposition. Neglecting reevaporation of the condensed phase from the relatively cold surface gives a net rate of condensation:

$$r_c = (2mk_B T)^{-1/2} p^*$$
$$= 3.513 \times 10^{22} (MT)^{-1/2} p^* \quad (9.10)$$

where M is the molar mass in grams and p is in torr.

Example 9.2. Meyerson *et al.* (1986) used a low-pressure (10^{-3} torr) CVD to deposit epitaxial silicon film on a silicon wafer at temperatures ranging from 750 to 850 °C. The average growth rate was reported to be 6 nm/min. Noting that the temperature range is similar to that for the MBE of silicon and that the growth rate can be used as that sustainable in the MBE, calculate the maximum pressure that can be allowed for film growth by MBE. Assume the gas temperature to be at 800 °C.

Solution. Since Eq. (9.10) is in atoms per square centimeter per second, the linear growth rate of 6 nm/min needs to be converted to the following:

$$\frac{G \rho_s N_A}{M_s} = \frac{(6 \times 10^{-7}/60)(2.33)(6.02 \times 10^{23})}{28}$$
$$= 5.01 \times 10^{14} \text{ atoms/(cm}^2\cdot\text{s)}$$

where G is the linear growth rate, ρ_s and M_s, respectively, are the silicon density and molecular weight, and N_A is Avogadro's number. For epitaxy by MBE, the condensation rate given by Eq. (9.10) should be less than or equal to the growth rate given in Example 9.1:

$$3.513 \times 10^{22} (MT)^{-1/2} p \leq 5.01 \times 10^{14}$$

or

$$p \text{ (torr)} \leq 2.47 \times 10^{-6}$$

As the example illustrates, the pressure required for MBE is much lower than that for CVD epitaxy, because the impingement rate cannot be larger than the maximum possible rate of epitaxial growth at a given temperature. This in turn means that the pressure for epitaxy by PVD should be very low but that the pressure for metallization (amorphous structure) can be relatively much higher. Typical pressures for silicon MBE are 10^{-8} torr, while those for metallization are in the 10^{-3} to 10^{-8} to 10^{-10} torr range. Impingement rates for a number of evaporants and background residual gases are given in Fig. 9-3. The broken lines in the figure indicate that a chromium condensation rate of 1 Å/s corresponds to 8×10^{14} atoms/(cm^2·s). It is necessary to minimize any residual gas species in the evaporation chamber, since they also deposit. As shown in Fig. 9-3, an oxygen partial pressure of 10^{-6} torr would result in an incorporation of oxygen at a competing rate of 4×10^{14} molecules/(cm^2·s). In contrast, a dopant source can be introduced to the chamber as the desired impurity for doping.

The relatively high temperature of the cold substrate surface has an important bearing on the defect level of an epitaxial film. The dislocation density of an MBE silicon decreases with increasing temperature, as shown in Fig. 9-4. This is

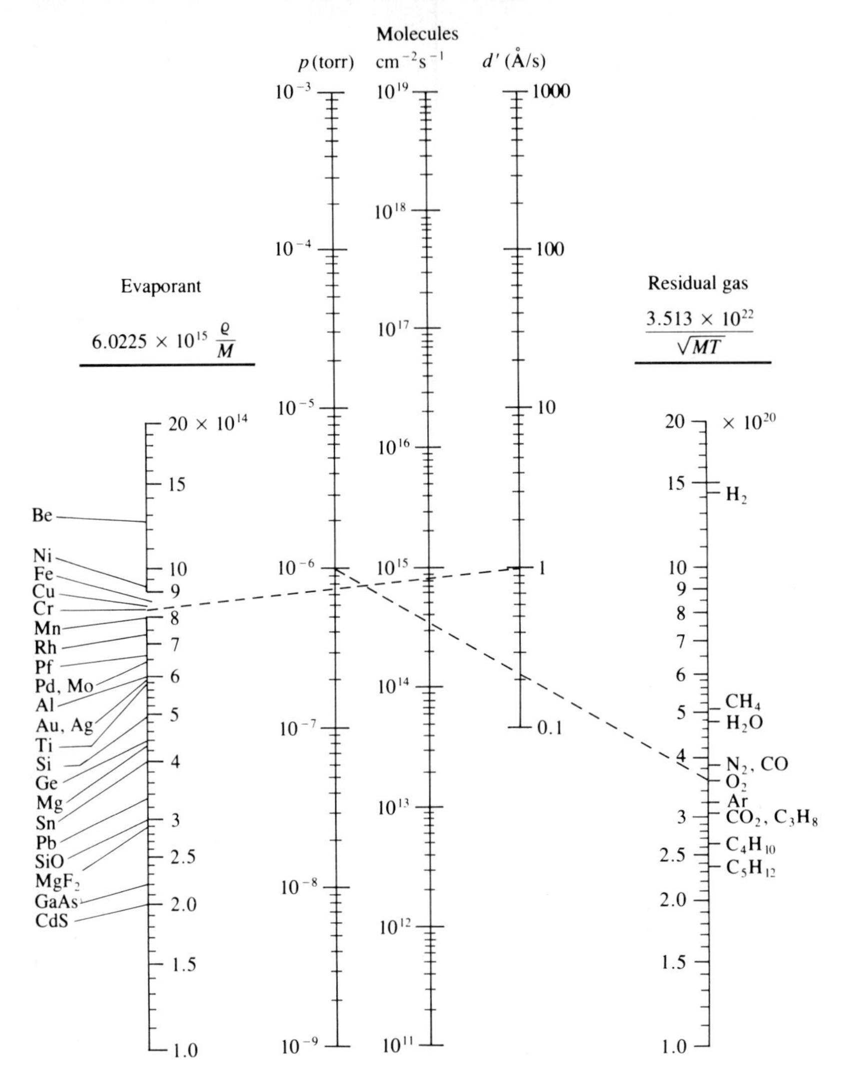

FIGURE 9-3
Nomogram to determine impingement rate. The appropriate point in the evaporant axis is to be connected with the observed (pure) deposition rate on the d' axis. For residual gases, points on the far right and on the pressure axes should be connected. The intersects with the center axis give the impingement rates (Glang, 1970).

expected since the surface migration rate increases with increasing temperature while the impingement rate decreases to some extent. One would also expect the defect level to decrease with decreasing pressure because of the corresponding decrease in the impingement rate, which favors more orderly epitaxial growth.

Vaporization of solid substances can also be caused by electron bombardment of the substance surface. The electron source is usually a hot cathode. The

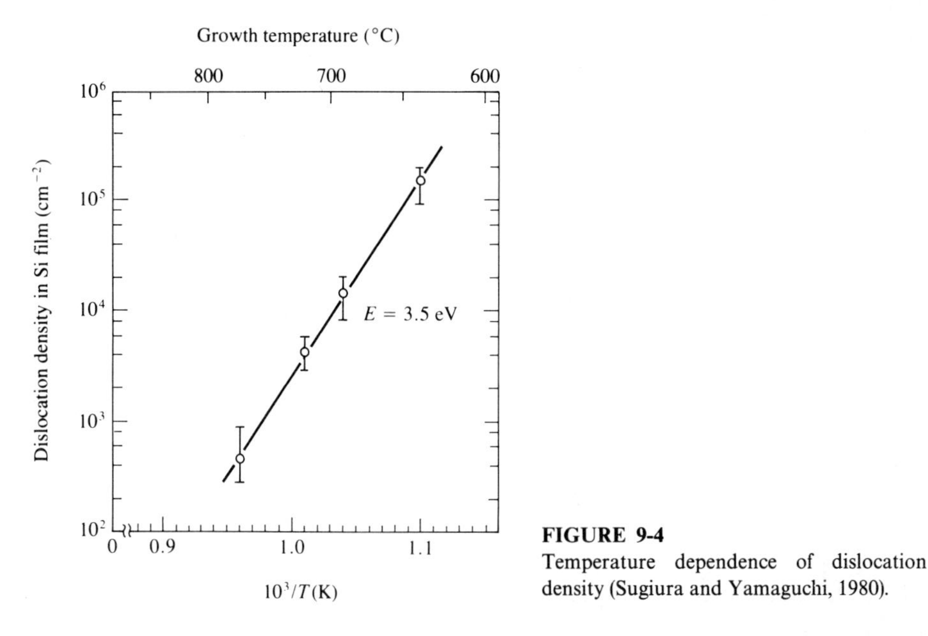

FIGURE 9-4
Temperature dependence of dislocation density (Sugiura and Yamaguchi, 1980).

electrons are accelerated through fields of 5 to 10 kV onto the evaporant surface. Since the energy is imparted by charged particles, only the surface is heated. The surface temperature can exceed 3000 °C (Glang, 1970) while the remainder of the solid is maintained at a lower temperature. Devices operating on the principle of electron-bombardment heating are referred to as electron guns. While solid substances have been the major evaporant, vapor substances are increasingly used in MBE, in particular for III-V compounds films (Cho, 1987).

Another major method of metallization for VLSI circuits is physical sputtering. An understanding of plasmas, however, is required to go into the sputtering.

9.3 PLASMA

A plasma is a collection of positively and negatively charged particles, including neutral ones. Charge neutrality requires that the number density of the positive charges is equal to that of the negative ones. Plasma constitutes the fourth state of matter. As a matter is heated, its state changes from solid to liquid, from liquid to gas, and then to plasma. The plasma state is in fact by far the most common form of matter (up to 99 percent of the universe). It is also the most energetic state: matter requires on the average 10^{-2} eV/particle to change its state from solid to liquid or from liquid to gas, but it requires 1 to 30 eV/particle to change its state from gas to plasma.

The essence of plasmas as applied to film deposition and etching lies in a self-sustaining discharge between cathode and anode, in particular glow discharge. Three types of self-sustaining discharges can result when electrons are emitted from a cathode (discharge). These are dark discharges, gas discharges,

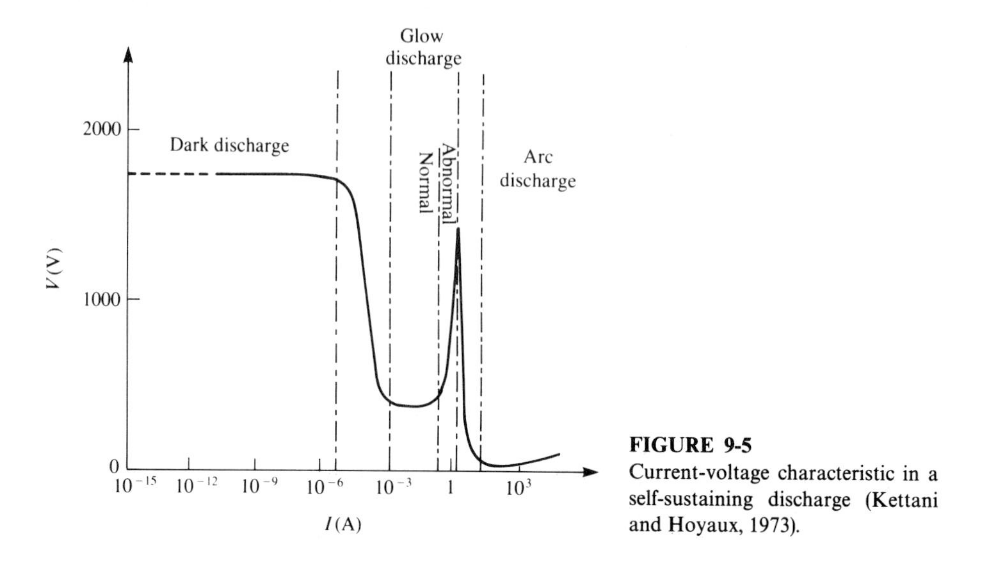

FIGURE 9-5
Current-voltage characteristic in a self-sustaining discharge (Kettani and Hoyaux, 1973).

and arc discharges. The types are determined by the current-voltage characteristics of the discharge as shown in Fig. 9-5. A current will flow between the separated electrodes in a gas at low pressure, provided the applied voltage is above the minimum, or breakdown, voltage. The breakdown voltage is that level at which the gas can be at least partially converted to plasma. As the electrons emitted from the cathode travel toward the anode, they make a fixed number of ionizing collisions. The ions that result from these collisions are also accelerated by the applied field and move toward the cathode. Some of these ions, on striking the cathode, eject secondary electrons from its surface. The glow that results is said to be self-sustained when the number of secondary electrons produced at the cathode is sufficient to maintain the discharge.

A self-sustained glow discharge has a fair degree of structure that is visible, as illustrated in Fig. 9-6. The Crookes dark space is where positive ions have accumulated. The dark space adjacent to an electrode (cathode) is often referred to as a plasma sheath or simply a sheath. Its thickness is approximately the mean distance traveled by an electron from the cathode before it makes an ionizing collision. Electrons traverse the dark space very rapidly because of the localized

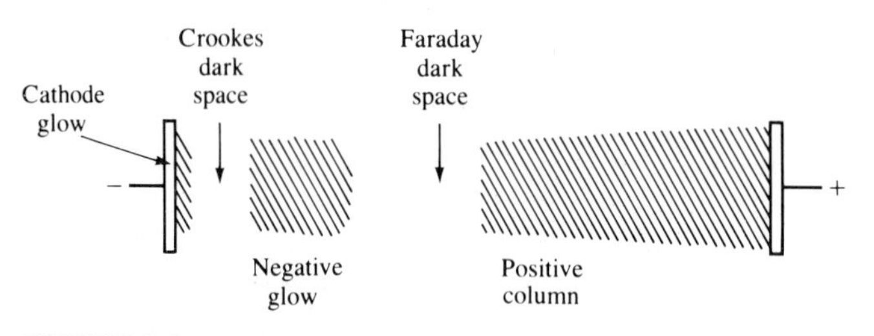

FIGURE 9-6
Cross-sectional view of glow discharge (Maissel, 1970).

space charge there. As the edge of the negative glow is reached, electrons begin to produce a significant number of ion-electron pairs, and the preponderance of the positive charge falls off very rapidly. A neutral region consisting of approximately equal numbers of ions and electrons begins, forming a plasma. At relatively low voltages the cross-sectional area of the glow is less than the available cathode area since there exists a minimum current density requirement for maintaining the glow. If additional power is applied to the discharge tube, the glow adjusts by increasing its cross-sectional area, thus raising the total current but keeping the current density at the cathode constant. As long as this current density does not increase, neither does the voltage across the dark space. The minimum voltage drop needed to maintain the glow is called the normal cathode fall, and the corresponding glow is referred to as the normal glow. When the power is increased beyond the point where the glow covers the entire available cathode area, the current density at the cathode must increase. This in turn implies an increase in the emission of secondary electrons from the cathode. The glow discharge operating in this mode is referred to as abnormal (Fig. 9-5). In practice, this is the only mode of interest for film deposition by physical sputtering. The current density in the normal glow is too low for atoms to be sputtered out of the cathode at a useful rate. At the same time, the corresponding voltage drop is quite low (Fig. 9-5) so that sputtering yields are similarly low.

When electrons enter the negative glow, they possess essentially the full cathode fall of potential. This energy is then lost through a series of either ionizing or excitation collisions. Eventually, the electrons' energy is reduced to a point where they are no longer able to produce additional ions. The region in the discharge where this happens defines the far edge of the negative glow, and since no more ions are being produced, the electrons begin to accumulate there, forming a region of slightly negative space charge. In this region, the electrons have insufficient energy to cause either ionization or excitation. Consequently, it is a dark region, known as the Faraday dark space. After passing through the Faraday dark space by diffusion, electrons are accelerated toward the anode. This region is called the positive column.

Since the self-sustaining feature of the discharge depends only on the emission of sufficient electrons at the cathode by positive ions from the negative glow, the exact location of the anode normally makes very little difference to the electrical characteristics of the glow. Thus, if the anode is moved closer and closer in toward the cathode, the positive column will be extinguished, the Faraday dark space will disappear, and, finally, a large fraction of the negative glow may be extinguished before any appreciable effect is seen in the electrical characteristics.

If an electrode is inserted into a glow and is biased with respect to the anode, the effect of negative bias is that a second cathode is created with its own ion sheath around it. However, since the flow is already being sustained by the flow of secondary electrons from the primary cathode, the second cathode can operate at as low a voltage as desired since the glow is not dependent on it for a supply of secondary electrons. The additional cathode is referred to as a probe. More details can be found in Maissel (1970).

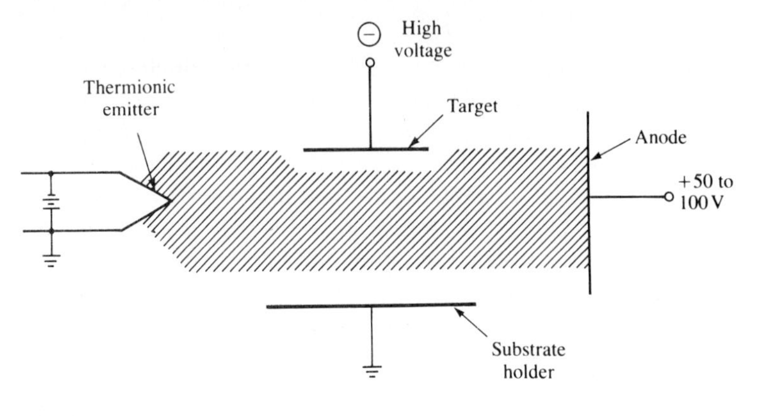

FIGURE 9-7
Schematic of thermionically supported glow discharge adapted for sputtering (triode sputtering) (Maissel, 1970).

The simplest discharge is the one produced by applying a dc potential between two metal electrodes in a partially evacuated enclosure. Typically, the discharge operates at pressures exceeding 3×10^{-2} torr and the applied voltage exceeding a few hundred volts. Instead of electrons by secondary emission, a hot cathode can be used to supply the electrons through thermionic emission, sometimes called a low-voltage arc. Such a system can provide finite current even at pressures lower than 3×10^{-2} torr. At still lower pressures ($\sim 10^{-3}$ torr), the mean free path for electrons exceeds the typical dimensions of discharge enclosure, and the probability of ionizing collisions is too small to maintain the discharge unless the electrons are confined by an external magnetic field (Bollinger and Fink, 1980) or by insertion of a probe (target), as shown in Fig. 9-7.

Suppose that an alternating voltage is applied to the two electrodes instead. At sufficiently low frequencies a dc discharge can be produced that alternates between the two electrodes at the same frequency. On the other hand, no dc discharge can be established if the frequency is high. The discharge simply oscillates between the electrodes and the electrons pick up sufficient energy during their oscillations (random motion) to cause ionization. The minimum pressure at which the discharge will occur is gradually reduced with increasing frequency. The frequency at which such a discharge is sustainable is usually in the radio-frequency (rf) range, the effect being detectable above about 50 kHz and leveling off for frequencies in excess of a few megahertz. This mode of discharge is referred to as rf discharge.

The rf discharge is the main method of producing a plasma in applications to IC processing because of several advantages it can offer. First, sputtering of an insulator is possible with an rf discharge. Simple substitution of an insulator for the metal target in dc discharge leads to failure because of the immediate buildup of a surface charge of positive ions on the front side of the insulator, preventing any further ion bombardment. This also permits the use of reactive gases typically used in plasma etching since the electrodes within the discharge can be covered with insulating material. Second, the rf discharge can be operated at pressures as low as 10^{-3} torr, where the mean free path of ions and of sputtered atoms become comparable with or larger than the enclosure dimensions. This reduces or eliminates many of the complications inherent in glow discharges such as diffusion of sputtered material back to the target, poorly defined bombarding-ion energies, charge exchange effects, etc. Third, the discharge can be sustained independently of the yield of secondary electrons from the walls and electrodes.

Low-pressure plasmas used in IC processing are generally characterized by a low degree of ionization (typically 0.1 to 1 percent) and an absence of thermal equilibrium between ions, electrons, and neutral gas molecules. The range of electron energies is between 1 and 20 eV. The ion densities are normally between 10^9 and 10^{12} cm^{-3}. This is set by fundamental limits. When the density is less than 10^9 cm^{-3}, the electrostatic force is weak enough for the charge to separate over a fairly large distance and neutrality is no longer maintained. This is usually unstable as an operating region. Plasma densities higher than 10^{12} cm^{-3} correspond to high currents and significant gas heating. High temperature can be harmful to substrate materials and leads to plasma instability and nonuniformity (Flamm and Herb, 1988). Considering that the density of gas molecules at 1 torr is about 10^{16} cm^{-3}, it can be seen that the discharges are weakly ionized. This results in a nearly ambient gas temperature, despite a mean electron temperature of about 10^4 to 10^5 K. The relatively low gas temperature permits the use of thermally sensitive materials such as organic resists.

Physical sputtering based on plasma can be used for both deposition and etching. In the case of deposition, the atoms sputtered out by the target (cathode) deposit onto a substrate as shown in Fig. 9-8. If the target itself is the substrate as shown in Fig. 9-8*b*, the sputtering results in etching of the substrate. The deposition is usually referred to as physical sputtering, although physical sputtering can also be used for etching as in ion milling (Bollinger and Fink, 1980).

9.4 PHYSICAL SPUTTERING

Two key quantities of interest in physical sputtering are the threshold energy and the sputtering yield. The threshold energy is defined as the minimum ion energy for sputtering to ensue from a target bombarded at normal incidence by an ion. The threshold energies for various metals with noble gases can be found in Wehner and Anderson (1970). The threshold values are roughly four times the metal sublimation energy for most inert gases used for physical sputtering. The sputtering yield is defined as the number of target atoms sputtered by one colliding ion. The sputtering yield S in the energy range of interest for film deposition (a few hundred to a few kiloelectronvolts) can be determined theoretically (Steinbruchel, 1985) from

$$S = \alpha(E^{1/2} - E_{th}^{1/2}) \tag{9.11}$$

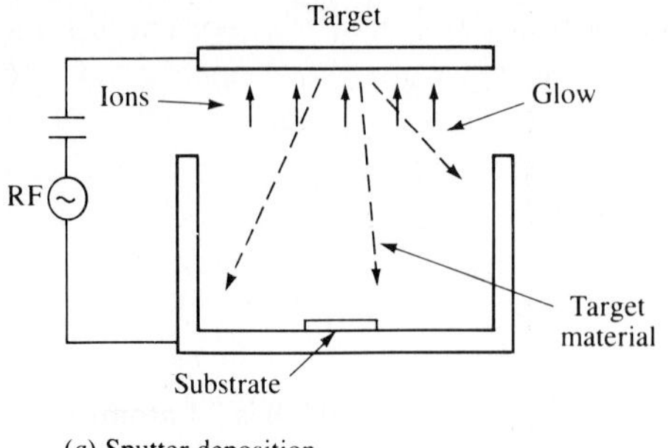

(a) Sputter deposition

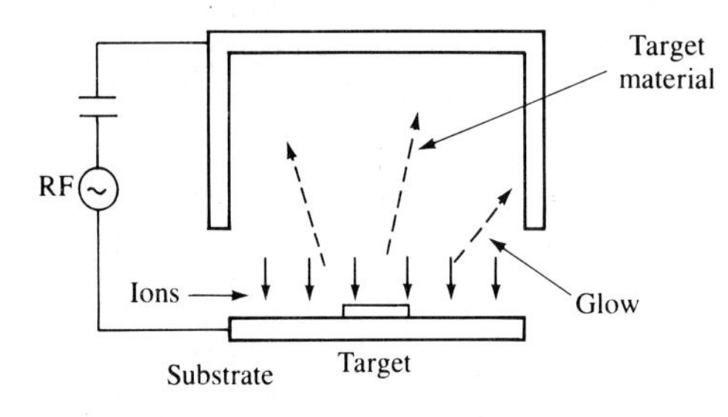

(b) Sputter etching

FIGURE 9-8
Two modes of operation of a two-electrode rf sputtering system (Horwitz, 1983).

where
$$\alpha = \frac{5.2}{U} \frac{Z_t}{(Z_t^{2/3} + Z_x^{2/3})^{3/4}} \left(\frac{Z_x}{Z_t + Z_x}\right)^{0.67}$$

E is the ion energy in kiloelectronvolts, E_{th} is the threshold energy in kiloelectronvolts, U is the surface binding energy in electronvolts, which can be taken as the sublimation energy, and Z_t and Z_x, respectively, are the atomic numbers of the target and the gas. When the gas is a molecule consisting of m atoms, the relationship becomes

$$S = m^{1/2}\alpha[E^{1/2} - (mE_{th})^{1/2}] \tag{9.12}$$

Example 9.3. Consider tungsten deposition in a dc discharge with Ar. The threshold energy is 33 eV. Calculate the sputtering yield for a cathode voltage of 100 volts. The heat of sublimation of tungsten is 191 kcal/mol. Use the sheath potential (voltage drop across the sheath) for the cathode potential. Note that 1 kcal/mol is equal to 0.0434 eV/molecule.

Solution. The ion energy is 100 eV for the cathode potential of 100 volts. Also, from the sublimation energy,

$$U = 0.0434 \times 191 = 8.29 \text{ eV/molecule (atom)}$$

For the target tungsten and the gas Ar,

$$Z_t = 74 \qquad \text{and} \qquad Z_x = 18$$

These values can be used in Eq. (9.11) for α:

$$\alpha = \frac{5.2}{8.29} \frac{74}{(74^{2/3} + 18^{2/3})^{3/4}} \left(\frac{18}{74 + 18}\right)^{0.67} = 1.41$$

and thus one has for the sputtering yield

$$S = 1.41(0.1^{1/2} - 0.033^{1/2})$$
$$= 0.19 \text{ atom/ion}$$

The experimental value for the yield (Wehner and Anderson, 1970) is 0.1 atom/ion.

The rate of sputtering, r_s, which can be defined as the number of atoms sputtered per unit time per unit area, can be obtained by multiplying the sputtering yield by the ion flux. Since ion current density (current flux) is equal to the ion flux times electron charge q, one has, for r_s,

$$r_s = \frac{Sj_i}{q} \tag{9.13}$$

where j_i is the ion current density. Although the rate of sputtering is given by Eq. (9.13), it is not a simple matter to relate the ion energy and the ion current density to pertinent electrical and physical characteristics of sputtering.

For deposition by physical sputtering, the pressure is usually low enough that collisions in the target sheath can be neglected. Under the "space charge limited condition," the ion current density can be expressed (Hasted, 1964) as

$$j_i = \frac{\beta V_{sp}^{3/2}}{d^2 M_i^{1/2}} \tag{9.14}$$

where β is a constant, which is equal to $0.85 p_0 q^{1/2}$ where p_0 is the permittivity of free space, V_{sp} is the sheath potential (voltage drop across the dark space adjacent to the cathode in dc sputtering and adjacent to the powered electrode in the case of rf sputtering), d is the sheath thickness, and M_i is the ion mass. For dc glow discharges, the sheath potential can be taken as the voltage applied to the cathode or the cathode potential. However, the situation is more complicated for rf discharges.

Example 9.4. The electrical quantities that are readily measurable in the dc sputtering are the cathode potential and the cathode current. Although the cathode current is mainly by ions, the secondary electrons emitted from the cathode by the incident ions do contribute to the current. If the number of secondary electrons emitted per incident ion is denoted by μ, the cathode current density j_c can be expressed (Brown, 1966) as

$$j_c = j_i(1 + \mu)$$

A typical number for μ is 0.1. Consider in this light tantalum deposition by dc sputtering as reported by Vratny (1967). The deposition rate reported is 8 nm/min with a cathode potential of 850 volts, and the corresponding cathode current density is 0.25 mA/cm^2 in an argon environment of 20 millitorr. Calculate the rate of sputtering, and compare it with the reported deposition rate. The heat of sublimation of Ta is 190 kcal/mol and the threshold energy is 26 eV. Discuss the results.

Solution. From the latent heat of sublimation,

$$U = 0.0434 \times 190 = 8.25 \text{ eV/atom}$$

For the target of tantalum and the gas Ar,

$$Z_t = 73 \qquad \text{and} \qquad Z_x = 18$$

As in Example 9.3, one can calculate α; it is 1.4. From Eq. (9.11),

$$S = 1.4(0.85^{1/2} - 0.026^{1/2})$$

$$= 1.065 \text{ atom/ion}$$

Since $j_c = j_i(1 + \mu)$ and $j_c = 2.5 \times 10^{-4}$ A/cm^2, one has, for μ of 0.1,

$$j_i = \frac{j_c}{1 + \mu} = \frac{2.5 \times 10^{-4}}{1.1} = 2.273 \times 10^{-4} \text{ A/cm}^2$$

From Eq. (9.13) for r_s,

$$r_s = \frac{1.065 \times 2.273 \times 10^{-4}}{1.6 \times 10^{-19}} = 1.51 \times 10^{15} \text{ atoms/(s·cm}^2)$$

To convert from a linear deposition rate (G) of 8 nm/min or 0.133 nm/s, one can use the following:

$$r_d \text{ [atoms/(s·cm}^2)] = \frac{G\rho N_A}{M}$$

$$= \frac{(0.133 \times 10^{-7})(16.6)(6.02 \times 10^{23})}{181}$$

$$= 7.34 \times 10^{14}$$

where ρ and M, respectively, are the density and atomic weight of tantalum and N_A is Avogadro's number.

The results show that the actual rate of deposition is approximately one half the rate of sputtering. Aside from the numbers, it should be understood that the actual rate of deposition is generally smaller than the intrinsic rate of sputtering, because of the effects involved in the transport of the sputtered atoms to the substrate located on the anode. The sputtered atoms can also deposit onto surfaces other than the substrate such as walls. On the other hand, the rate of etching is the same as the rate of sputtering, since then the target, substrate, is itself.

Consider rf discharge for the ion energy. Radio-frequency discharges consist of two parts, as shown in Fig. 9-9. These are the plasma body which has approximately equal numbers of electrons and ions and the two sheaths, one at each electrode, where the ion density is much larger than the electron density. Shown

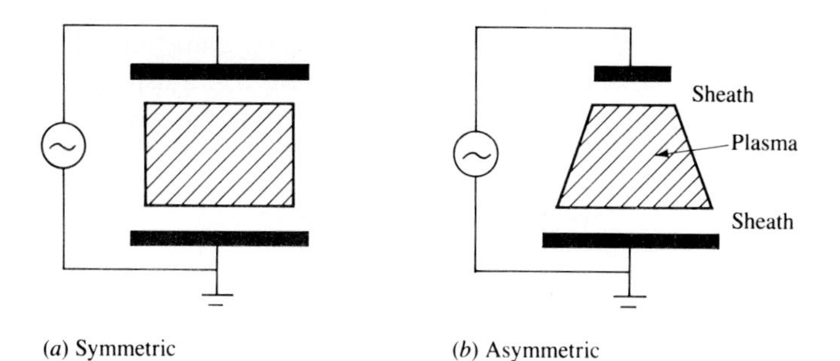

FIGURE 9-9
Symmetrical and asymmetrical electrode systems.

in Fig. 9-9*a* is the case where the powered electrode area (A_p) is the same as the grounded electrode area (A_g); Fig. 9-9*b* is for $A_g > A_p$. The former is referred to as a symmetrical electrode system; the latter is called asymmetrical. In the plasma sheaths, the current is dominated by ions. Further, no secondary electrons from the target are involved in maintaining the plasma. Therefore, the root mean square current measured at the powered electrode can be taken as the ion current density (when it is divided by A_p), j_i. In the plasma, however, electrons dominate as the major current carriers because of their higher mobility compared with the ions, even though the electron density is approximately equal to the cation density.

The plasma of an rf discharge develops an appreciable potential, V_p, which is positive with respect to both electrodes. This is shown in Fig. 9-10 for both symmetrical and asymmetrical cases. In the symmetric electrode system, the sheath potential V_{sp} is the same at both electrodes and is equal to the plasma potential V_p. Therefore, the average ion energy is that corresponding to V_p.

In the asymmetrical case, however, the current density varies because the electrode areas are different and yet the current is the same at both electrodes. This difference causes a bias of voltage at the powered electrode (small area). This is shown in Fig. 9-10*c* for time-varying voltage distributions. Note in this regard

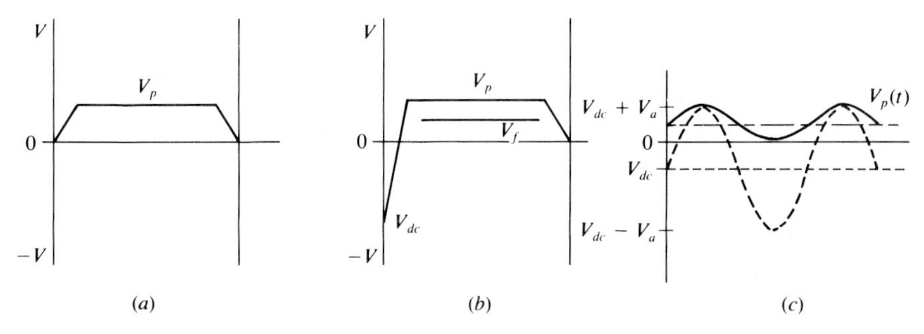

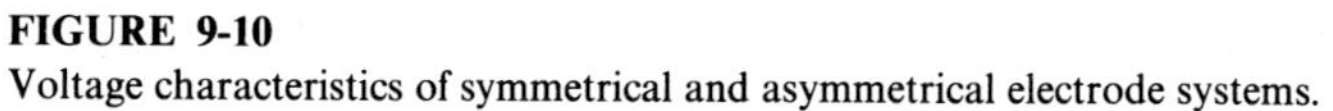

FIGURE 9-10
Voltage characteristics of symmetrical and asymmetrical electrode systems.

that the voltage distributions in Fig. 9-10*a* and *b* are for the time-averaged values. However, rf systems without a blocking capacitor do not have bias regardless of the area ratio. The average plasma potential can be determined (Chapman, 1980) from

$$V_p = 0.5(V_a + V_{dc}) \tag{9.15}$$

where V_a is the applied voltage amplitude (Fig. 9-10*c*). In a capacitively coupled rf discharge, the sheath potential is given (Coburn and Kay, 1972) by

$$V_{sp} = V_p - V_{dc} \tag{9.16}$$

Note that the reference voltage is the ground, such that the value of V_{dc} is negative.

Example 9.5. Consider an rf discharge operating at 13.56 MHz. Many rf discharge processes operate at this frequency because it is the one allotted to international communications. For an asymmetrical, parallel-plate system, suppose that the root mean square current and applied voltage amplitude are 0.1 ampere and 160 volts, respectively, and that the powered electrode area is 10 cm^2 whereas the grounded electrode area is 30 cm^2. The dc bias with respect to the ground is 80 volts. Calculate the aluminum sputtering rate for an aluminum target placed on the powered electrode in an argon environment. For aluminum, the threshold energy in argon is 13 eV/atom and the latent heat of sublimation is 3.25 eV/atom. Recalculate the rate for the case where the target is placed on the grounded electrode.

Solution. The sheath and plasma potentials follow from Eqs. (9.15) and (9.16):

$$V_p = 0.5(160 - 80) = 40 \text{ volts}$$

$$V_{sp} = 40 - (-80) = 120 \text{ volts}$$

Since ions dominate in the sheath, the ion current density at the powered electrode can be calculated as follows:

$$j_i = \frac{0.1}{10} = 0.01 \text{ A/cm}^2$$

The sputtering yield can be calculated from Eq. (9.11):

$$\alpha = \frac{5.2}{3.25} \frac{13}{(13^{2/3} + 18^{2/3})^{3/4}} \left(\frac{18}{13 + 18}\right)^{0.67} \qquad \text{for } Z_{\text{Al}} = 13$$

$$= 2.257$$

$$S = 2.257(0.12^{1/2} - 0.013^{1/2}) = 0.52 \text{ atom/ion}$$

since the ion energy is that corresponding to the sheath potential. The rate of sputtering follows directly from Eq. (9.13):

$$r_s = \frac{0.52(0.01)}{1.6 \times 10^{-19}}$$

$$= 3.25 \times 10^{16} \text{ atoms/(cm}^2\text{·s)}$$

If the target is placed on the grounded electrode, the ion energy is that corresponding to V_p, or 40 eV. The current density is

$$j_i = \frac{0.1}{30} = 0.0033 \text{ A/cm}^2$$

The sputtering yield is

$$S = 2.257(0.04^{1/2} - 0.013^{1/2}) = 0.194 \text{ atom/ion}$$

Thus,

$$r_s = \frac{0.194(0.0033)}{1.6 \times 10^{-19}}$$
$$= 4 \times 10^{15} \text{ atoms/(cm}^2\text{·s)}$$

This example shows that the intrinsic rate of sputtering, which is the maximum possible rate of deposition in the case of film deposition and the actual rate of etching in the case of etching, is much higher when the target is placed on the powered electrode than on the grounded electrode. Therefore, the asymmetrical system is usually used in rf sputtering and the area ratio, A_p/A_g, is an important design parameter. Since the rate of sputtering is entirely determined by ion energy and current density, it is necessary to relate these to the transport processes of electrons and ions in the sputtering apparatus. This is treated in the next chapter.

As discussed earlier, a simple substitution of an insulator for the metal target fails in a conventional dc sputtering system because of the immediate buildup of a surface charge of positive ions on the front side of the insulator. This prevents any further ion bombardment. However, in rf sputtering, the voltage alternates so that the target is alternately bombarded by cations and then electrons, which dissipates the charge buildup. Therefore, rf sputtering is the method of choice for both insulator etching and deposition.

An important feature of the sputtering process (Maissel, 1970) is that the chemical composition of a sputtered film is often the same as that of the target from which it was sputtered. This is true even though the components of the system may differ significantly in their relative sputtering rates, except when resputtering occurs. The very first time that sputtering is performed from a multicomponent target, the component with the highest sputtering rate comes off faster, but a so-called "altered region" soon forms at the surface of the target. This region becomes depleted in the higher sputtering rate component to compensate for its greater removal rate and later sputtering rates tend to even out. Subsequent deposits have the composition of the parent material. However, the stoichiometry of a film obtained from a compound material is not the same as the parent material. In this case, the target is sputtered both in the molecular and atomic forms (Coburn *et al.*, 1974). Often the film is deficient in gaseous or other volatile species. For example, an oxide film obtained by sputtering can be deficient in oxygen.

Example 9.6. Consider the example by Chapman (1980) for the sputtering of a 80:20 Ni-Fe alloy. For 1000-eV argon ions, the sputtering yield is 2.1 for Ni and 1.4 for Fe. After the initial period, the composition of the alloy film should be in the ratio of 80:20. Determine the composition of the target surface that would yield the desired film composition.

Solution. The initial ratio of the sputtered atoms will be (80 × 2.1):(20 × 1.4). As the iron enrichment continues, the sputtering rate of the iron atoms increases and of nickel atoms decreases until they are again leaving in the ratio 80:20. For the ratio of sputtered atoms to be 80:20, the following condition has to be satisfied:

$$\frac{2.1y}{1.4(1-y)} = \frac{80}{20}$$

where y is the fraction of Ni in the target surface after the initial period. The solution yields a value of 0.727 for y, meaning that the surface composition is 72.7 percent Ni and 27.3 percent Fe.

The deficiency in the stoichiometry of a deposited compound film can be compensated for by introducing a gas containing the deficient atoms, deficient gas, in the sputtering environment. For example, any oxygen deficiency in the deposited film can be corrected by introducing a suitable amount of oxygen to the gaseous environment such as argon (Erskine and Cserhati, 1978). This is turn means that any undesired impurities in the environment will also deposit onto the film. The trapping of the impurities can occur for inert ions such as Ar^+ if its energy is high enough to be implanted into the film as in ion implantation. For active ions such as $O_2{}^+$, adsorption is the typical mode of impurity trapping. For such ions, the fraction, f_i, of species i trapped in a film being deposited at the rate of r_d is given by

$$f_i = \frac{(r_a)_i}{(r_a)_i + r_d} \tag{9.17}$$

where $(r_a)_i$ is the rate of adsorption of species i.

The effect of target temperature on the rate of sputtering is minimal unless the target becomes too hot since only physical change is involved. The same can be said for the substrate temperature, since the only thermal effect is through evaporation of the deposited material. The pressure effect is on the transport of sputtered atoms, which is considered in the next chapter.

9.5 PLASMA DEPOSITION AND GAS-SOLID REACTION

Physical sputtering is a major method by which metallization is carried out. However, the fact that any reactive species can be generated in a plasma by simply introducing the desired gases to the electrode system has been exploited by others for film deposition other than metallization. If a film source gas is

TABLE 9.1
Comparisons between sputtering and plasma deposition (Catherine, 1985)

	Sputtering	Plasma
Source	Solid (gas)	Gas
Pressure, torr	$<10^{-2}$	$0.1 \sim 2$
Deposition temperature, °C	~ 25	<300
Deposition rate, nm/min	$0.1 \sim 10$	$1 \sim 50$
Crystallinity	Mostly amorphous (crystalline)	Mostly amorphous (polycrystalline)
DC bias, V	$500 \sim 3000$	<300
Power density, W/cm^2	0.5	0.5

introduced to an electrode system, the reactive species generated can lead to deposition at a lower temperature, because of its more reactive nature. Creation of a more active substrate surface by the impact of colliding ions is another enhancement factor in plasma deposition. This type of deposition is often referred to as plasma-assisted or plasma-enhanced deposition, or sometimes plasma-enhanced chemical vapor deposition. The difference between deposition by physical sputtering and plasma deposition may be described in terms of the source. The target material is the source in the case of physical sputtering, but the gas is the source for plasma deposition. Another major difference is that the substrate is usually immersed in the plasma in the case of plasma deposition. In this way, reactive atoms or molecules are more readily accessible to the substrate. Therefore, electrodes may not be present in some cases. An example is the glow generated by induction (rf) heating. Comparisons between sputtering and plasma deposition are given in Table 9.1.

When the substrate is immersed in a plasma, sheaths again form around it. Since the surroundings outside the sheaths is at the plasma potential V_p, and the substrate is at a "floating" potential V_f, which is lower than V_p to repel electrons, the energy barrier the electrons have to overcome is the difference, $(V_p - V_f)$. This floating potential is also shown in Fig. 9-10. Further, the net current to the substrate is zero, i.e., the ion current is the same as the electron current, because of the neutrality of plasma. This potential difference is approximately given (Chen, 1974) by

$$V_p - V_f = \frac{k_B T_e}{2q} \ln \left(\frac{M_i}{2.3 m_e} \right) \tag{9.18}$$

where T_e is the electron temperature and m_e is the mass of an electron. This potential difference is sufficiently small for the sputtering to be negligible, which is another reason for immersing the substrate in the plasma.

Example 9.7. The floating potential and electron temperature are determined experimentally by inserting a probe into the plasma. The electrical characteristics of a plasma with a probe (or equivalently immersed substrate) are completely defined by the Langmuir theory (Langmuir, 1923; Langmuir and Mott-Smith, 1924).

However, experimental realization is quite another matter (e.g., Morgan, 1985). Based on the following typical values given by Chapman (1980), calculate the potential drop across the plasma substrate sheath:

$$M_i = 6.6 \times 10^{-23}\ \text{g} \qquad m_e = 9.1 \times 10^{-28}\ \text{g}$$

$$T_e = 23{,}200\ \text{K} \qquad T_i\ (\text{gas temperature}) = 290\ \text{K}$$

It is useful to know that 1 eV corresponds to 11,600 K, which follows from the k_B value ($k_B = 8.62 \times 10^{-5}$ eV/K).

Solution. From Eq. (9.18),

$$\begin{aligned} V_p - V_f &= \frac{k_B T_e}{2q} \ln \frac{6.6 \times 10^{-23}}{2.3 \times 9.1 \times 10^{-28}} \\ &= \frac{8.62 \times 10^{-5} \times 23{,}200\ (\text{eV})}{2q}\ 10.358 \\ &= \frac{10.36}{q}\ \text{eV} \\ &= 10.36\ \text{volts} \end{aligned}$$

The unique features of plasma deposition are the generation of various reactive species at almost room temperature by colliding hot electrons in the plasma and the creation of energetically favorable substrate surface sites by colliding ions with the substrate surface. These unique features, which are the very reasons for employing the technique, are at the source of extraordinarily complex phenomena that have yet to be unraveled. For deposition kinetics, one needs an understanding of the reactions leading to the formation of reactive species in the plasma. The eventual adspecies are determined by the compatibility of the reactive species in the plasma with the energetics of the substrate surface. Therefore, a full description of the deposition kinetics requires not only the surface state as modified by colliding ions but also the heterogeneous reactions on the substrate. However, because of the complexity, such an understanding is not available for any plasma deposition process.

All ions (radicals) and neutrals in the plasma are originally generated by the impact of colliding electrons with molecules. When the electron impact leads to the formation of neutrals, the process is often referred to as dissociation (decomposition). Examples are:

$$e + H_2 \longrightarrow H + H + e$$

$$e + SiH_4 \longrightarrow SiH_2 + H_2 + e$$

$$e + CF_4 \longrightarrow CF_3 + F + e$$

As the examples show, dissociation leads to reactive neutrals, free radicals, such as H, SiH_2, F and CF_3 but it also leads to stable ("unreactive") neutrals such as

H_2. Ionization by electron impact can take several paths. Examples are:

$$e + \mathrm{Ar} \longrightarrow \mathrm{Ar}^+ + 2e$$

$$e + \mathrm{SiH_4} \longrightarrow \mathrm{SiH_3}^+ + \mathrm{H} + 2e$$

$$e + \mathrm{CF_4} \longrightarrow \mathrm{CF_3}^- + \mathrm{F}$$

The first reaction is simple ionization. The second is dissociative ionization, and the third is dissociative attachment, since the electron combines with CF_3 upon impact. These ions and neutrals can also form various products by reacting among themselves and with other species. Since radical-initiated reactions can lead to many products through propagation, recombination and disproportionation, it is not difficult to envisage a myriad of species present in plasma. Each reaction in the examples can be treated as an elementary reaction. For the reaction, $e + \mathrm{SiH_4} \rightarrow \mathrm{SiH_2} + \mathrm{H_2} + e$, for instance, one can write:

$$r_{\mathrm{SiH_2}} = k n_e n_{\mathrm{SiH_4}} \tag{9.19}$$

where k is the rate constant, n_e is the electron density, and n_{SiH4} is the silane density (typically in units of molecules/cm^3 or simply cm^{-3}).

For typical rf discharge systems, electron (and thus positive ion) densities in plasma are between 10^8 and 10^{12} cm^{-3}. Average electron energies are several electron volts while ion (and neutral) energies are at least two orders of magnitude lower. Further, the density of neutrals is greater than that of ions by three orders of magnitude and thus the neutral species are the primary contributors to film deposition (Hess, 1986). Because of the positive, i.e., $(V_p - V_f)$, sheath potential around the substrate, mostly positive ions and electrons but few negative ions reach the substrate, negative ions being much less mobile than electrons. Further, neutrals reach the substrate by diffusion across the sheath, and positive ions arriving at the substrate become immediately neutralized by electrons (Greene and Barnett, 1982). Thus, neutrals are the main adspecies precursors. A determination of the dominant reactive neutrals can be made on the basis of the dissociation energy required for their formation. The reactive neutrals can then become candidates for adsorption precursors.

Example 9.8. According to Turban (1984), the dissociation energies of silane are as follows:

		Dissociation energy, eV
$e + \mathrm{SiH_4} \longrightarrow$	$\mathrm{SiH_3} + \mathrm{H} + e$	4.04
	$\mathrm{SiH_2} + \mathrm{H_2} + e$	2.16
	$\mathrm{SiH_2} + 2\mathrm{H} + e$	6.7
	$\mathrm{SiH} + \mathrm{H} + \mathrm{H_2} + e$	5.7
	$\mathrm{Si} + 4\mathrm{H} + e$	13.2

Determine the ratio of the density of SiH_3 to that of SiH_2 in plasma, assuming Boltzmann statistics. Assume a plasma temperature of 300 K.

Solution. According to Boltzmann statistics, the probability of a molecule overcoming a dissociation energy barrier, E_{da}, and forming products is given by

$\exp(-E_{da}/k_B T)$. Thus, the ratio of SiH_3 density to SiH_2 density, n_{SiH_3} to n_{SiH_2}, is given by

$$\frac{n_{SiH_3}}{n_{SiH_2}} = \frac{\exp(-4.04/k_B T)}{\exp(-2.16/k_B T) + \exp(-6.7/k_B T)}$$

$$= \frac{1}{\exp[(4.04 - 2.16)/k_B T] + \exp[(4.04 - 6.7)/k_B T]}$$

$$= \frac{1}{\exp[1.88/(8.6 \times 10^{-5} \times 300)] + \exp[-2.66/(8.6 \times 10^{-5} \times 300)]}$$

$$= \frac{1}{\exp(72.7) + \exp(-102.9)}$$

$$= \exp(-72.7) = 2.67 \times 10^{-32}$$

An assumption here is that SiH_3 and SiH_2 are formed from SiH_4 only. Note that $k_B = 8.62 \times 10^{-5}$ eV/K (1 eV/molecule = 23.04 kcal/mol). It is seen that the SiH_3 density is negligible compared with n_{SiH_2}. Thus, the reactive neutral with the lowest dissociation energy dominates, particularly at room temperature, since a slight difference in the dissociation energy makes a large difference in the relative density at that temperature.

Deposition rate kinetics can be derived on the basis of assumed adspecies precursors, just as was done in Chap. 5. Although plasma deposition is mostly used on amorphous materials, crystalline structures can also be obtained. As in any type of deposition, the key to the crystallinity is the relative rate of precursor adsorption with respect to that of surface migration of adspecies. Either a polycrystalline or an amorphous material will result if the adsorption rate is higher than the migration rate. In plasma deposition, these rates are determined by the substrate temperature and the flux to the substrate of the adspecies precursor. Monocrystalline films occur at high substrate temperatures. For instance, an epitaxial silicon film has been grown by plasma deposition by heating the substrate to 750 °C (Donahue *et al.*, 1984). As discussed in detail in Sec. 5.4, first-order kinetics results when an amorphous film is formed by crosslinking:

$$r_a = KC_a = J_a \tag{9.20}$$

where r_a is the rate of amorphous film deposition per unit surface area, C_a is the adspecies concentration on the substrate surface, and J_a is the flux of an adspecies precursor, which is usually one of the reactive neutrals diffusing through the sheath around the substrate. If some level of crystallinity is present in the amorphous film, the rate is given (Sec. 5.4) by

$$r_a = \frac{k'C_a}{1 + KC_a} \tag{9.21}$$

The procedures of Sec. 5.4 can be applied to the kinetics of poly- and monocrystalline film growth. Any substrate surface modification made by colliding ions that promotes deposition would appear in the form of larger rate constants when compared with the film grown by conventional CVD.

Silicon nitride in the form of Si_xN_{1-x} is often grown as a passivating layer by plasma deposition. The reactants are usually either SiH_4/N_2 or SiH_4/NH_3 and the film is grown at temperatures below 300 °C (e.g., Gorowitz *et al.*, 1985). Silicides used as contact materials are also often grown by plasma deposition. For tungsten silicide, for instance, the reactants are WF_6 and SiH_4 (e.g., Akimoto and Wantanabe, 1981). For the deposition involving two species, one may consider competitive adsorption followed by incorporation into the solid structure. For the competitive adsorption, one may write:

$$A(g) + S \overset{K_A}{\rightleftharpoons} A \cdot S$$

$$B(g) + S \overset{K_B}{\rightleftharpoons} B \cdot S$$

where A and B are the adspecies precursors for the film atom ratio in the form of A_xB_{1-x}, and A·S and B·S are the corresponding adspecies. It follows (Sec. 5.4) from the adsorption steps that

$$\frac{C_{A \cdot S}}{C_t} = \frac{K_A C_A}{1 + K_A C_A + K_B C_B} \tag{9.22}$$

$$\frac{C_{B \cdot S}}{C_t} = \frac{K_B C_B}{1 + K_A C_A + K_B C_B} \tag{9.23}$$

where K_A and K_B are the adsorption equilibrium constants, and C_A and C_B are the concentrations of the precursors A and B. Each adspecies can become part of the amorphous structure independently. They can also combine into the film structure. Therefore, the steps for incorporation into the amorphous structure are

$$\begin{aligned} \alpha A \cdot S &\xrightarrow{k_A} A_\alpha(a) \\ \beta B \cdot S &\xrightarrow{k_B} B_\beta(a) \\ A \cdot S + \mu B \cdot S &\xrightarrow{k_c} AB_\mu(a) \end{aligned} \tag{9.24}$$

where $A_\alpha(a)$ and $B_\beta(a)$, respectively, are α number of adspecies A and β number of adspecies B incorporated into the amorphous structure, and $AB_\mu(a)$ is the part of amorphous incorporation involving one adspecies A and μ number of adspecies B. It follows from Eqs. (9.22), (9.23), and (9.24) that the atom ratio x is given by

$$x = \frac{\alpha r_A + r_c/\mu}{\alpha r_A + \beta r_B + r_c} \tag{9.25}$$

where r_A, r_B, and r_c are given by

$$r_A = \left(\frac{k_A K_A C_A}{1 + K_A C_A + K_B C_B}\right)^\alpha \tag{9.26}$$

$$r_B = \left(\frac{k_B K_B C_B}{1 + K_A C_A + K_B C_B}\right)^\beta \tag{9.27}$$

$$r_c = \frac{k_c K_A K_B^\mu C_A C_B^\mu}{(1 + K_A C_A + K_B C_B)^{1+\mu}} \tag{9.28}$$

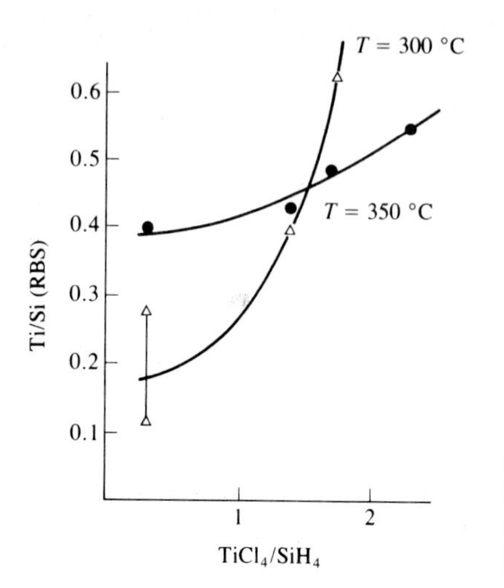

FIGURE 9-11
Film composition as a function of gas phase composition for plasma-enhanced CVD (Kemper *et al.*, 1985).

Example 9.9. Experimental results obtained by Kemper *et al.* (1985) are shown in Fig. 9-11 for Ti_xSi_{1-x} silicide film prepared by plasma deposition. The results show the Ti/Si atom ratio as a function of gas phase composition, $TiCl_4/SiH_4$. The concentration in Eqs. (9.25) through (9.28) are those at the substrate, whereas the concentration ratio in Fig. 9-11 is for the bulk gas phase. Therefore, the concentration ratio cannot be used in the equations directly. For the sake of illustration, assume the concentration ratio to be the adspecies concentration ratio at the surface. Reduce Eq. (9.25) for the following cases:

(*a*) r_c is negligible.
(*b*) r_c is negligible and $\alpha = \beta$.
(*c*) $\alpha = \beta = 2$, $\mu = 1$.

Compare the reduced results with the data in Fig. 9-11.

Solution

(*a*) If r_c is negligible, Eq. (9.25) reduces to

$$x = \frac{\alpha r_A}{\alpha r_A + \beta r_B} \tag{A}$$

$$= \frac{1}{1 + (\beta/\alpha)\{(k_B K_B C_B)^\beta / [(1 + K_A C_A + K_B C_B)^{\beta-\alpha}(k_A K_A C_A)^\alpha]\}}$$

(*b*) If $\alpha = \beta$, one has, from Eq. (A),

$$x = \frac{1}{1 + k_B K_B C_B/(k_A K_A C_A)} \tag{B}$$

(*c*) If $\alpha = \beta = 2$ and $\mu = 1$, it follows from Eqs. (9.26) through (9.28) that

$$x = \frac{2(k_A K_A C_A)^2 + k_c K_A K_A C_A C_B/\mu}{2(k_A K_A C_A)^2 + 2(k_A K_B C_B)^2 + k_c K_A K_B C_A C_B} \tag{C}$$

Under the assumption, $TiCl_4/SiH_4$ is equal to C_A/C_B. In terms of the atom ratio Ti/Si, x can be expressed as

$$x = \frac{\text{Ti/Si}}{1 + \text{Ti/Si}} \tag{D}$$

It can be deduced from Eqs. (B) and (D) that the data cannot be represented by Eq. (B). Additional information would be necessary for the other cases.

It should be recognized that the kinetics for plasma-enhanced CVD (PECVD) are in the same form as for the conventional CVD counterpart. The only difference is in the activation energy. Because of enhancement from the surface modification by colliding ions, the activation energy for PECVD is generally lower than that for conventional CVD and depends on the ion energy.

While plasma deposition can perhaps be traced to the work of Alt *et al.* (1963) for semiconductor processing, the advantage of low-temperature processing permitted by plasma was also recognized for the oxidation of native silicon (Ligenza, 1965). Reactions of gaseous species from a plasma with a native substrate, as in oxidation or nitridation of silicon, can be categorized as plasma gas-solid reactions. As in plasma deposition, reactive species generated in the plasma are utilized for the gas-solid reaction. The substrate can be immersed in a plasma or it can be placed outside. When the substrate is placed outside the plasma on an anode, the process is referred to as anodization, although it is usually limited to oxidation reactions.

The rate of oxidation or nitridation by plasma gas-solid reactions can be expressed in exactly the same manner as for conventional (thermal) gas-solid reactions (Sec. 5.5). Although the formulation is the same, some specifics are different. For one, it is much more difficult to identify which reactive species are primarily responsible for the plasma gas-solid reaction. For another, the transport mechanism from the bulk gas to the substrate surface is different. For thermal gas-solid reactions, film transport is by molecular diffusion. For plasma gas-solid reactions, transport across the sheath is by molecular diffusion for neutral species but can be dominated by drift for negative ions when the substrate is placed on an anode. In both cases, the distance of diffusion or drift, which is the sheath thickness, is dictated by the applied potential.

The reactive species generated in oxygen plasmas are

$$O_2 + e \longrightarrow O_2^+ + 2e$$

$$O_2 + e \longrightarrow 2O + e$$

$$O + e \longrightarrow O^-$$

$$O_2^+ + e \longrightarrow 2O$$

Although the reactive species resulting from electron collisions are O_2^+, O, and O^-, there is uncertainty as to which is the dominant reactive species (Friedel and Gourrier, 1983; Barlow *et al.*, 1985). However, certain inferences can be made on the basis of the type of discharge. Plasma anodization, which is by far the most commonly used method, involves placing the substrate on an anode. A negatively charged sheath forms around the anode due to the dc discharge. Therefore, negative ions can easily drift to the substrate whereas positive ions cannot because of an energy barrier. Thus, the dominant reactive species in this case are likely to be O and O^-. In the case of rf discharges, the transport of reactive species across the sheath is exactly the same as in plasma deposition. Thus, the dominant reactive species would be neutral, i.e., O.

The rate of oxidation as affected by transport processes can be derived in exactly the same manner as for thermal oxidation (Sec. 5.5). If neutral atomic oxygen is the dominant species during oxidation, the rate of oxidation is given by Eq. (5.89) with k_m replaced by D_o/d and C_b replaced by N_o. Here D_o is the molecular diffusivity of atomic oxygen, d is the sheath thickness, and N_o is the atomic oxygen concentration in the oxygen plasma. While the sheath thickness in rf discharges is determined by the applied voltage, the thickness in dc discharge can be varied for any voltage (Sec. 9.3) by simply moving the anode toward the plasma.

Example 9.10. Both neutral and negative oxygen ions can be the dominant species for plasma oxidation in dc discharges. However, assume for this example that the negative ion is primarily responsible for the oxidation. Because of the negative discharge, not only diffusion but also drift contributes to the flux of the ion to the substrate. For this case, one can write:

$$\mu_i n_i E + D_i \frac{dn_i}{dx} = 0 \tag{A}$$

for the density of the negative ion n_i in the negative sheath. Here μ_i and D_i are the mobility and diffusivity, respectively, of the ion, E is the electric field, and x is the distance normal to the substrate surface. Noting that $n_i = (n_i)_0$ at $x = 0$ and $-E = dV/dx$ where V is the potential, the solution of Eq. (A) yields

$$n_i(x) = (n_i)_0 \exp\left[\frac{q}{kT}(\Delta V)_x\right] \tag{B}$$

where $(\Delta V)_x$ is the voltage drop across a distance x and the Einstein relationship $(D/\mu = kT/q)$ has been used. At the surface of the substrate, $x = d$ and $n_i = (n_i)_s$. Thus,

$$(n_i)_s = (n_i)_0 \exp\left(\frac{q\,\Delta V}{kT}\right) = (n_i)_0\, v_i \tag{C}$$

where ΔV is now the voltage drop across the sheath. Using the approach in Sec. 5.5 and Eq. (C), derive a rate expression for the oxidation.

Solution. Rewriting Eq. (C) in terms of concentration, one has

$$N_s = N_p v_i \tag{D}$$

where N_s is $(n_i)_s$ and N_p is $(n_i)_0$ in terms of molar concentration. Equation (5.82) (Sec. 5.5) can be written as

$$-D_s \frac{dN}{dx} = D_s\left(\frac{N_s - N_i}{l}\right) \tag{E}$$

where N_i is the concentration at the oxide-solid reactant interface, D_s is the ion diffusivity in the oxide, and l is the sheath thickness. The equation corresponding to Eq. (5.83) is

$$r_c = k_i N_i \tag{F}$$

where k_i is the rate constant for the reaction by the ion. One then has, from Eqs. (D), (E), and (F) with the aid of Eqs. (5.86) and (5.87),

$$m\rho'_M \frac{dl}{dt} = \frac{k_i N_p v_i}{1 + k_i l/D_s} \tag{G}$$

which corresponds to Eq. (5.88). Therefore, the rate of oxidation is given by

$$\begin{aligned} l^2 + A_i l &= B_i t \\ A_i &= 2D_s/k_i \\ B_i &= \frac{2D_s v_i N_p}{m\rho'_M} \end{aligned} \tag{H}$$

Note that N_p is the negative ion concentration in the plasma.

When both neutral oxygen atoms and their negative ions are contributing to the oxidation, the overall oxidation rate can be written as

$$r_o = r_i + r_n$$

Furthermore, one can write, in place of Eq. (5.86),

$$\left(\frac{m}{A_i}\right)\frac{dN_m}{dt} = r_o = r_i + r_n \tag{9.29}$$

where r_i is the rate due to the ion and r_n is that for the neutral. A rate expression that is not in the form of Eq. (5.89) (Deal-Grove type of model) results (see Prob. 9.10).

Plasma nitridation (e.g., Petro *et al.*, 1986) is essentially the same as plasma oxidation in the framework of gas-solid reactions. Therefore, the rate expression derived for oxidation also applies to nitridation.

9.6 PLASMA ETCHING

The unique feature of plasma etching as compared to physical sputtering lies in the gasification of substrate at the surface by the reactive species generated in plasma. The modification of substrate surface by bombarding ions which significantly increases the rate of gas-solid reaction (gasification) is more important than the nature of the reactive species. This effect is clearly shown in Fig. 9-12. When only XeF_2 gas is used, the etch rate can be attributed mainly to the gasification of silicon by silicon fluoride whereas it is due to physical sputtering when only argon ions are used. When both are used, the etch rate becomes much greater than the sum of the individual rates, as evident from the figure. While the exact mechanism may be subject to dispute, there is no doubt that the modifi-

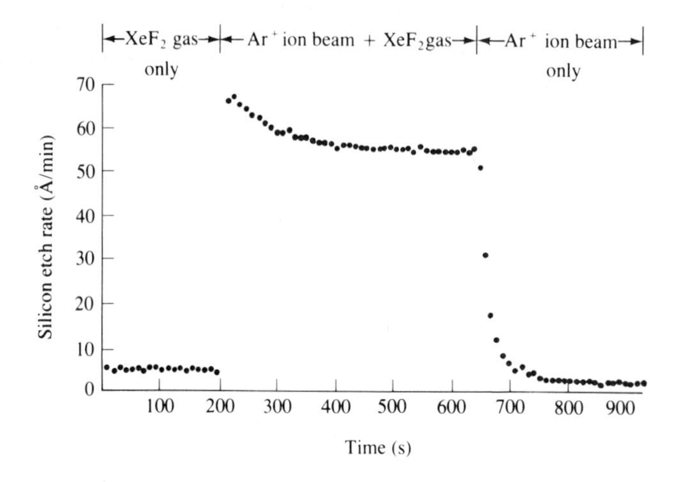

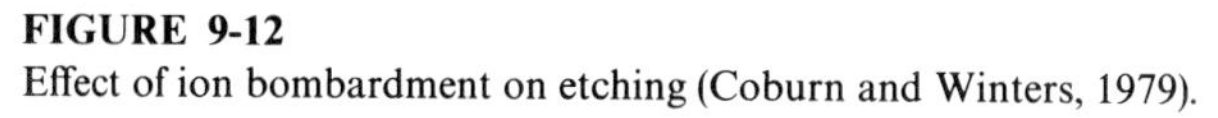

FIGURE 9-12
Effect of ion bombardment on etching (Coburn and Winters, 1979).

cation of the surface and/or surface species by bombarding ions is responsible for the synergistic effect. Although plasma was not involved in the etching, the results in the figure clearly show the role of ions.

Ions in plasma approach perpendicularly to the substrate surface under the usual operating conditions because of the directed electric field. The normal incidence is assured when the sheath thickness is much larger than the vertical dimension of window openings and the mean free path is larger than the sheath thickness. Ordinarily, therefore, anisotropy is easier to achieve than with wet etching. In this regard, two types of plasma etching are of interest. Plasma etching is ion-induced if no etching takes place in the absence of ions. In this case, anisotropy is normally expected since there cannot be any lateral etching with the normally incident ions. Plasma etching is ion-enhanced if etching can take place in the absence of ions, but it is greatly enhanced by ion bombardment. However, for ion-enhanced etching, the window to be etched out can be undercut. In either case, the physical momentum transfer from colliding ions is the initiator for plasma etching.

In plasma etching, the reactive neutral species in general are responsible for actual etching. The role of ions is to make the substrate surface more active with respect to the reactive neutral species and at the same time to provide the directionality of the etched surface. Typical gases used are halogen compounds such as CF_4 and Cl_2. Additional gases are also added such as H_2 or O_2 to provide desired selectivity and edge profile (Flamm and Donnelly, 1981). The dominant ion in CF_4 plasma is $CF_3{}^+$ (Harper *et al.*, 1981), which is primarily produced by the following:

$$e + CF_4 \longrightarrow CF_3{}^+ + F + 2e$$

The atomic fluorine can be produced by a number of other reactions. In chlorine plasma, the dominant ion is believed to be Cl_2^+. The reactive chlorine atom can be generated by

$$e + Cl_2 \longrightarrow 2Cl + e$$

The kinetics of plasma etching can be treated in the general framework of gasification taking place on the substrate surface. The neutral species formed in the plasma adsorbs on the substrate surface upon diffusing through the plasma sheath. The adsorbed species then goes through surface reaction(s) which lead to surface species that are precursors to the gasified species. The surface species then desorbs, resulting in the gasification of the substrate. As with any heterogeneous reaction sequence, one of these three steps can be rate-limiting. It has been found in plasma etching of InP in chlorine plasma, for instance, that the activation energy of the etching is close to the sublimation energy of $InCP_3$ (Donnelly *et al.*, 1982). This indicates that the rate-limiting step could be the desorption step.

The selectivity of an etchant with respect to two different materials is an important factor in pattern delineation. An example would be the selectivity of an etchant with respect to the silicon oxide layer and the underlying silicon. In opening a window through the oxide layer, an ideal situation is one in which the etchant etches the oxide but not the silicon. It is known in CF_4 plasma etching that addition of H_2 does not greatly affect the etch rate of SiO_2 but the Si etch rate decreases with increasing H_2 content (Eprath, 1979). This has to do with the number of fluorine atoms available for etching. In the plasma, hydrogen atoms can readily combine with fluorine atoms to form HF. The same process also takes place in the sheath. Therefore, the etch rate of silicon decreases with increasing H_2 content. When silicon dioxide is exposed to the plasma, oxygen liberated from the oxide surface competes with the fluorine atom for combination with hydrogen. Therefore, the etch rate of silicon oxide decreases slightly with increasing H_2 content. The effect of adding O_2 to CF_4 plasma, shown in Fig. 9-13, also has to do with the availability of fluorine atoms. One major source of disappearance of fluorine atoms is recombination with CF_x ions. As the oxygen concentration increases, oxygen combines with CF_x ions to produce species such as COF_2, CO, etc., thereby decreasing the availability of CF_x ions, resulting in increased fluorine atom levels in the plasma. When the oxygen concentration becomes too high, dilution of the plasma by oxygen dominates over enhancement and a maximum appears in the etch rate. If plasma reactions were solely responsible for the behavior in Fig. 9-13, similar trends would be observed for both Si and SiO_2. The observed difference has been attributed to the chemisorption behavior of oxygen on the substrates (Mogab *et al.*, 1978). On Si, oxygen chemisorbs whereas chemisorption on SiO_2 should be negligible. Inhibition by chemisorption of oxygen has been proposed as the reason for the more rapid decrease in the etch rate of Si at higher oxygen contents compared with the etch rate decrease of SiO_2.

When a halogen compound containing carbon is used, there is invariably carbon deposited on the substrate surface. This is caused by impact dissociation

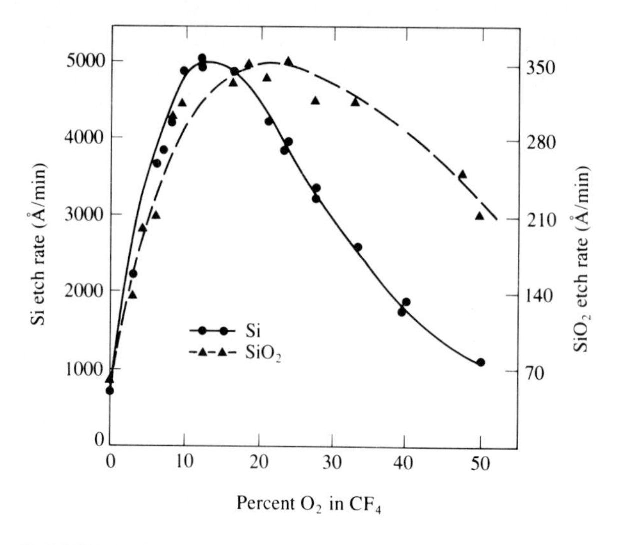

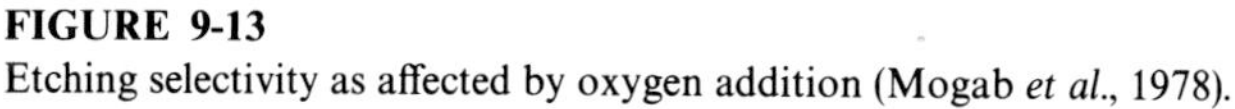

FIGURE 9-13
Etching selectivity as affected by oxygen addition (Mogab *et al.*, 1978).

of the ions on the surface. The carbon deposit has to be sputtered away, if the surface is to be etched. On the other hand, the deposit can be used as a means of achieving etching selectivity. Another problem is the formation of polymer both in the plasma and on the substrate surface in the form of $(CF_2)_n$, for example, when the ion concentration is relatively high. Here, again, the polymer formed can be used as a protective coating.

The etching kinetics for both ion-induced and ion-enhanced plasma etching can be treated in the same way. If one lets r_o be the observed intrinsic rate of etching and r_u the rate in the absence of ions, then the rate of etching r for both cases can be expressed as

$$r = r_o - r_u \tag{9.30}$$

For ion-induced etching, $r_u = 0$ and thus $r = r_o$. For ion-enhanced etching, r represents the rate enhancement due to the modification of surface and/or surface species by colliding ions. In general, adsorption processes of neutral species may involve successive adsorption steps (Flamm and Donnelly, 1981):

$$\mathrm{A + S} \rightleftharpoons \mathrm{A \cdot S}$$

$$\mathrm{A \cdot S + A} \rightleftharpoons \mathrm{(2A) \cdot S}$$

$$[(x-1)\mathrm{A}] \cdot \mathrm{S + A} \rightleftharpoons (x\mathrm{A}) \cdot \mathrm{S}$$

$$(x\mathrm{A}) \cdot \mathrm{S + A} \rightleftharpoons [(x+1)\mathrm{A}] \cdot \mathrm{S}$$

where A is the active neutral species. A·S is the substrate surface atom S with one adsorbed species and (2A)·S is the same but with two adsorbed neutral species, etc. Depending on the energetics of the surface sites, some of the adsorbed species may react to form product precursors in the adsorbed state by electron reconfiguration:

$$(2A)\cdot S \rightleftharpoons A_2S\cdot S$$
$$\vdots$$
$$(xA)\cdot S \rightleftharpoons A_xS\cdot S$$

where $A_xS\cdot S$ is the product precursor with x number of neutral species bonded to the underlying substrate surface atom S. Finally, the product precursors desorb, yielding the gasified product A_xS:

$$A_2S\cdot S \rightleftharpoons A_2S(g)$$
$$\vdots$$
$$A_xS\cdot S \rightleftharpoons A_xS(g)$$

Although rate expressions encompassing all the steps involved can be written, they are too complex and cumbersome to use. As discussed in Chap. 5, it is often sufficient to consider only the dominant surface species for the purpose of describing the kinetics. As an example, consider etching of silicon in CF_4 plasma. The main products of etching are SiF_2 and SiF_4. One may assume the dominant surface species to be the adsorbed SiF_2, as proposed by Donnelly and Flamm (1980), or in the above notation, $(2A)\cdot S$, where S is for surface Si atom and A symbolizes fluorine. Then, the surface reaction steps can be written as follows:

$$A(g) + S \underset{}{\overset{K_1}{\rightleftharpoons}} A\cdot S \qquad (F + S \rightleftharpoons F\cdot S)$$
$$A\cdot S + A(g) \overset{K_2}{\rightleftharpoons} (2A)\cdot S \qquad [F\cdot S + F \rightleftharpoons (2F)\cdot S]$$
$$(2A)\cdot S \rightleftharpoons A_2S\cdot S \qquad [(2F)\cdot S \rightleftharpoons F_2S\cdot S]$$
$$A_2S\cdot S \overset{K_d}{\rightleftharpoons} A_2S(g) + S \qquad [F_2S\cdot S \rightleftharpoons SiF_2(g) + S]$$

Further, one of the steps may be assumed to be the rate-controlling step. Suppose that the third step is the controlling step, for which one can write

$$(2A)\cdot S \xrightarrow{k_s} A_2S\cdot S$$

Note that this reaction represents the formation of the precursor for the gasified product A_2S in the form attached to the underlying silicon atom. Thus, $A_2S\cdot S$ is a notation for the state and does not involve any additional vacant site exposed to the gaseous environment. This fact is reflected in the surface reaction written above. Following the procedures detailed in Chap. 5, one has

$$\frac{C_v}{C_t} = \frac{1}{1 + K_1(C_A + K_2 C_A) + K_d C_{A_2S}} \tag{9.31}$$

and

$$r = k_s C_{2A\cdot S} = k_s K_1 K_2 C_v C_A^2 \tag{9.32}$$

where C_t is the surface concentration of total silicon atom sites and C_v is that of vacant silicon sites. Use of Eq. (9.31) in Eq. (9.32) yields

$$r = \frac{kC_{\mathrm{A}}^2}{1 + K_1(C_{\mathrm{A}} + K_2 C_{\mathrm{A}}) + K_d C_{\mathrm{A_2S}}} \tag{9.33}$$

In terms of the species F and SiF_2, the rate can be rewritten as

$$r = \frac{kC_{\mathrm{F}}^2}{1 + K_1(C_{\mathrm{F}} + K_2 C_{\mathrm{F}}) + K_d C_{\mathrm{SiF_2}}} \tag{9.34}$$

where C_{F} is the concentration of the neutral fluorine atom at the silicon surface and $C_{\mathrm{SiF_2}}$ is the surface concentration of SiF_2.

It is noted in this regard that the activation energies of the rate constants in Eq. (9.34) cannot be the same, even for the same plasma conditions, if the ion energy changes. This is due to the fact that surface modification and the resulting surface energetics are determined by the momentum of the colliding ions with the substrate surface. Thus, the activation energies depend on the ion energy.

The kinetics of plasma etching invariably involve terms for the neutral species concentration at the substrate surface. To relate this to feed conditions, the concentrations of the species in the plasma have to be known, which in turn has to be related to the surface concentration. Transport phenomena are the link to these relationships, which is treated in the next chapter.

NOTATION

A	Solid surface area (L^2)
A_i	Cross-sectional surface area (L^2)
C_a	Surface concentration of adspecies (mol/L^2)
C_{A}, C_{B}	Concentrations of species A and B, respectively (mol/L^3)
$C_{\mathrm{A\cdot S}}$, $C_{\mathrm{B\cdot S}}$	Concentrations of surface species A·S and B·S, respectively (mol/L^2)
C_v	Surface concentration of vacant sites (sites/L^2)
C_t	Surface concentration of total available sites (sites/L^2)
d	Sheath thickness (L)
D_i	Ion diffusivity (L^2/t)
E	Ion energy (E); electric field (V/L)
E_{th}	Threshold energy (E)
f_i	Fraction of species i trapped in film
G	Linear growth rate (L/T)
j_c	Cathode ion current density (I/L^2t)
j_i	Ion current density (I/L^2t)
J_a	Flux of adspecies (mol/L^2t)
k	Rate constant (units dependent on concentration units)
k_B	Boltzmann constant

K, K_A, K_B	Equilibrium constants (units dependent on concentration units)
L	Distance from center of substrate plane (L)
m	Mass of molecule (M)
m_e	Mass of electron (M)
M	Molecular weight (M/mol)
M_i	Ion mass (M)
M_s	Solid molecular weight (M/mol)
n_e	Electron density (electron/L^3)
n_i	Ion density (ions/L^3)
N	Number of molecules
N_A	Avogadro's number
N_m	Moles of species m
p	Partial pressure
p^*	Saturation pressure (P)
q	Electron charge (C)
r	Distance between source and substrate (L); rate of reaction (mol/L^2t)
r_c	Rate of condensation (atom/L^2t)
r_D	Rate of deposition (M/L^2t)
$(r_D)_0$	r_D directly above and r away from source
r_i	Rate due to ion (mol/L^2t)
r_n	Rate due to neutral (mol/L^2t)
r_s	Rate of sputtering (atom/L^2t)
S	Sputtering yield (atom/ion)
t	Time
T	Temperature
T_e	Electron temperature (T)
U	Surface binding energy (E)
V	Mass rate of evaporation (M/L^2t)
V_a	Applied voltage (V)
V_{dc}	dc bias in asymmetrical, capacitive rf plasma (V)
V_f	Floating potential (V)
V_p	Plasma potential (V)
V_{sp}	Sheath potential (V)
V_t	Mass rate of evaporation (M/t)
x	Spatial coordinate (L); atom fraction in binary alloy
Z_t	Atomic number of target material
Z_x	Atomic number of gas species

Greek letters

α	Constant; quantity defined in Eq. (9.11); angle in Fig. 9-1
β	Constant
μ	Constant; number of secondary electrons ejected per incident ion
μ_i	Ion mobility (L^2/Vt)

ρ_s	Solid density (M/L^3)
ω	Angle in Fig. 9-1
Ω	Angle in Fig. 9-1

Units

C	Coulomb
E	Energy
I	Current
L	Length
M	Mass
P	Pressure (M/Lt^2)
t	Time
T	Temperature
V	Voltage

PROBLEMS

9.1. Plot the relative rate, $r_D/(r_D)_0$, as a function of L/r for both point and small area evaporant sources up to L/r of 2. Comment on the uniformity of a film deposited onto a flat substrate for the two cases.

9.2. Suppose metallization is carried out for a metal strip 1 μm wide that is defined by a resist 1.5 μm thick. For the evaporation from a point source for metallization, which is 5 cm away from the center of the resist window, determine the maximum length of the groove that can be filled with a maximum thickness deviation less than 1 percent. For the maximum deposition rate of 1×10^{-5} g/(cm^2·s), calculate the time required to fill the resist window. Assume the metal density to be 4 g/cm^3.

9.3. For gold metallization at 300 K, suppose that the main residual gas is oxygen which constitutes 1 percent of the volume. Calculate the fraction of oxygen incorporated into the metal deposit. Use Fig. 9-3. The desired linear deposition rate is 1 nm/s.

9.4. For MBE, which is an epitaxial process, the rate of growth is dictated by the rate of surface migration of adatoms; i.e., the net rate of condensation should be less than the rate of surface migration. Otherwise, a monocrystalline structure cannot be obtained. The surface migration rate decreases exponentially with decreasing temperature. Suppose that the migration rate corresponds to 0.1 nm/min at 700 °C. Calculate the MBE chamber pressure required for Si MBE at 700 °C.

9.5. A dc discharge is used for tantalum deposition. The measured cathode voltage and current, respectively, are 700 volts and 4 mA in an argon environment. The cathode (target) area is 20 cm^2. Calculate the rate of sputtering, assuming that the contribution to the current by secondary electrons is 10 percent. The distribution of sputtered material deposited on the substrate surface also follows the cosine law but is dependent on the ratio of target diameter to electrode separation. When this ratio is larger than, say, 25, the rate of deposition is essentially the same as the rate of sputtering and the thickness distribution is uniform up to a value of 4 for the ratio of the distance from the center to electrode separation (Maissel, 1970). Determine the ratio of the target diameter to the distance from the center that satisfies the conditions. Discuss whether the electrode separation can be freely chosen.

9.6. Consider aluminum deposition by rf sputtering. For a symmetrical, parallel-plate system, the applied voltage is 120 volts and the electrode area is 20 cm^2. For aluminum, the sputtering yield S is given by

$$S(\text{atom/ion}) = 2.257(E^{1/2} - 0.013^{1/2})$$

Assuming that the root mean square current is 0.08 A, calculate the rate of sputtering.

9.7. Electrons accelerated by an applied field cause activation and formation of ions and reactive species in a plasma by collisions. Consider the following process:

$$e + SiH_4 \longrightarrow SiH_2 + H_2 + e$$

It is known that the activation energy is 2.16 eV. For a SiH_4 plasma at 1 torr, suppose that the ion density is 10^{11} cm^{-3}. Calculate the rate of formation of SiH_2 at 300 K in the plasma by the above process, assuming that the preexponential factor is 10^{24} cm^3/s.

9.8. Polymer deposition on p-type silicon substrates using plasma-enhanced CVD has been reported by Nguyen *et al.* (1985). Their data show that the polymer deposition rate increases exponentially with the power input, holding all other conditions constant. They show that the deposition rate is almost zero below a certain power input. To a first degree of approximation, one may assume the power input to be proportional to the ion energy. The enhancement of the deposition rate can be attributed to the change of the silicon surface by colliding ions, which could reduce the activation energy for the deposition. Based on the premises stated, postulate the dependence of the deposition rate on the ion energy. Assume the rates to be intrinsic. Use the rate in the form, $r_c = kf(C)$.

9.9. The rate of formation of SiH_2 in the plasma considered in Prob. 9.7 is 1.7×10^{15} molecules/(cm^3·s). Suppose the surface area of deposition for polysilicon is 5 cm^2. Determine the maximum constant linear deposition rate (in centimeters per minute) that is sustainable. Discuss what happens if the substrate temperature is high enough that the potential deposition rate is higher than that corresponding to 1.7×10^{15} molecules/(cm^3·s). Assume that the generation of adspecies SiH_2 is by Eq. (9.19) and that the adspecies for the polycrystalline silicon is SiH_2.

9.10. Plasma gas-solid reactions such as plasma oxidation or nitridation are due to both negative ions and neutrals supplied by the plasma. Show for such reactions that the rate of oxidation or nitridation, when both contribute to the reaction, can be determined from Eq. (9.29) written in the following form:

$$m\rho'_M \frac{dl}{dt} = \frac{k_i N_p v_i}{1 + k_i l/D_s} + \frac{kN_n}{1 + kD_0/d + kl/D_{sn}} \tag{A}$$

where d is the sheath thickness, k is the rate constant for the neutral, N_n is the neutral concentration in the plasma, D_0 is the gas phase diffusivity of the neutral, and D_{sn} is the diffusivity of the neutral in the solid. Show that Eq. (A) does not lead to the Deal-Grove type of relationship.

9.11. Ion-induced or ion-enhanced etching can be carried out in the absence of plasma. Separate sources of ion and gas species can be used for the etching. In fact, the result shown in Fig. 9-11 was obtained by feeding XeF_2 gas from the gas source directed to the substrate surface while a separate argon ion was directed to the same surface.

Consider the ion-enhanced etching by XeF_2 in the absence of plasma. According to Chuang (1980), XeF_2 adsorbs on silicon surface by dissociation, although the stoichiometry is not clear, and the dominant surface species is SiF_2. Assume the following adsorption/dissociation:

$$XeF_2(g) + S \rightleftharpoons S\cdot F_2 + Xe(g) \tag{A}$$

where S is the silicon site. Tu *et al.* (1981) have observed that about 75 percent of Si leaving the surface are in the form of SiF_4. Thus, one may postulate the following steps:

$$S\cdot F_2 + XeF_2(g) \rightleftharpoons S\cdot(SF_4) + Xe(g) \tag{B}$$

$$S\cdot(SF_4) \rightleftharpoons SiF_4(g) + S \tag{C}$$

Equation (B) represents the surface reaction leading to the formation of the surface species Si-SiF_4 and Eq. (C) represents desorption of the species SiF_4 attached to the underlying Si site, $S\cdot(SF_4)$. Derive the kinetics of etching for each of three cases in which one of the steps is rate-controlling.

9.12. For the etching of Si by XeF_2 with separate sources of ion and the etchant (Fig. P9-12), Gerlach-Meyer (1981) gives etching data for three types of ions as shown in the figure. The flow rate can be taken as equivalent to the concentration of XeF_2. The data have already been reduced such that the etching is solely for the ion-enhanced part, that is, r in Eq. (9.30). It should be clear from the results of Prob. 9.11 that none of the kinetics derived can describe the data, even when one assumes SiF_2 to be the dominating (sole) surface species. With additional information that the

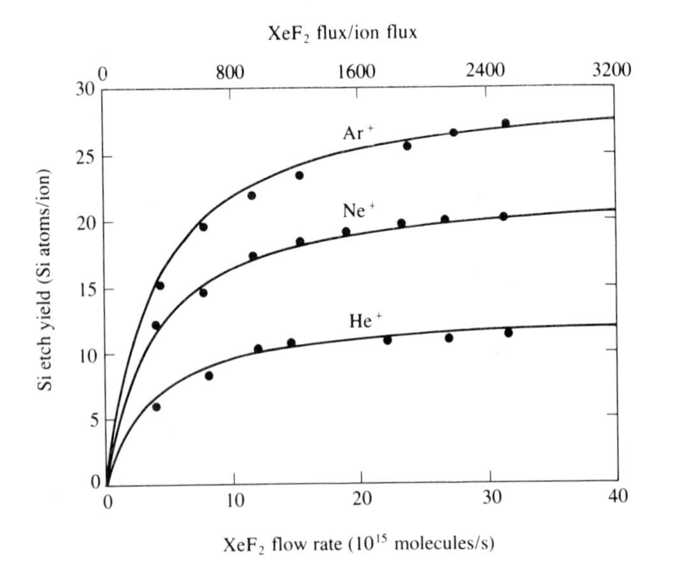

FIGURE P9-12
Etching data of Gerlach-Meyer (1981).

surface coverage by SiF_2 (to be more precise SiF_x) is almost complete and that the ion energy is the same for all ions in the figure, one may postulate the following mechanism:

$$XeF_2(g) + S \rightleftharpoons S \cdot F_2 + Xe(g); \qquad \text{complete coverage by } S \cdot F_2$$

$$S \cdot F_2 \underset{k_r}{\overset{k_i \text{ (ion)}}{\rightleftharpoons}} S \cdot F_2{}^*$$

$$S \cdot F_2{}^* + XeF_2(g) \xrightarrow{k^*} SiF_4(g) + Xe(g)$$

The second step is for the activation of the surface species by colliding ions. The rate of etching is then given by

$$r = k^* C_{S \cdot F_2{}^*} C_A \qquad \text{where } C_A = C_{XeF_2} \tag{A}$$

Invoke the pseudo steady-state approximation for the activated surface species $S \cdot F_2{}^*$ to arrive at the following:

$$r = \frac{k_i C_t k^* C_A}{k_r + k^* C_A} \qquad \text{where } C_{S \cdot F_2} = C_t \tag{B}$$

Noting that the only rate constant dependent on the ion type is k_i, offer a postulation for the differences due to the ion type.

REFERENCES

Akimoto, K., and K. Wantanabe: *Appl. Phys. Lett.*, vol. 39, p. 445, 1981.

Alt, L. L., S. W. Ing, and K. W. Laendle: *J. Electrochem. Soc.*, vol. 110, p. 465, 1963.

Barlow, K. J., A. Kiermasz, and W. Eccleston: in G. S. Mathad *et al.* (eds.), *Plasma Processing*, The Electrochemical Society, Pennington, N.J., 1985.

Bollinger, D., and R. Fink: *Solid State Technol.*, vol. 25, p. 79, 1980.

Brown, S. C.: *Introduction to Electrical Discharge in Gases*, McGraw-Hill, New York, 1966.

Catherine, Y.: in G. S. Mathad *et al.* (eds.), *Plasma Processing*, The Electrochemical Society, Pennington, N.J., 1985.

Chapman, B.: *Glow Discharge Processes*, Wiley, New York, 1980.

Chen, F. F.: *Introduction to Plasma Physics*, Plenum Press, New York, 1974.

Cho, A. Y.: *Proceedings of Compound Semiconductor Growth, Processing and Devices for the 1990s*, Japan/US Perspective, Gainesville, Fla., October 1987.

Chuang, T. J.: *J. Appl. Phys.*, vol. 51, p. 2614, 1980.

Coburn, J. W., and E. Kay: *J. Appl. Phys.*, vol. 43, p. 4965, 1972.

——— and H. F. Winters: *J. Appl. Phys.*, vol. 50, p. 3189, 1979.

———, E. Taglaner, and E. Kay: *Jap. J. Appl. Phys.*, suppl. 2, p. 501, 1974.

Donahue, T. J., W. R. Burger, and R. Reif: *Appl. Phys. Lett.*, vol. 44, p. 346, 1984.

Donnelly, V. M., and D. L. Flamm: *J. Appl. Phys.*, vol. 51, p. 5273, 1980.

Eprath, L. M.: *J. Electrochem. Soc.*, vol. 126, p. 1419, 1979.

Erskine, J. C., and A. Cserhati: *J. Vac. Soc. Technol.*, vol. 15, p. 1823, 1978.

Flamm, D. L., and V. M. Donnelly: *Plasma Chem. and Plasma Proc.*, vol. 1, p. 317, 1981.

——— and G. K. Herb: *A Short Course in Plasma Etching*, chap. 1, Academic Press, 1988.

———, V. M. Donnelly, and J. A. Mucha: *J. Appl. Phys.*, vol. 52, p. 3633, 1978.

Friedel, P., and S. Gourier: *J. Phys. Chem. Solids*, vol. 44, p. 353, 1983.

Gerlach-Meyer, U.: *Surface Sci.*, vol. 103, p. 524, 1981.

Glang, R.: in L. Maissel and R. Glang (eds.), *Handbook of Thin Film Technology*, chap. 1, McGraw-Hill, New York, 1970.

Gorowitz, B., T. B. Gorezyea, and R. J. Saia: *Solid State Technol.*, p. 179, June 1985.

Greene, J. E., and S. A. Barnett: *J. Vac. Sci. Technol.*, vol. 21, p. 285, 1982.

Harper, J. M. F., J. J. Cuomo, P. A. Leary, G. M. Summa, H. R. Kaufmann, and F. J. Bresnock: *J. Electrochem. Soc.*, vol. 128, p. 1077, 1981.

Hasted, J. B.: *Physics of Atomic Collision*, Butterworth, London, 1964.

Hess, D. W.: *Ann. Rev. Mater. Sci.*, vol. 16, p. 163, 1986.

Horwitz, C. M.: *J. Vac. Sci. Technol.*, vol. A1, p. 90, 1983.

Kemper, M. J. H., S. W. Koo, and F. Huizinga: in G. S. Mathad *et al.* (eds.), *Plasma Processing*, The Electrochemical Society, Pennington, N.J., 1985.

Kettani, M. A., and M. F. Hoyaux: *Plasma Engineering*, Wiley, New York, 1973.

Langmuir, I.: *General Electric Rev.*, vol. 26, p. 731, 1923.

——— and H. Mott-Smith: *General Electric Rev.*, vol. 27, pp. 449, 538, 616, 762, 810, 1924.

Ligenza, J. R.: *J. Appl. Phys.*, vol. 36, p. 2703, 1965.

Maissel, L.: in L. Maissel and R. Glang (eds.), *Handbook of Thin Film Technology*, chap. 4, McGraw-Hill, New York, 1970.

Meyerson, B. S., E. Ganin, D. A. Smith, and T. N. Nguyen: *J. Electrochem. Soc.*, vol. 133, p. 1232, 1986.

Mogab, C. J., A. C. Adams, and D. L. Flamm: *J. Appl. Phys.*, vol. 49, p. 3769, 1978.

Morgan, R. A.: *Plasma Etching in Semiconductor Fabrication*, Elsevier, London, 1985.

Nguyen, V. S., J. Underhill, S. Fridmann, and P. Pan: in G. S. Mathad *et al.* (eds.), *Plasma Processing*, The Electrochemical Society, Pennington, N.J., 1985.

Petro, W. G., B. R. Cairns, and K. V. Anand: in H. Huff *et al.* (eds.), *Semiconductor Silicon 1986*. The Electrochemical Society, Pennington, N.J. 1986.

Steinbruchel, C.: *J. Vac. Sci. Technol.*, vol. A3, p. 1913, 1985.

Sugiura, H., and M. Yamaguchi: *Jap. J. Appl. Phys.*, vol. 19, p. 583, 1980.

Tu, Y. Y., T. J. Chuang, and H. F. Winters: IBM Research Report RY 2810, 1981.

Turban, G.: *Pure and Appl. Chem.*, vol. 56, p. 215, 1984.

Vratny, F.: *J. Electrochem. Soc.*, vol. 114, p. 505, 1967.

Wehner, G. K., and G. S. Anderson: in L. Maissel and R. Glang (eds.), *Handbook of Thin Film Technology*, chap. 3, McGraw-Hill, New York, 1970.

CHAPTER 10

PHYSICAL VAPOR DEPOSITION APPARATUSES AND PLASMA REACTORS

10.1 INTRODUCTION

Apparatuses and reactors used for physical and physicochemical processes are diverse. They range from vacuum deposition apparatuses, sputtering equipment, elaborate molecular beam epitaxy (MBE) apparatuses, and various types of plasma reactors. Physical sputtering and MBE are the principal physical vapor deposition (PVD) processes. Physical sputtering remains a major method of metallization in which metals are deposited. MBE is the preferred choice for epitaxial deposition of thin films of III-V compounds, particularly when heterostructures with sharp doping profiles are desired. Any apparatus for PVD requires an elaborate vacuum system.

The phenomena taking place in a plasma reactor are quite complex. The added complications, compared to conventional CVD reactors, arise from the presence of plasma, a medium that literally determines the fate of deposition or etching. At the core of the complications are the collision processes in plasma leading to the formation of electrons, ions, and reactive neutral species, the transport phenomena within the plasma and in the sheaths, and their relation to the electrical characteristics of the plama. These phenomena interact with transport phenomena in the reactor and heterogeneous reactions on substrate surfaces.

For PVD, attention will be focused on apparatus used for sputtering and MBE in this chapter. Then diffusion of charged particles in plasma reactors will be considered. This will be followed by a section on the characteristics of plasma and their relation to measurable electrical variables. These can be combined with intrinsic kinetics and reactor conservation equations to describe plasma reactors, leading to design considerations for plasma reactors. Because of the complexities, many of the relationships for plasma are approximate, in particular those related to plasma characteristics. Further, only a few principal species, mainly responsible for plasma deposition or etching, can be considered. In spite of the complexities and approximations, certain limits and trends can be ascertained from the approximate relationships and the main species of interest.

10.2 PHYSICAL VAPOR DEPOSITION (PVD) APPARATUSES

An apparatus for vacuum deposition is shown in Fig. 10-1. As discussed in Chap. 9, a planetary surface receives the same amount of mass flux from a source. The arrangement of substrates in the figure takes advantage of this fact. The rest of the apparatus is for maintaining the desired vacuum level in the deposition chamber.

In contrast to the apparatus used for vacuum metal deposition (usually amorphous), the apparatus for MBE is much more complex and elaborate because of the epitaxy requirements. A schematic of a vacuum chamber in which MBE is carried out is shown in Fig. 10-2. Today the trend is toward the use of effusion cells to which liquid gas source species are introduced. Originally, melts were used exclusively for evaporant sources, and an electron gun was used to

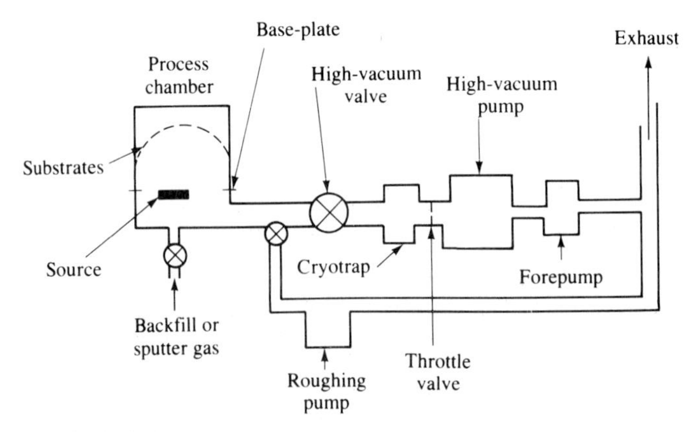

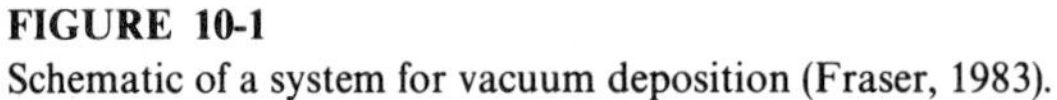
FIGURE 10-1
Schematic of a system for vacuum deposition (Fraser, 1983).

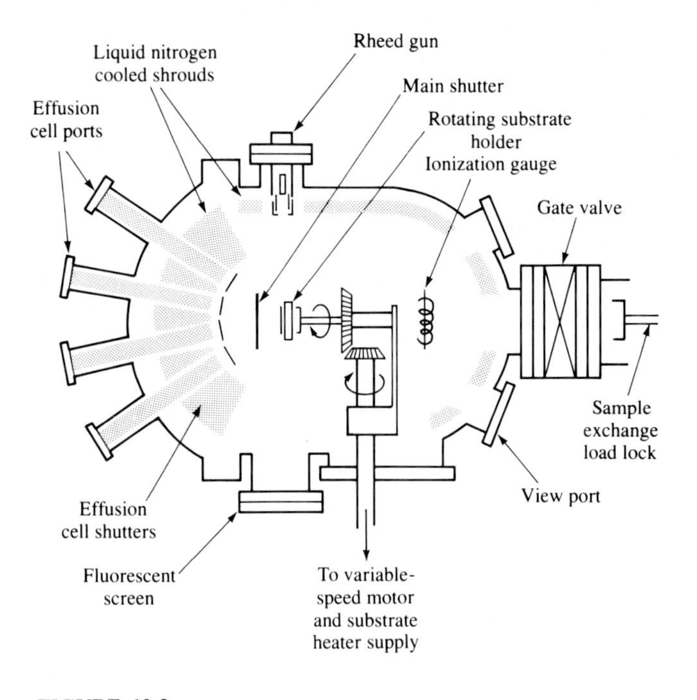

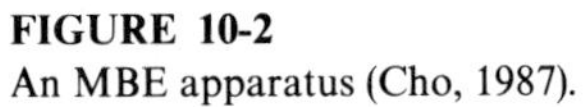

FIGURE 10-2
An MBE apparatus (Cho, 1987).

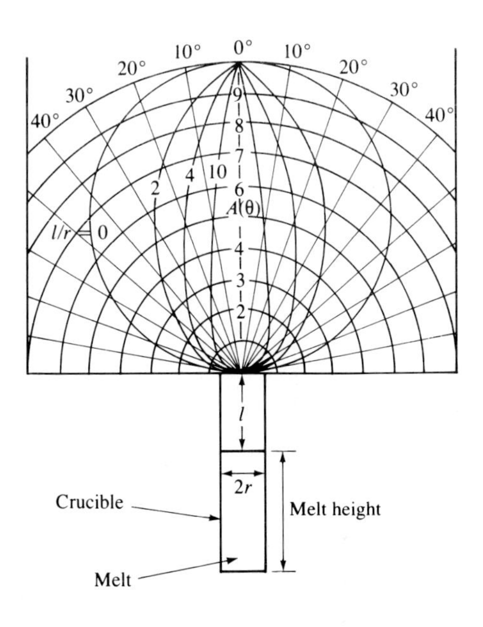

FIGURE 10-3
Angular distribution of flux from a crucible (Luscher and Collins, 1981).

produce source materials of high boiling point, such as silicon (see Fig. 10-2). There are several reasons for the change. As discussed in Chap. 9, MBE is most suitable for thin films with abrupt and well-defined doping profiles which can be realized at lower temperatures. Therefore, the use of MBE has been shifting to the preparation of such thin films for III-V compound semiconductors, typically used for optoelectronic applications. The shift has also been prompted by the fact that MBE film qualities are in general poorer than CVD film but the thin films with abrupt doping profiles necessary for optoelectronic devices are easier to obtain by MBE. For III-V compound semiconductors, liquid/gas source species are readily available and they are more convenient to use. Another reason for the change has to do with the ease with which film-thickness uniformity can be achieved with the liquid/gas source in "effusion" cells. With the melt source, the thickness distribution is determined by the cosine law. This is shown in Fig. 10-3, which gives angular distribution of flux by evaporation from a crucible as a function of time. In contrast, the source gas (vaporized, if liquid, by heating) is fed externally to the effusion cell as shown in Fig. 10-2. Because the external gas is fed into vacuum, the gas "jets" and as a result the trajectories of the gas molecules are well defined. This is shown in Fig. 10-4. As the figure illustrates, the trajectories are confined to the length (or diameter) of a substrate if the centerline of the effusion cell passes through the substrate center and if the extensions of the cell walls form straight lines with the substrate edges. Although the flux arriving at the substrate is dependent on the substrate position, uniform deposition is possible with the rotation of the substrate since then all parts of the surface receive equal exposure to the varying flux.

Although MBE used to be strictly by evaporation/condensation, such a distinction is no longer valid, particularly with the gas source material used in effusion cells. With the substrate temperatures higher than 400 °C and pressures in the 10^{-5} torr range, the epitaxial growth for III-V compound films is more likely due to heterogeneous surface reactions than to simple condensation. The similarity is particularly apparent in light of the fact that the substrate temperature in MBE is kept as high as the desired doping profile will allow to

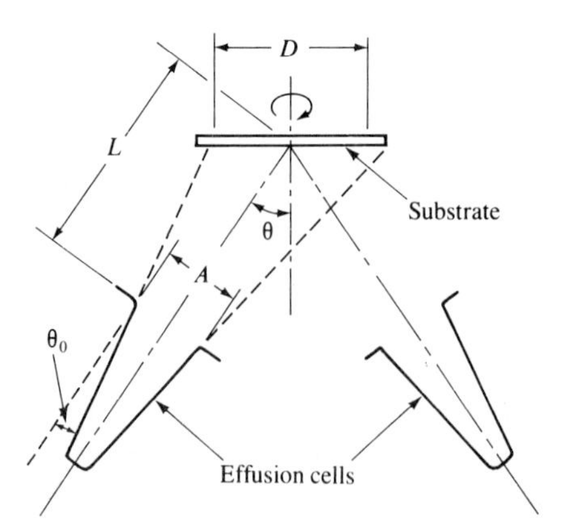

FIGURE 10-4
Trajectories of flux from effusion cells (Saito *et al.*, 1986).

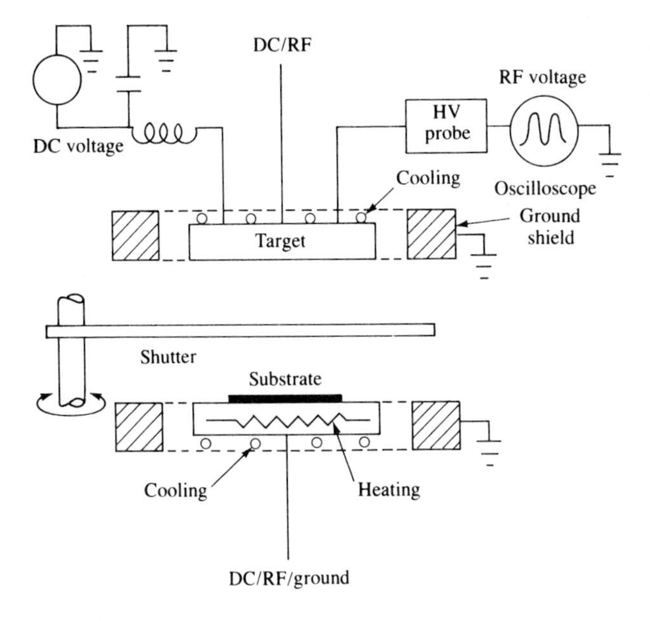

FIGURE 10-5
Schematic of a sputtering system (Chapman, 1980).

achieve better film qualities. In fact, the atomic layer epitaxy (ALE) discussed in Chap. 5 is essentially the same as the MBE with gaseous source species.

A schematic of a sputtering system is shown in Fig. 10-5. As shown in the figure, it can be operated as either a dc or rf discharge. The target is surrounded by a dark space shield, also known as a ground shield. This feature restricts ion bombardment and sputtering only to the target. To prevent ion bombardment of the protected regions, the space between the target and the ground shield must be less than the thickness of the dark space (sheath). Otherwise, a self-sustained discharge can form in the space. As will be shown in subsequent sections, the sheath thickness decreases with increasing pressure. Thus, the size of the gap between the target and shield sets an upper pressure limit for operating the system. The sheath thickness also decreases with increasing frequency in rf sputtering.

Another feature shown in Fig. 10-5 is a shutter that can be rotated into place between the electrodes. The shutter is typically used during a presputtering period when a few atomic layers of the target are sputtered away for removing any impurities that might be present on the target surface. In this way any contaminants introduced during loading, unloading, or *in situ* substrate cleaning can be removed from the target. Sputtering takes place entirely by momentum transfer and as such considerable heating of the target occurs. The localized heating can be excessive (up to 400 °C; Chapman 1980) and can lead to damage of the bonding between the target and the backing electrode or of the target itself. Thus, the target is usually cooled. However, various mechanical problems associated with cooling can be avoided if the power input is not too high. The desired substrate temperature is usually obtained by resistance heating. Here, again, cooling on the backside is required, as shown in Fig. 10-5.

The effect of current and voltage on the deposition rate by physical sputtering can be inferred directly from Eq. (9.13), which represents the maximum possible rate of deposition:

$$r_s = \frac{SJ_i}{q} \tag{9.13}$$

where S is the sputtering yield (atoms ejected per incident ion), J_i is the ion current density, and q is the electric charge. The effect of the current is larger than that of the voltage since the sputtering yield is proportional to the square root of the voltage [with a bias with respect to the threshold energy; see Eq. (9.11)] but J_i is directly proportional to the current. Therefore, for a given power, it is best to sputter at high current and low voltage for higher deposition rates. This can be achieved by using a thermionically supported glow or by going to higher pressure. As the pressure in a sputtering system is raised, the ion density and therefore the current density increases. The upper limit is set by the requirement of eliminating plasma formation in the space between the target and ground shield. The usual range for physical sputtering is from 20 to 130 mtorr, but the optimal range is 50 to 60 mtorr for dc discharges (Maissel, 1970). The rate of deposition decreases with increasing temperature. This is due to the fact that the deposition is by condensation. The cosine law also applies to deposition.

10.3 PLASMA REACTORS

Reactors for plasma processing were originally developed for etching. Some of these reactors were adapted for plasma deposition. Although the plasma reactors have been in use since the early 1970s, they are still in their infancy in terms of our ability to describe them mathematically. However, systematic studies have already been undertaken, particularly for etching (Alkire and Economou, 1985; Dalvie *et al.*, 1986; Stenger *et al.*, 1987).

Plasma reactors for etching are shown in Fig. 10-6. The so-called barrel reactor in Fig. 10-6*a*, which physically resembles a conventional LPCVD reactor, was the first plasma reactor in semiconductor processing. Initially it was mainly for stripping of photoresist in an oxygen plasma (Irving, 1971) and later for the etching of dielectric materials such as SiO_2, typically in CF_4/O_2 mixtures. As shown in the figure, the reactor consists of a cylindrical quartz chamber that has input gas manifolds and a vacuum pumping outlet to maintain the pressure at around 1 torr. Although the schematic shows the gas being fed from the top, it is also fed from the bottom in other arrangements. The plasma is sustained by an rf potential applied to two external electrodes. Usually, a perforated aluminum cylinder is placed coaxially around the wafers which are stacked on a boat as shown in the figure. In such an arrangement, the plasma is confined in the

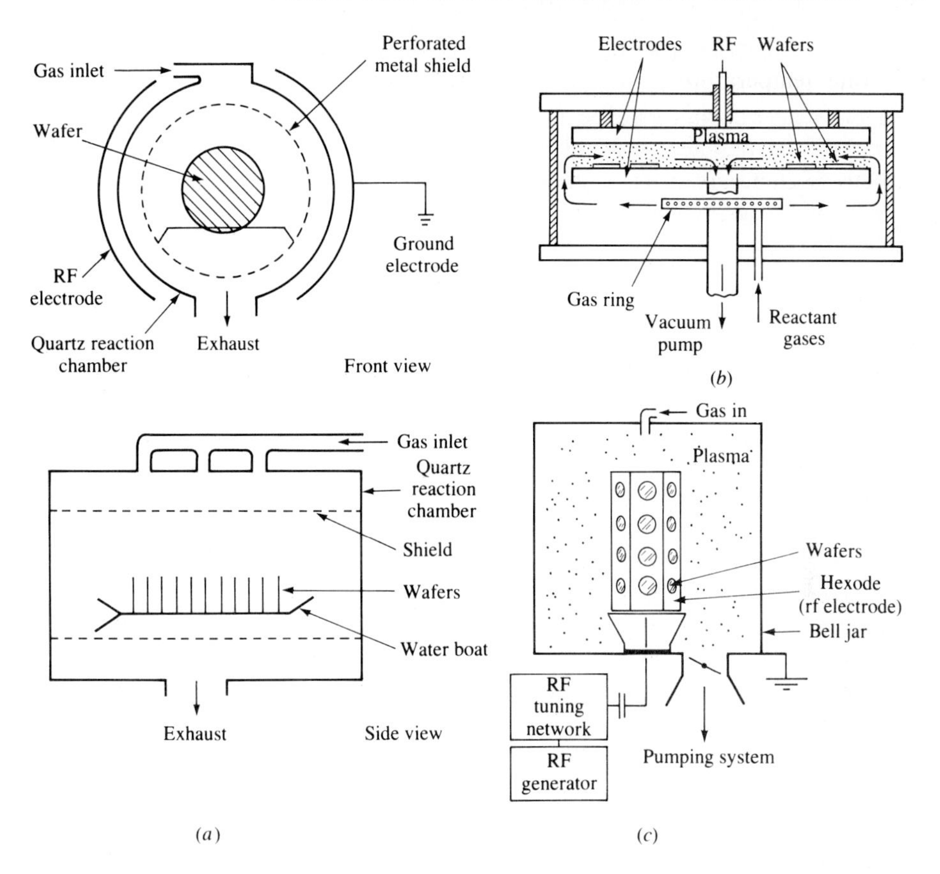

FIGURE 10-6
Plasma reactors for etching: (*a*) barrel reactor, (*b*) planar (radial) reactor (Reinberg, 1975; Heinecke, 1975), and (*c*) hexode (Bell Lab) reactor.

annular regions between the metal cylinder and the reactor wall. The mesh of the aluminum tunnel surrounding the wafers is sufficient to allow the passage of active neutral species while limiting the number of ions passing through. The aluminum tunnel is usually mounted in such a way that it electrically floats. With such an arrangement, a minimum of energetic ionized components are allowed to enter the tunnel from the plasma region (Morgan, 1985). This can partially prevent the overheating of the wafer surface by bombarding ions that can lead to poor uniformity of etching and poor selectivity. Before commencement of etching, some wafer temperature control can be attained by preheating the wafers in an inert gas plasma.

It should be evident from the reactor configuration that ion bombardment does not have any significant role in the barrel reactor. As such, the etching selectivity can be very high, as high as 100:1 for the selectivity of $Si:SiO_2$. The other consequence of such a configuration is that no anisotropy can be achieved.

Therefore, the barrel reactor cannot be used where linewidth control is critical. However, it does give a high throughput for applications not involving close linewidth control. Another aspect particular to the barrel reactor is that the concentration of active neutrals, which are responsible for etching, decreases as they travel from the tunnel mesh to the wafers because of recombination reactions.

Perhaps the most widely used plasma reactor for etching is the planar reactor shown in Fig. 10-6*b*. The reactor arrangement takes full advantage of the effects of ion bombardment. It allows for enhancement of etching due to the surface modification by bombarding ions and for anisotropy. The reactor typically consists of a top electrode that is powered by an rf supply and a bottom electrode that is normally grounded. The wafers are usually placed on the grounded electrode and are in "contact" with the plasma, as opposed to the barrel reactor. Note that a planar reactor will etch wafers located on either electrode surface. The wafer surface is bombarded by ions accelerated by the sheath potential. As shown in Fig. 10-6*b*, the etching gas is fed through a gas ring, which is in turn evacuated by a vacuum pump. In the arrangement shown in the figure, the wafers are immersed in the plasma.

In general, the disadvantages of the barrel reactor are the advantages of planar reactor and vice versa. The anisotropy attainable in a planar reactor is the major feature of the reactor.

One problem in planar reactors has to do with the physical sputtering that takes place on the electrode. The situation is aggravated by having the whole electrode surfaces facing each other. A reactor developed by Bell Laboratories to minimize the physical sputtering of electrodes is called the hexode (Bell Lab) reactor and is shown in Fig. 10-6*c*; it is similar in its physical appearance to the conventional CVD barrel reactor. The hexode on which wafers are placed is connected to an external rf generator as shown in the figure and the reactor wall is grounded. The hexode side walls face in different directions to partially prevent sputtered material from depositing on the wafer surface. In this arrangement, the wafers are on a powered electrode and thus are subject to a larger sheath potential.

Most plasma reactors are preconditioned for etching before wafers are placed to warm the reactor and to reduce wafer contamination. Wafers are introduced to the reactor through a load lock to minimize atmospheric contamination, especially water vapor. Optical emission spectra and mass spectroscopy are used to detect the endpoint of etching. The electrode materials used in reactors are usually anodized aluminum or titanium, which are relatively inert catalytically. Some atoms of electrode material are always present due to physical sputtering, and they in turn can catalyze polymer formation as in CCl_4 plasma (Tokauaga and Hess, 1980).

Reactors for plasma deposition are essentially the same as those for plasma etching. The barrel reactor in Fig. 10-6*a* has been used for plasma deposition and so has the planar reactor. A variation of the planar arrangement, particularly with respect to feeding, is shown in Fig. 10-7*a*. A distinct feature is the presence of a heater, since the deposition is carried out at much higher substrate tem-

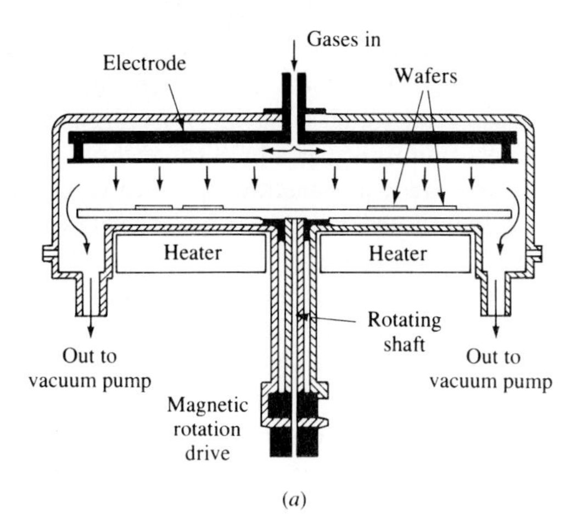

(a)

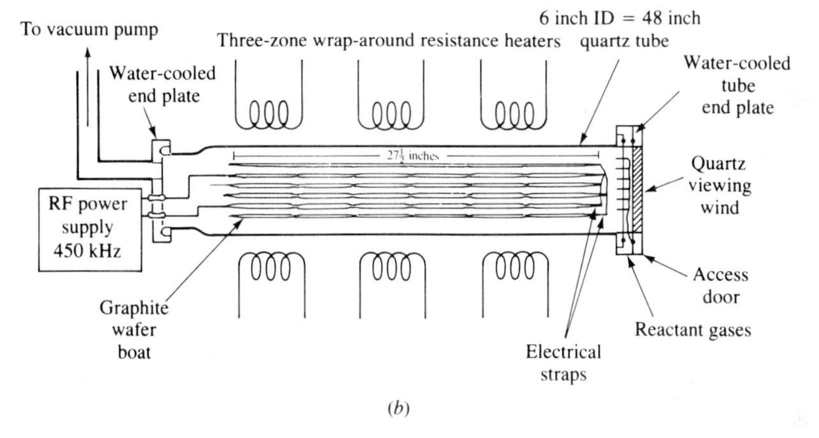

(b)

FIGURE 10-7
Plasma reactors for deposition: (*a*) planar reactor with perforated electrode (Applied Materials, Inc.) and (*b*) an LPCVD-type deposition chamber (Kumagai, 1984).

peratures than those for etching. A distinct departure from the reactors that were originally used for etching is shown in Fig. 10-7*b* for plasma deposition. The deposition chamber is in the form of a resistance-heated quartz furnace tube. Wafers are loaded onto both faces of an array of parallel graphite electrodes. Alternate electrodes in the array are connected to an external rf source. The reactant gas stream is directed along the axis of the chamber tube and between the electrodes. The plasma is ignited between adjacent electrodes. Another distinct feature is the use of pulsed rf excitation rather than the continuous excitation. Good thickness uniformity over large batch sizes and better temperature control are cited (Kumagai, 1984) for the reactor. More discussions on the reactors for plasma deposition can be found in the article by Weiss (1983).

The design objectives of plasma reactors for etching are high throughput, etching uniformity, high selectivity, and anisotropy (when linewidth control is critical). Contamination or side effects due to physical sputtering is another problem that such a reactor has to deal with. The design objectives of plasma reactors for deposition are essentially the same as those for the conventional CVD reactors, namely thickness uniformity and high throughput. As discussed in Chap. 5 for CVD reactors, there are also constraints. These constraints, such as those from film-quality considerations, provide a region of feasibility within which a reactor can be designed and operated.

For plasma reactors, the characteristics pertinent to the plasma have to be considered. They are determined by applied excitation power, excitation frequency, pressure, and a diffusion length characteristic of the reactor dimensions. These plasma characteristics are the same whether deposition or etching is involved. The only distinction is that a different aspect of the plasma characteristics is important for deposition (which involves deposition of reactive neutrals) versus etching (which involves gasification of substrate by reactive neutrals). Although the total pressure is important in CVD reactors, because of its role in determining flow regimes and enhancing diffusion, the pressure effects in plasma reactors should be viewed in terms of collisions. To begin with, a plasma is generated by collisions of electrons with neutral molecules and, as such, the pressure has a dominant effect on plasma characteristics. The typical total pressure in plasma reactors is at most 10 torr. The molecular flow regime is rarely encountered in plasma reactors since the plasma extinguishes when the electron mean free path approaches the magnitude of reactor dimension. It is noted in this regard that the mean free path of electrons is several times that of neutrals. The low limit on the pressure is dictated by the need to sustain a plasma. The limit decreases with increasing power and excitation frequency. The practical range is much higher than the low limit since an efficient discharge requires sufficient collisions. To gain an understanding of plasma characteristics, transport properties of charged particles (electrons and ions) are considered first.

10.4 TRANSPORT PROPERTIES OF IONS AND ELECTRONS

The transport properties of interest in plasma processing are diffusivities and mobilities of electrons and ions in a fluid medium. The mobility is the transport property that relates the external force (electric field) to the flux of charged particles due to the force. This flux is called "drift flux." An analogy in mass transport can be found where convective flux is caused by the external force of pressure. Thus, the drift flux is equivalent to the convective flux. The mobility relates the electric field to drift velocity, v_d, as follows:

$$v_d = \mu E \tag{10.1}$$

where μ is the mobility and E is the electric field. It has the unit of square centimetres per volt-second. Here again, the drift velocity is equivalent to fluid velocity in convective flow.

As in mass transport, the flux of charged particles consists of diffusive flux and drift (convective) flux:

$$j_i = -D_i \frac{dn_i}{dx} + \mu_i E n_i \tag{10.2}$$

$$j_e = -D_e \frac{dn_e}{dx} - \mu_e E n_e \tag{10.3}$$

which are the one-dimensional expressions for the (positive) ion flux j_i and the electron flux j_e. Here D_i and D_e are the diffusivities of ions and electrons, respectively, and μ_i and μ_e are the mobilities of ions and electrons. Note in Eq. (10.3) that the drift term is negative since electrons move in the direction opposite to the electric field.

According to the kinetic theory of Chapman-Enskog, the ion-atom mutual diffusion coefficient D_i is given to second order by

$$D_i = \frac{3\pi^{1/2}}{16}\left(\frac{2k_B T}{M_r}\right)^{7/2} \frac{1 + E_o}{(n_i + n_a)P_{ia}} \tag{10.4}$$

where M_r is the reduced mass given by $M_i M_a/(M_i + M_a)$, n_i is the ion and n_a the atom number densities, E_o is a second-order correction which may be taken as zero, and P_{ia} is an average of diffusion cross section over a Maxwellian velocity distribution. Often, it is of interest to find the dependence of the diffusivity on pressure and temperature. Because of the dependence of P_{ia} on the pressure and temperature, the diffusivity is given by

$$D_i = (\text{constant}) \frac{T^m}{M_r P} \tag{10.5}$$

where P is the total pressure and m can take on a value of 2 for low temperature or 1.66 for high temperature. In plasmas, n_i is much less than n_a and, if one assumes the elastic sphere model (Hirschfelder *et al.*, 1964), Eq. (10.4) reduces to

$$D_i = \frac{3}{8}\left(\frac{\pi k_B T}{2M_r}\right)(n_a \pi d_{ia}^2)^{-1} \qquad \text{where } d_{ia} = \frac{d_i + d_a}{2} \tag{10.6}$$

and d_i and d_a, respectively, are the ion and atom diameters. The mobility can then be obtained from the Einstein relation:

$$\frac{D_i}{\mu_i} = \frac{k_B T_i}{q} \tag{10.7}$$

The temperature and pressure dependence of the mobility can be written (McDaniel, 1964) as

$$\mu_i = (\text{constant}) \frac{T_i^n}{M_r^{1/2} P} \tag{10.8}$$

where n is zero at low temperature and $-\frac{1}{3}$ at high temperature. Here the temperature dependence of mobility is different from that expected from the Einstein relation because of the temperature dependence of velocity.

The Einstein relation, which is applicable to a system at thermal equilibrium, has to be modified for electrons in a fluid:

$$\frac{D_e}{\mu_e} = \frac{\lambda_t k_B T_e}{q} \tag{10.9}$$

where the Townsend coefficient λ_t (McDaniel, 1964) accounts for the nonequilibrium nature of the electrons. The coefficient increases nonlinearly with E/P. Use of Eq. (10.1) for electrons in Eq. (10.9) leads to

$$(v_d)_e = (\text{constant}) \frac{E/P}{\lambda_t T_e} \tag{10.10}$$

Experimental results for the drift velocity are usually correlated to E/P, typically on a log-log scale as shown in Fig. 10-8. For most gases, the drift velocity is of the order of 10^6 cm/s for electrons and of 10^5 cm/s for ions at E/P of 1 volt/(cm·torr). Once the drift velocity of electrons is known, the diffusivity can be calculated from Eq. (10.9).

The diffusivities are often calculated from experimental "ambipolar" diffusivities. When the density of electrons and ions becomes large (usually larger than

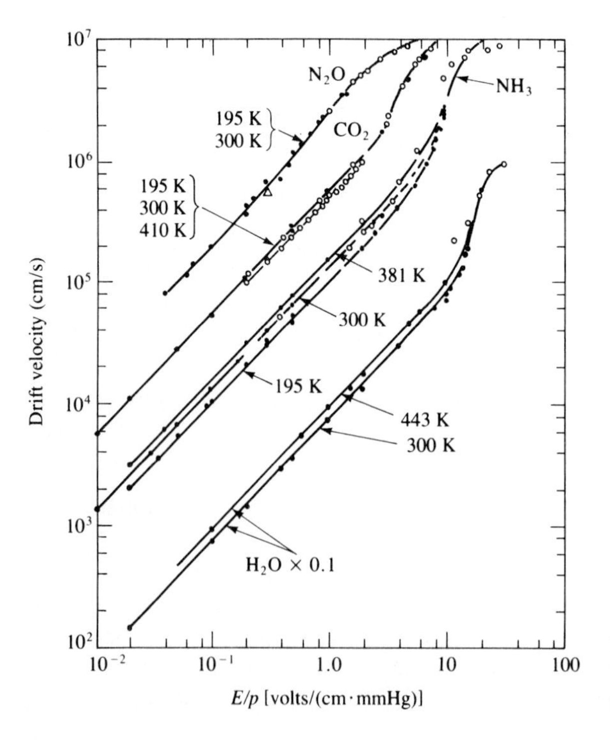

FIGURE 10-8
Drift velocity for several gases (McDaniel, 1964).

10^8 cm^{-3}), the ordinary diffusion does not hold because of their mutual Coulomb fields. The diffusivity of electrons is about four orders of magnitude higher than that of ions, a typical electron diffusivity being 10^6 cm^2/s. Thus, electrons attempt to diffuse more rapidly than ions toward regions of lower concentration, but their motion is impeded by the restraining space charge field thereby created. This same field has the opposite effect on the ions and causes them to diffuse at a faster rate than they would in the absence of the electrons. Both species of charged particles consequently diffuse with the same velocity, and since there is now no difference in the flow of the particles of opposite sign the diffusion is called ambipolar (McDaniel, 1964). Consider a fluid medium in which the density of electrons is the same as that of ions. Let this density be n and the velocity be v_a, which is the same for both species. Use of these in Eqs. (10.2) and (10.3), where now $j = j_i = j_e = v_a n$ and $n = n_i = n_e$, leads to the following definition of the (see Prob. 10.2) ambipolar diffusivity D_a:

$$D_a = \frac{D_e \mu_i + D_i \mu_e}{\mu_i + \mu_e} \tag{10.11}$$

Approximate relationships result from Eq. (10.11) and the Einstein relation (with Townsend coefficient of unity) when $\mu_i \gg \mu_e$ and $T_e \gg T_i$:

$$D_a = \begin{cases} \dfrac{k_B T_e}{q} \mu_i & (10.12) \\ D_i\left(1 + \dfrac{T_e}{T_i}\right) = D_i \dfrac{T_e}{T_i} & (10.13) \end{cases}$$

Note that $D_a P$ is a constant, and experimental values of the constant range from 100 to 900 cm^2·torr/s depending on the type of gas and ion.

The ambipolar diffusivity is obtained from an experimental determination of the rate of decay of the charged particle density in a cavity after the ionization source has been turned off. For this purpose, the particle number density is assumed to decay as $\exp(-t/t_r)$, where the decay constant t_r is the average lifetime of particles with respect to collision with the walls of a containing vessel. If the diffusion length appropriate for a given container is denoted by L, the ambipolar diffusivity is then given by

$$D_a = \frac{L^2}{t_r} \tag{10.14}$$

where the characteristic diffusion length L depends on the container geometry.

Example 10.1. The density of a charged particle, n, in a container decays according to $n = n_o \exp(-t/t_r)$. Here n_o is the density at the time the ionization source is turned off. Thus, a plot of ln n versus t yields the decay constant t_r. According to McDaniel (1964), the average collision lifetime, t_r, in a tube of 1 cm radius is 3×10^{-3} s at 1 torr of nitrogen and room temperature. The diffusion can be described by

$$\frac{\partial n}{\partial t} = \nabla(D_a \nabla n) \tag{A}$$

If n decays in an exponential way, one can write

$$n = N(r, z)e^{-t/t_r} \tag{B}$$

For the cylindrical coordinate, Eq. (A) with Eq. (B) can be rewritten as

$$\frac{d^2N}{dr^2} + \frac{1}{r}\frac{dN}{dr} + \alpha^2 N = 0 \qquad \text{where } \alpha = \frac{1}{D_a t_r} \tag{C}$$

where it has been assumed that the tube is infinitely long. The boundary conditions are the symmetry around the center ($dN/dr = 0$) and vanishment of n at the wall or $N = 0$ at $r = r_o$. The solution to Eq. (C) is

$$N(r) = AJ_0(\alpha r) \tag{D}$$

with the condition

$$AJ_0(\alpha r_o) = 0 \tag{E}$$

where J_0 is the Bessel function of the first kind of order zero. There are many solutions to Eq. (E) and the first zero of J_0 occurs at $\alpha_1 r = 2.405$. If α_i is the ith root of J, one can write the solution as

$$n(r, t) = \sum_{i=1} A_i J_0(\alpha_i r)e^{-t/t_i} \tag{F}$$

According to Eq. (10.14) and the definition of α_i in Eq. (C), the diffusion length L, which in this case is dependent on the roots α_i, and thus L_i, is given by

$$\frac{1}{L_i^2} = \frac{1}{D_a t_i} = \alpha_i^2$$

The diffusion length corresponding to the first root is called fundamental or first mode and it is given by

$$\frac{1}{L_1^2} = \frac{1}{D_a t_1} = \left(\frac{2.405}{r_o}\right)^2 \tag{G}$$

The first mode is dominant after the time corresponding to t_1 and thus L (and therefore t_r) is taken as that for the first mode for the analysis of the data and the subsequent determination of D_a.

(*a*) Using only the fundamental mode, calculate the ambipolar diffusivity.

(*b*) Show that $t_2/t_1 = 0.19$ and thus experimental data for time larger than t_1 will give an accurate value of D_a.

Solution

(*a*) Since $r_o = 1$ cm, one has, from Eq. (G),

$$\frac{1}{D_a t_r} = 2.405^2$$

or

$$D_a = \frac{1}{2.405^2 t_r} = \frac{1}{2.405^2 \times 3 \times 10^{-3}} = 57.6 \text{ cm}^2/\text{s}$$

(*b*) The second root of Eq. (E) is given by $\alpha_2 r_o = 5.5$, or $\alpha_2 = 5.5$. The ratio, α_1/α_2, is given by

$$\frac{\alpha_1}{\alpha_2} = \left(\frac{D_a t_2}{D_a t_1}\right)^{1/2} = \left(\frac{t_2}{t_1}\right)^{1/2} = \frac{1}{2.29}$$

Thus, $t_2 = 0.19t_1$. Using only two terms in Eq. (F),

$$n(r, t) = A_1 J_0\left(\frac{2.405}{r_o}\right)e^{-t/t_1} + A_2 J_0\left(\frac{5.5}{r_o}\right)e^{-t/0.19t_1}$$

When $t = t_1$, the first term decays by e^{-1} or 0.37 but the second by $e^{-5.26}$ or 0.005. It is seen that for $t > t_1$, the first term is sufficiently accurate for the time dependence of n or for the plot of ln n versus t. The same value of D_a should result, regardless of the mode used.

The characteristic diffusion length is often used to describe a plasma. For simple geometries, these are (e.g., McDaniel, 1964)

$$L^2 = \begin{cases} \dfrac{1}{\pi^2(1/a^2 + 1/b^2 + 1/c^2)} & \text{rectangular channel of width } a\text{, depth } b\text{, and length } c \\ \left(\dfrac{r_o}{\pi}\right)^2 & \text{sphere of radius } r_o \\ \left[\left(\dfrac{2.405}{r_o}\right)^2 + \left(\dfrac{\pi}{H}\right)^2\right]^{-1} & \text{cylinder of radius } r_o \text{ and height } H \end{cases} \tag{10.15}$$

Example 10.2. According to Edelson and Flamm (1984), the major ion of interest for plasma etching based on CF_4 is CF_3^+. They give the following diffusivity for the ion:

$$D_i\ (\text{cm}^2/\text{s}) = \frac{0.00146T_i^{1.75}}{P\ (\text{torr})}$$

For the plasma, they provide the following conditions:

$T_{gas} = 313$ K $T_i = 453$ K $T_e = 5.0$ eV (1 eV = 11,600 K)
Tube radius = 0.95 cm
Plasma length = 5 cm
$P = 0.5$ torr

Calculate the ion diffusivity and mobility. Obtain the ambipolar diffusivity. Calculate the characteristic diffusion length for the reactor and the average collision lifetime with reactor walls.

They also give the following conditions for the afterglow:

$T_g = T_i = 298$ K $T_i = 0.025$ eV
Flow = 80 cm^3/min (STP)

Calculate the distance down the tube after the plasma region at which the ion concentration reduces to 1 percent of the concentration in the plasma.

Solution. According to the expression for D_i,

$$D_i = \frac{0.00146(453)^{1.75}}{0.5}$$

$$= 130 \text{ cm}^2/\text{s}$$

Now the CF_3^+ ion mobility can be obtained from the Einstein relation [Eq. (10.7)]:

$$\mu_i = \frac{D_i q}{k_B T_i} = \frac{130}{8.62 \times 10^{-5} \times 453} = 3.33 \times 10^3 \text{ cm}^2/(\text{V·s})$$

where $k_B = 8.62 \times 10^{-5}$ eV/K. From Eq. (10.13),

$$D_a = D_i \frac{T_e}{T_i} = 130 \frac{5 \times 11{,}600}{453} = 16{,}630 \text{ cm}^2/\text{s}$$

For the cylindrical tube of length 5 cm, one has, from Eq. (10.15),

$$L^2 = \frac{1}{(2.405/r_o)^2 + (\pi/H)^2}$$

$$= \frac{1}{(2.405/0.95)^2 + (3.14/5)^2}$$

$$= 0.147 \text{ cm}^2$$

Therefore, the characteristic diffusion length is 0.383 cm. The lifetime can be obtained from Eq. (10.14):

$$t_r = \frac{L^2}{D_a} = \frac{0.147}{16{,}640} = 8.8 \times 10^{-6} \text{ s}$$

The average fluid velocity v_f is

$$v_f = \frac{80 \times 760/0.5}{3.14 \times 0.95^2} = 42{,}910 \text{ cm/s}$$

Since the concentration decays on the average by the factor of t_r, one has

$$\frac{n}{n_0} = \exp\left(-\frac{t}{t_r}\right)$$

or

$$\ln\left(\frac{n}{n_0}\right) = -\frac{t}{t_r}$$

where n_0 is the concentration in the plasma. Thus,

$$\ln(0.01) = -\frac{t}{8.8} \times 10^{-6}$$

$$t = 4.05 \times 10^{-5} \text{ s}$$

The length l is therefore given by

$$l = tv_f = 4.05 \times 10^{-5} \times 42{,}910 = 1.74 \text{ cm}$$

Thus, the plasma is extinguished to 1 percent of the plasma strength in the afterglow by the time it travels 1.74 cm.

10.5 PLASMA CHARACTERISTICS AND ELECTRICAL PROPERTIES

As indicated in the introduction, relationships for plasma and its electrical properties are approximate. Furthermore, they are by no means applicable to all cases. It is with this understanding and caution that the relationships given in this section should be used.

A plasma can be characterized by its Debye length defined by

$$\lambda_D = \left(\frac{p_0 k_B T_e}{q^2 n_e} \right)^{1/2} \tag{10.16}$$

where p_0 is the permittivity of free space. An ionized medium cannot be called a plasma when its Debye length is longer than any surrounding dimensions of its boundary. The Debye length is a measure of sheath thickness. This arises from the fact that the plasma potential decays approximately with the factor of $\exp(-x/\lambda_D)$, where x is the distance from the electrode.

Since collisions by electrons cause ionization, a basic quantity in plasmas is the electron mean free path, λ_e, which is the mean distance an electron can travel without encountering a collision. If one lets f_e be the electron collision frequency, which is the number of collisions an electron experiences per unit time, the electron velocity v_e is given by

$$v_e = \lambda_e f_e \tag{10.17}$$

There are a number of constraints that a plasma system has to meet for breakdown (discharge) to occur and more importantly for efficient breakdown. It is obvious, for instance, that no breakdown will occur if the electron mean free path is larger than the electrode spacing d or the characteristic diffusion length L, since then all electrons vanish at container surfaces. For a dc discharge, the breakdown voltage is often correlated to Pd (pressure times diameter) for infinite parallel plates but more appropriately to PL (pressure times characteristic dimension) for an arbitrary container shape. According to the breakdown voltage presented by Brown (1966), the breakdown voltage increases abruptly for most gases when PL is less than 1 cm·torr. As the pressure and the dimension decrease, an enormous amount of energy is required to maintain the necessary amount of secondary electrons, which results in an abrupt change in the breakdown voltage. Therefore, a condition for an efficient dc discharge may be written as follows:

$$PL = \text{constant} \tag{10.18}$$

where the constant is around 4 cm·torr. The constant is that corresponding to the pressure at which the mean free path is about equal to the electrode separation. For ac (rf) discharge, an additional factor related to the frequency of the exciting field has to be taken into consideration, which in turn introduces a number of constraints. The electron collision frequency f_e has to be larger than the angular frequency of the external field, ω, to avoid the excessive breakdown voltage that becomes necessary when the system goes through a transition from many collisions per oscillation of electron to many oscillations per collision. Another constraint is that the oscillation amplitude should be less than one half the characteristic diffusion length (one half the electrode spacing for infinite parallel plates) or else electrons can travel completely across the length and collide with the walls on every half-cycle. The combination of these constraints, including the mean free path constraint, that is, $L > \lambda_e$, leads to the following condition (Brown, 1966):

$$\frac{P^2 L}{\omega} = \text{constant} \tag{10.19}$$

where the constant is dependent on the external field applied. This condition is based on diffusion-controlled hydrogen plasma. Therefore, caution should be taken in applying it to other cases. Although the constants in Eqs. (10.18) and (10.19) are dependent on the plasma system, it is nevertheless very instructive to note that L or equivalently the electrode spacing is inversely proportional to P or P^2 for optimum breakdown voltage. This is the main reason why the electrode spacing is smaller for higher pressure plasmas. For example, for rf plasma systems at pressures higher than 1 torr the electrode spacing is typically less than 1 cm (Mathad, 1985).

Example 10.3. For hydrogen, the collision frequency is given (Brown, 1966) as follows:

$$f_e = 5.93 \times 10^9 P \text{ (torr)}$$

when the average electron energy is larger than a few electronvolts. Further, $P\lambda_e =$ 0.02 cm·torr for the average electron energy of 5 volts. Let m be the electron mass (9.1×10^{-28} g). Then

$$v_e^2 = \frac{8k_B T_e}{\pi m} \tag{A}$$

Calculate the electron temperature.

Solution. The velocity is given by Eq. (10.17):

$$v_e = f_e \lambda_e = 5.93 \times 10^9 P \frac{0.02}{P} = 1.19 \times 10^8 \text{ cm/s}$$

From Eq. (A),

$$T_e = \frac{\pi m v_e^2}{8k_B} = \frac{3.14 \times 9.1 \times 10^{-28} \times (1.19 \times 10^8)^2}{8 \times 1.38 \times 10^{-6}} = 3.7 \times 10^4 \text{ K}$$

Example 10.4. For the CF_4 plasma in Example 10.2, determine the constant in Eq. (10.19) assuming that they used the typical frequency of 13.56 MHz and that the constant evaluated in this way can be applied to the plasma system so long as the applied voltage does not change significantly around the value used in the experiment. Based on the result, calculate the characteristic diffusion length required for efficient ignition of plasma when the angular frequency changes from 13.56 to 3.0 MHz.

Solution. Since the frequency is equal to $2\pi\omega$, one has

$$13.56 \times 10^6 = 2\pi\omega$$

$$\omega = 2.16 \times 10^6$$

From Example 10.2, $L = 0.383$ cm and $P = 0.5$ torr. Thus,

$$\frac{P^2 L}{\omega} = \frac{0.5^2 \times 0.383}{2.16 \times 10^6}$$

$$= 4.43 \times 10^{-8} \text{ torr}^2\cdot\text{cm/s}$$

Thus, the criterion under the assumptions can be written as

$$\frac{P^2 L}{\omega} = 4.43 \times 10^{-8} \text{ torr}^2\cdot\text{cm/s} \tag{A}$$

When the angular frequency changes to 3.0 MHz, it has to satisfy Eq. (A):

$$L = \frac{4.43 \times 10^{-3}\omega}{P^2}$$

$$= \frac{4.43 \times 10^{-8} \times 3.0 \times 10^6/6.28}{0.25}$$

$$= 0.085 \text{ cm}$$

This length is too small to be practical.

Another quantity that characterizes a plasma is plasma frequency. It is the angular frequency with which the plasma will oscillate around its equilibrium position and is given by

$$\omega_p^2 = \frac{q^2 n_e}{m p_0} \tag{10.20}$$

where ω_p is the plasma frequency. In the sheath where the charge carriers are mainly ions, an equivalent quantity is "ion" frequency $(\omega_p)_i$, which can be defined the same way as with ω_p but with n_e and m_e replaced by n_i and M_i, respectively. An ac (rf) plasma system is often modeled as an equivalent electrical circuit for electrical characterization. The plasma frequency can be used to determine whether sheath and plasma can be treated as a resistor or a capacitor in the equivalent circuit. Although it is still subject to dispute, the following conditions due to Zarowin (1983) could be useful:

$$\begin{aligned} \omega &\gg \frac{\omega_p^2}{f} \quad \text{(capacitive)} \\ \omega &\ll \frac{\omega_p^2}{f} \quad \text{(resistive)} \end{aligned} \tag{10.21}$$

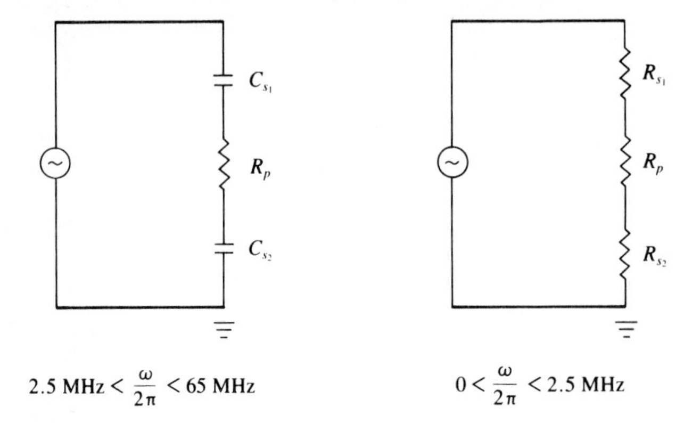

FIGURE 10-9
Equivalent circuit representation of plasma system (Zarowin, 1983).

It is understood in Eq. (10.21) that ω_p and f are $(\omega_p)_e$ and f_e for the plasma and they are $(\omega_p)_i$ and f_i for the sheath where f_i is the collision frequency of ions with neutrals in the sheath. The two equivalent circuits are shown in Fig. 10-9. According to Zarowin, the usual resistive plasma and capacitive sheaths result when

$$2.5\text{ MHz} < \frac{\omega}{2\pi} < 65\text{ MHz} \tag{10.22}$$

Example 10.5. Typical plasma-etching conditions are given (Zarowin, 1983) below:

$$n_e = 10^{10}\text{ cm}^{-3} \qquad f_e = 7 \times 10^{10}\text{ s}^{-1} \qquad f_i = 2 \times 10^{-7}\text{ s}^{-1}$$

$$n_i = 10^{10}\text{ cm}^{-3} \qquad M = 10^5\text{ m}$$

Show for the typical rf frequency of 13.56 MHz that the plasma is resistive and the sheath is capacitive according to the criteria of Eq. (10.21). Note that $p_0 = 8.85 \times 10^{-12}\text{ C}^2/(\text{N}\cdot\text{m}^2)$ and $q = 1.6 \times 10^{-19}$ C/electron.

Solution. For the plasma, Eq. (10.20) can be used:

$$\omega_p^2 = \frac{q^2 n_e}{m p_0}$$

$$= \frac{(1.6 \times 10^{-19})^2}{(9.1 \times 10^{-31})(8.85 \times 10^{-12})} n_e$$

$$= 3179 n_e \qquad (\text{m}^3/\text{s})$$

or

$$\omega_p = 56{,}380 n_e^{1/2} \qquad (n_e \text{ in cm}^{-3})$$

$$= 5.64 \times 10^9\text{ s}^{-1}$$

Thus, one has, for $f_e = 7 \times 10^{10}\ \text{s}^{-1}$,

$$\frac{\omega_p^2}{f_e} = \frac{(5.64 \times 10^9)^2}{7 \times 10^{10}} = 4.54 \times 10^8\ \text{s}^{-1} \tag{A}$$

For the excitation frequency, the angular frequency is

$$\omega = 2\pi \times 13.56 \times 10^6 = 8.52 \times 10^7\ \text{s}^{-1} \tag{B}$$

From Eqs. (A) and (B) and Eq. (10.21), it can be concluded that the plasma is resistive.

For the sheath,

$$\begin{aligned}(\omega_p)_i &= \frac{q^2 n_i}{M p_0} = \omega_p \left(\frac{m}{M}\right)^{1/2} \left(\frac{n_i}{n_e}\right)^{1/2} \\ &= 5.64 \times 10^9 (10^{-5})^{1/2} (1)^{1/2} \\ &= 1.78 \times 10^7\ \text{s}^{-1} \\ \frac{(\omega_p)_i^2}{f_i} &= \frac{(1.78 \times 10^7)^2}{2} \times 10^7 \\ &= 1.58 \times 10^7 \end{aligned} \tag{C}$$

Equations (B) and (C) show that the sheath is capacitive according to the criteria of Eq. (10.21).

The most prominent quantity in characterizing a plasma is the electron collision frequency or, equivalently, the mean free path. The collision frequency can be written as

$$f_e = Q_c n_0 v_e \tag{10.23}$$

where n_0 is the number density of neutrals, v_e is the electron velocity, and Q_c is called the collision cross section, which is a measure of the probability of collision between electrons and neutrals. The reason that the collision frequency cannot be readily determined is the strong nonlinear dependence of the collision cross section on the applied voltage. It is also strongly dependent on the type of gas molecules. Furthermore, many basic quantities required to characterize a plasma such as electron temperature cannot be measured directly, but must be inferred from electrical measurements. This is much more of a problem for ac discharges than for dc discharges.

Consider the simpler case of dc discharges. For deposition by physical sputtering, which is of major interest for dc discharges, the rate of sputtering is that given by Eq. (9.13). All quantities of interest can be determined from the measurements of voltage and current at the cathode. The requirement of eliminating plasma formation in the spacing between the target and ground shield can be satisfied by insisting that the sheath thickness corresponding to the cathode potential be smaller than the spacing. This gives a conservative value since the cathode potential is the highest in the container and thus the thickness is the maximum possible. This sheath thickness is that given by Eq. (9.14):

$$d^2 = \frac{0.85 p_0 q^{1/2} V_{sp}^{3/2}}{j_i M_i^{1/2}} \tag{10.24}$$

where V_{sp} is the cathode (sheath) potential, d is the sheath thickness, and j_i is the ion current density.

For ac discharges, two quantities that can be determined from electrical measurements have already been given in Chap. 9 (refer to Fig. 9-10):

$$V_p = 0.5(V_a + V_{dc}) \tag{10.25}$$

$$V_{sp} = V_p - V_{dc} \tag{10.26}$$

where V_a is the applied excitation voltage, V_{dc} is the corresponding dc bias, V_{sp} is the sheath potential, and V_p is the plasma potential. Measurement of the average (rms) current I_r yields the sheath current density (flux) at the powered electrode, J_{sp}:

$$J_{sp} = \frac{I_r}{A_p} \tag{10.27}$$

where A_p is the area of the powered electrode.

In rf discharges, the maximum power dissipation in the discharge occurs when the impedance of the external circuit is matched to the internal (discharge) impedance. Under this condition, and in the excitation frequency range given by Eq. (10.23) corresponding to the capacitive circuit model in Fig. 10-9, the sheath electric field E_{sp} is given by (Zarowin, 1983)

$$E_{sp} = \frac{J_{sp}}{\omega p_0} \tag{10.28}$$

It follows then that the sheath thickness (powered side) d_{sp} is given by

$$d_{sp} = \frac{V_{sp}}{E_{sp}} \tag{10.29}$$

where a linear voltage drop across the sheath has been assumed. Note that $d_{sp} = d_p$ and $V_{sp} = V_p$ when the powered electrode area is the same as the grounded electrode area.

The electron temperature can then be estimated from Eq. (9.18):

$$V_p - V_f = \frac{k_B T_e}{2q} \ln\left(\frac{M}{2.3m}\right) \tag{10.30}$$

where V_f is the floating potential with respect to a ground for a substrate (probe) immersed in a plasma.

Example 10.6. From the data and results given in Example 9.4, calculate the spacing between the target and ground shield required to eliminate plasma formation in the space.

Solution. From Example 9.4,

$$V_{sp} = 850 \text{ volts} = 850 \text{ kg·m}^2/(\text{s}^2\text{·C})$$

$$j_i = 2.27 \times 10^{-4} \text{ A/cm}^2 = 2.27 \times 10^{-4} \text{ C/(s·cm}^2)$$

Also, $q = 1.6 \times 10^{-19}$ C, $M = 0.018$ kg for Ar and $p_0 = 8.85 \times 10^{-12}$ $\text{C}^2\text{·s}^2/(\text{kg·m}^3)$. Therefore, Eq. (10.22) yields

$$d^2 = \frac{0.85 \times (8.85 \times 10^{-12})(1.6 \times 10^{-19})^{1/2}(850)^{3/2}}{(2.27 \times 10^{-4})(0.018)^{1/2}}$$

$$= 2.45 \times 10^{-12} \text{ cm}^2$$

or

$$d < 1.57 \times 10^{-6} \text{ cm}$$

It is seen that the ground shield in Fig. 10-5 should literally touch the target under the conditions.

Example 10.7. For an asymmetrical, parallel-plate rf discharge operating at 13.56 MHz, the root mean square current and applied voltage amplitude determined from electrical measurements are 0.1 A and 160 V, respectively. The dc bias with respect to the ground is (-40) V. The powered electrode area is 10 cm^2 and the grounded electrode area is 30 cm^2. For Ar plasma, suppose that the use of the probe shown in Fig. 10-10 with equal area electrodes yields a floating potential of 10 V with respect to the ground. Determine the following: V_p, V_{sp}, J_{sp}, d_{sp}, T_e, v_e, and the ion fluxes at both electrodes, assuming that ion collisions in the sheath are negligible.

Solution. From Eqs. (10.25) through (10.27), one has

$$V_p = 0.5(160 - 80) = 40 \text{ V}$$

$$V_{sp} = V_p - V_{dc} = 40 - (-40) = 80 \text{ V}$$

$$J_{sp} = \frac{I_r}{A_p} = \frac{0.1}{10} = 0.01 \text{ A/cm}^2$$

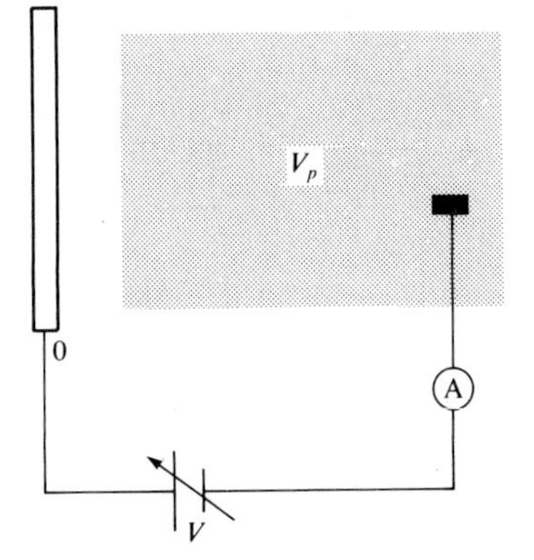

FIGURE 10-10
Schematic for probe measurements in a plasma (Chapman, 1980).

From Eqs. (10.28) and (10.29),

$$E_{sp} = \frac{J_{sp}}{\omega p_0}$$

$$= \frac{0.01 \times 10^4}{(2\pi \times 13.56 \times 10^6)(8.85 \times 10^{-12})}$$

$$= 1.31 \times 10^5 \text{ (s·A·N/C}^2\text{)}$$

$$= 1.31 \times 10^5 \text{ (N/C)} = 1.31 \times 10^5 \text{ V/m}$$

Thus,

$$d_{sp} = \frac{80}{1.31 \times 10^5} = 0.61 \times 10^{-3} \text{ m}$$

From Eq. (10.30),

$$V_p - V_f = 30 = \frac{k_B T_e}{2q} \ln\left(\frac{M}{2.3m}\right) = \frac{k_B T_e}{2q} \ln\left(\frac{18/6 \times 10^{23}}{2.3 \times 9.1 \times 10^{-28}}\right)$$

$$= 4.78\left(\frac{k_B T_e}{q}\right)$$

or

$$\frac{k_B T_e}{q} = \frac{30}{4.78} = 8.36 \text{ V}$$

Thus, the electron temperature is that corresponding to 8.36 eV or 97,000 K. For the electron velocity, Eq. (A) in Example 10.3 can be used:

$$v_e = \left(\frac{8k_B T_e}{\pi m}\right)^{1/2}$$

$$= \left(\frac{8 \times 1.38 \times 10^{-16} \times 97{,}000}{3.14 \times 9.1 \times 10^{-28}}\right)^{1/2} = 1.94 \times 10^8 \text{ cm/s}$$

The ion flux at the powered electrode is very close to that corresponding to J_{sp} [see Eq. (10.27)]. Thus,

$$j_{sp} = \frac{J_{sp}}{q} = \frac{0.01}{1.6 \times 10^{-19}} = 6.18 \times 10^{16} \text{ cm}^{-2}\text{·s}^{-1}$$

To calculate the ion flux, the fact that the current must be the same at both electrodes is used:

$$J_{sp} A_p = J_g A_g$$

Thus,

$$j_g = \frac{j_{sp} A_p}{A_g} = 6.18 \times 10^{16} \times 10/30 = 2.06 \times 10^{16} \text{ cm}^{-2}\text{·s}^{-1}$$

Example 10.8. Suppose that ambipolar diffusion experiments yield a D_a of 16,000 cm^2/s and a collision lifetime (time between collisions) of 8.8×10^{-14} s, when the same apparatus as in Example 10.7 is used. Based on the data here and those in Example 10.7, calculate μ_i and n_e. Also calculate the electron collision cross section. Assume the pressure is 0.5 torr and the plasma gas temperature is 300 K.

Use the following relationships:

$$J_p = \frac{C_d V_p}{H} \tag{A}$$

$$C_d = \frac{n_e q^2}{m f_e} \tag{B}$$

where C_d is the plasma conductivity and H is the electrode separation, which is 5 cm for this example.

Solution. The ion mobility can be obtained from Eq. (10.12):

$$\mu_i = \frac{q}{k_B T_e} D_a$$

$$= \frac{1}{(8.62 \times 10^{-5})(97{,}000)}(16{,}000) = 1.91 \times 10^3 \text{ cm}^2/(\text{V}\cdot\text{s})$$

where $k_B = 8.62 \times 10^{-5}$ eV/K. The collision lifetime is the inverse of the collision frequency. One has, from Eq. (10.23),

$$Q_c = \frac{f_e}{n_0 v_e} = \frac{1}{t_r n_0 v_e}$$

The number density of neutrals n_0 can be obtained from the ideal gas law:

$$n_0 = \frac{P}{k_B T} = \frac{(0.5/760)(6 \times 10^{23})}{82 \times 300} = 1.6 \times 10^{16} \text{ cm}^{-3}$$

Therefore,

$$Q_c = \frac{1}{(8.8 \times 10^{-14})(1.6 \times 10^{16})(1.94 \times 10^8)} = 3.66 \times 10^{-12} \text{ cm}^2$$

From Eq. (A), the conductivity C_d is

$$C_d = \frac{HJ_p}{V_p} = \frac{0.03 \times 0.05}{40} = 3.75 \times 10^{-5}\ \Omega^{-1}\cdot\text{m}^{-1}$$

From Eq. (B),

$$n_e = \frac{C_d m f_e}{q^2} = \frac{C_d m}{t_r q^2} = \frac{(3.75 \times 10^{-5})(9.1 \times 10^{-31})}{(8.8 \times 10^{-14})(1.6 \times 10^{-19})^2}$$

$$= 1.51 \times 10^{16} \text{ m}^{-3} = 1.51 \times 10^{10} \text{ cm}^{-3}$$

The above examples should reveal that most of the quantities of interest for plasma processing can be determined on the basis of electrical measurements and ambipolar diffusion experiments. However, these quantities are applicable only to the set of conditions under which the measurements are made. Therefore, for predictability, plasma characteristics have to be related to plasma design and operating conditions. These are: applied voltage, electrode areas, pressure, and reactor geometry for a given excitation frequency. A quantity representing the reactor geometry is the characteristic diffusion length. As was evident in Example 10.7, the electrode area ratio is another quantity that can be used to interpret plasma characteristics. The plasma electric field E_p is a measure of energy per unit distance over the potential. Further, pressure is a measure of the number of particles per unit volume. Thus, E_p/P is often used to represent plasma characteristics. Numerous examples of the use of E_p/P can be found, for example, in Brown (1966).

Another quantity often used in correlations for plasma characteristics is PL. This quantity can be viewed from the point of degree of ionization. If a beam of current i_0 is injected into a medium with a neutral number density N in a vessel with a characteristic diffusion length L, then the degree of ionization α, which can be defined as i_i/i_0, is given (McDaniel, 1964) by

$$\alpha = Q_i NL \tag{10.31}$$

where the ionization collision cross section Q_i can be expressed in terms of the collision probability P_i as follows:

$$Q_i = 0.283 \times 10^{-16} P_i \tag{10.32}$$

The probability P_i is in turn dependent on electron energy. As shown in Fig. 10-11, the probability increases almost linearly with electron energy and then goes through a maximum at around 90 eV. The probability is almost zero below

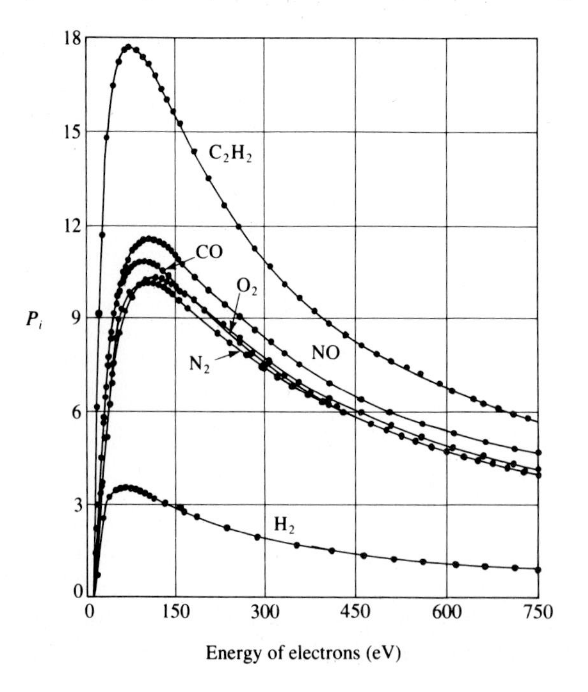

FIGURE 10-11
Collision probability for a number of gases (McDaniel, 1964).

a certain energy level, which is required to activate neutrals into ions. Note in this regard that N is proportional to P.

There are presently no *a priori* methods of relating the plasma characteristics to plasma operating and design conditions. The alternative is to correlate the plasma characteristics of interest to E_p/P and PL for a given plasma system through experiments. These results, in turn, can be used to describe or design a plasma reactor.

10.6 INTRINSIC KINETICS AND PLASMA AND TRANSPORT EFFECTS

There are two aspects unique to plasma deposition and etching. The first is that the rate constants in the intrinsic kinetics are dependent on ion energy (and ion flux) since the surface modification by ions is the major source of enhancement in the deposition and etching. The second is that, for ions, the transport effects are confined to the plasma sheath. The transport of active neutrals, which are responsible for the deposition or etching, is not restricted to the sheath but is operative throughout the reactor, as was the case in the conventional CVD.

Unlike the kinetics of conventional CVD, the intrinsic kinetics have to account for the enhancement caused by ion surface modification. The surface modification is mainly by momentum transfer with the colliding ions. Since the degree of modification is dependent on the ion energy, the activation energy in turn depends on the ion energy. The ion flux, on the other hand, determines the fraction of the substrate surface affected by the ion bombardment. One would expect that more than one site of the surface is affected per incident ion. However, if the ion flux is not sufficiently high, only a fraction of the surface would be affected. One would also expect the ion flux to have a small effect beyond a certain ion flux at which the whole surface is affected. This ion flux may be termed the critical ion flux. Therefore, all rate constants in the intrinsic kinetics are dependent not only on the ion flux but also on the ion energy after reaching the critical ion flux. Plasma reactors are usually designed and operated so as to increase the ion flux.

Therefore it is necessary to relate the rate constants to ion energy and ion flux after extracting intrinsic kinetic information from observed rates. With respect to transport effects, the buoyancy (free convection) effect is negligible due to low-temperature processing. As will be discussed shortly, the velocity profiles are well defined for a reactor that can be used to generate intrinsic kinetics. Thus, it is sufficient to consider only the mass transfer effect for the reactive neutral species of interest.

A criterion is available that can be used to determine whether or not the experimental data are free from the mass transfer effect. In conventional CVD, the concentration readily measurable via a sampling line is the bulk concentration. In plasma reactors, on the other hand, there can be two different concentrations that are relevant to the plasma processing: the bulk concentration if the sampling is directly from inside the plasma and the local concentration in the plasma. The point here is that any concentration measurement from outside the plasma does not represent the concentration of interest since the reactive neutrals can readily recombine outside the plasma. While the local concentrations can be measured readily, measurement of the bulk concentration would depend on reactor configuration. Therefore, the mass transfer coefficient for active neutral species, k_m, is defined as follows:

$$R_G = k_m(C_p - C_s) \tag{10.33}$$

where R_G is the observed rate of etching or deposition per unit area, C_p is the species concentration at the electrode opposite to the one where substrate is placed, and C_s is the surface concentration of the species at the substrate. Since the temperature gradient in plasma, and thus the mass flux due to the temperature gradient, is negligible, one can write

$$R_G = D \left. \frac{\partial C}{\partial y} \right|_0 \tag{10.34}$$

It can be shown from Eqs. (10.33) and (10.34) (see Prob. 10.15) that the Sherwood number is given by

$$\text{Sh} = \frac{k_m H}{D} = \begin{cases} 2 & \text{(no flux condition)} \\ 1 & \text{(constant wall concentration condition)} \end{cases} \tag{10.35}$$

where H is the electrode separation. The first is for the case where there is no consumption of the neutral species at the electrode opposite to the substrate electrode, which would be typical of plasma reactors. For the experimental reactors considered in this section, the concentration at the electrode opposite to the substrate electrode is kept constant. In this case, the Sherwood number of unity is used for the mass transfer coefficient. If the bulk concentration is to be used, the Nusselt numbers are

$$\text{Sh} = \begin{cases} \dfrac{20}{7} & \text{(no flux condition)} \\ 1 & \text{(constant wall concentration condition)} \end{cases} \tag{10.36}$$

One can readily determine whether the experimental data obtained are intrinsic or not on the basis of the mass transfer coefficient and the observed rate. As was the case in Chap. 6, this can be determined by calculating the magnitude of the following:

$$W = \frac{R_G H}{\text{Sh } DC_i} \qquad \text{for } i = b, p \tag{10.37}$$

where the subscripts b and p, respectively, are for the concentration in the bulk fluid and that in the fluid near the top wall.

Since the ratio W is equal to $(1 - C_s/C_p)$, which follows from Eq. (10.33), a W value close to unity means that the observed rate is severely affected by mass transfer resistance; a value close to zero means that the rate is almost unaffected by mass transfer limitations. Although one may conclude that plasma processing is typically free from mass transfer effects due to the high diffusivity (100 cm^2/s and higher) of neutrals at the low pressures (at most, 10 torr) typically employed, one should take note of the fact that the ratio W represents the relative magnitude of two rates. Thus, the relative rates and not the absolute values determine the extent to which the observed rate is affected by diffusion.

Example 10.9. In plasma etching in CF_4, fluorine atoms are the reactive neutral species. It is believed that the same is true in plasma etching in NF_3. For NF_3, Stenger *et al.* (1987) give the following conditions:

Etch rate = 0.15 μm/min	$T = 25\,°C$	$P = 40$ Pa (0.3 torr)
Inlet mole fraction of NF_3 = 0.4		Inlet flow rate = 25 sccm
Electrode separation = 1.5 cm		$D_F = 680\ cm^2/s$

Determine whether the kinetic data obtained under these conditions are affected by mass transfer resistance. Note in this regard that the use of NF_4 (as opposed to CF_4) does not lead to polymer formation. The intrinsic etching rate is first order with respect to F and the rate constant is 23.5 cm/s (Stenger *et al.*, 1987). Calculate the fraction of neutral F atoms in the plasma. Determine the electron density in the plasma assuming that all the rate constants involved are known. Assume for this part that the reactor is a simple planar reactor with symmetrical electrodes of 100 cm^2 and with no radial flow, i.e., flow in the direction parallel to the electrodes.

Solution. The etch rate can be calculated as follows:

$$R_G = \frac{\rho G}{M} = \frac{2.33 \times (0.2 \times 10^{-4}/60)}{28.1} = 2.764 \times 10^{-8}\ \text{mol/(cm}^2\text{·s)}$$

where M and ρ are the silicon atomic weight and density, respectively, and G is the linear (centimeters per second) etch rate. From the ideal gas law, the concentration of NF_4 is

$$C = \frac{p_i}{R_g T} = \frac{(0.3/760) \times 0.4}{82 \times 298} = 6.46 \times 10^{-9}$$

If the etch rate is first order with respect to F atoms and the rate constant is 23.5 cm/s, the fluorine atom concentration [F] is

$$[\text{F}] = \frac{R_G}{k} = \frac{2.764 \times 10^{-8}}{23.5} = 1.176 \times 10^{-9}\ \text{mol/cm}^3$$

Thus, the fraction of F atoms in the plasma f is

$$f = \frac{[\text{F}]}{C_t} = \frac{1.176 \times 10^{-9}}{6.46 \times 10^{-9}/0.4} = 0.073$$

where C_t is the concentration of all species in the plasma (note in this regard that concentrations of electrons and ions are negligible). According to the criterion of Eq. (10.37), one has

$$W = \frac{2.764 \times 10^{-8} \times 1.5}{\text{Nu}_m 680 \times 1.176 \times 10^{-9}} = \frac{0.052}{\text{Nu}_m} = \frac{0.052}{20/7} = 0.018$$

The ratio W in this case is $(1 - C_s/C_b)$. Thus, the data are almost free from mass transfer effects.

For part two, a steady-state balance for F atoms can be used:

$$\text{In} + \text{generation} = \text{out} + \text{consumption}$$

$$0 + v_p k_g n_e[\text{NF}_3] = Q[\text{F}] + k_1[\text{F}]A + k_2[\text{F}]v_p$$

where Q is the volumetric flow rate, A is the substrate area, v_p is the plasma volume, n_e is the electron concentration, and k_g, k_1, and k_2 are rate constants. Thus, the electron concentration is given by

$$n_e = \frac{[\text{F}](Q + k_1 A + k_2 v_p)}{k_g[\text{NF}_3]v_p}$$

One unique aspect of plasma kinetics is that a change in pressure causes not only a change in the concentration but also a change in the rate constants involved. Therefore, intrinsic data obtained at one pressure cannot be related to data at another pressure, even if all the other conditions remain the same, unless proper corrections are made. This is due to the fact that a pressure change also brings about changes in plasma characteristics, which in turn changes ion flux, ion energy, and the concentration of reactive neutral species, beyond the change expected from the pressure effect on concentration.

To examine how one can make a change in concentration without affecting the ion energy for given power input and rf frequency, examine the plasma and sheath (powered electrode) potentials. Depending on where the substrate is placed, one of the two potentials determines the ion energy, although some consideration should be given to the energy loss in the sheath due to collisions. When pressure is very low, the collision frequency of ions with neutrals is very low such that the potential is equivalent to the ion energy, as in a dc discharge. When the pressure is relatively high such that the mean free path of ions is not much larger than the sheath thickness, the potential is not equivalent to the ion energy. The probability of having no collision in travelling a distance x is given by $\exp(-x/\lambda)$, where λ is the mean free path (e.g., Chapman, 1980). The probability of having no collision in traveling the sheath thickness d, P_d, is therefore given by

$$P_d = \exp\left(-\frac{d}{\lambda}\right) \tag{10.38}$$

where the mean free path of ions in neutrals (McDaniel, 1964) is given by

$$\lambda = \left[\pi n_i d_i^2 (2)^{1/2} + \pi n_n \left(\frac{d_i + d_n^2}{2}\right)^{1/2} \left(1 + \frac{M_i}{M_n}\right)^{1/2}\right]^{-1} \tag{10.39}$$

The relationship gives good order of magnitude agreement with experiments for the slightly ionized gases, typical of plasma processing. Here n is the number density, d is the molecular diameter, M is the molecular weight, and the subscripts i and n, respectively, are for ions and neutrals. The mean free path decreases with increasing pressure. The sheath thickness also decreases with increasing pressure but at a much lower rate. Thus, as the pressure decreases, d/λ decreases and the probability of having no collision [Eq. (10.38)] approaches unity.

As apparent from Eq. (10.38), the probability of collision in the sheath approaches zero when d/λ is much less than unity, which occurs when the pressure is low, say less than 0.01 torr, as in a typical dc discharge. Thus, the sheath potential itself is the ion energy. In the other extreme case in which the sheath thickness is large relative to the mean free path, the average ion energy E_i (Davis and Vanderslice, 1963) is given by

$$E_i = \lambda_i q E_j \qquad \text{for } j = p \text{ or } sp \tag{10.40}$$

where E_j is the electric field for plasma or sheath. Note that the ion energy is proportional to E_j/P, since the mean free path is inversely proportional to pressure. Thus, the ion energy is the same if E_j/P is the same. The ion flux can be determined from the measured root mean square (rms) current although both electrons and ions contribute to the current in alternating frequencies (see Prob. 10.17 for a relationship for the ion flux). The ion flux is less than the current flux (density), but it is nevertheless proportional to the current density.

There are no universal relationships for E_j/P. Nevertheless, it is known that it can be correlated to PL and that at least in the regime where efficient ignition of plasma takes place [satisfying the condition of Eq. (10.18) for a dc discharge or that of Eq. (10.19) for an rf discharge], E_j/P is a monotonic function of PL: there is one and only one value of E_j/P for a given value of PL. This means that each time a pressure change is made in an experimental reactor, the characteristic diffusion length or the electrode spacing has to be changed to get experimental data for the same ion energy and thus the same rate constants, at least with a saturated ion flux. Therefore, it is necessary to establish experimentally a relationship between E_j/P and PL prior to kinetic experiments.

Example 10.10. For argon plasma with asymmetrical electrodes, Song (1988) gives the data in Fig. 10-12. The ranges of variables for the data are: 0.03 to 0.9 torr, 0.08 to 0.73 W/cm^2 power density, and electrode separation of 2 to 5 cm. For this example, however, suppose that the data are for symmetrical electrodes. Calculate the electrode separation required for 0.2 torr that will yield the same ion energy as that for 0.50 torr and electrode separation of 2 cm. The reactor is a quartz cylinder with a diameter of 10 cm and the electrodes are on the top and bottom of the cylinder.

Solution. For the reference conditions of 0.7 torr and 2 cm, one has, from Eq. (10.15),

$$\lambda_r = \left[\frac{1}{(2.405/5)^2 + 9.86/H^2}\right]^{1/2} = 0.61 \text{ cm} \tag{A}$$

Since $P = 0.5$ torr and $(PL)_r = 0.305$ cm·torr, the value of E_p/P corresponding to the value of PL in Fig. 10-12 is 30 V/(cm·torr). For the ion energy to be the same, E_p/P should remain the same. This in turn means that PL should be the same. Thus,

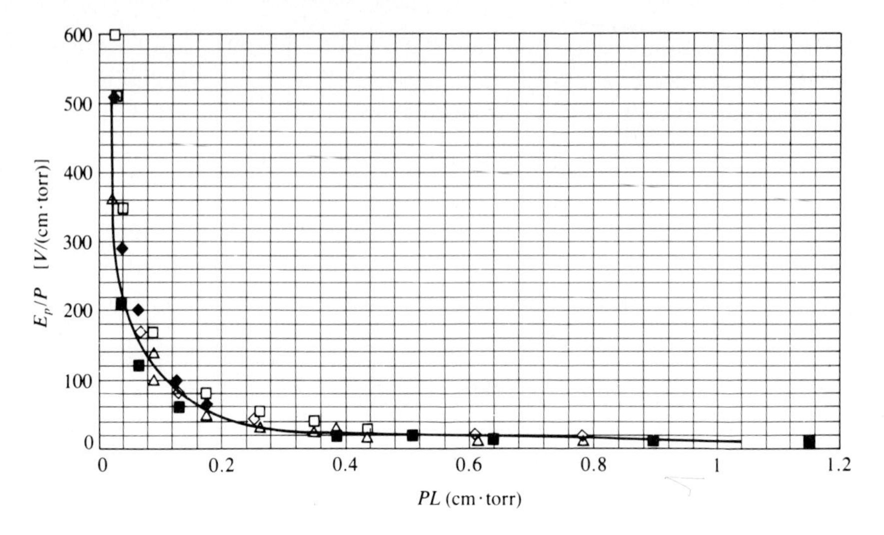

FIGURE 10-12
PL versus E_p/P (Song, 1988).

$PL = 0.305$. Solving this for L for 0.2 torr yields

$$L = \frac{0.305}{0.2} = 1.525 \text{ cm}$$

From Eq. (10.15),

$$1.525 = \left[\frac{1}{(2.405/5)^2 + 9.86/H^2}\right]^{1/2}$$

Solving this for H yields 6.98 cm.

As illustrated in the above example, kinetic data at various pressures (concentrations) but at the same ion bombardment energy can be obtained. There remain two items that must be resolved for the use of the data for intrinsic kinetics. The first is to properly account for mass transfer resistance, if any. The second has to do with the proper way of representing the effects of ion energy and ion flux on the rate. As discussed earlier in this section, the ion energy is a measure of the extent to which the activation energy is reduced by surface modification by colliding ions. Therefore, the activation energy is dependent on the ion energy. The ion flux is a measure of the fraction of the substrate surface affected by the colliding ions. Above a certain saturation value of the ion flux, any further increase should have negligible effects for a given ion energy. In analogy with the change of heat of adsorption with surface coverage for heterogeneous surfaces, one may postulate the following for the change of the activation energy E_a:

$$E_a = E_o - \frac{AJ_p}{1 + AJ_p} g[E_i] \tag{10.41}$$

where A is a constant representing the extent to which the ion flux approaches the saturation, E_o is the activation energy in the absence of ions, and g is a function representing the effect of the ion energy, E_i. It is noted that E_o can be determined by independent experiment. Although J_p is not the ion flux, it can be used as such since the ion flux is proportional to J_p and the proportionality constant can be absorbed into the constant A.

In general, intrinsic kinetics can be written as follows:

$$r_c = kf(C) \tag{10.42}$$

where r_c is the intrinsic rate, k is the rate constant, and $f(C)$ represents the concentration dependence. The rate constant, k, in plasma etching or deposition can be written in the following form with the aid of Eq. (10.41):

$$k = k_o \exp\left[\left(-E_o + \frac{AJ_p g}{1 + AJ_p}\right)\left(\frac{1}{R_g T}\right)\right] \tag{10.43}$$

where k_o is the preexponential factor and R_g is the gas constant. If the concentration dependence $f(C)$ is assumed for the determination of intrinsic kinetics from experimental data, the observed rate, R_G, is that given by

$$R_G = k_s f(C_s) \tag{10.44}$$

where the subscript s denotes evaluation of the quantity at the substrate surface. If first-order kinetics is assumed, as in Example 10.8, $r_c = k_s C_s$. The value of k_s is the rate constant at the substrate temperature; C_s can be determined from Eqs. (10.33) and (10.34) or Eq. (10.36). From the measured values of R_G and C_p(or C_b), the rate constant can be determined for an assumed form of $f(C_s)$, as illustrated in Chap. 5. The rate constant thus determined may be related to ion energy and ion flux in the form suggested by Eq. (10.43).

Since the intrinsic kinetics for etching or deposition involve the concentration of reactive neutral species, it is necessary to determine the homogeneous kinetics of the reactions leading to the formation of the neutral species. One complication with the homogeneous reactions is the formation of polymers and its deposition on the walls. For intrinsic kinetics, however, all that is required is steady state, i.e., constant reactive species concentration in the plasma. The steady-state concentrations can be used for the kinetics since an equilibrium exists between the wall polymer concentration and the neutral species concentration after the steady state is reached (Edelson and Flamm, 1984). A balance for the reactive neutral species for a flow-through reactor is

$$v_p r_G = QC_n + R_G A_s \tag{10.45}$$

where r_G is the net rate of generation of the reactive species, Q is the volumetric flow rate, C_n is the concentration of the reactive species exiting the plasma, and A_s is the substrate surface area. Since all quantities in the right-hand side of Eq. (10.45) are known, the net rate of generation can be determined from the measurements. The net rate of generation consists of the rate of generation of the reactive species and the rate of consumption due to recombination. Each rate process can be considered elementary and thus the rates can readily be written provided the major reactions are known. The rate of generation r_g in general can be written as

$$r_g = k_g n_e C_m = k'_g n_e n_m \tag{10.46}$$

where k_g is the rate constant and C_m is the concentration of the main neutral species in the plasma. For CF_4 plasma, for example, the concentration is that of CF_4. Depending on the recombination routes, the rate of consumption by recombination can be complicated. However, for CF_4 plasma, it appears that the main path is by recombination among the neutral F atoms (Edelson and Flamm, 1984), in which case the rate of recombination r_r can be written as $k_r C_F^2$. The electron concentration or the number density can be inferred from the fact that the plasma conductance, at least in the range of electron density of interest (10^9 to 10^{12} cm^{-3}), is proportional to the electron concentration. Thus, Eq. (10.46) is rewritten with the aid of Eq. (A) in Example 10.8 as follows:

$$r_g = k_G\left(\frac{HJ_p}{V_p}\right)C_m \qquad \text{where } k_G = k_g \alpha_g \tag{10.47}$$

and α_g is the proportionality constant. Given measurements of C_n, J_p and V_p, and C_m from the ideal gas law, the rate constants can be determined with the aid of Eq. (10.45) and the recombination rate.

An experimental reactor suitable for the determination of intrinsic kinetics is shown in Fig. 10-13, which is the same as that used by Song (1988) except that it has symmetrical electrodes. The gas is fed through a quartz distributor which is circular with a perforated bottom such that the gas is distributed onto the substrate located on the powered electrode, and then exits to the vacuum pump via the annulus formed by the distributor and the outside quartz walls. A ground shield is placed in the conduit to the pump. The figure also shows the measurement arrangement. Note in this regard that the reactor can be used for both etching and deposition.

The determination of intrinsic kinetics based on the experimental reactor in Fig. 10-13 can proceed as follows:

1. Establish a relationship between E_p/P and PL by varying pressure and electrode spacing for the desired excitation frequency and power input. In this process, the constant in Eq. (10.19) for efficient ignition of plasma can also be inferred. Also establish relationships between J_p and PL and those between V_p and PL. No substrates are involved in this phase of the experiment.
2. With the substrate placed on the powered electrode, run experiments in the ranges of substrate temperature and concentration of interest at constant ion energy levels.
3. Determine the intrinsic rate (kinetics) of etching or deposition with the aid of Eqs. (10.33) through (10.36) and Eqs. (10.41) through (10.44). Measure the main reactive neutral species concentration, J_p, and V_p.

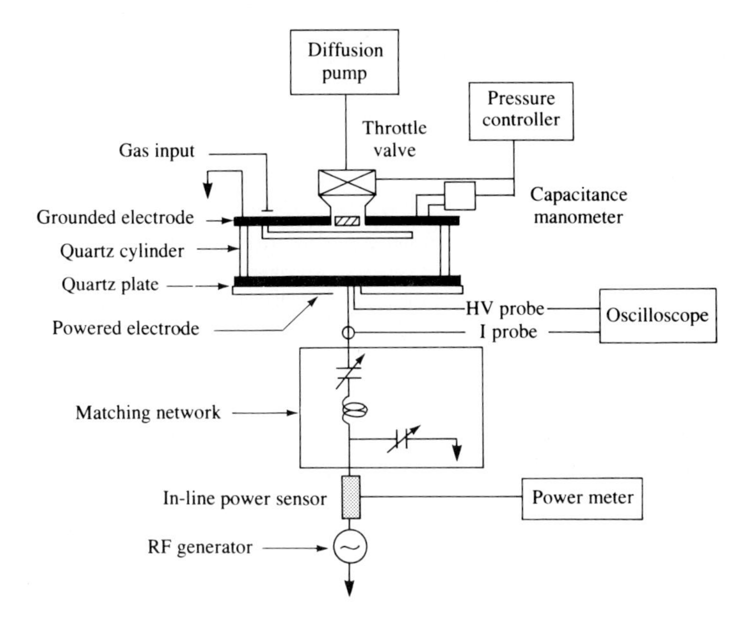

FIGURE 10-13
An experimental reactor for intrinsic kinetics.

4. Determine the homogeneous kinetics with the aid of Eqs. (10.45) through (10.47).

10.7 PLASMA REACTOR DESIGN AND ANALYSIS

There are no distinct configurational differences between plasma reactors for etching and those for deposition. The differences mainly lie in the operating conditions. This is not surprising since both take advantage of the surface modification by ions colliding with a substrate surface and of the generation of reactive neutral species in the plasma. In the case of deposition, the ion energy should be just sufficient to cause enhancement of the deposition rate; in etching, the same would be true except when anisotropy is critical, in which case a higher ion energy is required to obtain the desired directionality.

The main objectives in designing a plasma reactor are: uniformity in etching or deposition and minimum contamination by impurities at maximum possible throughput. In addition, achieving anisotropy is a major consideration in plasma etching. In both etching and deposition, but particularly in etching, there is a need to ensure that the substrate surface does not reach a high temperature, which causes temperature-related side effects. There are two main sources of contamination in plasma processing. The first is the undesired deposition of sputtered atoms on the substrate; the second is the undesired deposition from the gas phase as in polymer formation and subsequent deposition. Although electrodes are usually coated with a material with a low sputtering rate and they are inert to reactive species, minimizing the contamination is still a major concern.

The size of a plasma reactor is determined by the number of wafers to be processed per batch, which determines the area of the reactor, and the diffusion characteristic length L, which in turn fixes the other dimensions. On the other hand, the performance of the reactor with respect to uniformity and anisotropy (in the case of etching) is determined by the operating conditions. The outermost constraint that has to be satisfied is the condition for efficient ignition given by Eq. (10.18) for a dc discharge and Eq. (10.19) for an rf (ac) discharge. The constraints are given in terms of PL for a dc discharge and P^2L for an rf discharge. In general, the highest maximum possible ion flux is desired for the maximum rate and is accomplished by increasing the pressure. However, the power required to ignite plasma becomes too excessive as the pressure is increased unless the diffusion length (or the electrode separation) is reduced. The chosen conditions should reflect these conflicting factors. It should be noted in this regard that, at high pressures, the electrode separation can be very small (of the order of 1 cm), and therefore the degree to which two electrodes are parallel to each other can have a significant effect on the uniformity because of the difference in electric field from one point to another.

The flow in plasma reactors, in most cases, is in the laminar flow regime. Depending on the pressure and the electrode separation, the flow can be in the transition to the molecular flow regime, in which case the usual viscous laws do not apply. According to Eqs. (6.4) and (6.6), the flow will be in transition to molecular flow at 0.01 torr for an electrode separation of a few centimeters. At the higher pressures (at most 10 torr) typical of plasma reactors and with small temperature differences, buoyancy (free convection) effects, if any, are negligible and the flow is stable.

A necessary condition for etching or deposition uniformity is that there should be negligible depletion of source gases. To be more direct, the concentration of the main reactive neutral species should be almost the same everywhere over the wafers being processed. This is not possible with the usual feeding arrangements, unless the number of wafers to be processed is quite small. If the source gas is distributed as in Fig. 10-7*a* (similarly as in the experimental reactor in Fig. 10-13), all wafers would be exposed to the same concentration, ignoring minor variations due to the flow pattern. For such a reactor, Eq. (10.45) applies, and it can be rewritten with the aid of Eq. (10.47) as follows:

$$v_p\left[k_G\left(\frac{HJ_p}{V_p}\right)C_m - k_r C_n^n\right] = QC_n + R_G A_s \tag{10.48}$$

where k_r is the recombination rate constant, and the order for the reaction, n, is usually one or two. Given the intrinsic kinetics and the desired rate of etching or deposition, the volumetric flow rate is then given by Eq. (10.48). An inherent restriction in writing Eq. (10.48) is that the source gas must not be depleted.

Otherwise, the relationship does not hold. Since the source gas is mainly depleted through the consumption of the reactive neutral species generated [Eq. (10.47)], one has

$$Q\,\Delta C_m = r_g = k_G\left(\frac{HJ_p}{V_p}\right)C_m \tag{10.49}$$

where ΔC_m is the concentration change due to the consumption. If one insists on almost constant C_m, say $\Delta C_m/C_m < \beta$, where β is a small number, one can write:

$$Q > \frac{k_G(HJ_p/V_p)}{\beta} \tag{10.50}$$

For less than 1 percent change, the value of β would be 0.01. Thus, Eq. (10.48) can be used with the restriction of Eq. (10.50).

Example 10.11. Reconsider the etching in Example 10.9. The following kinetics were reported by Stenger *et al.* (1987), which may be considered intrinsic:

$$R_G = 23.5\ (\text{cm/s})C_n \qquad \text{for } C_n = [\text{F}]$$

$$r_r = 0.57\ (\text{s}^{-1})C_n$$

Although it was reported that a maximum variation of 8 percent occurred at 100 sccm, 0.1 torr, and 25 °C, assume for this example that true uniformity was attained under the conditions. The symmetric electrode area was 2450 cm^2, the plasma volume was 3850 cm^3, and the electrode spacing was 1.57 cm. Although the whole electrode area was not occupied by wafers, use the whole area in your calculation. Calculate the value of $k_G(J_p/V_p)$ for the etch rate of 0.15 μm/min. Also calculate the reactive fluorine atom concentration under the conditions.

Solution. From Eqs. (10.45) and (10.48) for the net rate of generation of reactive species, one has

$$r_G = k_G\left(\frac{HJ_p}{V_p}\right)\left(\frac{0.4P}{RT}\right) - 0.57C_n \tag{A}$$

where the ideal gas law has been used for C_m. Using Eq. (A) and R_G in Eq. (10.48) leads to

$$3850\left[\left(\frac{k_G J_p}{V_p}\right)\left(\frac{1.57 \times 0.4 \times 0.1}{760 \times 82 \times 298}\right) - 0.57C_n\right] = \left(\frac{760}{0.1}\right)\left(\frac{100}{0.60}\right)C_n + 2450R_G \tag{B}$$

From Example 10.9, $R_G = 2.764 \times 10^{-8}$ mol/(cm^2·s) and therefore $C_n = 1.176 \times 10^{-9}$ mol/cm^3. Substituting these values into Eq. (B) yields

$$\frac{k_G J_p}{V_p} = 5.49\ \text{cm}^{-1}$$

Note that the mass transport effect was negligible under the conditions.

It is instructive here to reexamine the role of V_p and J_p. It is well understood that a higher ion flux (J_p) but a relatively smaller ion energy (V_p or $V_p\lambda$) are desirable. A higher ion flux enhances the rate and assures local uniformity (activated versus nonactivated surface), and the lower ion energy also enhances the rate [see Eq. (10.43)] but only up to the level where physical sputtering is minimal. It is therefore desirable to maximize the ratio J_p/V_p subject to the constraint that the ion energy is sufficient for the activation. This is also the desired direction for increasing the rate of generation of the reactive neutral species, as Eq. (10.47) shows. The equation is rewritten as follows:

$$r_g = k_G\left(\frac{HJ_p}{V_p}\right)\left(\frac{Py_i}{R_g T}\right) \tag{10.51}$$

where y_i is the feed mole fraction of the source species. Noting that J_p and V_p are determined by PL for given excitation power and frequency, the design essentially reduces to specifying P, L, Q, and power input for a given number of wafers to be processed since the specification of these parameters also determines the rate of generation of the reactive species.

The specifications of P, L, and the power input (or density) are to be considered now that Q is specified by the necessary condition for the uniformity. The specification can proceed in a systematic way when the intrinsic kinetics are available. One consequence of determining the kinetics in the form of Eq. (10.43) is that the "sufficient" ion flux is that corresponding to $AJ_p \gg 1$, or

$$J_p \gg \frac{1}{A} \tag{10.52}$$

Further, the intrinsic kinetics should also reveal the minimum required ion energy [$g(E_i)$ in Eq. (10.43)]. Therefore, one has the following condition for the ion energy E_i:

$$\beta_1 < E_i < \beta_2 \tag{10.53}$$

where E_i is either qV_p or that given by Eq. (10.40), that is, $\alpha_u E_p/P$ where α_u is a constant. Here, β_1 is the minimum sufficient ion energy for the desired surface modification or rate enhancement and β_2 is the maximum allowable ion energy above which physical sputtering becomes a problem. In general, one would keep the ion energy close to β_1. This is also in line with the desire to maximize HJ_p/V_p to obtain the maximum possible rate of generation of reactive neutral species. From Eq. (10.51) one has

$$\text{Maximize}\left(\frac{HPJ_p y_i}{V_p}\right) \tag{10.54}$$

Finally the condition of efficient plasma ignition for an rf discharge [Eq. (10.19)] at a given excitation frequency may be used:

$$P^2L = \text{constant} \tag{10.55}$$

The conditions of Eqs. (10.52) through (10.55) can be used along with the relationships for J_p and V_p, given in terms of PL, for the specification of the design and operating parameters. It is seen that the determination of intrinsic kinetics leads to the reactor specification.

Example 10.12. As discussed in Chap. 9, the ion energy has to be in excess of the threshold energy for physical sputtering to take place. For most metals, the threshold energy is approximately 20 eV, which means that physical sputtering would be minimal for the ion energy less than 15 eV (see Prob. 10.5). No one in the literature has yet provided all the information suitable for illustrating the above design procedures. Therefore, the data reported by Stenger *et al.* (1987) and Song (1988) will be adapted and combined for this example, although they studied two different plasma systems. Shown in Figs. 10-14 and 10-15 are the J_p and V_p adapted from the experimental data of Song. Specify Q, P, and L using the kinetic parameters of Stenger *et al.* and other information in Example 10.11. Assume a k_G value of 10^6 (cm·V/A), a value of 0.92 cm·torr^2 for the constant in Eq. (10.54), and β_1 and β_2 values of 10 and 15 eV. Also use 0.01 for β in Eq. (10.50).

The following relationships can be used:

$$\frac{E_p}{P} = \frac{1000}{PL} \tag{A}$$

$$\lambda_i \text{ (cm)} = \frac{0.008}{P \text{ (torr)}} \tag{B}$$

Solution. As evident from Fig. 10-14, J_p increases with PL and then reaches a plateau at PL of 0.6. Thus, an equivalent of Eq. (10.52) is

$$PL > 0.6 \text{ cm·torr} \tag{C}$$

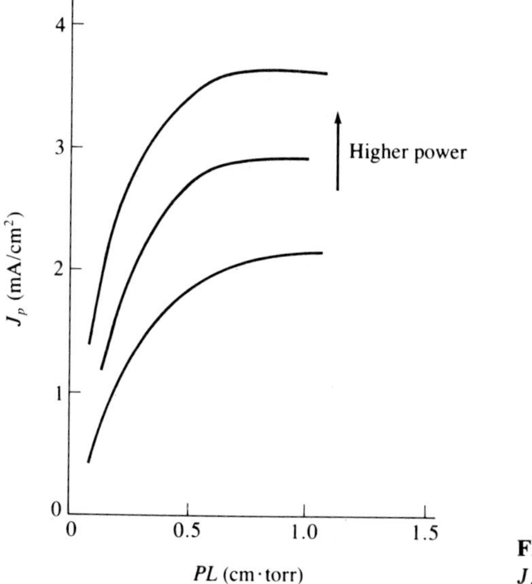

FIGURE 10-14
J_p versus PL.

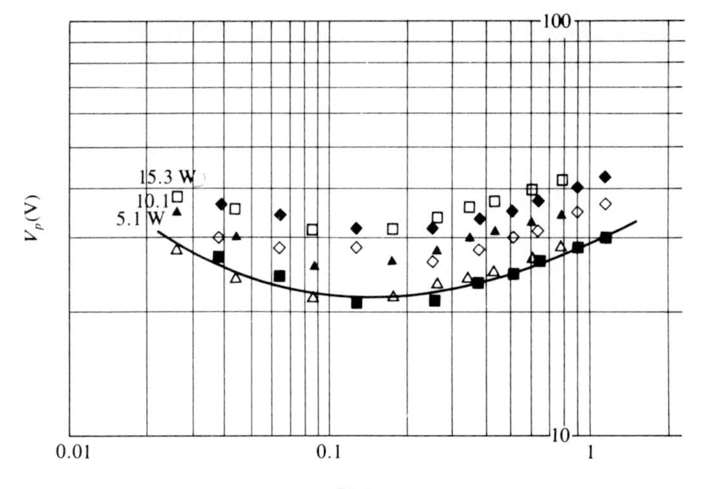

FIGURE 10-15
V_p versus PL.

According to the problem statement, $10 < E_i < 15$ eV. From Eqs. (B) and (10.40), $\alpha_u = 0.008q$ (cm·torr). Therefore,

$$10 < \frac{0.008E_p}{P} < 15 \tag{D}$$

Using Eq. (A) in Eq. (D), one has

$$10 < \frac{8}{PL\ (\text{cm·torr})} < 15 \text{ eV}$$

This, when solved for PL, yields

$$0.53 < PL < 0.8 \text{ cm·torr} \tag{E}$$

Equations (C) and (E) yield

$$0.6 < PL < 0.8 \text{ cm·torr} \tag{F}$$

since Eq. (C) satisfies the lower limit in Eq. (E). The conditions imposed by Eq. (F) are shown shaded in Fig. 10-16. Since y_i is specified (0.4) in Example 10.11, Eq. (10.54) becomes

$$\text{Maximize}\left(\frac{HPJ_p}{V_p}\right) \tag{G}$$

It is useful to note that L is essentially H/π since the radius is much larger than 2.405 cm [refer to Eq. (10.15)]. Therefore, Eq. (G) can be rewritten as

$$\text{Maximize}\left(\frac{PLJ_p}{V_p}\right) = \text{Maximize } I \tag{H}$$

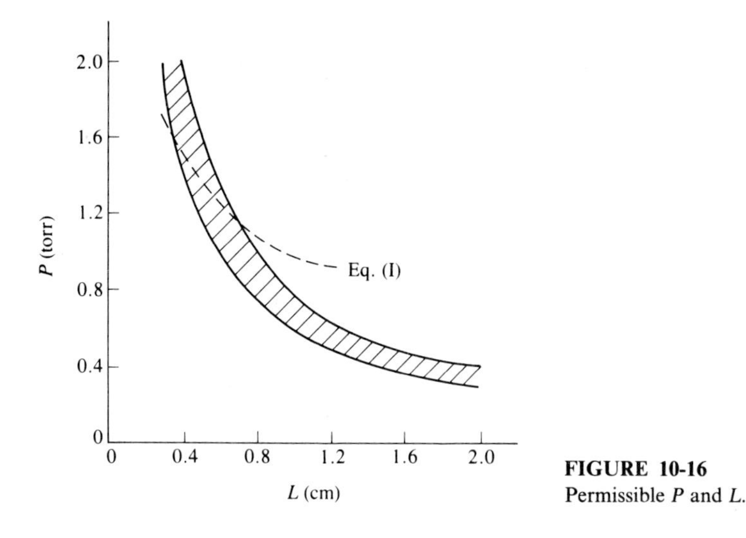

FIGURE 10-16
Permissible P and L.

Since J_p is already at its maximum possible, one is left with the maximum of PL/V_p. In the range given by Eq. (F), Fig. 10-15 reveals that V_p increases from approximately 24 to 28 volts when PL increases from 0.6 to 0.8 cm·torr. Now, one has, for Eq. (10.55),

$$P^2L = 0.92 \text{ cm·torr}^2 \tag{I}$$

Equation (I) is shown in Fig. 10-16 as the dashed line. Since the maximum of I in Eq. (H) is that corresponding to $PL = 0.8$, the intersection between the line of PL of 0.8 and the dashed line gives the desired operating conditions, which correspond to P of 1.16 torr and L of 0.72 cm. To be reasonably within the boundaries specified by Eq. (F), one may choose a P of 1.1 torr and an L of 0.70 cm. Thus, the specifications are:

Pressure = 1.1 torr
Electrode spacing = 0.7 × 3.14 = 2.2 cm

Now that the specifications are available, the flow rate required for uniformity, with a gas distribution arrangement designed for uniform source concentration, can be obtained from Eq. (10.48):

$$3850\left[10^6\left(\frac{2.2 \times 2.0 \times 10^{-3}}{27}\right)\left(\frac{0.4 \times 1.1}{760 \times 82 \times 298}\right) - 0.57C_n\right]$$
$$= QC_n + 2450 \times 23.5C_n \tag{J}$$

or

$$\frac{1.48 \times 10^{-2}}{C_n} = Q + 59{,}770 \tag{K}$$

where it has been assumed that the mass transfer effect is negligible, which can be checked once the desired rate is selected. Note here that $J_p = 2$ mA/cm^2 and $V_p =$ 27 volts from the chosen conditions and Figs. 10-14 and 10-15. A plot of Q versus R_G $(=23.5C_n)$ is given in Fig. 10-17. The vertical dotted line represents the restriction imposed by Eq. (10.50). As indicated by the solid line, specification of C_n

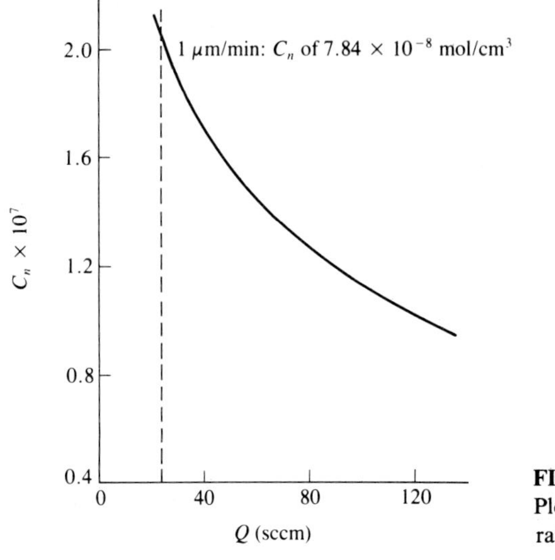

FIGURE 10-17
Plots for the desired volumetric feed rate.

(or R_G) fixes the flow rate, or vice versa. Suppose that an etch rate of 1.27 μm/min (or C_n of 10^{-7} mol/cm^3) is desired. Then the corresponding flow rate should be 127 sccm. As examined in Example 10.9, the etching is not mass transfer limited. Thus, no further calculations are required. Note that the maximum rate is that corresponding to the flow rate of 24 sccm (Fig. 10-17).

This example illustrates the essence of the procedures one can follow in designing a plasma reactor, although details may vary depending on the system of interest. The design involves satisfying a number of conditions while maximizing the rate. It is made clear here that there are no essential differences between the design of plasma reactors for etching and those for deposition unless the former requires anisotropic etching.

The same procedures can be followed for asymmetrical systems, the only differences being the ion flux and ion energy. In etching, where anisotropy is critical, asymmetrical electrodes are used. The substrates are placed on the powered electrode, which is smaller than the grounded electrode. In this manner the voltage drop across the powered sheath is higher than the plasma potential drop, the ion flux is increased by the factor of A_g/A_p for the substrate on the powered electrode, and contamination can be minimized due to the smaller area exposed relative to a symmetrical system. As discussed in Example 10.7, the current flux ratio can be written as

$$\frac{J_{sp}}{J_p} = \frac{A_g}{A_p} \tag{10.56}$$

where A_g is the grounded electrode area and A_p is the powered electrode area. Although the ratio of potentials has been reported to vary with the fourth power of the area ratio (Koenig and Maissel, 1970), it can be shown (Zarowin, 1983; see

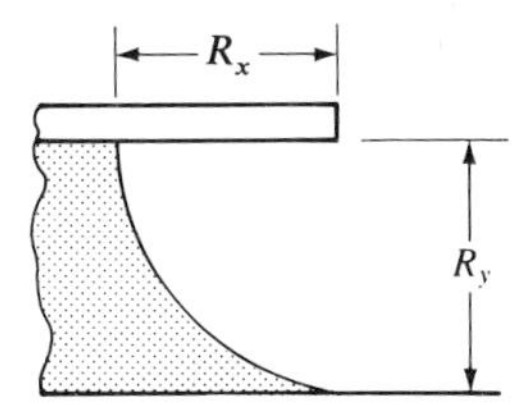

FIGURE 10-18
Definition of etch directionality.

Prob. 10.9) that the following holds for capacitive plasma:

$$\frac{V_p}{V_{sp}} = \begin{cases} A_p/A_g & L < \lambda \\ \left(\dfrac{A_p}{A_g}\right)\left(\dfrac{d_{sp}}{d_p}\right) & \text{otherwise} \end{cases} \tag{10.57}$$

where d_{sp} is the sheath thickness at the powered electrode and d_p is the sheath thickness at the grounded electrode. These values can be calculated using electrical measurements (Prob. 10.9). The same procedures used for symmetrical electrode design can be followed for the asymmetrical reactor design with the aid of Eqs. (10.56) and (10.57) for the additional parameter, A_g/A_p.

An additional factor that must be considered for directional etching is the desired degree of anisotropy. Referring to Fig. 10-18, the etch directionality, U, can be defined as

$$U = \frac{R_x}{R_y}$$

Complete anisotropy results when $U = 0$; complete isotropy occurs when $U = 1$. According to Zarowin (1983), the etch directionality is inversely proportional to the following:

$$\frac{1}{U} \propto \left(\frac{E_{sp}}{P}\right)\left(\frac{1 + M_i/M_n}{Q_i}\right) \tag{10.58}$$

where M_i is the ion mass, M_n is the mass of dominant neutral species and E_{sp} is the electric field across the sheath of the powered electrode where the substrate is placed. For a given plasma, the desired anisotropy can be attained if

$$\frac{E_{sp}}{P} > \text{constant} \tag{10.59}$$

An equivalent way of expressing this condition is to use P_w/P in place of E_{sp}/P for a given system where P_w is the applied rf power. The constant in this case, reported by Zarowin, is 1.4 W/torr for polysilicon etching by Cl_2-He plasma. As indicated by Eq. (10.58), the constant in Eq. (10.59) becomes larger when the dominant neutral species has a higher molecular weight. Thus, one would expect the constant to increase with decreasing content of helium in Cl_2-He plasma, as the experimental results of Zarowin show. Note that an equivalent form of Eq. (10.59) can be written as $J/P >$ constant. This means that a higher current flux results in a higher anisotropy.

For design purposes, the anisotropy requirement places the uppermost constraint on the range of PL allowed since E_{spl}/P can be correlated to PL, as shown in Fig. 10-12. The electrode area ratio gives an additional degree of freedom to work with the added constraint. The design procedures remain essentially the same as those for symmetric electrodes.

Analysis of a plasma reactor involves the same considerations as its design. However, the reactor transport of reactive and source species has to be considered in the analysis when the mode of gas distribution does not guarantee equal exposure of the wafers to the same concentrations of the species. This is typically the case with reactors being used in practice, such as the barrel, planar, and hexode reactors in Fig. 10-6. The transport of species in a planar plasma reactor (Stenger *et al.*, 1987) will be considered first. Mass balances for the reactive neutral species and the source species are

$$v_r \frac{\partial C_n}{\partial r} = D\left[\frac{1}{r}\frac{\partial}{\partial r}\left(r\frac{\partial C_n}{\partial r}\right) + \frac{1}{r^2}\frac{\partial^2 C_n}{\partial \theta^2} + \frac{\partial^2 C_n}{\partial z^2}\right] + k_g n_e C_m - r_r \qquad (10.60)$$

$$v_r \frac{\partial C_m}{\partial r} = D\left[\frac{1}{r}\frac{\partial}{\partial r}\left(r\frac{\partial C_m}{\partial r}\right) + \frac{1}{r^2}\frac{\partial^2 C_m}{\partial \theta^2} + \frac{\partial^2 C_m}{\partial z^2}\right] - mk_g n_e C_m \qquad (10.61)$$

where radial flow parallel to the electrodes has been assumed and m is a stoichiometric coefficient. The configuration and coordinates are shown in Fig. 10-19*a*. If the concentration is averaged over the electrode spacing and the wafers are

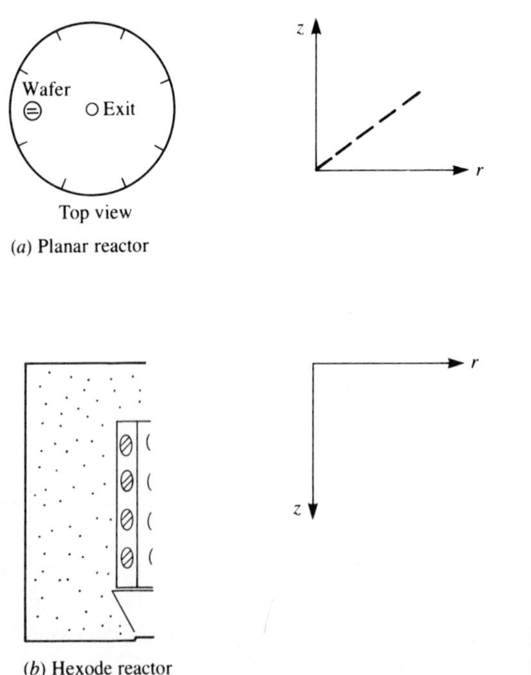

FIGURE 10-19
Coordinate and geometry for two reactors.

placed symmetrically, the mass balances (see Prob. 10.16) reduce to

$$\frac{v_o r_o}{r}\frac{d\bar{C}_n}{dr} = \frac{D}{r}\frac{d}{dr}\left(r\frac{d\bar{C}_n}{dr}\right) - \frac{A_s r_c}{H} + k_g n_e \bar{C}_m - R_r \tag{10.62}$$

$$\frac{v_o r_o}{r}\frac{d\bar{C}_m}{dr} = \frac{D}{r}\frac{d}{dr}\left(r\frac{d\bar{C}_m}{dr}\right) - mk_g n_e \bar{C}_m \tag{10.63}$$

where v_o is the average bulk velocity at the outside radius r_o of the reactor boundary and R_r is the rate of recombination written in terms of $\bar{C}_n$. For the system considered in Example 10.9, C_n is the concentration of F atoms, C_m is the concentration of NF_3, $r_c = kC_n$, and $r_r = k_r C_n$. As shown in Prob. 10.16, the concentration difference between any normalized radial position x and the outlet $(x = 0)$ is proportional to the following:

$$\Delta\bar{C}_m \propto x^{\mu} I_{\mu}(\alpha x) \tag{10.64}$$

where $$\mu = \frac{v_o r_o}{2D} = \frac{\text{Pe}}{2}$$

$$\alpha = \left(\frac{mk_g n_e r_o^2}{D}\right)^{1/2}$$

and Pe is a Peclet number and I_{μ} is the modified Bessel function of order μ. One conclusion from Eq. (10.64) is that the concentration of the source species cannot be uniform unless the Peclet number approaches infinity, in which case the feed velocity approaches infinity. In such a case, the reactive neutral species are swept away by the flow and the rate of etching or deposition can become unacceptably low. An optimization for the uniformity leads only to a small window of acceptable operating conditions, and this is the reason why feeding arrangements that lead to uniform concentration over the wafers are being sought in practice. Note that the reactor transport, such as that represented by Eqs. (10.60) and (10.61), need not be considered if a feeding arrangement provides a uniform concentration over the wafers of interest.

The barrel reactor in Fig. 10-6 has a feeding arrangement similar to that for an LPCVD reactor. Here again, however, the trend is toward a feeding arrangement for uniform exposure of wafers to the same source species concentration. In such a case, the reactive species concentration at the edges of the wafers is the same for all wafers. This system was analyzed by Alkire and Economou (1985) in detail for the limiting case of no recombination. Under that restriction, the uniformity is guaranteed if the following condition is satisfied:

$$\frac{RG\rho_s}{1.5B_r C_o D M_w} < \tanh\left(\frac{1.203s}{R}\right) < 1 \tag{10.65}$$

which is Eq. (6.49*a*) derived earlier for an LPCVD reactor. Here R is the wafer radius, G is the desired etch or deposition rate in length per time, ρ_s is the density of the material being etched or deposited, M_w is its molecular weight, C_o is the concentration of the reactive neutral species at the wafer edges, B_r is the degree to which the uniformity is desired (0.01 for 1 percent deviation, for example), and s is the interwafer distance. When the reactor has a feeding arrangement similar to that for an LPCVD reactor (Fig. 10-6*b*), mass balance equations have to be solved first for the annular region between the wall and the metal grid for the plasma confined in the annulus, and then between the metal grid and the wafer edges for the reactive neutral species. The concentration of the reactive neutral species at the wafer edges can then be used for the transport between two adjacent wafers.

The hexode plasma reactor is essentially the same as the barrel CVD reactor. The only difference in the physical layout is that the inner hexode with side walls (on which the wafers are placed) is not inclined in the case of the plasma reactor to assure uniform electric field. If one treats the inner cylinder ideally with side walls as a perfect cylinder, the conservation equations, similar to Eqs. (10.60) and (10.61) for the hexode reactor, are

$$v_z \frac{\partial C_n}{\partial z} = D\left[\frac{1}{r}\frac{\partial}{\partial r}\left(r\frac{\partial C_n}{\partial r}\right) + \frac{\partial^2 C_n}{\partial z^2}\right] + k_g n_e C_m - r_r \tag{10.66}$$

$$v_z \frac{\partial C_m}{\partial z} = D\left[\frac{1}{r}\frac{\partial}{\partial r}\left(r\frac{\partial C_m}{\partial r}\right) + \frac{\partial^2 C_m}{\partial z^2}\right] - mk_g n_e C_m \tag{10.67}$$

where v_z is the velocity in the direction of flow and r is now the radial coordinate perpendicular to the flow direction (Fig. 10-19*b*). As discussed in Prob. 10.17, the equations, after averaging over the radial direction, reduce to

$$\frac{d\bar{C}_n}{dy} = \frac{1}{P_m}\frac{d^2\bar{C}_n}{dy^2} + t_f\left(k_g n_e \bar{C}_m - R_r - \frac{2R_i \bar{r}_c}{R_o^2 - R_i^2}\right) \tag{10.68}$$

$$\frac{d\bar{C}_m}{dy} = \frac{1}{P_m}\frac{d^2\bar{C}_m}{dy^2} - t_f(k_g n_e \bar{C}_m) \tag{10.69}$$

where the Peclet number P_m is given by Zv_z/D, $t_f = Z/v_z$, R_o and R_i are the radius for the outer and the inner cylinder, respectively, Z is the inner cylinder length, and y is the length coordinate normalized with respect to Z. The solutions to Eqs. (10.68) and (10.69) can be written for linear rate expressions as discussed in Prob. 10.17. As was the case in the planar plasma reactor, only a small window of operating conditions is available for near uniformity. A radial feeding arrangement along the inner reactor wall would be desirable. Another alternative is to introduce an inclined divider coaxially with the inner cylinder wall.

It should be clear from this section that the main design considerations are those for obtaining the plasma characteristics that are desired for the best possible performance of the plasma reactor. The design, therefore, involves satisfying a number of conditions for the desired plasma characteristics. The reactor transport effect is primarily determined by the way the source gas is distributed along the reactor. This effect can be eliminated by devising a feeding arrangement that leads to equal exposure of all wafers to the same source species concentration.

Under such an arrangement, the gas-solid interface transport effect can still be present but can be compensated for to avoid a nonuniformity problem.

Finally, it should be kept in mind in using approximate relationships, in particular those related to plasma characteristics, that there are always exceptions to the relationships.

NOTATION

a, b, c	Width, depth, and length of a channel, respectively (L)
A	Constant in Eq. (10.41)
A_g, A_p	Area of ground and powered electrode, respectively (L^2)
A_s	Substrate area (L^2)
C	Concentration (mol/L^3)
C_b	Concentration of source species in bulk fluid; bulk concentration (mol/L^3)
C_d	Conductivity
C_m	Concentration of main neutral species (mol/L^3)
C_n	Concentration of reactive neutral species (mol/L^3)
C_p	Concentration of source species at a point near cold wall (mol/L^3)
C_s	Concentration of source species at substrate surface (mol/L^3)
d_a, d_i	Diameter of atom and ion, respectively (L)
D	Diffusivity; diffusivity of source species (L^2/t)
D_a	Ambipolar diffusivity (L^2/t)
E	Electric field (V/L); energy (E)
E_a	Activation energy (E)
E_i	Average ion energy (E)
E_o	Activation energy in absence of ions (E)
f	Frequency ($1/t$)
$f(C)$	Concentration dependence of rate
G	Linear growth rate (L/t)
H	Electrode spacing (L); cylinder length in Eq. (10.15) (L)
I_r	Root mean square current (I)
j	Ion flux (ion/L^2t)
J	Current density (current/L^2t)
k	Rate constant
k_B	Boltzmann constant
k_g	Rate constant for rate of generation of ions
k_m	Mass transfer coefficient (L/t)
k_o	Preexponential factor for rate constant
k_r	Rate constant for rate of recombination of reactive neutral species
L	Characteristic diffusion length (L)
m	Electron mass (M); constant exponent in Eq. (10.5); stoichiometric constant in Eq. (10.61)
M	Ion mass (M); molecular weight (M)
M_r	Reduced mass (M)

n	Number density (number/L^3); constant exponent in Eq. (10.8)
n_a	Atom number density (atom/L^3)
N_n	Number density of neutral species (neutral molecule/L^3)
p_0	Permittivity of free space
P	Pressure (P)
P_d	Probability of no collision
P_i	Collision probability
P_m	Peclet number in Eq. (10.69)
P_{ia}	Average of diffusion cross section in Eq. (10.4)
P_w	Power (W)
Pe	Peclet number defined in Eq. (10.64)
q	Elementary charge (1.6×10^{-19} C)
Q	Volumetric flow rate (L^3/t)
Q_c	Collision cross section
Q_i	Ionization collision cross section
r	Radial coordinate (L)
r_c	Intrinsic rate
r_G	Net rate of generation
r_r	Recombination rate
r_o	Radius of a sphere in Eq. (10.15); radius of electrode on which substrates are placed
r_s	Rate of sputtering (atoms/L^2t)
R_g	Gas constant
R_G	Observed rate
R_r	Normalized r_r
S	Sputtering yield (atoms/ion)
Sh	Sherwood number defined in Eq. (10.35)
t_f	Fluid residence time in a reactor (t)
t_r	Average lifetime (t)
T	Temperature (T)
U	Etch directionality
v	Velocity (L/t)
v_d	Drift velocity (L/t)
v_f	Average fluid velocity (L/t)
v_o	Average bulk velocity at outside radius r_o in Eq. (10.62)
v_p	Plasma volume (L^3)
v_r	Radial velocity (L/t)
v_z	Axial velocity (L/t)
V	Potential (V)
V_a	Applied voltage (V)
V_f	Floating potential in Eq. (10.30) (V)
V_{dc}	DC bias (V)
W	Quantity defined by Eq. (10.37)
x	r/H
y	Mole fraction

z	Axial coordinate (L)
Z	Hexode reactor length (L)

Greek letters

α	Quantity defined in Example 10.1; quantity defined in Eq. (10.64)
β	Small number in Eq. (10.50)
β_1, β_2	Constant ion energies in Eq. (10.53)
λ	Mean free path (L)
λ_t	Townsend coefficient in Eq. (10.9)
λ_D	Debye length (L)
μ	Mobility; constant in Eq. (10.64)
ρ	Density (M/L^3)
ω	Angular frequency (radian/t)
ω_p	Plasma frequency defined by Eq. (10.20) ($1/t$)

Subscripts

a	Atom
e	Electron
i	Ion
g	Ground electrode
m	Main (source) neutral species
n	Neutral species
p	Plasma; ground electrode sheath
sp	Powered electrode sheath

Superscript

—	Average

Units

E	Energy (ML^2/t^2)
L	Length
M	Mass
P	Pressure (M/it^2)
t	Time
T	Temperature
V	Volt

PROBLEMS

10.1. Show that Eq. (10.10) follows from Eqs. (10.1) and (10.9), assuming that $D_e P$ is constant. The log-log plot in Fig. 10-8 is almost linear for N_2O for E/P less than 1 volt/(cm·torr). Determine the form of Eq. (10.10) in this range. Comment on the functional dependence of λ_t on E/P. The temperature may be combined into the constant term. Calculate the electron mobility of N_2O plasma at 1 torr under an electric field of 1 volt/cm.

10.2. Consider a medium in which the density of electrons is the same as that of ions. Let this density be n and velocity be v_a, which are the same for both species. Since $j = j_i = j_e = v_a n$ and $n = n_i = n_e$, Eqs. (10-2) and (10.3) yield

$$j_i = -D_i \frac{dn}{dx} + \mu_i En \tag{A}$$

$$j_e = -D_e \frac{dn}{dx} - \mu_e En \tag{B}$$

Combining these to solve for En yields

$$En = \frac{D_i - D_e}{\mu_i - \mu_e} \frac{dn}{dx} \tag{C}$$

Now $v_a n = j = -D_a(dn/dx)$, which defines the ambipolar diffusivity D_a. Using this definition, derive Eq. (10.11).

10.3. The characteristic diffusion length (fundamental mode) for hexode and barrel (with metal grid) plasma reactors with outer radius r_o and inner radius r_i can be expressed as

$$L^2 = \left[\left(\frac{B}{r_o}\right)^2 + \left(\frac{\pi}{H}\right)^2\right]^{-1}$$

where B is given in the following table:

r_o/r_i	1.2	1.5	2.0	2.5	3.0	3.5	4.0
B	15.701	6.270	3.123	2.073	1.549	1.234	1.024

For planar reactors, r_o is large and thus L is well approximated by H/π, where H is the electrode spacing. For hexode and barrel reactors, H, which is now the cylinder length, is large. Thus L is well approximated by r_o/B. Noting that H is of the order of 1 cm and r_o is of the order of at least one wafer radius, make conclusions on the relative magnitude of ion energies for the same operating pressure and excitation power density.

10.4. In some applications, plasma is generated outside a reactor and then fed to the reactor. For an average lifetime of ions of 10^{-3} s, calculate the residence time in the reactor for the exit ion concentration to be at least 10 percent of the inlet concentration. Calculate the time for both once-through (plug-flow) and well-mixed reactors.

10.5. Davis and Vanderslice (1963) give the following expression for ion energy (E_i) density distribution when the sheath thickness is large relative to the mean free path:

$$f(E_i) = (\lambda q E)^{-1} \exp\left(-\frac{E_i}{\lambda q E}\right)$$

where E is the electric field. The mean energy for this distribution is λqE. Since the mean free path λ is inversely proportional to P, λE is proportional to E/P. Show that the distribution can be rewritten as follows:

$$f(E_i) = \left(\frac{\alpha_r qE}{P}\right)^{-1} \exp\left(-\frac{E_i P}{\alpha_r qE}\right) \qquad \text{where } \alpha_r = \lambda_r P_r$$

and the subscript r is for a reference case. For the same applied voltage, how would you expect the pressure to affect the ion energy distribution? Suppose for an electrode material the threshold energy is 20 eV. Determine the fraction of ions that have an energy level above the threshold energy for an electrode spacing of 1 cm at 1 torr.

10.6. In plasma processing, it is desirable to minimize sputtering so that the impurities due to sputtered atoms are minimized. For Prob. 10.5, calculate the applied voltage at which only 1 percent of the ions have an energy level higher than 20 eV. Do the calculations for 1 and 2 torr. Discuss your result.

10.7. For the hexode reactor, the length equivalent to H in Eq. (10.35) is that given by Eq. (6.32):

$$H_{eq} = R_o \ln\left(\frac{R_o}{R_i}\right)$$

For a hexode reactor with an outside radius of 25 cm (R_o) and R_o/R_i of 1.08, determine the concentration of F atoms at the substrate surface for etching as per Example 10.9. Determine the total pressure at which the ratio of surface to bulk concentration becomes larger than 0.99.

10.8. According to the experimental results of Coburn and Winters (1979), the rate of silicon etching is 0.5 nm/min when only XeF_2 gas is used. When an argon beam is introduced, the rate increases to 5 nm/min under identical conditions. Assuming surface temperature to be 600 K, calculate the change in the activation energy due to the ion bombardment.

10.9. For capacitive plasmas (excitation frequency higher than, say, 3 MHz), Zarowin (1983) gives the following relationship:

$$\frac{E_p}{E_{sp}} = \frac{A_p}{A_g} \tag{A}$$

where E_p is the electrode field across the sheath of the grounded electrode, E_{sp} is the electrode field across the sheath of the powered electrode, A_p is the area of the powered electrode, and A_g is that of the grounded electrode. The average energy ions acquire in arriving at an electrode is equal to the force (qE_j, $j = p$ or sp) multiplied by the distance they travel. Thus, the ion energy E_i, when there are negligible collisions in the sheath, is given by

$$(E_i)_j = qE_j d_j \qquad \text{for } j = p \text{ or } sp \tag{B}$$

where d is the sheath thickness. It follows from Eq. (B) that

$$(E_i)_j = qV_j \tag{C}$$

When the pressure is relatively high, such that the sheath thickness is large relative to the mean free path, the distance ions travel is λ instead of d. Thus, the ion energy is given by

$$(E_i)_j = qE_j\lambda = \frac{qV_j\lambda}{d_j} \tag{D}$$

Show that the following relationships (Song, 1988) hold:

$$\frac{(E_i)_p}{(E_i)_{sp}} = \begin{cases} \dfrac{V_p}{V_{sp}} = \left(\dfrac{A_p}{A_g}\right)\left(\dfrac{d_p}{d_{sp}}\right) & \text{when Eq. (C) holds} \quad \text{(E)} \\ \left(\dfrac{V_p}{V_{sp}}\right)\left(\dfrac{d_{sp}}{d_p}\right) = \dfrac{A_p}{A_g} & \text{when Eq. (D) holds} \quad \text{(F)} \end{cases}$$

According to Song, the ion flux at the powered electrode j_i is given by

$$j_i = \frac{I_r}{2\pi A_p[1 - (1/\pi)\cos^{-1}(-V_{dc}/V_a)]q} \tag{G}$$

10.10. For an asymmetric system in which the rms current is 100 mA at the powered electrode, determine the sheath potentials and current fluxes at both electrodes. Also obtain the sheath thicknesses at both electrodes. Use the following information:

$$-V_{dc} = 10 \text{ volts} \qquad V_a = 20 \text{ volts}$$

$$A_p = 10 \text{ cm}^2 \qquad A_g = 40 \text{ cm}^2$$

$$\text{Excitation frequency} = 13.56 \text{ MHz}$$

Assume that the pressure is high enough for the sheath thickness to be large relative to the ion mean free path.

10.11. For Prob. 10.10, calculate the ion energies at both electrodes and the ion flux at the powered electrode for the plasma at 1 torr and 300 K. Assume that ions and neutrals have the same mass and diameter and that the diameter is 0.5 nm, for which Eq. (10.39) reduces to

$$\lambda = (2^{1/2}\pi d^2 N)^{-1}$$

10.12. Show that a relationship between E_{sp}/P versus PL along with one between V_{sp} versus PL are sufficient, given the electrode areas, to obtain information on ion energy and ion flux for the design of a plasma reactor. List the equations necessary.

10.13. As the minimum dimension of devices decreases, it becomes increasingly more important to attain anisotropy at a lower ion energy so that the damage caused by ion bombardment does not lead to serious defect problems. Discuss whether an additional freedom can be gained by operating in the pressure region where the sheath thickness is smaller than the mean free path.

10.14. The rate of etching in a reactor decreases as the number of wafers placed increases. This effect is known as the loading effect (Mogab, 1977). The effect of volumetric flow rate on the etch rate is known to initially increase with increasing flow and then decrease, going through a maximum (Chapman and Minkiewicz, 1978). Although these effects have not been well documented for plasma deposition, the same effects are expected to occur in deposition as well. Although these effects cannot be described readily without solving reactor conservation equations for neutral and source species, they can be fully described for those reactors where the gas distribution assures the equal exposure to the same concentration for all wafers. For these reactors, Eqs. (10.48) and (10.49) apply:

$$v_p\left[k_G\left(\frac{HJ_p}{V_p}\right)C_m - k_r C_n\right] = QC_n + R_G A_s \tag{A}$$

$$Q[(C_m)_{\text{in}} - C_m] = k_G\left(\frac{HJ_p}{V_p}\right)C_m \tag{B}$$

where the order n in Eq. (10.48) has been assumed to be unity. Assuming that the mass transfer effect is negligible and that the rate of etching or deposition is first order with respect to the concentration of the reactive neutral species, that is, $R_G = kC_n$, find the flow rate at which the etching rate becomes the maximum. Also, illustrate the loading effect.

10.15. The concentration profile in the direction perpendicular to the flow can be approximated by a parabolic profile:

$$C = a_0 + a_1 y + a_2 y^2 \tag{A}$$

where the a_i values are constants to be determined and y is the distance from the substrate. Since $C = C_s$ at $y = 0$, Eq. (A) can be rewritten as

$$C = C_s + a_1 y + a_2 y^2$$

When there is no feed from the upper electrode (opposite the electrode on which wafers are placed) $D(dC/dy) = 0$ at $y = H$. Further, $C = C_p$ at $y = H$. Using these conditions in Eq. (A) yields

$$C = C_s + \frac{2(C_p - C_s)}{H}y - \frac{C_p - C_s}{H^2}y^2 \tag{B}$$

Based on Eq. (B) and Eqs. (10.33) and (10.34), derive the first part of Eq. (10.35). Discuss why the Sherwood number is unity when the source gas is fed through the upper electrode such that the concentration there is constant.

10.16. Show that Eqs. (10.60) and (10.61) reduce to Eqs. (10.62) and (10.63) when the concentrations averaged over the electrode spacing are used as follows:

$$\bar{C} = \frac{1}{H}\int_0^H C\,dz$$

For the solution of Eq. (10.63), write Eq. (10.63) as follows:

$$\frac{d^2\bar{C}_m}{dx^2} + \frac{1 - a_1}{x}\frac{d\bar{C}_m}{dx} - a_2\bar{C}_m = 0 \tag{A}$$

where

$$a_1 = \text{Pe} = 2\mu$$

$$a_2 = \frac{mk_g n_e r_o^2}{D} = \alpha^2$$

By substitution, show that the following is a solution:

$$\bar{C}_m = Bx^{\mu}I_{\mu}(\alpha x) \tag{B}$$

where B is an integration constant. For the boundary conditions of $\bar{C}_m = C_i$ at $x = 1$ and $\bar{C}_m = \bar{C}_{out}$ at $r = 0$, show that the solution is

$$\frac{\bar{C}_m}{C_{in}} = \frac{C_{out}}{C_{in}} + \left(1 - \frac{C_{out}}{C_{in}}\right) \frac{x^\mu I_\mu(\alpha x)}{I_\mu(\alpha)} \tag{C}$$

Thus, the concentration change is proportional to $x^\mu I_\mu(\alpha x)$.

10.17. Define the average concentrations in Eqs. (10.68) and (10.69) by

$$\bar{C} = \frac{\int_{R_i}^{R_o} rC\, dr}{\int_{R_i}^{R_o} r\, dr}$$

Multiply both sides of the equations by r and integrate from R_i to R_o to obtain Eqs. (10.68) and (10.69) using the above definition. Use appropriate radial boundary conditions. For $r_r = k_r C_n$ and first-order deposition kinetics, obtain the solutions for $\bar{C}_n$ and $\bar{C}_m$. The boundary conditions are

$$\frac{d\bar{C}_i}{dy} = \frac{1}{P_m}[\bar{C}_i - (C_i)_f] \qquad \text{at } y = 0; \quad i = m \text{ or } n$$

$$\frac{d\bar{C}_i}{dy} = 0 \qquad \text{at } y = 1; \quad (C_i)_f = \text{feed concentration}$$

10.18. The electrical properties of an rf or dc plasma can be simulated by solving electron and ion conservation equations, Poisson's equation for the electric field, and the electron energy equation, if one assumes that the pressure is high enough for such a continuum model to be used (Thompson and Sawin, 1986). The electron and ion conservation equations are:

$$\frac{\partial n_e}{\partial t} = \nabla(\mu_e E n_e) + D_e \nabla^2 n_e + (k_+ - k_-)N n_e \tag{A}$$

$$\frac{\partial n_i}{\partial t} = \nabla(\mu_i E n_i) + D_i \nabla^2 n_i + (k_+ - k_-)N n_e \tag{B}$$

where n is the number density, μ is the mobility, D is the diffusivity, N is the number density of source gas, E is the electric field, k_+ and k_- are the ionization and attachment rate constants, respectively, and the subscripts i and e are for ions and electrons, respectively. Poisson's equation for this case is

$$\nabla^2 V = \frac{q(n_e - n_i)}{p_0} \tag{C}$$

The boundary conditions are

$$\begin{aligned} n_e = n_i &= 0 && \text{at electrodes} \\ V &= 0 && \text{at grounded electrode} \\ V &= V_{dc} + V_a \sin(\omega t) && \end{aligned} \tag{D}$$

The first condition follows from the assumption that charged particles are neutralized at the electrodes. Suppose solutions are sought in the following forms

(Thompson and Sawin, 1986):

$$n_e = k_e(t)g_e(x, t)$$
$$n_i = k_i(t)g_i(x, t) \quad \text{(E)}$$

Show that, based on Eqs. (A), (D), and (E) for n_e, under certain conditions g_e is almost time invariant.

REFERENCES

Alkire, R. C., and D. J. Economou: *J. Electrochem. Soc.*, vol. 132, p. 648, 1985.

Brown, S. C.: *Introduction to Electrical Discharges in Gases*, Wiley, New York, 1966.

Chapman, B. N.: *Glow Discharge Processes*, Wiley, New York, 1980.

——— and V. J. Minkiewicz: *J. Vac. Sci. Technol.*, vol. 15, p. 239, 1978.

Cho, A.: *Conference Proceedings, Compound Semiconductor Growth, Processing and Devices for the 1990s*, University of Florida, Gainesville, October 1987.

Coburn, J. W., and H. F. Winters: *J. Appl. Phys.*, vol. 50, p. 3189, 1979.

Dalvie, M., K. F. Jensen, and D. B. Graves: *Chem. Eng. Sci.*, vol. 41, p. 653, 1986.

Davis, W. D., and T. A. Vanderslice: *Phys. Rev.*, vol. 131, p. 219, 1963.

Edelson, D., and D. L. Flamm: *J. Appl. Phys.*, vol. 56, p. 1552, 1984.

Fraser, D. B.: in S. M. Sze (ed.), *VLSI Technology*, chap. 9, McGraw-Hill, New York, 1983.

Golan, V. E., *et al.*: in S. Brown (ed.), *Fundamentals of Plasma Physics*, Wiley, New York, 1979.

Heinecke, R. A.: *Solid State Elect.*, vol. 18, p. 1146, 1975.

Hirschfelder, J. O., C. F. Curtis, and R. B. Bird: *Molecular Theory of Gases and Liquids*, Wiley, New York, 1964.

Irving, S. M.: *Solid State Technol.*, vol. 14(6), p. 47, 1971.

Koenig, H. R., and L. I. Maissel: *IBM J. Res. Dev.*, vol. 14, p. 168, 1970.

Kumagai, H. Y.: in McD. Robinson *et al.* (eds.), *Chemical Vapor Deposition, 1984*, p. 198, The Electrochemical Society, Pennington, N.J., 1984.

Luscher, P. E., and D. M. Collins: in B. R. Pamplin (ed.), *Design Considerations for MBE Systems*, Pergamon, London, 1981.

McDaniel, E. W.: *Collision Phenomena in Ionized Gases*, Wiley, New York, 1964.

Maddox, R. L., and H. L. Parker: *Solid State Technol.*, p. 107, April 1978.

Maissel, L.: in L. Maissel and R. Glang (eds.), *Handbook of Thin Film Technology*, chap. 3, McGraw-Hill, New York, 1970.

Mathad, G. S.: *Solid State Technol.*, p. 221, April 1985.

Mogab, C. J.: *J. Electrochem. Soc.*, vol. 124, p. 1262, 1977.

Morgan, R. A.: *Plasma Etching in Semiconductor Fabrication*, Elsevier, Amsterdam, 1985.

Reinberg, A. R.: U.S. patent 3,757,733, 1975.

Rosler, R. S., and G. M. Engle: *Solid State Technol.*, p. 172, April 1981.

Saito, J., T. Igareski, T. Nakamura, K. Kondo, and A. Shibatoni: 4th International Conference on MBE, York, England, September 1986.

Song, M. K.: "Application of Impedance Analysis to Reactive Ion Etching of Silicon and Teflon," PhD Thesis, Drexel University, 1988.

Stenger, H. G., Jr., H. S. Caram, C. F. Sullivan, and W. M. Russo: *Ass. Ind. Chem. Engrs J.*, vol. 33, p. 1187, 1987.

Thompson, B. E., and H. H. Sawin: *J. Appl. Phys.*, vol. 60, p. 89, 1986.

Tokauaga, C., and D. W. Hess: *J. Electrochem. Soc.*, vol. 127, p. 928, 1980.

Weiss, A. D.: *Semicond. Int.*, vol. 6, p. 88, 1983.

Zarowin, C. B.: *J. Electrochem. Soc.*, vol. 130, p. 1144, 1983.

CHAPTER 11

PACKAGING

11.1 DIE BONDING AND PACKAGING

When wafer device fabrication is finished, the chips on the wafers, called "dice," are separated by scribing with a diamond-tipped tool. The electrically good dice are then selected for packaging in the final product form of ICs. The first step in packaging is to bond the back of a die to an appropriate mounting media (substrate). The bond pads on the circuit side of the chip are then electrically interconnected to the package. Then the whole package is either molded into final product form or encapsulated and sealed to protect the circuit from the environment. An example is shown in Fig. 11-1 for a die in a plastic material (DIP). As shown in the figure, the backside of the die is bonded to a substrate (die support paddle), the die circuits are connected via bond wires to the leadframe, and the whole package is then molded into the final product form using a plastic molding compound. The final product is then suitable for plugging into a circuit board.

The two most common types of material for die bonding are hard solder (eutectic) and polymers. The bonding material should offer a low electrical resistance between the die and the package plus good thermal conductance. Eutectic die bonding, shown in Fig. 11-2, involves mechanically attaching a die to either a metal or ceramic substrate or to a package. The package is typically either a metal leadframe made of a Cu or Fe-Ni alloy or a ceramic such as Al_2O_3. The backside of the die is metallized to make it wettable by the die bonding preform. The substrate material is usually metallized with plated Ag (leadframes) or Au (leadframes or ceramic). The solder-preform materials such as 98% Au–2% Si react to dissolve the silicon at temperatures above the eutectic temperature (370 °C, for example, for 98% Au–2% Si). As the eutectic composition (96.4%

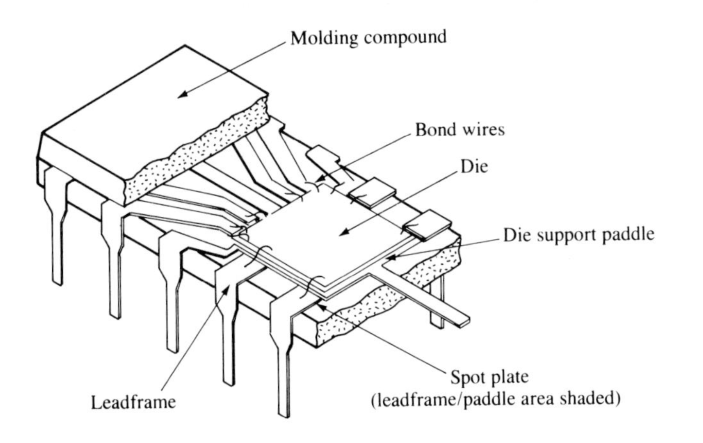

FIGURE 11-1
Ball- and wedge-bonded silicon die in a plastic material (DIP) (Howell, 1981).

Au–3.6% Si, for example) is reached and then exceeded, the composition of the composite structure becomes more silicon rich. The die bond is completed (Steidel, 1983), when the composite freezes. Polymer bonding materials are usually silver-filled epoxy or polyimide adhesives. The silver serves as a good conducting medium for both heat and electricity. Properties of materials used in packaging are given in Table 11.1.

The packages to which a die (or dice) is attached come in a variety of forms. They are made of either ceramic or plastic with metal leads, which are used for electrical interconnections between the die and the leads. The usual lead material for ceramic packages is an Fe-Ni-Co alloy called Kovar. The packages can have brazed pins or leads and edge or array pinouts for making the interconnections. Figure 11-3 shows various types of packages with different interconnection arrangements.

The electrical interconnection between die and package is usually carried out by one of the following methods: wire bonding, tape-automated bonding (TAB), or the flip-chip method. The first two methods are illustrated in Fig. 11-4. The flip-chip method involves making gold bumps on the circuit side of the die,

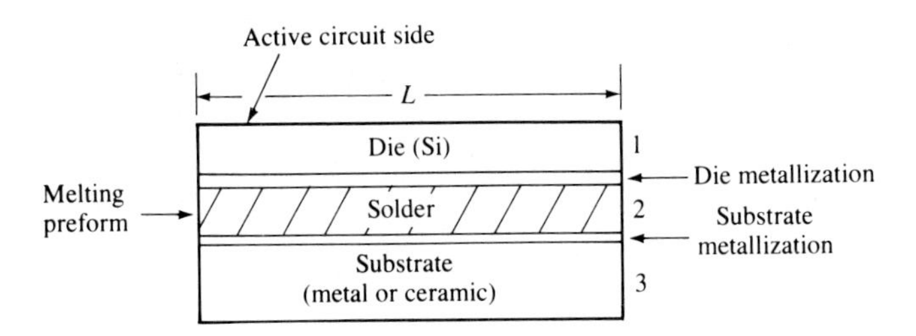

FIGURE 11-2
The basic structure of a silicon device die-bonded with a metal platform (Steidel, 1983).

TABLE 11.1
Properties of packaging materials (Sloan, 1985; Steidel, 1983)

Material	Expansion coefficient, cm/(cm °C × 10^6)	Elastic modulus, GPa	Thermal, W/(cm °C)	Density, g/cm^3
Metal				
Al	23.4	69	1.38	2.71
Ag	19.3	73	4.20	10.5
Au	14.2	77	3.06	19.29
Au-20% Sn	15.9	59	0.57	—
Au-3% Si	12.3	83	0.27	—
Be	12.2	290	1.63	1.82
Be-Cu	39.6	128	1.06	8.21
Cd	13.1	68	0.91	8.62
Cu	16.7	119	3.68	8.90
K_{ovar}	13.9	135	0.21	8.84
Mg	26.5	45	0.86	1.80
Ni	13.1	206	0.75	8.84
Pb-5% Si	29.0	7	0.63	—
Si	2.6	13.03	1.47	2328
Steel carbon (1010)	11.5	207	0.58	8.02
Stainless (304)	16.0	196	0.16	8.02
Ceramics				
Alumina (Al_2O_3)	5.6	373	0.34	3.59
Beryllia (BeO)	6.8	317	1.84	2.90
Mira	7.6	69	0.0035	2.90
Quartz	0.5	72	0.40	2.60
Magnesia (MgO)	8.8	69	1.46	2.79
Plastics				
Epoxy glass (G10) (*X*/*Y*)	9.9	16	0.0026	1.96
(*Z*)	72.0	16	0.0026	1.96
Lexan	67.5	2.6	0.0019	1.30
Nylon	90.0	1.5	0.0023	1.13
Teflon	90.0	1.0	0.0021	2.13
Mylar	16.9	3.8	0.0014	1.38

flipping the chip, and bonding to a substrate. Thus, the method can accomplish both die bonding and interconnection.

The ultimate objective of any packaging is to provide an environment for circuits to be reliable and have a long lifetime with good electrical performance. Trace amounts of water vapor and metals such as sodium and residuals such as Cl can cause corrosion of the circuit. The circuit is either encapsulated entirely or sealed with a suitable material for protection from corrosion. The circuits also generate heat and, in the tightly sealed environment of the package, overheating can cause circuit malfunctioning unless heat dissipation is efficient. In the process of bonding and interconnection, the interconnection bumps or wires are subjected to heating and relaxation. The thermal- and fatigue-induced stress can cause the "interconnects" to fail. Some basics of these subjects are considered in

(*a*)

(*b*)

(*c*)

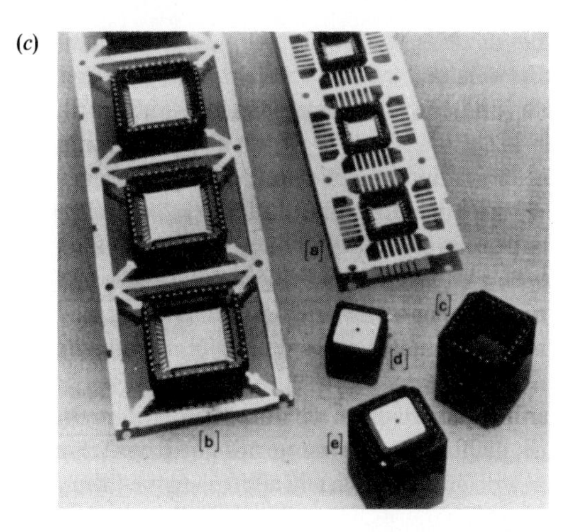

FIGURE 11-3

Package types with different interconnection arrangements: (*a*) ceramic package types (Kyocera), (*b*) glass-sealed Cerdip packages (Steidel, 1983), and (*c*) premolded packages (Levinthal, 1979).

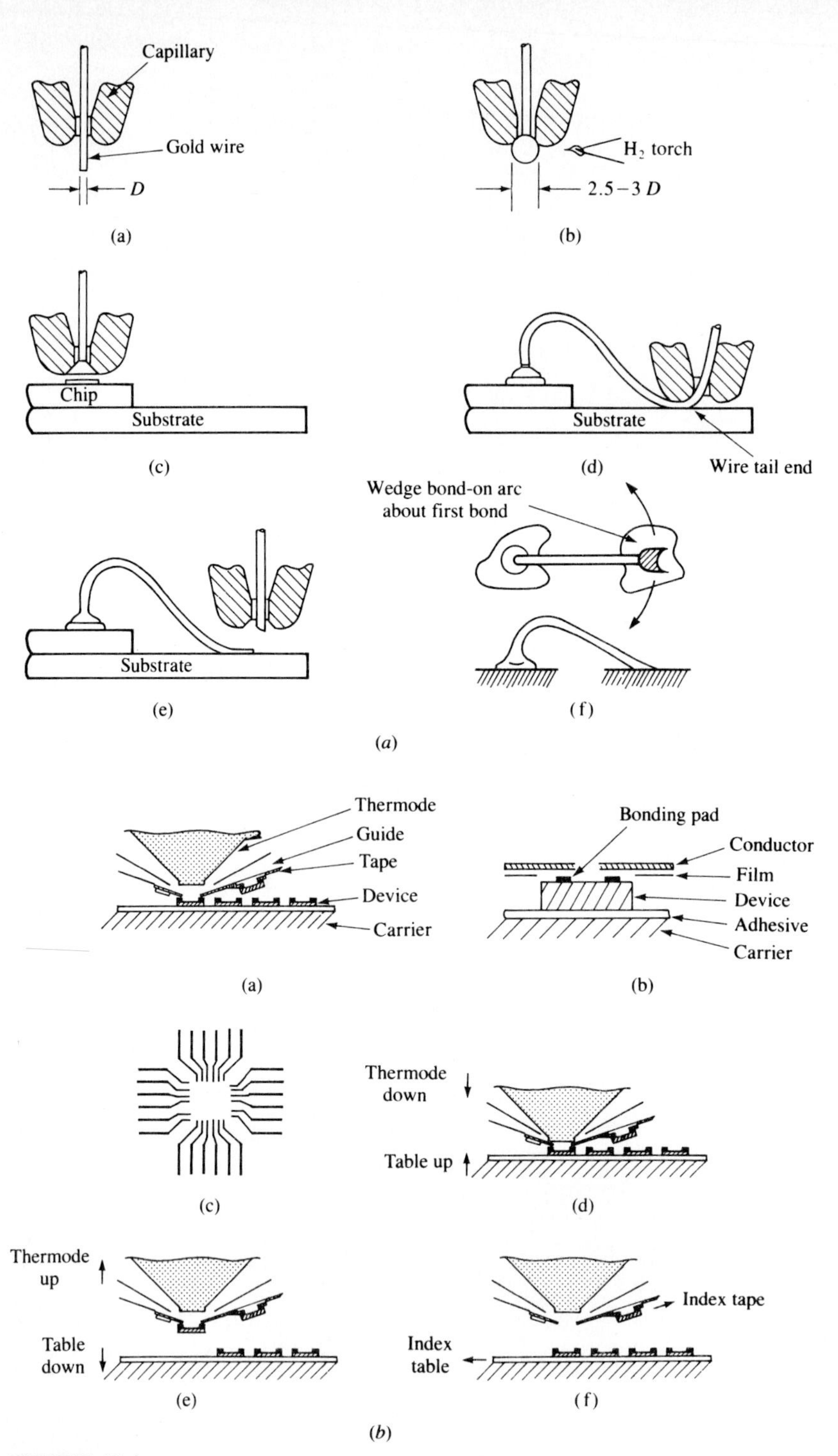

FIGURE 11-4
(*a*) Thermosonic ball-wedge bonding of a gold wire (Stafford, 1982), and (*b*) TAB inner lead-bonding process (Keizer and Brown, 1978).

subsequent sections. More details on packaging can be found in a chapter by Steidel (1983). Packaging materials are discussed first.

11.2 PACKAGING MATERIALS

Typical packages are made of either polymeric or ceramic materials. A polymer package has certain advantages over the use of a hermetically sealed ceramic package. The wire bonds used for interconnections are more reliable since the injection molding process removes weak bonds and shrinkage after molding keeps the bonds in compressive loading. Failures from handling damage are much less frequent for plastic packages than for more fragile glass and ceramic packages. Mechanized handling is simplified since the molding process produces devices with more consistent dimensions, and the possibility of high-volume manufacture allows less expensive devices to be produced (Amerasekera and Campbell, 1987).

The most common polymer encapsulants are novolac epoxies and silicones. The novolac epoxies are produced by reacting novolac resin with epichlorohydrin and a base. Sodium and chlorine ions must be carefully washed from the resulting epoxy since ionic contamination can alter the electrical properties of the device and open circuit interconnections.

Silicones consist of a polysiloxane backbone with various organic substituents at the silicon atom. The main chain of silicon and oxygen provides thermal stability, while the substituents add other properties such as moisture repulsion from alkyl groups and solvent resistance from fluorinated alkyl groups. The polymers are usually produced by the Rochow process, which is initiated by passing a stream of alkyl or aryl chloride through a heated bed of pure silicon alloyed with copper. The major product, dichlorosilane, is hydrolyzed to form a hydroxy end-blocked (HEB) siloxane polymer. The molecular weight of the HEB siloxane is dictated by the reaction conditions and its end-blocking can be altered if desired. For the room temperature vulcanizing (RTV) silicones used by the electronics industry, airborne moisture is used to initiate curing with an alkoxy crosslinker. The reliability problems caused by ionic contamination of epoxy resins may be avoided with silicones as high-purity electronic grade silicones are commercially available (Wong and Rose, 1983).

Ceramic packages for integrated circuits are produced in a variety of forms, from the low-cost and relatively unsophisticated Cerdip to high-performance multilayer multichip modules. The advantages of ceramics over plastics as packaging materials are hermetic sealing, higher thermal conductivity, and thermal expansion coefficients closer to that of silicon. Ceramic packages tend to cost more than their plastic analogs, but the advantages listed above tend to make the ceramic packages more reliable in most environments and offset the expense.

The Cerdip (Fig. 11-5) is a widely used integrated circuit package that provides a hermetically sealed environment at low cost. The package is prepared by pressing a ceramic cap and substrate. The chip and leadframe assembly is then sandwiched between the two halves of the package and the unit is sealed with

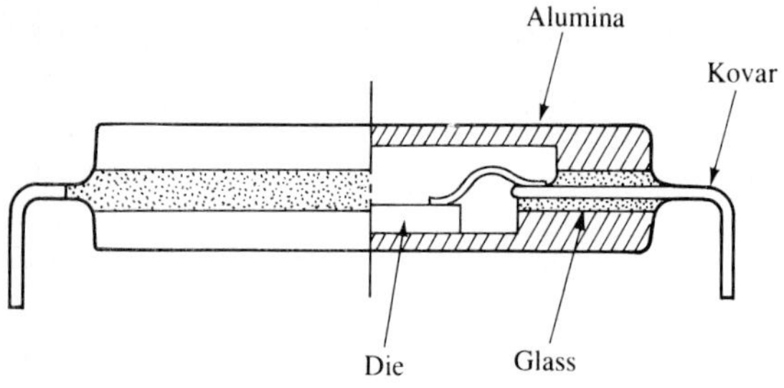

FIGURE 11-5
Schematic of a Cerdip glass-sealed ceramic package (Huatek, 1987).

low-temperature glass, typically a $PbO/ZnO/B_2O_3$ mixture. The sealing step is completed by heating the unit to at least 400 °C in an oxidizing environment. The ease with which this process can be automated allows the Cerdip to compete with plastics for low-cost packaging.

Multilayer ceramic (MLC) technology allows the package to act as a functional part of the electronic system. The MLC package provides not only interconnections and environmental protection but also intraconnections and power distribution. The more advanced modules contain over one hundred chips and approximately thirty layers of interconnections in a single package (see Fig. 11-6).

Although some modules are produced using a sequential process, lamination is the more common technique. The sequential process is begun with a fired substrate for the base. Alternating layers of dielectric and metallurgy are deposited on the base and fired, building a three-dimensional structure. The dielectrics are typically glass-ceramics or devitrifiable glasses, with Ag, Au, Pt, Pd, and Cu used for metallization. The module is typically fired each time a layer is added.

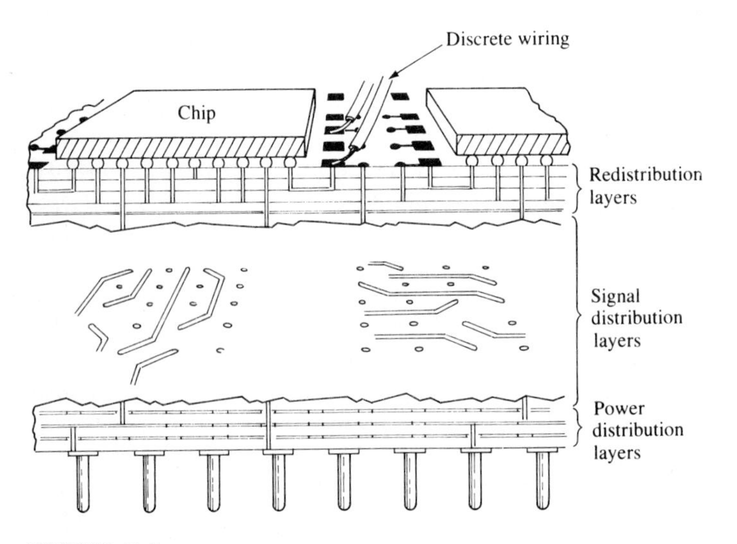

FIGURE 11-6
Schematic of a multilayer, multichip ceramic module (Blodgett, 1980).

In the laminated process, the module is assembled and then fired once. The process is begun by casting sheets of unfired (green) ceramic. These green sheets are punched with via holes, screened with metallurgy, and laminated to form the desired structure. The laminated structure is then fired, simultaneously densifying the ceramic and metallurgy.

The green sheets are cast by passing a mixture of ceramic powder solvents, plasticizers, and binders under a doctor blade. The green sheet is then cut into single sheets. The via holes, which provide for interconnection between the layers, are punched on a mechanical press. After the vias have been punched, electrical patterns are screened onto the sheets with molybdenum or tungsten pastes and the vias are filled. The screened sheets are then dried, laminated, and cut to their final size. The structure is then sintered, removing the remaining volatile components and producing the desired monolithic body (Schwartz, 1986).

The material properties of the ceramic substrate have a direct effect on the electrical, thermal, and mechanical performance of the module. Beyond the obvious requirement of strength, an ideal substrate material will have a low dielectric constant, high thermal conductivity, and a thermal expansion coefficient equal to that of silicon, $2.6 \times 10^{-6}\,°\text{C}^{-1}$.

The value of the dielectric constant controls the signal delay time, a measure of signal transmission speed, as follows:

$$t_0 = \frac{(K_D)^{1/2} D}{C} \tag{11.1}$$

where D is the distance the signal travels, K_D is the dielectric constant, and C is the speed of light. The delay time could be reduced by decreasing the dimensions of the package, but this approach can be limited by properties of the ceramic also. The width and thickness of the metal interconnections must decrease as the total package size decreases, increasing the resistance of the lines. Although silver and gold are usually used for high conductivity wiring, their melting points are too low for cofiring with most ceramics. Platinum and palladium have high melting points, but unfortunately have low conductivities as well. The refractory metals, molybdenum and tungsten, are an adequate compromise with intermediate conductivities and sufficiently high melting points for firing with most ceramics (Schwartz, 1984).

The thermal expansion mismatch between substrate and chip imposes a limit on the size of the chip. Chips are typically attached to MLC substrates using flip-chip technology, in which aluminum bonding pads on the chip are attached to solder balls that have been deposited on the substrate. This procedure allows an array of electrical interconnections to be produced simultaneously. Each time the chip is turned on and off, however, the resulting heating and cooling produces a strain, E_h, on the solder pads given by

$$E_h = \frac{(\Delta\alpha)(\Delta T) D_{np}}{H} \tag{11.2}$$

where $\Delta\alpha$ is the difference between the thermal expansion coefficients of the substrate and silicon, ΔT is the temperature change, D_{np} is the distance from the neutral point of the chip, and H is the height of the pad. The number of temperature cycles the package can undergo before a bond is broken is proportional to E_h^2. As the number of interconnections per chip continues to increase, the current stragegy of concentrating the bonding pads near the center of the chip will lose its effectiveness. Substrates with thermal expansion coefficients closer to that of silicon will be needed (Schwartz, 1984).

The major path of heat flow from an operating chip is through the solder pads and the substrate. Since the failure rate of the bonds is increased by making them shorter and wider, increased cooling for high-power devices can only be obtained by increasing the thermal conductivity of the substrate and solder pads or providing for heat removal from the back of the chip. Increasing the thermal conductivity of the substrate not only improves cooling but also reduces the mechanical stresses at the chip-substrate interface and within the substrate by reducing the temperature gradients.

Alumina, Al_2O_3, is currently the dominant ceramic material for integrated circuit packaging. The aluminas used are typically 90 to 96 percent purity, with moderate strength and thermal conductivity but high thermal expansion coefficients and dielectric constants.

The most commonly used method for preparing alpha-alumina ceramics is calcination of Bayer aluminum trihydrate. The Bayer process (Fig. 11-7) consists of five major steps: raw materials preparation, digestion, clarification, precipitation, and calcination. The bauxite ore that serves as the raw material for most

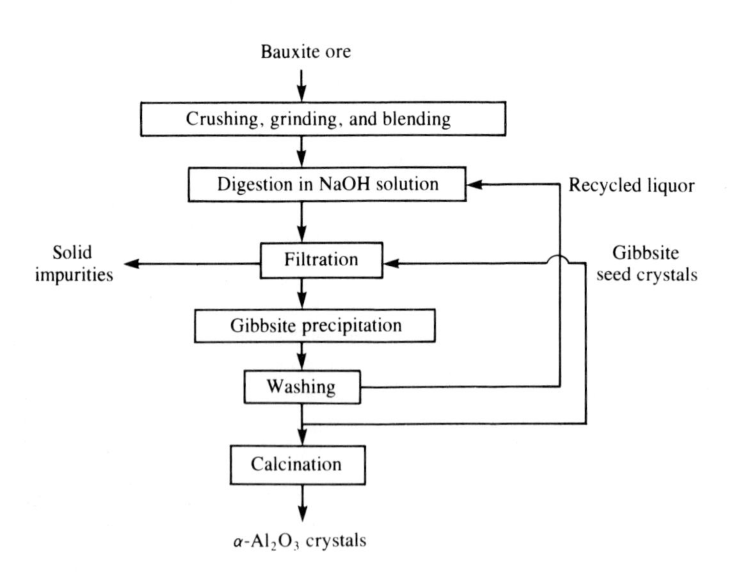

FIGURE 11-7
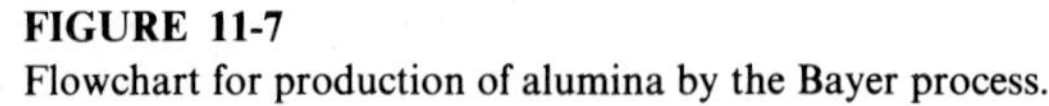
Flowchart for production of alumina by the Bayer process.

alumina produced must be crushed, blended, and ground to provide as uniform an end product as possible.

The alumina from the bauxite ore is then dissolved in a heated, pressurized caustic soda solution. This produces a sodium aluminate solution from which insoluble iron, silicon, and titanium are removed by settling and filtration. After cooling, gibbsite (α-$Al_2O_3 \cdot 3H_2O$) seed crystals from a previous cycle are used to precipitate about one-half of the alumina.

The seeding conditions, temperature, agitation, and time allowed for precipitation are used to control process economics, particle size distribution, texture, and purity of the precipitated gibbsite. The crystals are classified and washed to remove impurities. Since agglomerates of crystals often form, special processing conditions are required to remove soda and other impurities that may be trapped between crystals.

The purified gibbsite crystals are then calcined in large rotary drums, driving off the combined water. At approximately 350 to 400 °C, the dehydration rate reaches a plateau and recrystallization of gibbsite to alumina without shrinkage produces high surface-area transition aluminas. These activated aluminas, having a porous structure and high surface area, are used as drying agents and as catalysts and catalyst supports. As the temperature increases, the surface area and water content of the crystals continue to decrease. At 900 to 950 °C the formation of alpha alumina commences, and conversion to alpha alumina is complete within one hour at temperatures of 1200 to 1300 °C (McLeod *et al.*, 1985). The aluminas used for electronic packaging are high-purity forms that sinter at approximately 1600 °C and have a much smoother surface after sintering than normal aluminas.

Beryllia, BeO, is useful for high-power devices in which heat dissipation is important. Beryllia has a slightly lower dielectric constant than alumina, a thermal conductivity approximately ten times greater, and slightly less strength. Unfortunately, the toxicity of respirable berryllium requires special processing precautions that have limited the use of beryllia.

The production of beryllia begins with the removal of beryllium from beryl ore, 11% BeO, and bertrandite ore, $<1\%$ BeO. Sulfuric acid is used to leach the beryllium from the ores; then the beryllium is extracted from the leach liquor by solvent extraction. Stripping of the loaded organic is followed by hydrolysis to produce relatively pure beryllium hydroxide.

The beryllium hydroxide must be further purified for production of beryllia ceramics. This is accomplished by dissolving the beryllium hydroxide in pure sulfuric acid and filtering. The filtrate is heated to evaporate water and then cooled in the presence of beryllium sulfate seed crystals to precipitate beryllium sulfate tetrahydrate. The crystals are recovered and washed in a centrifuge.

Berryllium oxide is produced by thermal decomposition of the beryllium sulfate crystals (Fig. 11-8). The hotter and longer the decomposition cycle is, the lower the surface area, the lower the sulfur content, and the larger is the crystal size. The beryllia used most often by the semiconductor industry has a surface area of 9 to 12 m/g and has a fired density greater than 2.85 g/cm^3 (the theoreti-

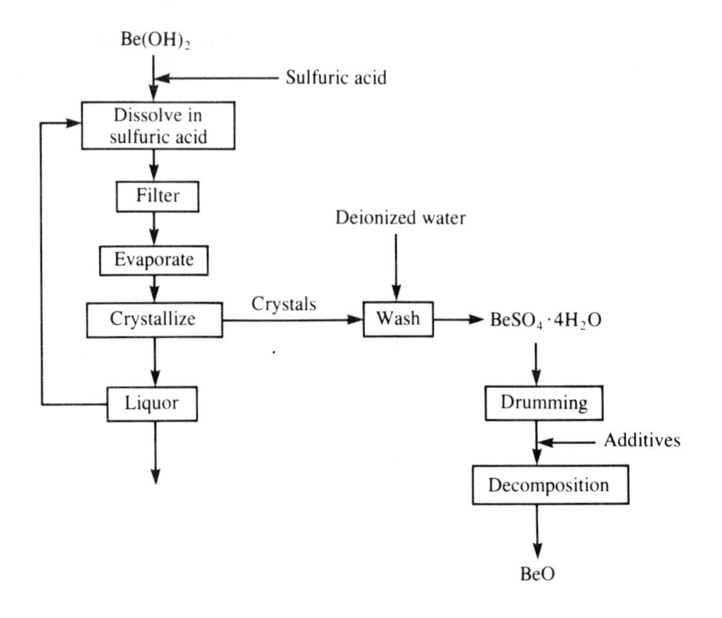

FIGURE 11-8
Schematic diagram for production of beryllium oxide from beryllium hydroxide (Kaczynski and Walsh, 1985).

cal maximum density of BeO is 3.01 g/cm^3).

The toxicity of beryllium presents a health hazard, but has been successfully controlled for some 30 years. The only form of exposure that causes problems is inhalation of respirable beryllium, dust particles smaller than 10 μm. Although susceptibility to the exposure depends on the individual, some people exhibit an immunological response. For this reason, special precautions are in order whenever beryllia or beryllium oxide dust, fumes, or mists may be created (Kaczynski and Walsh, 1985).

A mixture of 55 percent alumina and 45 percent lead borosilicate glass powders has been employed to overcome some of the disadvantages of aluminum. The alumina-glass material can be sintered at 900 °C, allowing the use of gold, for high conductivity, and of silver-palladium, for lower cost, instead of molybdenum or tungsten metallurgy. The alumina-glass substrate also has a lower dielectric constant than alumina, reducing signal delay times (Shimada *et al.*, 1983).

Silicon carbide, SiC, has been used for high-power devices because of its high thermal conductivity and coefficient of thermal expansion near that of silicon. The primary disadvantage of silicon carbide is the high dielectric constant, approximately four times that of silicon (Kohara *et al.*, 1986).

The thermal properties of aluminum nitride make it an excellent potential substrate material. Although it has a high sintering temperature, substrates have been prepared with tungsten metallurgy using both normal and hot-press sintering. Silicon nitride has high strength, but it also has to be sintered or hot-pressed at high temperatures. The thermal expansion coefficient and dielectric constant of silicon nitride are within reasonable limits for packaging of silicon devices. The glass-ceramics and mullites are also promising, but they are still in the developmental stage.

11.3 CORROSION

Most corrosion of materials refers to the chemical attack of metals, which is caused by oxidation of metals by the surrounding environment. Most corrosion reactions are electrochemical in nature. A piece of iron placed in a beaker of dilute hydrochloric acid dissolves or corodes in the acid producing iron chloride and hydrogen gas, as indicated by the following overall reaction:

$$\mathrm{Fe} + 2\mathrm{HCl} \longrightarrow \mathrm{FeCl_2} + \mathrm{H_2} \tag{11.3}$$

In a simplified ionic form, omitting the chloride ions, the reaction is

$$\mathrm{Fe} + 2\mathrm{H}^+ \longrightarrow \mathrm{Fe}^{2+} + \mathrm{H_2}$$

which consists of two half-cell reactions:

$$\begin{aligned} &\mathrm{Fe} \longrightarrow \mathrm{Fe}^{2+} + 2e^- &&\text{(oxidation half-cell reaction: anodic reaction)} \\ &2\mathrm{H}^+ + 2e^- \longrightarrow \mathrm{H_2} &&\text{(reduction half-cell reaction: cathodic reaction)} \end{aligned} \tag{11.4}$$

In the first half-cell reaction in Eq. (11.4), the metal forms the cation that goes into aqueous solution and the oxidation reaction is called the anodic reaction. The local regions on the metal surface where the oxidation reaction takes place are called local anodes. In the second half-cell reaction, the hydrogen ion is reduced. This reduction reaction in which a metal or nonmetal changes to a more negative (less positive) valence level is called the cathodic reaction. The local regions on the metal surface where metal or nonmetal ions are reduced are called local cathodes. There is a production of electrons in the anodic reaction whereas there is a consumption of electrons in the cathodic reaction. The anode is the one that is said to corrode.

The tendency of a metal to donate electrons, become anodic, is determined by its electrochemical potential relative to hydrogen ionization, which is taken to have a zero half-cell potential as a reference. The electrode potential for dilute solution E is given by

$$E \text{ (volts)} = E_0 + \frac{0.0257 \ln C}{n} \tag{11.5}$$

TABLE 11.2
Standard electrode potentials of metals

Metal and oxidation (corrosion) reaction	Electrode potential (volts)
$Au \longrightarrow Au^{3+} + 3e^-$	1.498
$2Cl^- \longrightarrow Cl_2 + 2e^-$	1.300
$2H_2O \longrightarrow O_2 + 4H^+ + 4e^-$	1.229
$Pt \longrightarrow Pt^{2+} + 2e^-$	1.200
$Ag \longrightarrow Ag^+ + e^-$	0.799
$2Hg \longrightarrow Hg^{2+} + 2e^-$	0.788
$Fe^{2+} \longrightarrow Fe^{3+} + e^-$	0.771
$4(OH)^- \longrightarrow O_2 + 2H_2O + 4e^-$	0.401
$Cu \longrightarrow Cu^{2+} + 2e^-$	0.337
$Sn^{2+} \longrightarrow Sn^{4+} + 2e^-$	0.150
$H_2 \longrightarrow 2H^+ + 2e^-$	0
$Pb \longrightarrow Pb^{2+} + 2e^-$	−0.126
$Sn \longrightarrow Sn^{2+} + 2e^-$	−0.136
$Ni \longrightarrow Ni^{2+} + 2e^-$	−0.250
$Co \longrightarrow Co^{2+} + 2e^-$	−0.277
$Cd \longrightarrow Cd^{2+} + 2e^-$	−0.403
$Fe \longrightarrow Fe^{2+} + 2e^-$	−0.440
$Cr \longrightarrow Cr^{3+} + 3e^-$	−0.744
$Zn \longrightarrow Zn^{2+} + 2e^-$	−0.763
$Al \longrightarrow Al^{3+} + 3e^-$	−1.662
$Mg \longrightarrow Mg^{2+} + 2e^-$	−2.363
$Na \longrightarrow Na^+ + e^-$	−2.714

where E_0 is the standard potential at 25 °C and 1 molar solution, C is the solution concentration in moles per liter (less than 1), and n is the number of electrons removed per ion. The standard electrode potentials are given in Table 11.2 for various metals.

The tendency for corrosion increases with decreasing electrode potential, such that more anodic metals tend to have higher corrosion (oxidation) rates. Aluminum will corrode, for example, in relation to nickel if a small amount of water is present since the electrode potential of aluminum is more negative than that of nickel, as Table 11.1 shows. This is the reason why metals are often gold-plated to protect them from corrosion, since the electrode potential of gold (Table 11.1) is very high. The worst impurity in packaged ICs from the standpoint of corrosion is sodium, since it is one of the metals that has a very low electrode potential and thus corrodes very readily. A galvanic cell forms when two dissimilar metals are immersed in a solution of their own ions. In such a cell, the metal with the lower electrode potential corrodes.

The galvanic cells leading to corrosion are not restricted to two dissimilar metals. A galvanic cell can form within a metal when a particular part is subjected to stress or is isolated, e.g., by covering. The part subjected to stress or isolation becomes anodic in relation to the other part which acts as the cathode.

A local corrosion attack is more likely to occur along grain boundaries, pointed parts of a metal, cracks, or crevices, where the surface energy is locally higher.

The major consideration in packaging to prevent corrosion is the preservation of the package from water vapor. Only an adsorbed layer of moisture can act as an electrolyte. Therefore, even a minute amount of water initially present in the package or permeated through the package from its environment can cause corrosion. Passivating layers are used to protect the underlying circuits and the whole package is hermetically sealed during high-performance IC production. With the advent of plasma processing, interconnect metal corrosion has become a greater problem, particularly when chlorine compounds are involved (Chang and Chao, 1985).

11.4 HEAT DISSIPATION

Integrated circuits generate heat. Unless the heat is dissipated effectively, the circuits deteriorate and their lifetime is shortened. Therefore, one objective of package design is to provide efficient heat dissipation paths. Heat dissipation for

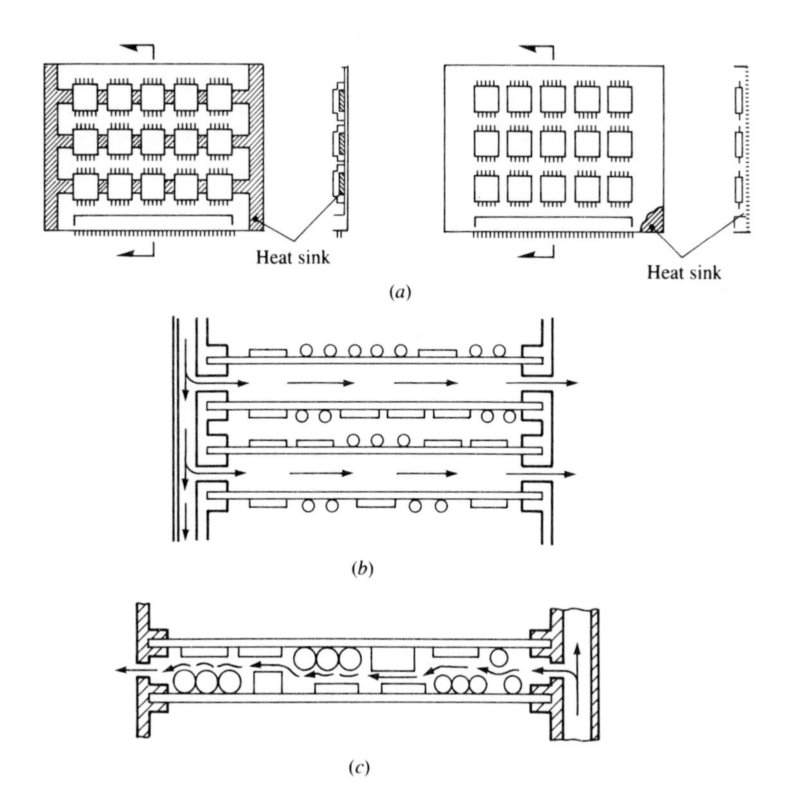

FIGURE 11-9
(*a*) Circuit card and heat sink configurations, (*b*) indirect cooling of printed wiring boards, and (*c*) impingement cooled circuit cards (Sloan, 1985).

an assembly of circuit boards and its enclosure in the form of a chasis requires complicated and detailed consideration. Heat dissipation at the printed wiring board is the only aspect considered here.

Heat dissipation at the printed wiring board level involves heat conduction from the circuits to a heat sink or a cold plate, which is in turn cooled by either natural or forced convection. Examples of heat sinks and cold plates are shown in Fig. 11-9. When no convection is involved, the heat dissipated to the sink is generally transferred to constant-temperature walls by conduction.

For the purpose of determining temperature distribution in a board, it is convenient to divide heat sources into two groups: a concentrated (line) heat source and a distributed (area) heat source. A concentrated heat source is either an individual heat-dissipating component or a small local cluster of several such components. The junctions of a chip are an example of a concentrated heat source. A distributed heat source may consist of many nearly equivalent thermal sources distributed over a defined area.

Consider heat dissipation by conduction only. Typical arrangements for this type of heat dissipation involve attaching a circuit board to supporting structures along opposite edges. The heat is removed by conduction at the edge supports in this arrangement. An example is shown in Fig. 11-6. Since the conduction from the board to the edges is by physical contact, the edge support should provide a large contacting surface area. Further, the edge surface should be smooth and in intimate contact with the board surface to promote effective heat conduction. Thus, the thermal resistance decreases with smoothness and increasing pressure with which the edge surface is held to the board surface. If the line and the area heat sources are uniform in the y direction with width w and if the thickness h is small, then the heat conduction is mainly in the x direction (Fig. 11-10).

The steady-state heat balance for these assumptions is

$$kA \frac{d^2T}{dx^2} = -Q \qquad \text{where } A = bh \tag{11.6}$$

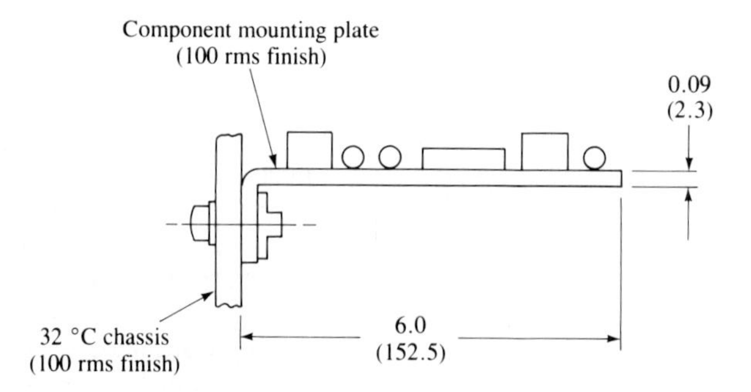

FIGURE 11-10
Edge-cooled component assembly (Sloan, 1985).

and k is the thermal conductivity, A is the cross-sectional area perpendicular to the main heat conduction direction, and Q is the rate of heat generation from the heat sources both concentrated and uniform. The boundary conditions are

$$kA\frac{dT}{dx} = \begin{cases} g_0(T - T_0) - q_0 & \text{at } x = 0;\ g_0 = kA/L_0 \quad (11.7a) \\ -g_l(T - T_l) + q_l & \text{at } x = l;\ g_l = kA/L_l \quad (11.7b) \end{cases}$$

where T_0 and T_l, respectively, are the edge support temperatures at $x = 0$ and $x = l$, L_0 and L_l, respectively, are the contact lengths at $x = 0$ and $x = l$, and q_0 and q_l, respectively, are the heat sources, if any, at $x = 0$ and $x = l$. Note that Q consists of concentrated (line) and uniform (area) heat sources such that for the line source Q is denoted by q_i (energy per time) and for the area source Q is denoted by q_j (energy per time per length) for i and j numbers of heat sources located at various positions along x, say x_i and x_j. The solution of Eq. (11.6) (Sloan, 1985) is

$$T(x) = T(0) + [g_0(T(0) - T_0) - q_0]\frac{x}{kA} - \sum_i^n T_i(x - x_i)H(x - x_i) \quad (11.8)$$

where $H(x - x_i)$ is a Heavyside function such that $H = 0$ for $x < x_i$ and $H = 1$ for $x > x_i$, and $T(0)$ and $T_i(x - x_i)$ are given by

$$T_i(x - x_i) = \begin{cases} q(x - x_i)/kA & \text{(concentrated heating)} \\ \dot{q}(x - x_i)^2/2kA & \text{(distributed heating)} \\ \dfrac{1}{kA}\displaystyle\int_0^x q_b(x)\,dx & \text{[continuously varying heating, } q_b(x)] \end{cases} \quad (11.9)$$

$$T(0) = \frac{(1 + g_l l/kA)(g_0 T_0 + q_0) + g_l T_l + q_l + g_l \sum_{i=1}^n T_i(l - x_i) + Q}{g_0 + g_l + g_0 g_l l/kA} \quad (11.10)$$

The total heat generated by the assembly, excluding the edge sources q_0 and q_l, is given by

$$Q = \sum_{i=1}^n [q_i + \dot{q}_i(l - x_i) + q_{bi}(l)] \quad (11.11)$$

Example 11.1. Components dissipating 20 watts are mounted to a 2-mm thick, 120-mm wide aluminum plate which is fastened to an aluminum chassis. The pressure and the contact area of 2900 mm^2 (50 mm long) are such that the plate-to-chassis interface resistance is 400 mm^2·°C/W. Determine the temperature distribution for the following: 10 W uniformly distributed for the whole plate ($x_i = 0$), 5 W concentrated 100 mm from the chassis interface, and 5 W linearly distributed with its maximum at the plate end. The chassis temperature is at 32 °C and the length of the plate is 152.5 mm. As shown in Fig. 11-10, the system is symmetric geometrically with respect to the contacting at the edges. One watt is 14.34 cal/min.

Solution. Since the system is symmetric geometrically,

$$g_0 = g_l = \frac{2900 \text{ mm}^2}{400 \text{ mm}^2 \cdot {}^\circ\text{C/W}} = 7.25 \text{ W/}^\circ\text{C} \qquad \text{(A)}$$

For the uniform heat source, $x_i = 0$. Thus, Eq. (11.11) yields

$$Q = 5 + 10 + 5 = 20 \text{ W} \qquad \text{(B)}$$

Note in Eq. (11.11) that $\dot{q} = 10/l$ but that $x_i = 0$ such that $\dot{q}_i l = 10$. For the heating sources, one has, from Eq. (11.9),

$$T_i = \begin{cases} 5(x - 100)/kA & \text{(concentrated)} \\ 10x^2/2kAl & \text{(distributed)} \\ \dfrac{1}{kA}\displaystyle\int_0^x \frac{q_{\max} x}{l}\, dx = \frac{x^2 q_{\max}}{2kAl} & \text{(linearly varying source)} \end{cases} \qquad \text{(C)}$$

For the linearly varying heat source, $q = q_{\max}\, x/l$.
Since $5 = [l_0\, q_{\max}\, x/l\, dx]/l$, it follows that $q_{\max} = 10$ W.
Since $g_i = kA/L_i$ and $L_i = 50$ mm,

$$kA = 7.25 \times 50 = 362.5 \text{ mm} \cdot \text{W/}^\circ\text{C} \qquad \text{(D)}$$

Using Eqs. (C) and (D) in Eq. (11.10), one has

$$T(0) = \frac{[1 + 7.25(152.5)/362.5](7.25 \times 32 + 0) + 7.25 \times 32 + 0 + 7.25\Sigma T_i(l) + 20}{7.25 + 7.25 + 7.25^2(152.5)/362.5}$$

$$\sum T_i(l) = \frac{1}{362.5}\left[5(152.5 - 100) + \frac{10(152.5)}{2} + \frac{10(152.5)}{2}\right]$$

$$= 4.93$$

Thus,

$$T(0) = 33.52\,^\circ\text{C}$$

From Eq. (11.8), one has

$$T(x) = 33.52 + [7.25(33.59 - 32) - 0]\frac{x}{362.5} - \sum T_i(x - x_i)H(x - x_i)$$

$$= 33.52 + 0.032x - 0.0138(x - 100)H(x - 100) - 18.1(10^{-5})x^2$$

where x is in millimeters and T is in degrees Celsius.

The one-dimensional temperature distribution that results when a conducting plate is cooled by natural convection (Sloan, 1985) to ambient temperature T_a is given by

$$T(x) = T(0) \cosh Mx + [\beta_1 + T(0)\beta_2] \sinh Mx - T_a(\cosh Mx - 1) - \sum T_i(x - x_i)H(x - x_i) \qquad (11.12)$$

where

$$T_i(x - x_i) = \begin{cases} (Mq/N_s hb) \sinh M(x - x_i) \\ (q/N_s hb)[\cosh M(x - x_i) - 1] \end{cases} \tag{11.13}$$

$$T(0) = \frac{B}{(g_0 + g_l) \cosh Ml + (N_s hb + g_0 g_l/kA) \sinh Ml/M} \tag{11.14}$$

where

$$\begin{aligned} B &= (q_0 + g_0 T_0 + g_l T_a) \cosh Ml + (kAMT_a - g_l \beta_1) \sinh Ml \\ &\quad + q_l + g_l(T_l - T_a) + g_l \sum [T_i(l)]_l + kA \sum [T_i(l)]_d \\ N_s &= \text{convective heat transfer area}/bl \\ M^2 &= hN_s/ka \\ \beta_1 &= -(q_0 + g_0 T_0)/MkA \\ \beta_2 &= g_0/MkA \end{aligned} \tag{11.15}$$

Here h is the heat transfer coefficient for natural convection. Note that the distribution can be obtained by letting g_0 and g_l approach infinity, when the edge temperatures are held at T_0 and T_l.

Component mounting surfaces with nonuniform heat distributions and irregular geometries have to be analyzed using a numerical technique based on either a finite difference or finite element method (Baker, 1983). A general expression for the heat transfer problem is

$$k\nabla^2 T_s = -\sum_i Q_i \tag{11.16}$$

$$\nabla T_a = \frac{hb}{\dot{m}C_p}(T_s - T_a) \tag{11.17}$$

where Q_i represents various types of heat sources with appropriate dimensions, T_s is the plate temperature, $\dot{m}$ is the coolant mass flow rate, and C_p is the coolant's specific heat. The boundary relationship is

$$-k\frac{\partial T_s}{\partial n} = h(T_s - T_a) \tag{11.18}$$

where n is an outward vector normal to the surface. Since the plate thickness is very small, it is usually sufficient to treat the problem as two dimensional.

11.5 STRESS-STRAIN AND FATIGUE

Mechanical stress develops during fabrication, testing, and equipment operation and is due mainly to temperature variations caused by heating and cooling. The stress occurs on assemblies or at attachments where materials exhibiting different thermal expansion coefficients are used. Heating and cooling cycles, which are a consequence of normal equipment operation or of variations in ambient temperature, cause alternating expansions and contractions proportional to the temperature difference relative to the original assembly temperature. Sometimes only a few cycles are sufficient to cause failure of a solder point, fastened connection, or electrical interconnection. The failures usually emerge as cracked solder joints, cracked interface material of bonded or joined surfaces, and fractured or pulled component leads. The primary reason for the failures is fatigue induced by thermal cycling.

The (tensile) stress S of a rod of length l and cross-sectional area A_0 subjected to a uniaxial tensile force F is defined by

$$S = \frac{F}{A_0} \tag{11.19}$$

By definition, the stress has units of pressure. The corresponding strain E is defined as the fractional change in the length:

$$E = \frac{L - L_0}{L_0} \tag{11.20}$$

where L is the extended length due to the stress. A typical stress-strain relationship is shown in Fig. 11-11. The region in which the stress is proportional to the strain is called the elastic region. In this region, the material returns to its original length when the load is removed, and Hooke's law applies:

$$S = \mu E \tag{11.21}$$

where μ is the modulus of elasticity or Young's modulus. Outside the elastic region, a material can exhibit plastic deformation as shown in Fig. 11-11. The stress at which the material shows significant plastic deformation is called the yield strength (stress). It is difficult to precisely pinpoint where elastic strain ends

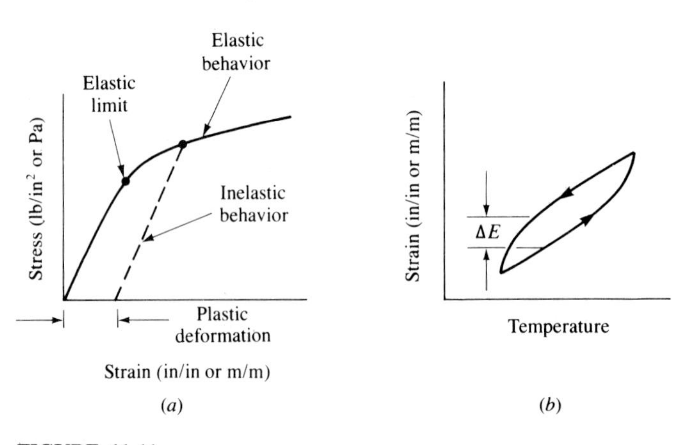

FIGURE 11-11
Stress-strain and strain hysteresis.

and plastic strain begins on the stress-strain curve. The yield strength is chosen to be that strength at which a definite amount of plastic strain has occurred, typically 0.2 percent. The ultimate tensile strength is the maximum strength reached in the stress-strain curve. When a material is under a constant stress, it may undergo progressive plastic deformation over a period of time. This time-dependent strain is called creep. Materials can rupture from creep. The yield strength and creep rupture stress decrease with increasing temperature. For shear stress, the proportionality constant in Eq. (11.21) is called the shear modulus. The shear strain is defined as the shear displacement divided by the distance over which the shear acts.

When any element or lead of an assembly deforms plastically during a thermal cycle but returns to its original length when the assembly is returned to its initial temperature, the strain-temperature curve shows hysteresis. The mean number of cycles to failure N_f for this plastic flow cycle is proportional to the square of the hysteresis width (deviation), which is the constant plastic strain in each cycle:

$$N_f \propto \frac{1}{\Delta E} \tag{11.22}$$

where ΔE is shown in Fig. 11-11*b*. The life of the assembly is strongly dependent upon the magnitude of the cyclic strain. The cyclic strain of component leads in a soldered assembly increases if stress relaxation in the solder joints occurs, since the leads carry a greater portion of the load. Plastic deformations are likely to occur in areas of stress concentration or geometric irregularities.

The stresses generated in an assembly by thermal expansion due to a temperature difference can be determined from Hooke's law, a force balance, and from the definition of the thermal expansion coefficient. Consider the assemblies in Fig. 11-12. The first one represents a parallel system, which can be used, for example, as an idealized model for a component mounted on a circuit board. Because of symmetry, only half the system is depicted. The system undergoes an actual displacement of d due to the load W and thermal expansion. For each lead, the definition of the thermal expansion coefficient α_i allows one to write the free displacement in the absence of the attachment v_i as follows:

$$v_i = \alpha_i L_i \, \Delta T \tag{11.23}$$

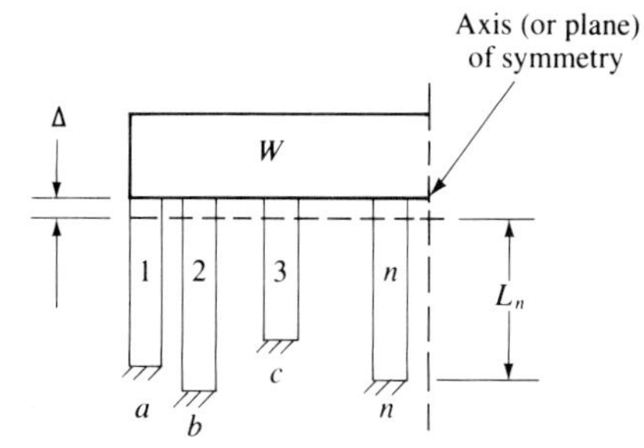

FIGURE 11-12
Thermal-elastic component models (Sloan, 1985).

where L_i is the ith lead length and T is the temperature difference. A force balance on the system is

$$d\sum_i^n K_i + \frac{W}{2} = \sum_i^n K_i v_i \qquad \text{where } K_i = \frac{A_i \mu_i}{L_i} \tag{11.24}$$

where A_i is the lead cross-sectional area. The left-hand side represents the downward force; balancing this force is the force due to constrained free displacement. The force load on each lead F_i is the net difference between the downward and upward forces:

$$F_i = K_i(d - v_i) \tag{11.25}$$

Since $S_i = F_i/A_i$ from Eq. (11.14), use of Eqs. (11.23) through (11.25) for S_i leads to

$$S_i = \left(\frac{\mu}{L}\right)_i \left[\frac{T \sum (A\mu\alpha)_i - W/2}{\sum_i (A\mu/L)_i} - \Delta T(\alpha L)_i\right] \tag{11.26}$$

It is noted that $F = (A\mu/L)(L - L_0) = K(L - L_0)$, which follows from Eqs. (11.19) through (11.21).

Example 11.2. Determine the stress in the copper leads of Fig. 11-13 for an epoxy encapsulated component subject to a temperature difference of 50 °C (Sloan, 1985). The length of the copper leads including that part in the encapsulated component and circuit board is 1.9 mm. Use the material properties given in Table 11.1 for copper and epoxy.

Solution. This system is equivalent to that in Fig. 11-12 with two leads: one for the copper lead and the other for the epoxy between the copper leads. When a line of symmetry is used as shown in Fig. 11-13, the cross-sectional areas are

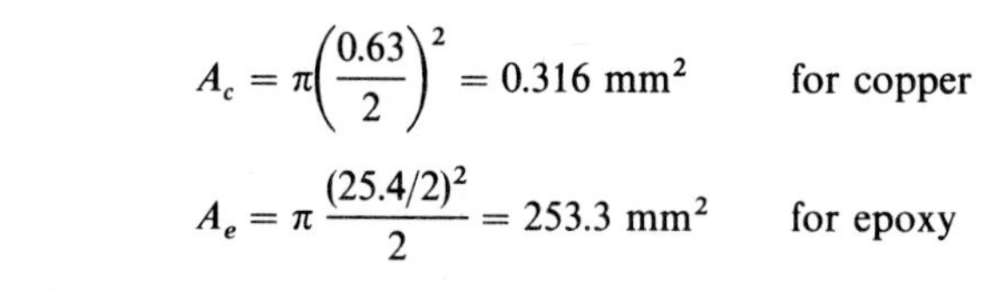

$$A_c = \pi\left(\frac{0.63}{2}\right)^2 = 0.316 \text{ mm}^2 \qquad \text{for copper}$$

$$A_e = \pi\,\frac{(25.4/2)^2}{2} = 253.3 \text{ mm}^2 \qquad \text{for epoxy}$$

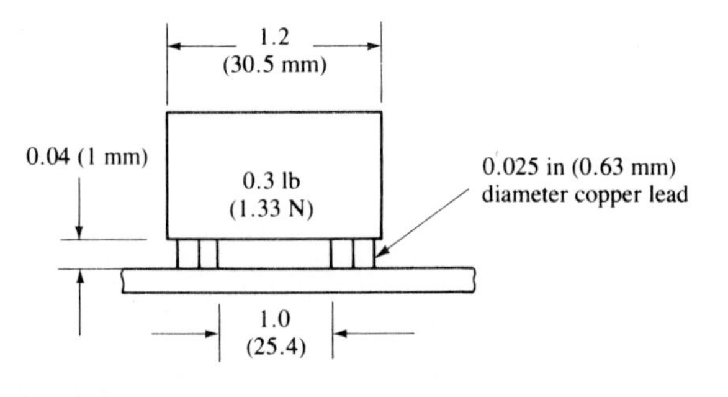

FIGURE 11-13
A copper leads system.

Note that A_e is half the cross-sectional area of the epoxy because of the line of symmetry. The load W is

$$W = 1.33N(10^3 \text{ mm/m})^2 = 1.33 \times 10^{-3} \text{ GPa·mm}^2$$

From Table 11.1, the moduli and coefficients of thermal expansion are:

	Copper	**Epoxy**
μ, GPa	119	16.3
α, m/(m · °C)	1.68×10^{-5}	7.2×10^{-5}

Using these values in Eq. (11.21) for the copper lead, one has

$$S_c = \frac{119}{1.9}\left[\frac{50(632 \times 10^{-4} + 0.297) - 665 \times 10^{-4}}{19.79 + 4130} - 1.596 \times 10^{-3}\right]$$

$$= 0.1246 \text{ GPa}$$

Bonded joints are often used in packaging. A change in temperature frequently encountered in the bonding process introduces stresses in joined materials of dissimilar thermal expansion coefficients. The thermal stress caused by cooling from high bonding temperatures can cause fracture and joint failure. Some insight into the effects of various physical parameters on the stress can be gained from force balance equations for the basic bonding geometry in Fig. 11-12. Consider a one-dimensional force balance. If f is defined as the axial force per unit width of the bonding material (shaded in Fig. 11-14), the force balances are (Sloan, 1985)

$$\frac{df_1}{dx} - U(x) = 0 \tag{11.27}$$

$$\frac{df_2}{dx} - U(x) = 0 \tag{11.28}$$

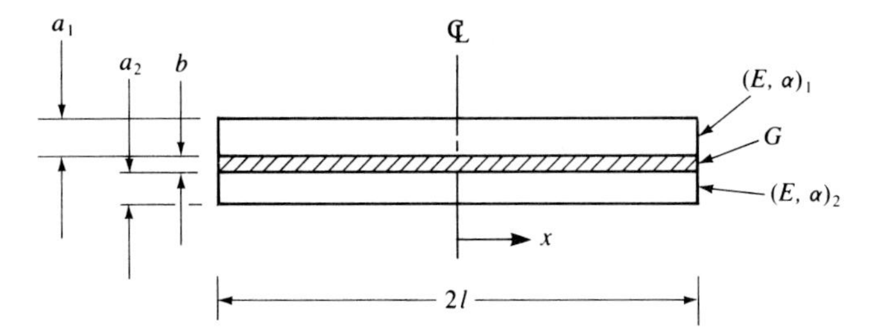

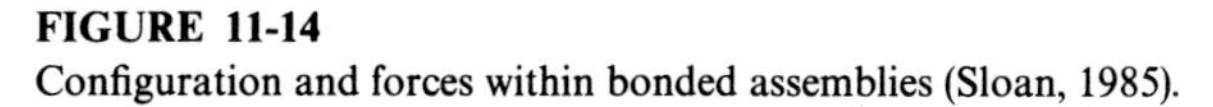

FIGURE 11-14
Configuration and forces within bonded assemblies (Sloan, 1985).

where U is the joint shear stress. The resulting strain of each of the bonded materials 1 and 2 in Fig. 11-14 is a combination of the strain due to the axial force $[(\Delta v_i)_f = (f/\mu a)_i \, \Delta x]$ and that due to the thermal expansion $[(\Delta v_i)_t = \alpha_i \, \Delta T \, \Delta x]$:

$$\frac{dv_i}{dx} = \left(\frac{f}{\mu a}\right)_i + \alpha_i \, \Delta T \qquad \text{for } i = 1, 2 \tag{11.29}$$

From the definition of shear modulus G, one has

$$U = G\left(\frac{v_1 - v_2}{b}\right) \tag{11.30}$$

Differentiating Eq. (11.30) with respect to x with the aid of Eq. (11.29) and differentiating again the resulting expression with the aid of Eqs. (11.27) and (11.28) yields

$$\frac{d^2U}{dx^2} = A^2U \tag{11.31}$$

where

$$A^2 = \frac{G}{b}\left[\left(\frac{1}{\mu a}\right)_1 + \left(\frac{1}{\mu a}\right)_2\right] \tag{11.32}$$

At the ends, the forces are relieved and thus one has

$$f_i = 0 \qquad \text{at } x = l \tag{11.33}$$

Further, the shear stress is zero at the center, i.e.,

$$U = 0 \qquad \text{at } x = 0 \tag{11.34}$$

The solution of Eq. (11.31) with these boundary conditions is

$$U = \frac{G \, \Delta T(\alpha_1 - \alpha_2) \sinh Ax}{bA \cosh Al} \tag{11.35}$$

It is understood that material 1 has a higher coefficient of thermal expansion than material 2 and that the joint bonding material is more pliant than the bonded materials 1 and 2. The axial force F at equilibrium ($P_2 = -P_1$) follows directly from Eqs. (11.27), (11.28), and (11.35):

$$F = fb = \frac{bG \, \Delta T(\alpha_1 - \alpha_2)(\cosh Al - \cosh Ax)}{bA^2 \cosh Al} \tag{11.36}$$

Example 11.3. Slip occurs when the shear stress exceeds a certain value. Determine the position at which the shear stress is the maximum. Noting that the bonding thickness b is the only variable that can be manipulated for given materials and configurations, determine the dependence of the maximum shear stress on the thickness.

Solution. The hyperbolic sine function is a monotonically increasing function of x. Thus, the maximum occurs at the largest value of Ax or at $x = l$. The maximum shear stress U_m is therefore given by

$$U_m = \frac{G\,\Delta T(\alpha_1 - \alpha_2)\tanh Al}{bA} \tag{A}$$

For $Al > 3$, $\tanh Al$ is close to unity. For most applications of interest, $Al > 3$. Thus, it follows from Eq. (11.32) and Eq. (A) that

$$U_m \propto b^{-0.5} \tag{B}$$

11.6 OTHER CONSIDERATIONS

Adhesion of a solid to another solid, although not unique to packaging, is an important consideration in packaging. There are no widely recognized or used theories for the adhesion of solid materials. However, an extensive body of theoretical and experimental results based on electrostatic interaction is available (Deryagin *et al.*, 1978). The semi-phenomenological approach views adhesion as the surface interaction between the two solid surfaces and treats the interactions as those of an electrical double layer. For modeling purposes, the electrical double layer is regarded as a parallel-plate microcapacitor with a charge density Ω. The force of interaction F is then equal to the force of attraction of the capacitor plates:

$$F = 2\pi\Omega^2 \tag{11.37}$$

The effect of permittivity is neglected for simplicity. A more detailed account of the adhesion force utilizes a donor-acceptor reaction that extends along the entire surface. In this view, then, the force of an adhesive bond between two solids is given by

$$F = \frac{q\Omega}{r^2} = \frac{Nq^2}{r^2} \tag{11.38}$$

where q is the electric charge, N is the number of charge pairs per unit area, and r is the separation between the solids, which is on the order of a few angstroms. Although the approach discussed here needs fuller development to be useful, it is nevertheless instructive that the adhesive bonding strength is proportional to the square of the surface charge density.

Electrical performance at the IC package level is another consideration that is attracting more attention as circuit speed and noise reduction requirements increase. This aspect is beyond the scope of this book and interested readers can refer to Hill *et al.* (1969) and O'Neill (1981).

An optimal packaging design requires integration of all the considerations treated so far. Depending on final application, some considerations will weigh more heavily than others during optimization. A systematic and quantitive approach to optimal packaging is a challenge that has not yet been fully explored.

NOTATION

a	Thickness (L)
A	Cross-sectional area (L^2) perpendicular to heat flow; quantity defined in Eq. (11.29)
A_0	Cross-sectional area of a rod (L^2)
b	Width (L); thickness of bonding material in Eq. (11.30) (L)
B	Quantity defined in Eq. (11.14) (EL/t)
C	Solution concentration (mol/L^3); speed of light (L/t)
C_p	Specific heat
d	Actual displacement in Eq. (11.24) (L)
D	Distance a signal travels (L)
D_{np}	Distance from neutral point of chip (L)
E	Electrode potential (V) in Eq. (11.3); strain in Eq. (11.20)
E_h	Strain caused by cycles of heating and cooling
E_0	Standard electropotential (V)
ΔE	Hysteresis width in Fig. 11-11b
f_1, f_2	Axial forces per unit width (F/L)
F	Force (F)
g_0, g_l	Quantities defined in Eq. (11.7) (EL/tT)
G	Shear modulus (F/L^2)
h	Height (L); heat transfer coefficient in Eq. (11.13)
H	Step function in Eq. (11.8); height of pad (L)
k	Thermal conductivity (E/tT)
K_D	Dielectric constant
K_i	Quantity defined in Eq. (11.24) (F/L)
l	Length (L)
L	Length of rod (L)
L_0	Original length of a rod in Eq. (11.20)
L_0, L_l	Contact lengths in Eq. (11.7) (L)
L_i	ith lead length (L)
$\dot{m}$	Mass rate (M/t)
M	Quantity defined in Eq. (11.15)
n	Number of electrons removed per ion
N	Number of charge pair per unit area ($1/L^2$)
N_f	Number of cycles resulting in fatigue
N_s	Quantity defined in Eq. (11.15)
q	Electric charge
q_0, q_l	Heat sources at $x = 0$ and $x = l$, respectively (E/t)
q_i	Line heat source (E/t)
$\dot{q}_i$	Area heat source (E/tL)

q_b	Continuously varying heat source with axial distance (E/L)
Q	Rate of heat transfer from heat sources (E/t)
r	Separation between solids (L)
S	Shear stress (F/L^2)
t_D	Signal delay time (t)
T	Temperature (T)
T_a	Ambient temperature (T)
T_i	Temperature defined in Eq. (11.9) (T)
T_0	Temperature at $x = 0$ (T)
T_l	Temperature at $x = l$ (T)
T_s	Solid temperature (T)
x	Axial coordinate (L)
U	Shear stress (F/L^2)
U_m	Maximum shear stress
v_i	Free displacement (L)
W	load (F)

Greek letters

α_i	Thermal expansion coefficient ($1/T$)
β_1, β_2	Quantities defined in Eq. (11.15)
μ	Modulus of elasticity (F/L^2)
Ω	Charge density (L^{-3})

Units

E	Energy (L^2M/t^2)
F	Force (LM/t^2)
L	Length
M	Mass
t	Time
T	Temperature
V	Voltage

PROBLEMS

11.1. A gold wire is connected to a copper lead. Which one is more likely to corrode?

11.2. When each component on an assembly dissipates approximately the same amount of heat, the whole assembly may be considered as a uniformly heated plane. A heat balance for such a rectangular plane with width b and length l can be written as

$$k\left(\frac{\partial^2 T}{\partial x^2} + \frac{\partial^2 T}{\partial y^2}\right) + \frac{Q}{w} = 0$$

where Q is the rate of heat generation from the uniform source and w is the thickness. If the edges are kept at T_0, an approximate solution is (Sloan, 1985):

$$T = \frac{5Qbl}{2kw(b^2 + l^2)}\left[\frac{x}{l}\left(1 - \frac{x}{l}\right)\right]\left[\frac{y}{b}\left(1 - \frac{y}{b}\right)\right] + T_0$$

Suppose the temperature difference between the center and the edges is 10 °C, the thickness is 0.3 cm, b/l is 2, and the thermal conductivity is 1 W/(cm · °C). Determine the rate of uniform heat generation, Q.

11.3. To decrease the stress on the copper lead in Example 11.2, either the lead cross-sectional area or the lead length can be increased. Discuss why it is more effective to increase the lead length. Determine the increase in the lead length required to decrease the stress to one-half the value calculated in Example 11.2.

11.4. Discuss the effect of the thickness of bonded material on the maximum shear stress considered in Example 11.3. Determine the thickness of bonding material required for the same maximum shear stress.

11.5. Reflow soldered planar component leads usually fail by a peeling action that fractures the solder joint (Fig. P11-5). According to Sloan (1985), the peeling occurs when the force normal to the board surface equals or exceeds that given by

$$F_{\text{peel}} = 0.38WS_u\left[ht^3\left(\frac{\mu_l}{\mu_s}\right)\right]^{1/4}$$

where S_u is the ultimate solder strength, and μ_l and μ_s, respectively, are the lead and solder moduli of elasticity. Discuss the effect of temperature on the peeling force.

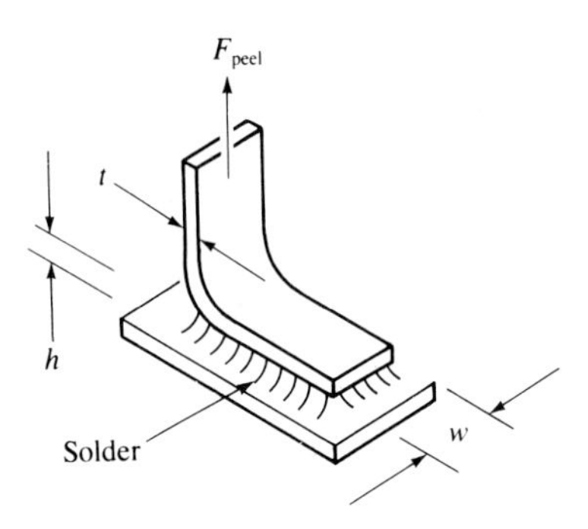

FIGURE P11-5
Peeling of solder joint.

11.6. Mismatch between the thermal expansion coefficient of a plastic molding compound and of a silicon die or gold wire is directly responsible for the stresses induced after molding and also during temperature cycling. Figure P11-6 shows a DIP cross section and an enlarged view of the wire bond. For modeling purposes, chip and leadframe may be combined as a structural member that resists the shrinkage of the plastic (Steidel, 1983; Usell and Smiley, 1981). The strain in the chip-leadframe combination, E_c, and the strain in the plastic, E_p, at temperatures below the mold and cure temperature are given by

$$E_i = \left(\frac{F}{\mu A}\right)_i + \alpha_i\,\Delta T \quad \text{for } i = c, p \tag{A}$$

Show that

$$E_c = \frac{(\alpha_p - \alpha_c)\,\Delta T}{1 + (\mu A)_c/(\mu A)_p} + \alpha_c\,\Delta T \tag{B}$$

Note that $E_p = E_c$ since they are joined together and $F_p = F_c$ at equilibrium.

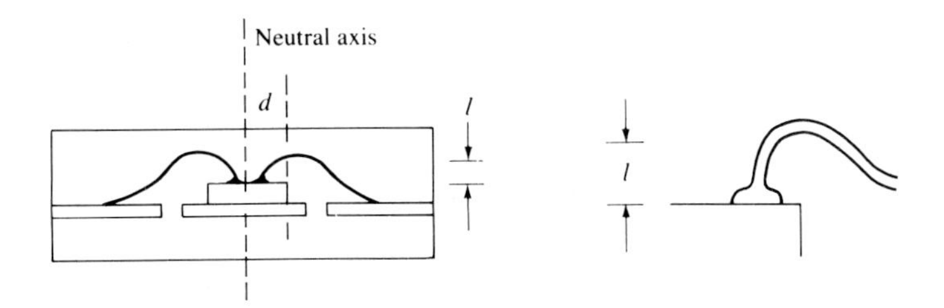

FIGURE P11-6
DIP cross section.

REFERENCES

Ameraskera, E. A., and D. S. Campbell: *Failure Mechanisms in Semiconductor Devices*, Wiley, New York, 1987.

Baker, A. J.: *Finite Element Computational Fluid Mechanics*, McGraw-Hill, New York, 1983.

Blodgett, A. J., Jr.: *IEEE Trans. Comp. Hyb. Mfg. Technol.*, vol. CHMT-3 (4), p. 634, 1980.

Chang, P. C., and K. K. Chao: *Plasma Processing*, The Electrochemical Society, Pennington, N.J., 1985.

Deryagin, B. V., N. A. Krotova, and V. P. Smilga: *Adhesion of Solids*, Plenum Publishing Company, New York, 1978.

Hill, Y. M., N. O. Reckford, and D. R. Winner: *IBM J. Res. Dev.*, vol. 13, p. 314, 1969.

Howell, J. R., *Proc. Int. Rel. Phys. Symp.*, p. 104, 1981.

Huatek, E. R.: *Integrated Circuit Quality and Reliability*, Marcel Deffer, Inc., New York, 1987.

Kaczynski, D. J., and K. A. Walsh: *Cer. Engng Sci. Proc.*, vol. 6, p. 1261, 1985.

Keizer, A., and D. Brown: *Solid State Technol.*, p. 59, March 1978.

Kohara, M., M. Harra, H. Genyo, H. Shibata, and H. Nakata: *IEEE Trans. Comp. Hyb. Mfg. Technol.*, vol. CHMT-9, p. 386, 1986.

Kyocera: "Design Guidelines for Multilayer Ceramics," Publication A-125E.

Levinthal, D. S.: *Semicond. Int.*, p. 33, April 1979.

McLeod, C. T., J. W. Kastuer, T. J. Carbone, and J. P. Starr: *Cer. Engng. Sci. Proc.*, vol. 6, p. 1233, 1985.

O'Neill, T. G.: *Cer. Engng Sci. Proc.*, p. 43, March 1981.

Schwartz, B.: *Am. Cer. Soc. Bull.*, vol. 63 (4), p. 577, 1984.

———: *Am. Cer. Soc. Bull.*, vol. 65 (7), p. 1032, 1986.

Shimada, Y., *et al.*: *IEEE Trans. Comp. Hyb. Mfg. Technol.*, vol. CHMT-6, p. 382, 1983.

Sloan, J. L.: *Design and Packaging of Electronic Equipment*, Van Nostrand Reinhold Company, New York, 1985.

Stafford, J. W.: *Semicond. Int.*, p. 82, May 1982.

Steidel, C. A.: in S. M. Sze (ed.), *VLSI Technology*, chap. 13, McGraw-Hill, New York, 1983.

Usell, R. J., Jr., and S. A. Smiley: *Proc. Int. Rel. Phys. Symp.*, p. 65, 1981.

Wong, C. P., and D. M. Rose: *IEEE Trans. Comp. Hyb. Mfg. Technol.*, vol. CHMT-6, p. 485, 1983.

APPENDIX A

TABLES

TABLE A.1
Periodic table

Period	Group IA	Group IIA	Group IIIB	Group IVB	Group VB	Group VIB	Group VIIB	Group VIII			Group IB	Group IIB	Group IIIA	Group IVA	Group VA	Group VIA	Group VIIA	Group 0
1 1*s*	1 H																1 H	2 He
2 2*s*3*p*	3 Li	4 Be											5 B	6 C	7 N	8 O	9 F	10 Ne
3 3*s*3*p*	11 Na	12 Mg											13 Al	14 Si	15 P	16 S	17 Cl	18 Ar
4 4*s*3*d*4*p*	19 K	20 Ca	21 Sc	22 Ti	23 V	24 Cr	25 Mn	26 Fe	27 Co	28 Ni	29 Cu	30 Zn	31 Ga	32 Ge	33 As	34 Se	35 Br	36 Kr
5 5*s*4*d*5*p*	37 Rb	38 Sr	39 Y	40 Zr	41 Nb	42 Mo	43 Tc	44 Ru	45 Rh	46 Pd	47 Ag	48 Cd	49 In	50 Sn	51 Sb	52 Te	53 I	54 Xe
6 6*s*(4*f*)5*d*6*p*	55 Ca	56 Ba	57* La	72 Hf	73 Ta	74 W	75 Re	76 Os	77 Ir	78 Pt	79 Au	80 Hg	81 Tl	82 Pb	83 Bi	84 Po	85 At	86 Rn
7 7*s*(5*f*)6*d*	87 Fr	88 Ra	89† Ac															

*Lanthanide series 4*f*	58 Ce	59 Pr	60 Nd	61 Pm	62 Sm	63 Eu	64 Gd	65 Tb	66 Dy	67 Ho	68 Er	69 Tm	70 Yb	71 Lu
†Actinide series 5*f*	90 Th	91 Pa	92 U	93 Np	94 Pu	95 Am	96 Cm	97 Bk	98 Cf	99 Es	100 Fm	101 Md	102 No	103 Lw

TABLE A.2
International system of units

Quantity	Unit name	Symbol	Dimensions
Fundamental			
Length	meter	m	L
Mass	kilogram	kg	M
Time	second	s	t
Temperature	kelvin	K	T
Current	ampere	A	I
Derived			
Capacitance	farad	F	$I^2t^4M^{-1}L^{-2}$ (C/V)
Conductance	siemens	S	$ML^2t^{-3}I^{-2}$ (A/V)
Electric charge	coulomb	C	It
Energy	joule	J	ML^2t^{-2}
Force	newton	N	MLT^{-2}
Frequency	hertz	Hz	t^{-1}
Induction	henry	H	$I^2t^2M^{-1}L^{-2}$ (Wb/A)
Magnetic induction	tesla	T	$Mt^{-2}I^{-1}$ (Wb/L^2)
Magnetic flux	weber	Wb	$ML^2t^{-2}I^{-1}$ (V·t)
Potential	volt	V	$ML^{-2}t^{-3}I^{-1}$
Power	watt	W	ML^2t^{-3} (VI, J/t)
Pressure	pascal	Pa	$Mt^{-2}L^{-1}$
Resistance	ohm	Ω	$ML^2t^{-3}I^{-2}$ (V/A)

TABLE A.3
Physical constants and unit prefixes

Avogadro constant	N	6.022×10^{23} mol^{-1}
Boltzmann constant	k	1.3807×10^{-23} J/K
		8.62×10^{-5} eV/K
Elementary charge	q	1.6022×10^{-19} C
Electron rest mass	m_e	9.11×10^{-28} g
Gas constant	R	1.987 cal/(mol·K)(=kN)
Permittivity of free space	ϵ_0	8.85×10^{-14} F/cm
Planck constant	h	6.63×10^{-34} J·s
Proton rest mass	m_p	1.6726×10^{-24} g
Speed of light	c	2.998×10^{10} cm/s

1 eV = 1.6×10^{-19} J = 23.053 kcal/mol
Wavelength of 1 μm = 1.24 eV
Room temperature (300 K) value of kT = 0.0259 eV
Torr = 1 mmHg = 1333.2 Pa
Calorie = 4.184 J
1 μm = 10^{-4} cm

Unit prefixes

Multiple	Prefix	Symbol	Multiple	Prefix	Symbol
10^{15}	peta	P	10^{-3}	milli	m
10^{12}	tera	T	10^{-6}	micro	μ
10^{9}	giga	G	10^{-9}	nano	n
10^{6}	mega	M	10^{-12}	pico	p
10^{3}	kilo	k	10^{-15}	femto	f

TABLE A.4
Properties of semiconductor materials (Streetman, 1980) (all values at 300 K)

	E_g, eV	μ_n, cm²/(V·s)	μ_p, cm²/(V·s)	ρ, Ω·cm	Transition	Doping	Lattice	a, Å	ϵ_r	Density, g/cm³	Melting point, °C
Si	1.11	1350	480	2.5×10^5†	i	n, p	D	5.43	11.8	2.33	1415
Ge	0.67	3900	1900	43	i	n, p	D	5.66	16	5.32	936
SiC(α)	2.86	500		10^{10}	i	n, p	W	3.08	10.2	3.21	2830
AlP	2.45	80		10^{-5}	i	n, p	Z	5.46		2.40	2000
AlAs	2.16	180		0.1	i	n, p	Z	5.66	10.9	3.60	1740
AlSb	1.6	200	300	5	i	n, p	Z	6.14	11	4.26	1080
GaP	2.26	300	150	1	i	n, p	Z	5.45	11.1	4.13	1467
GaAs	1.43	8500	400	4×10^8†	d	n, p	Z	5.65	13.2	5.31	1238
GaSb	0.7	5000	1000	0.04	d	n, p	Z	6.09	15.7	5.61	712
InP	1.28	4000	100	8×10^{-3}	d	n, p	Z	5.87	12.4	4.79	1070
InAs	0.36	22600	200	0.03	d	n, p	Z	6.06	14.6	5.67	943
InSb	0.18	10^5	1700	0.06	d	n, p	Z	6.48	17.7	5.78	525
ZnS	3.6	110		10^{10}	d	n	Z, W	5.409	8.9	4.09	1650‡
ZnSe	2.7	600		10^9	d	n	Z	5.671	9.2	5.65	1100‡
ZnTe	2.25		100	100	d	p	Z	6.101	10.4	5.51	1238‡
CdS	2.42	250	15		d	n	W, Z	4.137	8.9	4.82	1475
CdSe	1.73	650		10^5	d	n	W	4.30	10.2	5.81	1248
CdTe	1.58	1050	100	10^{10}	d	n, p	Z	6.482	10.2	6.20	1098
PbS	0.37	575	200	5×10^{-3}	i	n, p	H	5.936	161	7.6	1119
PbSe	0.27	1000	1000	10^{-3}	i	n, p	H	6.147	280	8.73	1081
PbTe	0.29	1600	700	10^{-2}	i	n, p	H	6.452	360	8.16	925

† Intrinsic resistivity.

‡ Vaporizes.

Definitions of symbols: ρ is resistivity of high-purity material; i is indirect; d is direct; D is diamond; Z is zinc blende; W is wurtzite; H is halite (NaCl). Values of mobility and resistivity are for material of available purity; these values are considered approximate (exception: Si and GaAs resistivities are extrapolated to intrinsic material). Most of the values in this table were taken from publications of the Electronic Properties Information Center (EPIC), Hughes Aircraft Company, Culver City, California; also, M. Neuberger, *III-V Semiconducting Compounds—Data Tables*, published with permission from Plenum Publishing Corporation, copyright 1970.

Crystals in the wurtzite structure are not described completely by the single lattice constant given here since the unit cell is not cubic. Several II-VI compounds can be grown in either the zinc blende or wurtzite structures.

Many values quoted here are approximate or uncertain, particularly for the II-VI and IV-VI compounds.

Source: B. G. Streetman, *Solid State Electronic Devices*, 2d ed., Prentice-Hall, New York, 1980.

TABLE A.5
Properties of Si and GaAs at 300 K

Properties	Si	GaAs
Atoms/cm^3	5.0×10^{22}	4.42×10^{22}
Atomic weight	28.09	144.63
Breakdown field, V/cm	$\sim 3 \times 10^5$	$\sim 4 \times 10^5$
Crystal structure	Diamond	Zinc blende
Density, g/cm^3	2.328	5.32
Distance between neighboring atoms, nm	0.2328	0.5320
Effective density of states		
Valence band, cm^3	1.04×10^{19}	7×10^{18}
Conduction band, cm^{-3}	2.8×10^{19}	4.7×10^{17}
Effective mass, m^*/m_0		
Electrons	$m_l^* = 0.98$	0.067
	$m_t^* = 0.19$	
Holes	$m_{lh}^* = 0.16$	$m_{lh}^* = 0.082$
	$m_{hh}^* = 0.49$	$m_{hh}^* = 0.045$
Electron affinity, V	4.05	4.07
Energy gap, eV	1.12	1.424
Index of refraction	3.42	
Intrinsic carrier concentration, cm^{-3}	1.45×10^{10}	1.79×10^6
Intrinsic Debye length, μm	24	2250
Intrinsic resistivity, Ω·cm	2.3×10^5	10^8
Lattice constant, nm	0.5431	0.5646
Linear coefficient of thermal expansion, cm/(cm·K)	2.6×10^{-6}	6.86×10^{-6}
Melting point, °C	1415	1238
Minority-carrier lifetime, s	2.5×10^{-3}	$\sim 10^{-8}$
Mobility, cm^2/(V·s)		
Electron	1500	8500
Hole	475	400
Optical phonon energy, eV	0.063	0.035
Phonon mean free path, nm	7.6 (electron)	
	5.5 (hole)	
Poisson's ratio	0.42	
Relative permittivity	11.9	13.1
Specific heat, J/(g·K)	0.7	0.35
Thermal conductivity, W/(cm·K)	1.5	0.46
Thermal diffusivity, cm^2/s	0.9	0.44
Vapor pressure, Pa	1 at 1659 °C	100 at 1050 °C
	10^{-6} at 900 °C	1 at 900 °C
Young's modulus, g/cm	1.089×10^9	

TABLE A.6
Properties of SiO_2 and Si_3N_4 at 300 K

Properties	SiO_2	Si_3N_4
Density, g/cm^3	2.2	3.1
DC resistivity, μm·cm	10^{14}–10^{16}	$\sim 10^{14}$
Dielectric strength, V/cm	$\sim 10^7$	$\sim 10^7$
Energy gap, eV	9	~ 5
Inframed absorption band, μm	9.3	11.5–12.0
Melting point, °C	~ 1600	—
Refractive index	1.46	2.05
Relative permittivity (dielectric constant)	3.9	7.5
Structure	Amorphous	Amorphous
Thermal expansion coefficient, K^{-1}	5×10^{-7}	—
Thermal conductivity, W/(cm·K)	0.014	—

TABLE A.7
Error function values

z	erf (z)	z	erf (z)	z	erf (z)	z	erf (z)
0.00	0.000 000	0.50	0.520 500	1.00	0.842 701	1.50	0.966 105
0.01	0.011 283	0.51	0.529 244	1.01	0.846 810	1.51	0.967 277
0.02	0.022 565	0.52	0.537 899	1.02	0.850 838	1.52	0.968 413
0.03	0.033 841	0.53	0.546 464	1.03	0.854 784	1.53	0.969 516
0.04	0.045 111	0.54	0.554 939	1.04	0.858 650	1.54	0.970 586
0.05	0.056 372	0.55	0.563 323	1.05	0.862 436	1.55	0.971 623
0.06	0.067 622	0.56	0.571 616	1.06	0.866 144	1.56	0.972 628
0.07	0.078 858	0.57	0.579 816	1.07	0.869 773	1.57	0.973 603
0.08	0.090 078	0.58	0.587 923	1.08	0.873 326	1.58	0.974 547
0.09	0.101 281	0.59	0.595 936	1.09	0.876 803	1.59	0.975 462
0.10	0.112 463	0.60	0.603 856	1.10	0.880 205	1.60	0.976 348
0.11	0.123 623	0.61	0.611 681	1.11	0.883 533	1.61	0.977 207
0.12	0.134 758	0.62	0.619 411	1.12	0.886 788	1.62	0.978 038
0.13	0.145 867	0.63	0.627 046	1.13	0.889 971	1.63	0.978 843
0.14	0.156 947	0.64	0.634 586	1.14	0.893 082	1.64	0.979 622
0.15	0.167 996	0.65	0.642 029	1.15	0.896 124	1.65	0.980 376
0.16	0.179 012	0.66	0.649 377	1.16	0.899 096	1.66	0.981 105
0.17	0.189 992	0.67	0.656 628	1.17	0.902 000	1.67	0.981 810
0.18	0.200 936	0.68	0.663 782	1.18	0.904 837	1.68	0.982 493
0.19	0.211 840	0.69	0.670 840	1.19	0.907 608	1.69	0.983 153
0.20	0.222 703	0.70	0.677 801	1.20	0.910 314	1.70	0.983 790
0.21	0.233 522	0.71	0.684 666	1.21	0.912 956	1.71	0.984 407
0.22	0.244 296	0.72	0.691 433	1.22	0.915 534	1.72	0.985 003
0.23	0.255 023	0.73	0.698 104	1.23	0.918 050	1.73	0.985 578
0.24	0.265 700	0.74	0.704 678	1.24	0.920 505	1.74	0.986 135
0.25	0.276 326	0.75	0.711 156	1.25	0.922 900	1.75	0.986 672
0.26	0.286 900	0.76	0.717 537	1.26	0.925 236	1.76	0.987 190
0.27	0.297 418	0.77	0.723 822	1.27	0.927 514	1.77	0.987 691
0.28	0.307 880	0.78	0.730 010	1.28	0.929 734	1.78	0.988 174
0.29	0.318 283	0.79	0.736 103	1.29	0.931 899	1.79	0.988 641
0.30	0.328 627	0.80	0.742 101	1.30	0.934 008	1.80	0.989 091
0.31	0.338 908	0.81	0.748 003	1.31	0.936 063	1.81	0.989 525
0.32	0.349 126	0.82	0.753 811	1.32	0.938 065	1.82	0.989 943
0.33	0.359 279	0.83	0.759 524	1.33	0.940 015	1.83	0.990 347
0.34	0.369 365	0.84	0.765 143	1.34	0.941 914	1.84	0.990 736
0.35	0.379 382	0.85	0.770 668	1.35	0.943 762	1.85	0.991 111
0.36	0.389 330	0.86	0.776 100	1.36	0.945 561	1.86	0.991 472
0.37	0.399 206	0.87	0.781 440	1.37	0.947 312	1.87	0.991 821
0.38	0.409 009	0.88	0.786 687	1.38	0.949 016	1.88	0.992 156
0.39	0.418 739	0.89	0.791 843	1.39	0.950 673	1.89	0.992 479
0.40	0.428 392	0.90	0.796 908	1.40	0.952 285	1.90	0.992 790
0.41	0.437 969	0.91	0.801 883	1.41	0.953 852	1.91	0.993 090
0.42	0.447 468	0.92	0.806 768	1.42	0.955 376	1.92	0.993 378
0.43	0.456 887	0.93	0.811 564	1.43	0.956 857	1.93	0.993 656
0.44	0.466 225	0.94	0.816 271	1.44	0.958 297	1.94	0.993 923
0.45	0.475 482	0.95	0.820 891	1.45	0.959 695	1.95	0.994 179
0.46	0.484 655	0.96	0.825 424	1.46	0.961 054	1.96	0.994 426
0.47	0.493 745	0.97	0.829 870	1.47	0.962 373	1.97	0.994 664
0.48	0.502 750	0.98	0.834 232	1.48	0.963 654	1.98	0.994 892
0.49	0.511 668	0.99	0.838 508	1.49	0.964 898	1.99	0.995 111

Table A.7 (*continued*)

z	erf (z)	z	erf (z)	z	erf (z)	z	erf (z)
2.00	0.995 322	2.50	0.999 593	3.00	0.999 977 91	3.50	0.999 999 257
2.01	0.995 525	2.51	0.999 614	3.01	0.999 979 26	3.51	0.999 999 309
2.02	0.995 719	2.52	9.999 634	3.02	0.999 980 53	3.52	0.999 999 358
2.03	0.995 906	2.53	0.999 654	3.03	0.999 981 73	3.53	0.999 999 403
2.04	0.996 086	2.54	0.999 672	3.04	0.999 982 86	3.54	0.999 999 445
2.05	0.996 258	2.55	0.999 689	3.05	0.999 983 92	3.55	0.999 999 485
2.06	0.996 423	2.56	0.999 706	3.06	0.999 984 92	3.56	0.999 999 521
2.07	0.996 582	2.57	0.999 722	3.07	0.999 985 86	3.57	0.999 999 555
2.08	0.996 734	2.58	0.999 736	3.08	0.999 986 74	3.58	0.999 999 587
2.09	0.996 880	2.59	0.999 751	3.09	0.999 987 57	3.59	0.999 999 617
2.10	0.997 021	2.60	0.999 764	3.10	0.999 988 35	3.60	0.999 999 644
2.11	0.997 155	2.61	0.999 777	3.11	0.999 989 08	3.61	0.999 999 670
2.12	0.997 284	2.62	0.999 789	3.12	0.999 989 77	3.62	0.999 999 694
2.13	0.997 407	2.63	0.999 800	3.13	0.999 990 42	3.63	0.999 999 716
2.14	0.997 525	2.64	0.999 811	3.14	0.999 991 03	3.64	0.999 999 736
2.15	0.997 639	2.65	0.999 822	3.15	0.999 991 60	3.65	0.999 999 756
2.16	0.997 747	2.66	0.999 831	3.16	0.999 992 14	3.66	0.999 999 773
2.17	0.997 851	2.67	0.999 841	3.17	0.999 992 64	3.67	0.999 999 790
2.18	0.997 951	2.68	0.999 849	3.18	0.999 993 11	3.68	0.999 999 805
2.19	0.998 046	2.69	0.999 858	3.19	0.999 993 56	3.69	0.999 999 820
2.20	0.998 137	2.70	0.999 866	3.20	0.999 993 97	3.70	0.999 999 833
2.21	0.988 224	2.71	0.999 873	3.21	0.999 994 36	3.71	0.999 999 845
2.22	0.998 308	2.72	0.999 880	3.22	0.999 994 73	3.72	0.999 999 857
2.23	0.998 388	2.73	0.999 887	3.23	0.999 995 07	3.73	0.999 999 867
2.24	0.998 464	2.74	0.999 893	3.24	0.999 995 40	3.74	0.999 999 877
2.25	0.998 537	2.75	0.999 899	3.25	0.999 995 70	3.75	0.999 999 886
2.26	0.998 607	2.76	0.999 905	3.26	0.999 995 98	3.76	0.999 999 895
2.27	0.998 674	2.77	0.999 910	3.27	0.999 996 24	3.77	0.999 999 903
2.28	0.998 738	2.78	0.999 916	3.28	0.999 996 49	3.78	0.999 999 910
2.29	0.998 799	2.79	0.999 920	3.29	0.999 996 72	3.79	0.999 999 917
2.30	0.998 857	2.80	0.999 925	3.30	0.999 996 94	3.80	0.999 999 923
2.31	0.998 912	2.81	0.999 929	3.31	0.999 997 15	3.81	0.999 999 929
2.32	0.998 996	2.82	0.999 933	3.32	0.999 997 34	3.82	0.999 999 934
2.33	0.999 016	2.83	0.999 937	3.33	0.999 997 51	3.83	0.999 999 939
2.34	0.999 065	2.84	0.999 941	3.34	0.999 997 68	3.84	0.999 999 944
2.35	0.999 111	2.85	0.999 944	3.35	0.999 997 838	3.85	0.999 999 948
2.36	0.999 155	2.86	0.999 948	3.36	0.999 997 983	3.86	0.999 999 952
2.37	0.999 197	2.87	0.999 951	3.37	0.999 998 120	3.87	0.999 999 956
2.38	0.999 237	2.88	0.999 954	3.38	0.999 998 247	3.88	0.999 999 959
2.39	0.999 275	2.89	0.999 956	3.39	0.999 998 367	3.89	0.999 999 962
2.40	0.999 311	2.90	0.999 959	3.40	0.999 998 478	3.90	0.999 999 965
2.41	0.999 346	2.91	0.999 961	3.41	0.999 998 582	3.91	0.999 999 968
2.42	0.999 379	2.92	0.999 964	3.42	0.999 998 679	3.92	0.999 999 970
2.43	0.999 411	2.93	0.999 966	3.43	0.999 998 770	3.93	0.999 999 973
2.44	0.999 441	2.94	0.999 968	3.44	0.999 998 855	3.94	0.999 999 975
2.45	0.999 469	2.95	0.999 970	3.45	0.999 998 934	3.95	0.999 999 977
2.46	0.999 497	2.96	0.999 972	3.46	0.999 999 008	3.96	0.999 999 979
2.47	0.999 523	2.97	0.999 973	3.47	0.999 999 077	3.97	0.999 999 980
2.48	0.999 547	2.98	0.999 975	3.48	0.999 999 141	3.98	0.999 999 982
2.49	0.999 571	2.99	0.999 976	3.49	0.999 999 201	3.99	0.999 999 983

For a more complete table, see L. J. Comrie, *Chambers Six Figure Mathematical Tables*, vol. 2, W. & R. Chambers, Edinburgh, 1949.

ACKNOWLEDGEMENTS

Chapter 1

Figure 1-1. From S. M. Sze (ed.), VLSI Technology, McGraw-Hill, New York (1983). Used with permission.

Figure 1-2. From D. A. Hodges and H. G. Jackson, Analysis and Design of Digital Integrated Circuits, McGraw-Hill, New York (1983). Used with permission.

Figure 1-3. From W. F. Smith, Principles of Materials Science and Engineering, McGraw-Hill, New York (1986). Used with permission.

Figure 1-4. B. G. Streetman, Solid State Electronic Devices, 2e, © 1980. Reprinted by permission of Prentice-Hall, Inc., Englewood Cliffs, New Jersey.

Figure 1-7 & Figure 1-8. A. S. Grove, Physics and Technology of Semiconductor Devices, © 1967. Reprinted by permission of John Wiley & Sons, Inc., New York.

Figures 1-9 through 1-10 and Figures 1-12 through 1-20. B. G. Streetman, Solid State Electronic Devices, 2e, © 1980. Reprinted by permission of Prentice-Hall, Inc., Englewood Cliffs, New Jersey.

Figures 1-22 through 1-24. R. A. Colclaser, Microelectronics Processing and Device Design, © 1980. Reprinted by permission of John Wiley & Sons, Inc., New York.

Figure 1-25. With permission, Bell Syst. Tech. J., *41*, 2 (1962).

Figure 1-26. With permission, Bell Syst. Tech. J., *39*, 3 (1960).

Chapter 2

Figure 2-2 & Table 2-1. From C. P. Khattak and K. V. Ravi (eds.), Silicon Processing for Photovoltaics, vol. II, North-Holland, New York (1987). Used by permission of Elsevier Science Publishers.

Figures 2-3 through 2-6. From H. R. Huff and E. Sirtl (eds.), Semiconductor Silicon 1977, The Electrochemical Society, Inc., Pennington, New Jersey (1977). Used by permission of the Electrochemical Soc.

Figures 2-7 through 2-10. From C. P. Khattak and K. V. Ravi (eds.), Silicon Processing for Photovoltaics, vol. I, North-Holland, New York (1985). Used by permission of Elsevier Science Publishers.

Chapter 3

Figure 3-1. From S. K. Ghandhi, VLSI Fabrication Principles, © 1983. Reprinted by permission of John Wiley & Sons, Inc., New York; With permission, S. M. Sze (ed.), VLSI Technology, McGraw-Hill, New York (1983).

Figure 3-2. From H. R. Huff and E. Sirtl (eds.), Semiconductor Silicon 1977, The Electrochemical Society, Inc., Pennington, New Jersey (1977). Used by permission of the Electrochemical Soc.; With permission, S. M. Sze (ed.), VLSI Technology, McGraw-Hill, New York (1983).

Figures 3-5, 3-9, 3-11, 3-12, and 3-14. From D. R. Askeland, The Science and Engineering of Materials, © 1985. Reprinted by permission of PWS-KENT Publishing Company, Boston.

Figure 3-8. From S. M. Sze, Physics of Semiconductor Devices, © 1981. Reprinted by permission of John Wiley & Sons, Inc., New York.

Table 3.1. From S. M. Sze (ed.), VLSI Technology, McGraw-Hill, New York (1983). Used with permission.

Table 3.2. From A. G. Milnes, Deep Impurities in Semiconductors, © 1973. Reprinted by permission of John Wiley & Sons, Inc., Englewood Cliffs, New Jersey.

Figure 3-15. With permission, J. Appl. Phys., *39*, 5205 (1968).

Figure 3-16. With permission, J. Crystal Growth, *50*, 865 (1980); With permission, J. Crystal Growth, *54*, 267 (1981).

Figure 3-18. With permission, J. Crystal Growth, *65*, 189 (1983).

Chapter 4

Figures 4-1 and 4-2. B. G. Streetman, Solid State Electronic Devices, 2e, © 1980. Reprinted by permission of Prentice-Hall, Inc., Englewood Cliffs, New Jersey.

Tables 4.1 and 4.2 and Figure 4-6. S. M. Sze, Physics of Semiconductor Devices, © 1981. Reprinted by permission of John Wiley & Sons, Inc., New York.

Figure 4-3. With permission, IBM J. Res. Dev., *22*, 72 (1978).

Figure 4-4. With permission, Phys. Rev. Lett., *24*, 256 (1981).

Figure 4-5. With permission, Solid State Electron., *13*, 1011 (1970).

Table 4.3. From S. M. Sze (ed.), VLSI Technology, McGraw-Hill, New York (1983). Used with permission.

Figure 4-10. With permission, J. Electrochem. Soc., Abstract 191, p452 (1975).

Figure 4-11. With permission, 18th Proceedings on Reliability Symposium, New York, p165, IEEE (1980).

Figure 4-12. With permission, 23rd Annual Proceedings on Reliability Physics, Orlando, Fla., p81, IEEE (1985).

Figure 4-14. With permission, IEEE Trans. Elect. Dev., ED-27, 606 (1980).

Figure 4-15. M. J. O. Strutt, Semiconductor Devices, 2e, © 1966. Reprinted by permission of John Wiley & Sons, Inc., New York.

Figure 4-16. With permission, Appl. Phys. Lett., *8*, 111 (1966).

Figure 4-18. With permission, IEEE Trans. Elect. Dev., ED-21, 667 (1974).

Figure 4-19. With permission, J. Vac. Sci. Tech., *14*, 17 (1977).

Chapter 5

Figure 5-4. With permission, J. Crystal Growth, *50*, 581 (1980).

Figure 5-5. With permission, J. Crystal Growth, *63*, 493 (1983).

Figure 5-6. With permission, J. Electrochem. Soc., *127*, 194 (1980).

Figure 5-7. With permission, J. Vac. Sci. Tech., *4*, 209 (1967).

Figure 5-13. From R. P. Wayne, Photochemistry, American Elsevier, New York (1970). Used by permission.

Table 5.2. With permission, J. Vac. Sci. Tech., *B1*, 969 (1983).

Figure P5-19. With permission, IEEE J. Quantum Electr., *16*, 1233 (1980).

Chapter 6

Figure 6-3. With permission, J. Vac. Sci. Tech., *15*, 13 (1973).
Figure 6-7. With permission, J. Electrochem. Soc., *129*, 634 (1982).
Figure 6-8. With permission, J. Electrochem. Soc., *125*, 317 (1973).
Figure 6-9. With permission, J. Fluids Engng., *105*, 5 (1983).
Figure 6-15. From S. M. Sze (ed.), VLSI Technology, McGraw-Hill, New York (1983). Used with permission.
Figure 6-16. With permission, Appl. Phys. Lett., *48*, 1681 (1986).

Chapter 7

Figure 7-2. H. Ryssel and I. Ruge, Ion Implantation, © 1986. Reprinted by permission of John Wiley & Sons, Inc., New York.
Table 7.1. From F. F. Y. Wang (ed.), Impurity Doping Process in Silicon, North-Holland, New York (1981). Used by permission of Elsevier Science Publishers.
Table 7.2. With permission, D. Shaw (ed.), Atomic Diffusion in Semiconductor, Plenum, New York (1973).
Table 7.3. and Figure 7-5. From S. M. Sze (ed.), VLSI Technology, McGraw-Hill, New York (1983). Used with permission.
Figure 7-13. With permission, J. Electrochem. Soc., *127*, 1334 (1980).
Figure 7-15. With permission, J. Appl. Phys., *35*, 2695 (1964).

Chapter 8

Figures 8-1 and 8-2. R. A. Colclaser, Microelectronics Processing and Device Design, © 1980. Reprinted by permission of John Wiley & Sons, Inc., New York.
Figures 8-3, 8-4, 8-8, 8-14, 8-27, 8-30, 8-31 and Table 8.1. From G. R. Brewer (ed.), Electron-Beam Technology in Microelectric Fabrication, Academic Press, New York (1980). Used with permission.
Figures 8-5 and 8-17. From S. M. Sze (ed.), VLSI Technology, McGraw-Hill, New York (1983). Used with permission.
Figures 8-6, 8-7, and 8-10. From R. M. Eisberg and L. S. Lerner, Physics: Foundations and Applications, McGraw-Hill, New York (1981). Used with permission.
Figure 8-9. With permission, IEEE Trans. Elect. Dev. ED-22, 445 (1975).
Figures 8-11, 8-12 and 8-19. L. F. Thompson et al. (eds.), Introduction to Microlithography, ACS Symposium Series 219, ACS, Washington (1983). Copyright by the American Chemical Soc. Used with permission.
Figure 8-13. With permission, IEEE Trans. Elect. Dev., ED-22, 1305 (1975).
Figures 8-15, 8-16, and 8-18. From R. Newman (ed.), Fine-Line Lithography, North-Holland, Amsterdam (1980). Used by permission of Elsevier Science Publishers.
Figure 8-28. With permission, J. Vac. Sci. Tech., *12*, 1275 (1975).
Figure 8-29. With permission, IEEE Trans. Elect. Dev., ED-22, 456 (1975).
Figure 8-32. With permission, I. Brodie and J. J. Muray, The Physics of Microfabrication, Plenum, New York (1982).

Chapter 9

Figures 9-1, 9-2, 9-3, 9-6 and 9-7. From L. Maissel and R. Glang (eds.), Handbook of Thin Film Technology, McGraw-Hill, New York (1970). Used with permission.
Figure 9-5. M. A. Kettani and M. F. Hoyaux, Plasma Engineering, © 1973. Reprinted by permission of John Wiley & Sons, Inc., New York.
Figure 9-8. With permission, J. Vac. Sci. Tech., A1, 90 (1983).

Table 9.1 and Figure 9-11. From G. S. Mathad et al. (eds.), Plasma Processing, The Electrochemical Society, Pennington, New Jersey (1985). Used with permission.
Figure 9-12. With permission, J. Appl. Phys., *50*, 3189 (1979).
Figure 9-13. With permission, J. Appl. Phys., *49*, 3769 (1978).
Figure P9-12. With permission, Surface Sci., *103*, 524 (1981).

Chapter 10

Figure 10-1. From S. M. Sze (ed.), VLSI Technology, McGraw-Hill, New York (1983). Used with permission.
Figures 10-5 and 10-10. B. N. Chapman, Glow Discharge Processes, © 1980. Reprinted by permission of John Wiley & Sons, Inc., New York.
Figure 10-8 and 10-11. E. W. McDaniel, Collision Phenomena in Ionized Gasses, © 1964. Reprinted by permission of John Wiley & Sons, Inc., New York.
Figure 10-9. With permission, J. Electrochem. Soc., *130*, 1144 (1983).

Chapter 11

Figure 11-2. From S. M. Sze (ed.), VLSI Technology, McGraw-Hill, New York (1983). Used with permission.
Table 11.1, Figures 11-9, 11-10, 11-12, and 11-14. From J. L. Sloan, Design and Packaging of Electronic Equipment, Van Nostrand Reinhold Company, New York (1985). Used with permission.
Figure 11-4. With permission, Solid State Tech., p59, March (1978).
Figure 11-5. From E. R. Huatek, Integrated Circuit Quality and Reliability, Marcel Dekker, Inc., New York (1987). Reprinted by courtesy of Marcel Dekker, Inc.
Figure 11-6. With permission, IEEE Trans. Comp. Hyb. Mfg, Tech., CHMT-3, 634 (1980).
Figure 11-8. With permission, Cer. Engng, Sci. Proc., *6*, 1261 (1985).

INDEX